# Oblique Derivative Problems for Elliptic Equations

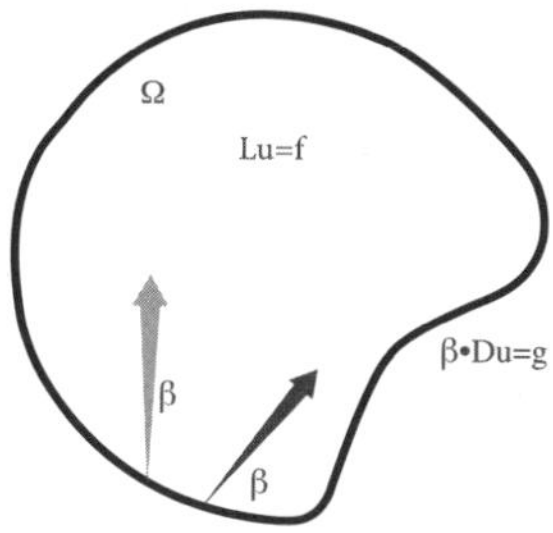


**Gary M Lieberman**
*Iowa State University, USA*

World Scientific

NEW JERSEY • LONDON • SINGAPORE • BEIJING • SHANGHAI • HONG KONG • TAIPEI • CHENNAI

*Published by*

World Scientific Publishing Co. Pte. Ltd.

5 Toh Tuck Link, Singapore 596224

*USA office:* 27 Warren Street, Suite 401-402, Hackensack, NJ 07601

*UK office:* 57 Shelton Street, Covent Garden, London WC2H 9HE

**British Library Cataloguing-in-Publication Data**
A catalogue record for this book is available from the British Library.

**OBLIQUE DERIVATIVE PROBLEMS FOR ELLIPTIC EQUATIONS**



ISBN 978-981-4452-32-8

Printed in Singapore by World Scientific Printers.

# Oblique Derivative Problems for Elliptic Equations

This book is dedicated to my wife Linda, our four grandchildren (Eliott, Leo, Thisbe, and Edie), and to the many people who helped make this book possible; in particular, Xing-Bin Pan, Chunquan Tang, and Hien Nguyen.

# Preface

Oblique derivative problems are those in which the boundary condition has the form $b(x, u, Du) = 0$ when $b$ depends on the gradient $Du$ of the unknown function $u$ in a suitable way. The model boundary condition is the normal derivative condition $\gamma \cdot Du = f(x)$, where $\gamma$ denotes the unit inner normal to the given domain $\Omega$ in which the boundary value problem is studied. When the boundary condition is linear, that is, $b(x, u, Du) = \beta(x) \cdot Du + \beta^0(x)u + g(x)$, then the boundary condition is oblique if $\beta \cdot \gamma > 0$. For nonlinear boundary conditions, we write $p$ as dummy variable for $Du$, and then the condition is oblique if $\beta = \partial b/\partial p$ satisfies $p \cdot \gamma > 0$.

When I first started to study oblique derivative problems over thirty years ago, I had a grand ambition: to put the theory of such problems in the same shape as the theory for the Dirichlet problem. It was clear even then that there was a major difference in attitudes towards the two problems in the eyes of most researchers in partial differential equations. The book [91] of Ladyzhenskaya and Ural′tseva devotes nine chapters to the Dirichlet problem and then one chapter to "other boundary value problems", and the book [64] of Gilbarg and Trudinger only devoted a few pages to oblique derivative problems. Fortunately, I was aware of some work on the topic: for example, the papers of Fiorenza [55, 54, 52, 53] suggested that there was some interest in the oblique derivative problem and the work of Concus and Finn [31, 32] indicated that there was even some practical interest in these problems.

In writing this book, I started with a similarly grand ambition: to write a complete description of the theory of oblique derivative problems. With the books [91, 64] as models (and my own book [114]), I felt that such an ambition might be within reach. Unfortunately, the theory is not yet in the complete state that I had hoped, and the theory is more technical than that for the Dirichlet problem. Nevertheless, what I present here is a first attempt at delineating the ideas for oblique derivative problems in a systematic fashion. An important point at the heart of my choice of topics is to find elements of the theory which are not immediate consequences of the corresponding theory of the Dirichlet problem. Hence, I have made a conscious choice not to duplicate the content of [91, 64] as much as possible. Specifically, the proofs of properties of Sobolev spaces and of much of the measure theory are not given here; I only refer to [64] or [114] for the proofs. Furthermore, I have given alternative proofs of the classical Schauder theory in Chapter 2; my point is not that I consider the original proof using potential theory to be obsolete but rather that the modern proof given there allows us to derive the appropriate Schauder theory for oblique derivative problems in Chapter 4. For the most part, I have included proofs of results from real analysis and functional analysis whenever they are not in [64, 114] or in standard graduate texts on those topics. Hence, the Weierstrass approximation theorem is proved here because the form I use is not the one most familiar to modern audiences. On the other hand, I have chosen not to prove the Fredholm-Schauder-Riesz theorem (Theorem 2.29) because the proof is essentially contained in Section 5.3 of [64].

It is also important to recognize that the oblique derivative problem is best understood not in a vacuum but, rather, in conjunction with the theory of the Dirichlet problem. For this reason, I have abbreviated some of the proofs since they are very close to results from [91, 64]; any reader who has difficulty understanding the proofs here is recommended to peruse these books for additional assistance. Moreover, even though I don't expect prospective readers to have a detailed knowledge of these books, it is extremely helpful to have some familiarity with their contents. I have provided quite a few detailed references to these books, too, at appropriate points in the text.

There are also other books which describe the theory of oblique derivative problem in some detail: [151, 155, 205], but they are really interested in what is more properly called the tangential-oblique derivative problem. When they consider the boundary condition $\beta(x) \cdot Du = g(x)$ on $\partial\Omega$ where the vector $\beta$ satisfies the condition $\beta \cdot \gamma \geq 0$, where $\gamma$ is the unit inner normal to $\partial\Omega$. For the problem to be oblique, we need to have $\beta \cdot \gamma > 0$ everywhere, and the tangential-oblique derivative problem has several features not present in the oblique derivative problem (or the Dirichlet problem); the most important ones are that the solution need not be as smooth as the data would suggest and that the difference between the dimensions of the kernel and cokernel of the operator need not be zero (or even finite in more than two space dimensions).

To keep this book from becoming even larger, I have not discussed a number of topics that are relevant to the general theory of oblique derivative problems. For example, the theory presented in this book has played a major role in the study of reflected shocks in transonic flow. The first application of the theory was made in [27] (for steady transonic shocks), which led to more papers than we care to mention here; MathSciNet lists 68 papers referring to [27]. Another important example is the capillary problem; although we prove some simple estimates for this problem, there is a vast theory which provides a complete description of the regularity of the solutions under various hypotheses. The book [46] by Finn gives a description of this theory as was known in 1986, and his more recent survey articles [49, 48, 47] present some more recent advances. In addition, Finn suggested in [50] that a slight variation of the historical capillary problem provides a more accurate model of the underlying physical situation, and the work in [51, 4] suggests that the theory in this book can be applied to this problem.

I have essentially ignored the two-dimensional theory of oblique derivative problems. The basic theory is quite old and appears in [10, 52], and many of the results in this book can be proved much more easily (or under weaker hypotheses) when there are only two dimensions. For example, a good part of the theory in this book is easily proved in two dimensions in [105], which is the source of the ideas in Chapter 4.

I also cannot resist mentioning one application of oblique derivative problems which affects my life outside mathematics: overtone series for musical instruments, specifically, three of the instruments that I play regularly. The overtone series of the clarinet differs from that of most other instruments (such as the flute and oboe) in that the even-numbered overtones are extremely faint, and hence the timbre of the clarinet is very different from those instruments. This difference in overtones comes from oblique boundary conditions, but there is no space here to justify this claim.

Here, I thank all the people who have helped with this book. First and foremost are Xing-Bin Pan, who invited me (several times) to present lecture series on oblique derivative problems at East China Normal University, my Ph.D. student Chunquan Tang, who gave me new impetus to examine oblique derivative problem more closely and who helped proofread parts of this book, and Hien Nguyen, who helped with the proofreading. Of course,

I take full responsibility for any errors in this book, and a website for errata will be available at lieb.public.iastate.edu/ODPBOOK/errata.pdf. I also thank the editors at World Scientific. In addition, I thank Iowa State University, East China Normal University, and the Centre for Mathematics and its Applications at the Australian National University for their support.

# Contents

# Chapter 1

# Pointwise Estimates

## Introduction

This chapter examines some basic estimates for solutions of linear oblique derivative problems, although some of the techniques used are quite sophisticated. The heart of all the estimates in this chapter (and, by extension, most of the estimates in the entire book) is the maximum principle, which is quite straightforward for the Dirichlet problem but not so for the oblique derivative problem. To illustrate the differences between these two problems, we look first at a simple one-dimensional example. If $u'' \geq 0$ on the interval $(0, 1)$, then $u$ attains its maximum at one of the endpoints, so prescription of the values for $u$ at these points leads to a simple estimate of the maximum over the interval. If we prescribe values for the first derivative of $u$ at the endpoints, then we get no information about the maximum value for $u$ because we can add an arbitrary constant (positive or negative in this case) to $u$ without changing the differential equation or boundary conditions. The control of the maximum in this case is achieved via one of three modifications, and we shall see instances of the higher-dimensional analog of each one at various points within the text. First, we can modify the boundary conditions to read $u'(0) + au(0) \geq -A$ and $-u'(1) + bu(1) \geq -B$ for negative constants $a$ and $b$, and nonnegative constants $A$ and $B$. (The sign conventions here are designed to match our normalization of the boundary condition in higher dimensions.) It follows that the maximum of $u$ is bounded by $\max\{A/|a|, B/|b|\}$. Second, we can modify the boundary conditions to $u'(0) \geq -A$ and $u(1) \leq B$ for nonnegative constants $A$ and $B$. (The case $u(0) \leq B$ and $u'(1) \leq A$ is handled similarly.) In this case, the maximum of $u$ is no greater than $B + A$. Our third modification is to the differential equation; if we assume that $u'' + cu \geq 0$ for some negative constant $c$ and that $u'(0) \geq -A$ and $u'(1) \leq A$ for some nonnegative constant $A$, then it is possible to estimate the maximum of $u$ in terms of $A$ and $c$. Since the details are not immediately obvious, we defer a discussion of this case to Section 1.3 where the higher-dimensional version is examined.

In fact, there is another important point for the oblique derivative problem which does not cause major problems for the Dirichlet problem. The classical definition of the oblique derivative condition is that some directional derivative of the solution is prescribed on the boundary under the assumption that the direction is nowhere tangential. If the boundary of the domain is sufficiently smooth, then this definition makes sense; however, in many applications, the domain is not assumed smooth. Such is the case if we look at the Neumann problem for harmonic functions in a square, and it will appear in many other contexts as well. For this reason, we devote Section 1.2 to a definition of obliqueness appropriate to more general, usually Lipschitz, domains.

Although it is relatively simple to develop a number of pointwise estimates for solutions by the classical maximum principle (which estimates the solution in terms of simple properties of the coefficients of the differential equation and boundary condition and the maximum of the inhomogeneous term in the differential equation and boundary condition), there are some significant results which allow us to study a large variety of nonlinear equations but for which the classical maximum principle is inadequate. A particular instance of such a result is a modulus of continuity estimate for the solution in terms of bounds of the coefficients in the differential equation and boundary condition. For oblique derivative problems, the appropriate starting point is a maximum principle, first presented by Bakel′man [5], Aleksandrov [2], and Pucci [157] for the Dirichlet problem in the 1960's, which estimates the maximum of the solution of the boundary value in terms of the $L^n$ norm of the right hand side of the differential equation, where $n$ is the number of spatial dimensions in the problem. We present several forms of this estimate which fit the oblique derivative context very well, and we prove several important estimates from this maximum principle.

We also prove a simple gradient estimate for solutions of equations with constant coefficients in Section 1.12.

## 1.1 The maximum principle

The most fundamental estimate for solutions of elliptic equations is an estimate on the maximum of the solution in terms of data. We shall provide several different estimates based on different information.

Our first estimate is a simple one which will be generalized very quickly. To apply it to oblique derivative problems, we need to introduce a basic concept that will be used repeatedly in this book. Let $\Omega$ be an open subset of $\mathbb{R}^n$ with boundary $\partial\Omega$ and let $\beta$ be a vector field defined on $\partial\Omega$. We say that a vector $\beta$ is *inward pointing* at $x \in \partial\Omega$ if there is a positive number $h_0$ such that $x + h\beta \in \overline{\Omega}$ for any $h \in (0, h_0)$. We also write $\beta \cdot Du(x) \le g$ if

$$\limsup_{h \to 0^+} \frac{u(x + h\beta) - u(x)}{h} \le g.$$

If $\beta$ is a vector field defined on some subset $S$ of $\partial\Omega$ such that $\beta(x)$ is inward pointing at $x$ for each $x \in S$, then we say that $\beta$ is *inward pointing on* $S$ and we write $\beta \cdot Du \le g$ on $S$ (for some function $g$ defined on $S$) if $\beta \cdot Du(x) \le g(x)$ for all $x \in S$. In addition, we define $\beta \cdot Du(x) \ge g$ to mean that

$$\liminf_{h \to 0^+} \frac{u(x + h\beta) - u(x)}{h} \ge g,$$

and we write $\beta \cdot Du(x) = g$ if $\beta \cdot Du(x) \le g$ and $\beta \cdot Du \ge g$.

Note that, if $u$ is differentiable at a point $x \in \partial\Omega$, then for any inward pointing vector field $\beta$, our definitions of $\beta \cdot Du(x) \le g$, $\beta \cdot Du(x) \ge g$, and $\beta \cdot Du(x) = g$ agree with the inequalities (or equality) satisfied by the expressions involving the gradient of $u$. More precisely, our inequalities on $\beta \cdot Du$ are equivalent to inequalities on the one-sided derivate of $u$ in the direction parallel to $\beta$ and any equation involving $\beta \cdot Du$ provides (at least for nonzero $\beta$) a prescription of the directional derivative of $u$ in the direction of $\beta$.

The most important fact we use is that, if $u$ has a maximum at $x \in \partial\Omega$ and $\beta$ is an inward pointing vector at $x$, then $\beta \cdot Du(x) \le 0$.

In addition, we introduce two operators to simplify our notation. Let $[a^{ij}]$ be a positive definite matrix-valued function defined on some bounded open set $\Omega$, let $b = (b^1, \dots, b^n)$ be a vector-valued function defined on $\Omega$, and let $c$ be a function defined on $\Omega$. We then write $L$ for the operator defined on $C^2(\Omega)$ by

$$Lu = a^{ij} D_{ij} u + b^i D_i u + cu.$$

If $\beta$ is a vector field defined on $\partial\Omega$ and $\beta^0$ is a scalar field defined on $\partial\Omega$, we define the operator $M$ by

$$Mu = \beta \cdot Du + \beta^0 u.$$

If $u$ is continuous on $\overline{\Omega}$, we take the inequality $Mu \geq g$ on some subset $S$ of $\partial\Omega$ to mean

$$\beta(x) \cdot Du(x) \geq -\beta^0(x)u(x) + g(x)$$

for all $x \in S$ and we define $Mu \leq g$ or $Mu = g$ similarly. Unless otherwise stated, we always assume that the functions $[a^{ij}]$, $b$, $c$, $\beta$, and $\beta^0$ are given, subject to any explicitly stated hypotheses.

**Lemma 1.1.** *Suppose $[a^{ij}]$ is positive semidefinite matrix and $\beta$ is inward pointing on $\partial\Omega$. Suppose also that $c < 0$ in $\Omega$ and $\beta^0 < 0$ on $\partial\Omega$. If $u \in C^2(\Omega) \cap C^0(\overline{\Omega})$ satisfies*

$$Lu \geq 0 \text{ in } \Omega, \quad Mu \geq 0 \text{ in } \partial\Omega, \tag{1.1}$$

*then $u \leq 0$ in $\overline{\Omega}$.*

***Proof.*** Let $x_0$ be a point at which $u$ attains its maximum value. If $x_0 \in \Omega$, then $a^{ij}D_{ij}u(x_0) \leq 0$ and $b^i D_i u(x_0) = 0$, so $cu(x_0) \geq 0$. Hence $u(x_0) \leq 0$.

On the other hand, if $x_0 \in \partial\Omega$, then $\beta \cdot Du \leq 0$, so $\beta^0 u(x_0) \geq 0$ and hence $u(x_0) \leq 0$ again.

Since $u(x_0) \leq 0$, it follows that $u \leq 0$ in $\overline{\Omega}$. □

The restrictions $c < 0$ and $\beta^0 < 0$ can be relaxed somewhat but it is important to recognize that they cannot be eliminated completely. If $c \equiv 0$ and $\beta^0 \equiv 0$, then any constant will satisfy the inequalities in (1.1). Moreover, the case $\beta \equiv 0$ (which is an inward pointing vector field) gives a maximum principle for the Dirichlet problem. We shall take advantage of this observation several times.

Our next step is a variant on our original maximum principle in which $c \leq 0$ and $\beta^0 < 0$. In this case, we need to make some assumptions about the relationship between $[a^{ij}]$ and $b$.

**Lemma 1.2.** *Suppose that there is some nonnegative function $\lambda$ such that*

$$a^{ij}\xi_i\xi_j \geq \lambda|\xi|^2 \tag{1.2}$$

*for all $\xi \in \mathbb{R}^n$ and suppose that, for any compact subset $K$ of $\Omega$, there is a nonnegative constant $\mu_1(K)$ such that $|b| \leq \mu_1\lambda$ on $K$. Suppose also that $c \leq 0$ in $\Omega$, $\beta$ is inward pointing vector field on $\partial\Omega$, and $\beta^0 < 0$. If (1.1) holds, then $u \leq 0$ in $\Omega$.*

***Proof.*** Fix $\varepsilon > 0$ and let $x_0 \in \Omega$. Setting $M_0 = \sup_{\partial\Omega} u$, we see that there is a compact set $K \subset \Omega$ such that $x_0$ is an interior point of $K$ and $u \le M_0 + \varepsilon$ on $\partial K$. We now set $\gamma = 1 + \mu_1(K)$, and we define the function $\varphi$ by

$$\varphi(x) = \exp(-\gamma x^1).$$

A simple calculation yields

$$\begin{aligned} a^{ij} D_{ij}\varphi + b^i D_i\varphi &= (a^{11}\gamma^2 - b^1\gamma)\exp(-\gamma x^1) \\ &\ge \lambda(\gamma^2 - \mu_1\gamma)\exp(-\gamma x^1) \\ &> 0. \end{aligned}$$

Now let $\Omega^+$ denote the subset of $K$ on which $u > 0$, and suppose $u + \varepsilon\varphi$ has a positive maximum at some $x_1 \in \overline{\Omega^+}$. If $x_1 \in \Omega^+$, then we have

$$a^{ij} D_{ij}(u + \varepsilon\varphi) + b^i D_i(u + \varepsilon\varphi) = a^{ij} D_{ij}(u + \varepsilon\varphi) \le 0$$

at $x_1$. However, in $\Omega^+$, we have

$$a^{ij} D_{ij}(u + \varepsilon\varphi) + b^i D_i(u + \varepsilon\varphi) \ge Lu + \varepsilon(a^{ij} D_{ij}\varphi + b^i D_i\varphi) > 0.$$

It follows that $u + \varepsilon\varphi$ can't have an interior positive maximum, so

$$u(x_0) + \varepsilon\varphi(x_0) \le \sup_{\partial K} u^+ + \varepsilon\varphi.$$

From our choice of $\varphi$, it follows that

$$u(x_0) \le M_0 + 2\varepsilon,$$

and hence

$$\sup_{\Omega} u \le \sup_{\partial\Omega} u^+.$$

In other words, if $u$ has a positive maximum, it must occur on $\partial\Omega$, say at some $x_2 \in \partial\Omega$. Then

$$\beta \cdot Du \le 0$$

at $x_2$ and the boundary condition implies that

$$\beta \cdot Du \ge -\beta^0 u$$

at $x_2$. Hence $u(x_2) \le 0$. It follows that $u$ cannot attain a positive maximum and hence $u \le 0$. □

An analogous result is true if $c < 0$ and $\beta^0 \le 0$ but several new ideas are needed. The most important one is our definition of obliqueness of a vector field, and that definition is the topic of the next section.

## 1.2 The definition of obliqueness

We now turn to the definition of oblique vector fields, which will be a special case of inward pointing vector fields. Our description of oblique vector fields starts with the same scheme as for inward pointing vector fields. We say that a vector $\beta$ is *oblique at a point* $x \in \partial\Omega$ if there is a cone of finite height, vertex at $x$ and axis parallel to $\beta$ which lies inside $\Omega$. We also say that a vector field $\beta$ defined on a subset $S$ of $\partial\Omega$ is oblique on $S$ if $\beta(x)$ is oblique at $x$ for every $x \in S$.

Note that every oblique vector field on $S$ is inward pointing on $S$ but an inward pointing vector field need not be oblique. Two special situations are worth noting. First, if $\Omega$ has an outward pointing cusp at $x \in \partial\Omega$, then the axis of the cusp is inward pointing but not oblique. Second, if $\beta$ is tangential to $\partial\Omega$, it will be inward pointing but not oblique. In particular, if

$$\Omega = \{x = (x^1, x^2) : |x| < 1,\ x^2 > 0\},$$

then the vector $(1, 0)$ is inward pointing but not oblique at $(0, 0)$.

For many of our results, a quantitative measure of obliqueness is useful. For $\varepsilon \in (0, 1)$, we say that a vector field $\beta$ defined on a subset $S$ of $\partial\Omega$ has *modulus of obliqueness at least* $\varepsilon$ *on* $S$ if there are positive constants $R$ and $\omega_0$, and, for each $s \in S$, a function $\omega : B_{n-1}(0, R)$ and a coordinate system $(y', y^n) = (y^1, \dots, y^n)$ centered at $s$ such that

$$\Omega \cap B(s, R) = \{(y', y^n) : y^n > \omega(y'), |y| < R\} \tag{1.3a}$$

$$|\omega(y_1') - \omega(y_2')| \le \omega_0 |y_1' - y_2'|, \tag{1.3b}$$

$$\omega_0 |\beta'(x)| \le (1 - \varepsilon)\beta^n(x) \tag{1.3c}$$

for all $x \in S$ with $|x - s| < R$. (Here $\beta^n$ is the component of $\beta$ in the $y^n$ direction, and $\beta'$ is the projection $\beta$ onto the $y'$ hyperplane.) If we don't wish to emphasize the number $\varepsilon$, we say only that $\beta$ is uniformly oblique on $S$.

We say that $\partial\Omega \in C^1$ if there is a continuously differentiable function $g\colon \mathbb{R}^n \to \mathbb{R}$ such that $\Omega = \{x \in \mathbb{R}^n : g(x) > 0\}$ and $Dg$ is nowhere zero on $\partial\Omega$. Our definition of obliqueness on $S \subset \partial\Omega$ in this case is equivalent to the condition that $\inf_S \beta \cdot \gamma/|\beta| > 0$, where $\gamma$ is the unit inner normal to $\partial\Omega$ (so $\gamma = Dg/|Dg|$). One of the main points in this book is that the more general definition, which allows domains which are not $C^1$, still leads to a rich and useful theory.

## 1.3 The case $c < 0$, $\beta^0 \leq 0$

We now consider the case $c < 0$ and $\beta^0 \leq 0$ in proving a maximum principle. An important point for the maximum principle is that, under our general definition, there is a smooth function $\rho$ on $\overline{\Omega}$ such that $\beta \cdot D\rho > 0$ on $\partial\Omega$. ( Such a function essentially plays the role of $g$ in the case of $\partial\Omega \in C^1$.) Such a result can be proved once we have two simple results concerning continuous functions. The first result is a simple extension theorem. To state the theorem, we recall that a *modulus of continuity* is an increasing function $\omega\colon [0,\infty) \to [0,\infty)$ such that $\omega(s+t) \leq \omega(s) + \omega(t)$ for all $s, t > 0$ and $\omega(0) = 0$.

**Lemma 1.3.** *Let $S \subset \mathbb{R}^n$ and let $f : S \to \mathbb{R}$ be bounded from below. Suppose that there is a modulus of continuity $\mu_f$ such that*

$$|f(s_1) - f(s_2)| \leq \mu_f(|s_1 - s_2|) \tag{1.4}$$

*for all $s_1$ and $s_2$ in $S$. Then there is a function $\hat{f}\colon \mathbb{R}^n \to \mathbb{R}$ such that*

$$|\hat{f}(x_1) - \hat{f}(x_2)| \leq \mu_f(|x_1 - x_2|) \tag{1.5a}$$

*for all $x_1$ and $x_2$ in $\mathbb{R}^n$, and*

$$\hat{f}(s) = f(s) \tag{1.5b}$$

*for all $s \in S$. In addition, if there are constants $m$ and $M$ such that $m \leq f \leq M$ on $S$, then $\hat{f}$ can be chosen so that $m \leq \hat{f} \leq M$ on $\mathbb{R}^n$.*

***Proof.*** We define $g : \mathbb{R}^n \to \mathbb{R}$ by

$$g(x) = \inf_{y \in S}\{f(y) + \mu_f(|x - y|)\}.$$

Now let $x_1$ and $x_2$ be points in $\mathbb{R}^n$. For any $\varepsilon > 0$, there is a point $y \in S$ such that $g(x_2) \geq f(y) + \mu_f(|x_2 - y|) - \varepsilon$. Then we have

$$\begin{aligned} g(x_1) - g(x_2) &\leq f(y) + \mu_f(|x_1 - y|) - f(y) - \mu_f(|x_2 - y|) + \varepsilon \\ &= \mu_f(|x_1 - y|) - \mu_f(|x_2 - y|) + \varepsilon \\ &\leq \mu_f(|x_1 - x_2|) + \varepsilon. \end{aligned}$$

Since $\varepsilon$ is arbitrary, we conclude that $g(x_1) - g(x_2) \leq \mu_f(|x_1 - x_2|)$ and, by reversing the roles of $x_1$ and $x_2$, we conclude that

$$|g(x_1) - g(x_2)| \leq \mu_f(|x_1 - x_2|).$$

In addition, if $s \in S$, then it's easy to see that $g(s) \leq f(s)$. On the other hand, for any $y \in S$, we have

$$f(s) \leq f(y) + \mu_f(|s - y|),$$

so $g(s) = f(s)$. For the first part of the lemma, we take $\hat{f} = g$.

For the second part of the lemma, we first observe that, if $f \geq m$ on $S$, then $g \geq m$ on $\mathbb{R}^n$. The proof is completed by taking

$$\hat{f}(x) = \min\{M, g(x)\}.$$

□

Our next step is the Weierstrass Approximation Theorem.

**Lemma 1.4.** *Let $f$ be a uniformly continuous function defined on some bounded set $K$ and suppose there is a modulus of continuity $\mu_f$ such that $|f(x) - f(y)| \leq \mu_f(|x-y|)$ for all $x$ and $y$ in $K$. Then there is a sequence $(P_m)$ of polynomials which converges uniformly to $f$ on $K$ and such that*

$$|P_m(x) - P_m(y)| \leq \mu_f(|x-y|)$$

*for all $x$ and $y$ in $K$.*

***Proof.*** First, we note that there is a positive number $R$ such that $|x| \leq R$ for all $x \in K$. Then we define $\bar{f} = f - \inf f$ on $K$ and we set $\bar{f}(x) = 0$ if $|x| \geq 2R$. Note that $\bar{f}$ satisfies the inequality $|\bar{f}(x) - \bar{f}(y)| \leq \mu_f(|x-y|)$ for all $x$ and $y$ in the domain of $\bar{f}$. We now extend $\bar{f}$ to a function $\hat{f}$ defined on all of $\mathbb{R}^n$ by using Lemma 1.3.

For our next step, we define the constants $a_m$, for $m$ a nonnegative integer, by

$$a_m = \int_{B(0,4R)} [(4R)^2 - |x|^2]^m \, dx,$$

and we define $\varphi_m$ by

$$\varphi_m(x) = \frac{[(4R)^2 - |x|^2]^m}{a_m}.$$

Our polynomials $P_m$ are then given by the equation

$$P_m(x) = \int_{\mathbb{R}^n} \varphi_m(y)\hat{f}(x-y)\, dy + \inf f.$$

We note that $P_m$ is well-defined because $\bar{f}$ has compact support. It follows from a change of variables that

$$P_m(x) = -\int_{\mathbb{R}^n} \varphi_m(x-y)\hat{f}(y)\, dy + \inf f,$$

and hence $P_m$ is a polynomial.

Furthermore, if $x_1$ and $x_2$ are in $K$, then $|x_1-y|\geq 2R$ and $|x_2-y|\geq 2R$ if $|y|\geq 4R$, so

$$\begin{aligned}|P_m(x_1)-P_m(x_2)| &= |\int_{\mathbb{R}^n}\varphi_m(y)(\hat{f}(x_1-y)-\hat{f}(x_2-y))\,dy| \\ &= |\int_{B(0,4R)}\varphi_m(y)(\hat{f}(x_1-y)-\hat{f}(x_2-y))\,dy| \\ &\leq \int_{B(0,4R)}\varphi_m(y)\mu_f(|x_1-x_2|)\,dy = \mu_f(|x_1-x_2|).\end{aligned}$$

To see that $(P_m)$ converges uniformly to $f$ on $K$, we examine the difference

$$D_m(x)=\hat{f}(x)-\int_{\mathbb{R}^n}\varphi_m(y)\hat{f}(x-y)\,dy$$

for $|x|\leq R$. We first observe that

$$D_m(x)=\int_{B(0,4R)}\varphi_m(y)[\bar{f}(x)-\bar{f}(x-y)]\,dy.$$

With $\delta\in(0,4)$ to be chosen, we now rewrite

$$D_m(x)=I_{1,m}(x)+I_{2,m}(x)$$

with

$$\begin{aligned}I_{1,m}(x) &= \int_{B(0,\delta R)}\varphi_m(y)[\bar{f}(x)-\bar{f}(x-y)]\,dy, \\ I_{2,m}(x) &= \int_{B(0,4R)\setminus B(0,\delta R)}\varphi_m(y)[\bar{f}(x)-\bar{f}(x-y)]\,dy.\end{aligned}$$

Then

$$|I_{1,m}(x)|\leq\int_{B(0,\delta R)}\varphi_m(y)\mu_f(\delta R)\,dy\leq\mu_f(\delta R)$$

because $\delta\leq 4$, and

$$|I_{2,m}(x)|\leq 2\sup\bar{f}\int_{B(0,4R)\setminus B(0,\delta R)}\varphi_m(y)\,dy.$$

We now observe that

$$\lim_{m\to\infty}\frac{(1-s^2)^m}{(1-(s/4)^2)^m}=0$$

for any $s\in(0,1)$. It follows that

$$\lim_{m\to\infty}\frac{\sup_{B(0,4R)\setminus B(0,\delta R)}\varphi_m}{\inf_{B(0,\delta R/4)}\varphi_m}=0,$$

and hence

$$\lim_{m\to\infty} \frac{\int_{B(0,4R)\setminus B(0,\delta R)} \varphi_m(y)\,dy}{\int_{B(0,\delta R)} \varphi_m(y)\,dy} = 0,$$

so

$$\lim_{m\to\infty} \int_{B(0,4R)\setminus B(0,\delta R)} \varphi_m(y)\,dy = 0.$$

Hence, if $\varepsilon > 0$ is given, we can find first $\delta$ sufficiently small so that $|I_{1,m}(x)| \le \varepsilon/2$ for all $x$ with $|x| \le R$ and then $m$ sufficiently large so that $|I_{2,m}(x)| \le \varepsilon/2$. Hence $|D_m(x)| \le \varepsilon$ if $m$ is sufficiently large, so $(P_m)$ converges uniformly to $f$ on $K$. □

We can now prove the existence of the function $\rho$ described earlier in this section.

**Lemma 1.5.** *Let $\varepsilon \in (0,1)$ be given and let $\Omega$ be a domain which possesses a vector field with modulus of obliqueness at least $\varepsilon$. Then there is a function $\rho \in C^\infty(\overline{\Omega})$ such that $\beta \cdot D\rho > |\beta|$ for any vector field $\beta$ with modulus of obliqueness at least $\varepsilon$.*

***Proof.*** With $R$ as in the definition of the modulus of obliqueness, we cover $\partial\Omega$ by finitely many balls $B_1, \ldots, B_N$ with radius $R$ centered on $\partial\Omega$, and we let $(\psi_j)_{j=1}^N$ be a partition of unity on $\partial\Omega$ subordinate to this cover. In each ball $B_j$, we use a coordinate system $Y$ (centered at the center of $B_j$) so that

$$\Omega \cap B_j = \{y : y^n > \omega_j(y'), |y| < R\}$$

for some Lipschitz function $\omega_j$. Then, with $\eta > 0$ to be chosen, the Weierstrass Approximation Theorem provides a function $\omega_{j,\eta} \in C^\infty(B_{n-1}(0,R))$ such that $|\omega_j(y') - \omega_{j,\eta}(y')| < \eta$ and $|\omega_{j,\eta}(x') - \omega_{j,\eta}(y')| < \omega_0|x' - y'|$ for all $x'$ and $y'$ in $B_{n-1}(0,R)$. We define $w_j$ on $B_j$ by

$$w_j(x) = y^n - \omega_{j,\eta}(y').$$

Next, we set $\kappa = 1/(1+\omega_0^2)^{1/2}$ and note that $\beta^n \ge \kappa|\beta|$, using the $y$ coordinates in any ball $B_j$. It follows that $w = \sum \psi_j w_j$ satisfies

$$\begin{aligned}\beta \cdot Dw &= \sum \psi_j \beta \cdot Dw_j + \sum w_j \beta \cdot D\psi_j \\ &\ge [\varepsilon\kappa - \eta \sum |D\psi_j|]|\beta|.\end{aligned}$$

Since the cover is finite, we choose

$$\eta = \frac{\varepsilon\kappa}{2\sup\sum|D\psi_j|}$$

to infer that $\beta \cdot Dw \ge \frac{1}{2}\varepsilon\kappa|\beta|$ on $\partial\Omega$. Hence the desired function is

$$\rho = \frac{2}{\varepsilon\kappa} w.$$

□

With these preliminaries, the maximum principle follows easily.

**Lemma 1.6.** *Suppose that there is a constant $C_2$ such that $|a^{ij}| + |b^i| < -C_2 c$ in $\Omega$. Suppose also that $\beta$ is uniformly oblique and that $\beta^0 \le 0$ on $\partial\Omega$. If $u$ satisfies (1.1), then $u \le 0$ in $\Omega$.*

***Proof.*** Since $\beta$ is uniformly oblique, there is $\varepsilon \in (0,1)$ such that $\beta$ has modulus of obliqueness less than $\varepsilon$. For any $K \in \mathbb{R}$, we set $w_K = K - \rho$ with $\rho$ the function from Lemma 1.5. If $K \ge \sup \rho$, then $\beta \cdot Dw_K + \beta^0 w_K < 0$ on $\partial\Omega$ and, if $K$ is sufficiently large (determined only by $C_2$ and $\sup |D^2\rho| + |D\rho| + |\rho|$), we also have

$$a^{ij} D_{ij} w_K + b^i D_i w_K + c w_K = -a^{ij} D_{ij}\rho - b^i D_i \rho + c(K - \rho) < 0.$$

As before, it follows that $u - \delta w_K$ can't have a positive maximum for any $\delta > 0$ and hence $u \le 0$. □

## 1.4 A generalized change of variables formula

For our next collection of maximum principles, we need an expression which generalizes the classical change of variables formula. Since the result is not generally known in the United States, we provide a simple proof. In fact, a more precise version of this theorem can be proved by similar methods (see the Notes for details), but we give a result which suffices for our purposes.

**Theorem 1.7.** *Let $\Omega$ be an open subset of $\mathbb{R}^n$ and let $F\colon \Omega \to \mathbb{R}^n$ be a $C^1$ map. Then*

$$\int_\Omega u \circ F(x) |\det DF(x)|\, dx \ge \int_{F(\Omega)} u(y)\, dy \tag{1.6}$$

*for any nonnegative measurable function $u$.*

***Proof.*** First, we use Sard's Theorem to observe that the set of $y$ such that $F(x) = y$ for some $x \in \Omega$ with $\det DF(x) = 0$ has measure zero. Hence, if we write $\Omega^*$ for the subset of $\Omega$ on which $\det DF \ne 0$, we have

$$\int_\Omega u \circ F(x) |\det DF(x)|\, dx = \int_{\Omega^*} u \circ F(x) |\det DF(x)|\, dx, \tag{1.7a}$$

$$\int_{F(\Omega)} u(y)\, dy = \int_{F(\Omega^*)} u(y)\, dy. \tag{1.7b}$$

Suppose now that $u$ is bounded above. Now let $D$ be a compact subset of $\Omega^*$. Then, by the Inverse Function Theorem, for each $y \in D$, there is an open subset $U_y$ of $\Omega^*$ such that $F$ is a homeomorphism from $U_y$ to $F(U_y)$ and $y \in U_y$. Since $D$ is compact, we can cover $D$ by finitely many of these open sets, which we label as $U_1, \dots, U_M$ for convenience. It follows that there is a positive constant $\delta$ such that any subset of $D$ with diameter less than $\delta$ is contained in some $U_j$. We now fix $\zeta > 0$ and take a countable collection $\{D_j\}$ of open subsets of $\Omega^*$ with diameter $\delta/2$ such that $D \subset \cup D_j$ and $\sum |D_j| \le |D| + \zeta$.

The classical change of variables formula implies that

$$\int_{D_j} u \circ F(x) |\det DF(x)|\, dx = \int_{F(D_j)} u(y)\, dy,$$

and hence

$$\sum \int_{D_j} u \circ F(x) |\det DF(x)|\, dx \ge \sum \int_{F(D_j)} u(y)\, dy.$$

Since $F(D) \subset \cup F(D_j)$, it follows that

$$\sum \int_{F(D_j)} u(y)\, dy \ge \int_{F(D)} u(y)\, dy.$$

In addition there is a constant $K$ such that $u \circ F |\det(DF)| \le K$ on $\cup D_j$, and this constant can be chosen independent of $\zeta$. It follows that

$$\int_D u \circ F(x) |\det DF(x)|\, dx + K\zeta \ge \int_{F(D)} u(y)\, dy.$$

Since this inequality is true for all $\zeta > 0$, we conclude that

$$\int_D u \circ F(x) |\det DF(x)|\, dx \ge \int_{F(D)} u(y)\, dy.$$

Next, we note that there are an increasing sequence of bounded measurable functions $(u_j)$ with $u_j \to u$ a.e. and an increasing sequence of compact sets $(D_j)$ such that $\cup D_j = \Omega^*$, so the Monotone Convergence Theorem gives

$$\int_{\Omega^*} u \circ F(x) |\det DF(x)|\, dx \ge \int_{F(\Omega^*)} u(y)\, dy$$

for any nonnegative measure $u$. The proof is completed by combining this inequality with (1.7). □

## 1.5 The Aleksandrov-Bakel′man-Pucci maximum principles

There are times when the maximum principles described so far are inadequate for important estimates. In this section, we present results in which the maximum of the solution is estimated in terms of some integral norm of data. For these estimates, we introduce an important set and an important function. If $u$ is a continuous function defined on some open set $\Omega$, we write $\Gamma(u)$ for the *upper contact set of* $u$, that is, the set of all $x \in \Omega$ such that there is a vector $p$ with

$$u(y) \le u(x) + p \cdot (x - y) \tag{1.8}$$

for all $y \in \Omega$. In addition, we write $\chi_u$ for the *normal mapping*, that is, $\chi_u(x)$ is the set of all vectors $p$ such that (1.8) holds for all $y \in \Omega$. In particular, $\chi_u(x)$ is non-empty if and only if $x \in \Gamma(u)$.

As a first step, we present a simple version of the main result.

**Lemma 1.8.** *Let $\Omega$ be a bounded open set, let $\beta$ be inward pointing on $\partial\Omega$, and let $\beta^0 < 0$ on $\partial\Omega$. Suppose that there is a constant $B_0$ such that*

$$|\beta| \le -B_0\beta^0 \tag{1.9}$$

*on $\partial\Omega$. Let $h$ be a positive function in $L^1_{loc}(\mathbb{R}^n)$ and define $H$ by*

$$H(t) = \int_{\{|p|<t\}} h(p)\, dp. \tag{1.10}$$

*If $u \in C^2(\Omega) \cap C(\overline{\Omega})$ and if $g$ is a nonnegative constant with*

$$Mu \ge \beta^0 g$$

*on $\partial\Omega$, then*

$$u \le g + (B_0 + \operatorname{diam}\Omega)H^{-1}\left(\int_{\Gamma^+(u)} h(Du)|\det D^2u|\, dx\right), \tag{1.11}$$

*where $\Gamma^+(u)$ is the subset of $\Gamma(u)$ on which $u \ge g$.*

***Proof.*** Let $p$ be a vector such that $p \notin \chi_u(\Gamma^+(u))$. Our first step is to look at a point $x_0$ at which $w$, defined by $w(x) = u(x) - p \cdot x$, attains its maximum.

For all $y \in \Omega$, we have $w(y) \le w(x_0)$, so

$$u(y) = w(y) + p \cdot y \le w(x_0) + p \cdot y = u(x_0) + p \cdot (y - x_0) \tag{1.12}$$

for all $y \in \Omega$, and hence $x_0 \in \Gamma(u)$. If $x_0 \in \Omega$, then $u(x_0) \le g$, while, if $x_0 \in \partial\Omega$, it follows that

$$0 \ge \beta \cdot Dw = \beta \cdot Du - \beta \cdot p \ge -\beta^0[u - B_0|p| - g]$$

at $x_0$. Hence $u(x_0) \le B_0|p| + g$ and then equation (1.12) implies that

$$\sup u \le g + (B_0 + \operatorname{diam}\Omega)|p|. \tag{1.13}$$

We now estimate $|p|$ in terms of the norm of $\det(D^2u)$. To this end, we set

$$R = H^{-1}\left(\int_{\Gamma^+(u)} h(Du)|\det D^2u|\,dx\right). \tag{1.14}$$

Because $Du$ is continuously differentiable, $\chi_u(x) = \{Du(x)\}$ if $x \in \Gamma^+(u)$, and then the change of variables theorem (Theorem 1.7) implies that

$$\int_{\chi_u(\Gamma^+(u))} h(y)\,dy \le \int_{\Gamma^+(u)} h(Du)|\det D^2u|\,dx. \tag{1.15}$$

It follows that, for any $\varepsilon > 0$, there is a vector $p$ with $|p| < R + \varepsilon$ and $p \notin \Gamma^+(u)$. Hence (1.13) gives

$$\sup u \le g + (B_0 + \operatorname{diam}\Omega)(R + \varepsilon),$$

and hence

$$\sup u \le g + (B_0 + \operatorname{diam}\Omega)R.$$

Combining this inequality with (1.14) and (1.15) yields (1.11). □

We point out that, if $h \in L^1(\mathbb{R}^n)$, then (1.11) only provides a bound for the supremum if

$$\int_{\Gamma^+(u)} h(Du)|\det D^2u|\,dx < \int_{\mathbb{R}^n} h(p)\,dp.$$

To continue, we study the corresponding estimate when $u$ satisfies a linear differential inequality in $\Omega$.

**Theorem 1.9.** *Suppose $[a^{ij}]$ is positive definite and set $\mathcal{D} = \det[a^{ij}]$. Suppose $c \le 0$ in $\Omega$. Suppose $\beta$ is inward pointing and $\beta^0 < 0$ and suppose that there is a nonnegative constant $B_0$ such that (1.9) holds. Suppose also that there is a nonnegative constant $B_1$ such that*

$$\int_\Omega \frac{|b|^n}{\mathcal{D}}\,dx \le B_1. \tag{1.16}$$

*If $u \in C^2(\Omega) \cap C(\overline{\Omega})$ and if*

$$Lu \geq -f \text{ in } \Omega, \quad Mu \geq g\beta^0 \text{ on } \partial\Omega \tag{1.17}$$

*for some nonnegative function $f$ and some nonnegative constant $g$, then*

$$u \leq g + C(n, B_1)(B_0 + \operatorname{diam}\Omega) \left\| \frac{f}{\mathcal{D}^{1/n}} \right\|_{L^n(\Gamma^+(u))}, \tag{1.18}$$

*where $\Gamma^+(u)$ is the subset of $\Gamma(u)$ on which $u > g$.*

***Proof.*** With $\mu$ a constant to be further determined, we define

$$h(p) = \frac{1}{\mu^n + |p|^n}$$

and note that

$$H(t) = \omega_n \ln\left(1 + \frac{t^n}{\mu^n}\right).$$

Hence Lemma 1.8 gives

$$\sup u \leq g + H^{-1}\left(\int_{\Gamma^+(u)} h(Du)|\det D^2u|\,dx\right).$$

Next, we estimate the integral on the right. For simplicity, we abbreviate $\Gamma^+(u)$ to $G$. First observe that, for any $x \in G$ and all sufficiently small $h$, we have

$$u(x+h) + u(x-h) - 2u(x) \leq 0$$

so $[D^2u]$ is negative semidefinite on $G$. It follows that $|\det D^2u| = \det(-D^2u)$, and the matrix inequality

$$\det A \det B \leq \frac{\operatorname{tr}(AB)^n}{n}$$

(valid for all positive semidefinite matrices $A$ and $B$) allows us to conclude that

$$\det(-D^2u) \leq \frac{(-a^{ij}D_{ij}u)^n}{n\mathcal{D}}.$$

Since $u > 0$ on $G$, we infer that

$$\det(-D^2u) \leq \frac{(f + |b||Du|)^n}{n\mathcal{D}}$$

on $G$. Now we observe that

$$\begin{aligned} f + |b||Du| &\leq 2\max\{\frac{f}{\mu}, |b|\}\max\{\mu, |Du|\} \\ &\leq 2(\frac{f^n}{\mu^n} + |b|^n)^{1/n}(\mu^n + |Du|^n)^{1/n}, \end{aligned}$$

and therefore

$$h(Du)\det(-D^2u) \le \frac{2^n}{n}\left(\frac{f^n}{\mu^n \mathcal{D}} + \frac{|b|^n}{\mathcal{D}}\right).$$

Therefore

$$\int_G h(Du)|\det(D^2u)|\,dx \le \frac{2^n}{n}\left(\int_G \frac{f^n}{\mu^n \mathcal{D}}\,dx + B_1\right).$$

We now take $\mu$ to satisfy

$$\mu > \left(\int_G \frac{f^n}{\mathcal{D}}\,dx\right)^{1/n}$$

to conclude that

$$\sup u \le g + (B_0 + \operatorname{diam}\Omega)H^{-1}(1+B_1)\mu,$$

Sending

$$\mu \to \left(\int_G \frac{f^n}{\mathcal{D}}\,dx\right)^{1/n}$$

completes the proof. □

Note that we can replace the integral of $|b|^n/\mathcal{D}$ over $\Omega$ with the integral over $G$ in this result. For our applications to strong solutions in Chapter 6, we point out a further variant.

**Corollary 1.10.** *Theorem 1.9 remains valid if we take $G$ to be the set of all $x \in \Gamma(u)$ for which $u(x) > g$ and (1.8) holds with some $p$ satisfying $u(x) \ge B_0|p| + g$.*

***Proof.*** We only need to observe that, if $p$ is chosen not in $\chi_u(G)$ and $x \in \Gamma(u)$ is a point at which $u(x) > g$ and (1.8) holds, then we have $u(x) \le B_0|p| + g$. Hence (1.13) holds. The remainder of the proof is unchanged. □

Another variation can be obtained if the vector $\beta$ is further restricted to a cone.

**Theorem 1.11.** *Suppose $[a^{ij}]$ is positive definite with determinant $\mathcal{D}$ and that $c \le 0$. Suppose $\beta$ is inward pointing and $\beta^0 \le 0$. Suppose further there is a cone $K$ with opening angle $\theta^*$ such that*

$$\inf_{v\in K} \frac{\beta\cdot v}{|\beta||v|} = B_0 > 0.$$

*If $u \in C^2(\Omega) \cap C(\overline{\Omega})$ satisfies $Lu \geq -f$ in $\Omega$ for some nonnegative function $f$ and $Mu \geq -g|\beta|$ for some nonnegative constant $g$ on the subset of $\partial\Omega$ on which $u > 0$, then*

$$\sup u \leq C(n, \theta^*, B_1) \operatorname{diam} \Omega \left( \frac{g}{B_0} + \left\| \frac{f}{\mathcal{D}^{1/n}} \right\|_{L^n(G)} \right),$$

*where $G$ is the subset of $\Gamma(u)$ on which $u > 0$ and (1.8) holds for some $p \in -K$ with $|p| > g/B_0$.*

***Proof.*** This time, we take $p$ to be a vector in $-K \setminus \chi_u(G)$ with $|p| > g/B_0$ and we define $w$ by $w(x) = u(x) - p \cdot x$. Just as before, we take $x_0 \in \overline{\Omega}$ to be a point at which $w$ attains its maximum. The proof of Lemma 1.8 shows that $x_0$ can't be in $\Omega$ unless $u(x_0) \leq 0$, and $u(x_0) \leq 0$ implies that

$$\sup u \leq \operatorname{diam} \Omega |p|. \tag{1.19}$$

If $x_0 \in \partial\Omega$ and $u(x_0) > 0$, then we have

$$\begin{aligned} 0 &\geq \beta \cdot Dw = \beta \cdot Du - \beta \cdot p \geq -g|\beta| - \beta \cdot p \\ &\geq |\beta|(-g + B_0|p|). \end{aligned}$$

Hence $|p| \leq g/B_0$, which contradicts our assumption on $p$. Hence $x_0$ can't be in $\partial\Omega$ if $u(x_0) > 0$, so we obtain (1.19).

We now define $\omega_n(K)$ to be the measure of the unit ball intersected with $K$ (note that this positive number is completely determined by $n$ and $\theta^*$), and we choose $R$ so that

$$\omega_n(K) \int_0^R h(r) r^{n-1} \, dt = \int_{\chi_u(G)} h(|y|) \, dy$$

with $h$ from the proof of Theorem 1.8. The proof of that theorem now shows that

$$R \leq C(n, \theta^*, B_1) \left( \int_G \frac{f^n}{\mathcal{D}} \, dx \right)^{1/n}.$$

We now note that, for any $\varepsilon > 0$, $p$ can be chosen so that $|p| \leq \max\{g/B_0, R\} + \varepsilon$. It follows that (after sending $\varepsilon \to 0$) that

$$\sup u \leq \operatorname{diam} \Omega \max\{g/B_0, R\},$$

and this inequality along with our estimate for $R$ finishes the proof. □

Just as before, we can repeat the proof with the integral of $|b|^n/\mathcal{D}$ over $\Omega$ with the integral over $G$, and we can take $G$ to be an even smaller set.

**Corollary 1.12.** *Theorem 1.11 remains valid if we take $G$ to be the set of all $x \in \Gamma(u)$ for which $u(x) > 0$ and (1.8) holds with some $p$ satisfying the additional conditions $u(x) \geq \operatorname{diam} \Omega |p|$ and $|p| \geq g/B_0$.*

## 1.6 The interior weak Harnack inequality

We can prove a number of useful estimates on solutions of oblique derivative problems from the pointwise estimates of the previous section. One of the most important of these is known as the weak Harnack inequality because it is related to the classical inequality for harmonic functions proved by Harnack (see paragraph 19, page 2 from [70]). Although we could prove this result directly for solutions of oblique derivative problems, it will be more helpful to recall the result for solutions of elliptic equations. We follow the basic plan from Section 9.8 of [64] with a slight variation. Our first step is a pointwise estimate for solutions of elliptic equations.

**Lemma 1.13.** *Suppose that there are positive constants $\mu$, $\mu_1$ and $\mu_2$, and a positive function $\lambda$ such that*

$$\lambda|\xi|^2 \le a^{ij}\xi_i\xi_j \le \mu\lambda|\xi|^2 \tag{1.20a}$$

*for all $\xi \in \mathbb{R}^n$,*

$$|b| \le \mu_1\lambda, \tag{1.20b}$$

$$c \ge -\mu_2\lambda. \tag{1.20c}$$

*Let $x_1 \in \mathbb{R}^n$ and let $r$ be a positive constant. Then, for any constants $\theta_1$ and $\theta_2$ in $(0,1)$ with $\theta_1 < \theta_2$, there are positive constants $C_1$ and $\zeta$, determined only by $n$, $\theta_1$, $\theta_2$, $\mu$, $\mu_1 r$, and $\mu_2 r^2$, such that, if $u \in C^2(B(x_1,r)) \cap L^\infty(B(x_1,r))$ satisfies*

$$Lu \le \lambda f \text{ in } B(x_1,r) \tag{1.21}$$

*for some nonnegative function $f \in L^n(B(x_1,r))$, with $\|f\|_n = F$, then, for any nonnegative constant $h$ such that*

$$|\{x \in B(x_1,\theta_1 r) : \bar{u}(x) < h\}| \le \zeta^n r^n, \tag{1.22}$$

*we have*

$$C_1\bar{u} \ge h \text{ in } B(x_1,\theta_2 r), \tag{1.23}$$

*where $\bar{u} = u + Fr$.*

***Proof.*** We define $\eta$ by

$$\eta(x) = 1 - \frac{|x-x_1|^2}{r^2},$$

and we let $q > 2$ be a constant at our disposal.

Because $u \geq 0$, we may assume without loss of generality that $c \leq 0$ and then

$$L(\eta^q) = q\eta^{q-2}[\eta a^{ij} D_{ij}\eta + (q-1)a^{ij} D_i\eta D_j\eta + \eta b^i D_i\eta + \frac{c}{q}\eta^2].$$

Since $|D\eta| \geq \theta_1/r$ in $B(x_1, r) \setminus B(x_1, \theta_1 r)$ and $a^{ij} D_{ij}\eta \geq -\mu\lambda/r^2$, it follows that there is a number $q_0 > 2$ such that

$$L(\eta^q) \geq 0 \text{ in } B(x_1, r) \setminus B(x_1, \theta_1 r)$$

if $q \geq q_0$.

We now choose $q = q_0$ and we set $v = h\eta^q - u$. It follows that there are positive constants $C_2$ and $C_3$ such that $L(\eta^q) \geq -C_2\lambda/r^2$ in $B(x_1, r)$ and

$$\left\| \frac{(Lv)^-}{\mathcal{D}^{1/n}} \right\|_{L^n(G)} \leq C_3 h\zeta/r + F,$$

where $G$ is the subset of $\Gamma(v)$ on which $v \geq 0$. Theorem 1.9 now implies that

$$v \leq C_4(h\zeta + Fr)$$

in $B(x_1, r)$. Next, we set $C_5 = \frac{1}{2}\inf_{B(x_1, \theta_2 r)} \eta^q$ and choose $\zeta$ so that $C_3\zeta \leq C_5$. Then

$$\bar{u} \geq -C_4 rF + C_5.$$

Since $\bar{u} \geq rF$, this inequality implies (1.23) with $C_1 = C_5/(1 + C_5)$. □

Our proof of the weak Harnack inequality rests on two results from harmonic analysis. We refer the reader to Section 9.2 of [64] for proofs; specifically, our Lemma 1.14 is Lemma 9.2 from [64] and our Lemma 1.15 is Lemma 9.23 from [64].

**Lemma 1.14.** *Let $f$ be a measurable function on some open subset $\Omega$ of $\mathbb{R}^n$ such that $|f|^p \in L^1(\Omega)$ for some $p > 0$, and set*

$$\mu(t) = \{x \in \Omega : |f(x)| \geq t\}.$$

*Then*

$$\int_\Omega |f|^p\, dx = p\int_0^\infty t^{p-1}\mu(t)\, dt. \tag{1.24}$$

For our next lemma, we write $K_r(z)$ for the cube

$$\{y \in \mathbb{R}^n : |y^i - z^i| < r \text{ for } i = 1, \dots, n\}.$$

**Lemma 1.15.** *Let $K_0$ be a cube in $\mathbb{R}^n$ and let $w \in L^1(K_0)$. For $k \in \mathbb{R}$, set*

$$\Gamma_k = \{z \in K_0 : w(z) < k\},$$

*and suppose that there are positive constants $\delta < 1$ and $C_0$ such that*

$$\sup_{K_0 \cap K_{3\rho}(z)} (w - k) \le C_0 \tag{1.25}$$

*whenever $k$, $z$, and $\rho$ satisfy $K_\rho(z) \subset K_0$ and*

$$|\Gamma_k \cap K_\rho(z)| \ge \delta |K_\rho(z)|. \tag{1.26}$$

*Then, for any $k \in \mathbb{R}$, we have*

$$\sup_{K_0}(w - k) \le C_0 \left(1 + \frac{\ln(|\Gamma_k|/|K_0|)}{\ln \delta}\right).$$

Our weak Harnack inequality takes the following form.

**Theorem 1.16.** *Let $L$ satisfy the hypotheses from Lemma 1.13 and suppose $u \in C^2(\Omega)$ satisfies $Lu \le \lambda f$, for some $f \in L^n(\Omega)$, and $u \ge 0$ in $\Omega$. Then there are positive constants $\kappa$ and $C$ such that*

$$\left(R^{-n} \int_{B(R)} u^\kappa \, dx\right)^{1/\kappa} \le C \left(\inf_{B(R)} u + R\|f\|_{L^n(B(2R))}\right)$$

*for any $R$ such that $B(2R) \subset \Omega$.*

***Proof.*** We set

$$F = \|f\|_{L^n(B(2R))}, \quad \bar{u} = u + FR, \quad w = -\ln \bar{u}.$$

We then let $\zeta$ and $C_1$ be the constants from Lemma 1.13 corresponding to $\theta_1 = 1/(3n)$ and $\theta_2 = 1/\sqrt{n}$.

By using $h = e^{-k}$, we conclude from Lemma 1.13 that (1.26) (with $r = 3\rho$) implies (1.22) provided $K_0 = K(x_0, 3R/\sqrt{n})$. It follows that (1.25) is satisfied with $C_0 = -\ln C_1$.

We now use Lemma 1.15 to infer that

$$\mu(t) = |\{z \in K_0 : \bar{u}(z) > t\}|$$

(which is equal to $|\Gamma_k|$ for $k = e^{-t}$) satisfies the inequality

$$\mu(t) \le C(\inf u/t)^\kappa R^n \tag{1.27}$$

for some positive constants $C$ and $\kappa$ determined by the same quantities as for $C_0$. If we multiply this equation by $t^{-1+\kappa/2}$, integrate $t$ from $\inf \bar{u}$ to $\infty$, and set $p = \kappa/2$, we see that

$$\int_{\inf \bar{u}}^{\infty} t^{p-1} \mu(t) \, dt \le C(\inf \bar{u})^p R^n.$$

On the other hand, $\mu(t) \le CR^n$, so we also have

$$\int_0^{\inf \bar{u}} t^{p-1}\mu(t)\,dt \le C(\inf \bar{u})^p R^n.$$

Adding these inequalities and using Lemma 1.14 tells us that

$$\left(R^{-n}\int_{B(R/3n)} \bar{u}^p\,dx\right)^{1/p} \le C(\inf_{B(R/3n)} \bar{u}),$$

and a simple chaining argument yields

$$\left(R^{-n}\int_{B(R/2)} \bar{u}^p\,dx\right)^{1/p} \le C(\inf_{B(R/2)} \bar{u}).$$

The proof is completed by recalling the definition of $\bar{u}$. □

Although we can derive a number of important results from the weak Harnack inequality, we defer their statements (and proofs) to a later section.

## 1.7 The weak Harnack inequality at the boundary

In deriving the interior weak Harnack inequality, a crucial role was played by the function $\eta$ in Lemma 1.13. The corresponding function for the oblique derivative boundary condition is more complicated. To see why this is so, let us first consider a special case. Suppose $\beta = (0, \dots, 0, 1)$, and set $S = \{x : |x| < 1, x^n = 0\}$ and $B^+ = \{x : |x| < 1, x^n > 0\}$. Then $w(x) = 1 - |x|^2$ satisfies $\beta \cdot Dw = 0$ on $S$, $w > 0$ on $B^+$ and $w$ vanishes on $\partial B^+ \setminus S$. (Note that $w$ is exactly the function $\eta$ from Lemma 1.13.) For more general uniformly oblique $\beta$, it's easy to modify this function to obtain a function satisfying the following conditions: $w$ is a function such that $\beta \cdot Dw \le 0$ in some neighborhood $N^0$ of 0 in $S$ and $w > 0$ in some open set $\Omega^*$ with $w = 0$ on $\partial\Omega^* \setminus N^0$ and $N^0 \subset \partial\Omega^*$. Specifically, set $x_\tau = (0, \dots, 0, -\tau)$ for $\tau \in (0,1)$ and define $w(x) = 1 - |x - x_\tau|^2$. By choosing $\tau$ sufficiently close to 1 and choosing $\Omega^*$ to be the subset of $B^+$ on which $w > 0$, we obtain the desired function. In fact, if $\partial\Omega$ is sufficiently smooth, it is not hard to adapt this construction. Our main point here is to provide such a function for more general domains.

We begin by taking $\alpha_1$, $\varepsilon$, and $\omega_0$ to be positive constants with $\alpha_1 < 1$ and $\varepsilon < 1$. We then set $A = 2^{1+2\varepsilon}\varepsilon^{-4\varepsilon}$, and we define $G\colon \mathbb{R}^n \times (0,\infty) \to \mathbb{R}$

by

$$G(y,r) = \left(\frac{|y'|^2}{r^2} + \alpha_1^2\right)^{(1+\varepsilon)/2} + \frac{(y^n)^2}{(A\omega_0 r)^2}. \tag{1.28}$$

We also introduce, for any $x_1 \in \mathbb{R}^n$, the set

$$E(x_1,r) = \{x : G(x - x_1, r) < 1\}. \tag{1.29}$$

With these definitions, we can state the following lemma.

**Lemma 1.17.** *Let $R > 0$ and suppose*

$$\{x : x^n > \omega_0|x'|, |x| < R\} \subset \Omega, \tag{1.30a}$$

$$\omega_0|\beta'| \le (1-\varepsilon)\beta^n \text{ on } \partial\Omega[R], \tag{1.30b}$$

*where $\partial\Omega[R]$ is the intersection of $\partial\Omega$ with the ball of radius $R$ centered at the origin. Let $r \in (0,R)$ and suppose $x_1 = (0,\dots,0,x_1^n)$ satisfies the conditions*

$$x_1^n \ge \left(A - \frac{\varepsilon^6}{4}\right)\omega_0 r \tag{1.31}$$

*and $E(x_1,r) \subset B(0,R)$. Then $\eta$ defined by*

$$\eta(x) = 1 - G(x - x_1, r)$$

*satisfies the inequality*

$$\beta \cdot D\eta \ge \varepsilon^4 \frac{\beta^n}{A\omega_0 r}$$

*on $S$, the subset of $\partial\Omega$ on which $\eta < 1$.*

***Proof.*** It follows from (1.30a) that

$$x_1^n - x^n \ge \left(A - \frac{\varepsilon^6}{4}\right)\omega_0 r - \omega_0|x'|$$

for all $x \in S$. Since $\varepsilon^6/4 < 1$ and $A > 2$, we have $A - \varepsilon^6/4 > 1$. Moreover, because $|x'| < r$ on $S$, we have $x_1^n - x^n > 0$ on $S$.

We now set

$$N = \frac{x_1^n - x^n}{A\omega_0 r}, \quad P = \frac{|x'|}{r}.$$

We also set $\alpha = \varepsilon^6/4$. Our previous inequality then becomes

$$N \ge 1 - \frac{\alpha}{A} - \frac{P}{A}, \tag{1.32}$$

and the definition of $G$ implies that

$$(P^2 + \alpha_1^2)^{(1+\varepsilon)/2} + N^2 \leq 1$$

on $S$. Combining these inequalities yields

$$P^{1+\varepsilon} + \left(1 - \frac{\alpha}{A} - \frac{P}{A}\right)^2 \leq 1,$$

and hence

$$P^{1+\varepsilon} - \frac{2P}{A} \leq \frac{2\alpha}{A}. \tag{1.33}$$

We now claim that

$$P \leq \left(\frac{2 + 2\varepsilon^2}{A}\right)^{1/\varepsilon}. \tag{1.34}$$

If this claim is false, then we have

$$P^{1+\varepsilon} - \frac{2P}{A} = [AP^\varepsilon - 2]\frac{P}{A} > [(2 + 2\varepsilon^2) - 2]\frac{P}{A} = \frac{2\varepsilon^2 P}{A}.$$

But $P \geq (2/A)^{1/\varepsilon} = \varepsilon^4/4$, so $\varepsilon^2 P \geq \alpha$, and hence

$$P^{1+\varepsilon} - \frac{2P}{A} > \frac{2\alpha}{A}.$$

It follows that (1.33) does not hold if (1.34) does not hold. Hence (1.34) is valid.

Our next step is a suitable lower bound for $N$. From (1.34), we have

$$\alpha + P \leq \frac{\varepsilon^6}{4} + \varepsilon^4 \frac{(1 + \varepsilon^2)^{1/\varepsilon}}{4}.$$

A straightforward calculation shows that the function $H$ defined by

$$H(s) = (1 + s^2)^{1/s}$$

satisfies

$$H'(s) = \frac{H(s)}{s^2} g(s^2)$$

with

$$g(\sigma) = \frac{2\sigma}{1 + \sigma} - \ln(1 + \sigma).$$

Since $g'(\sigma) = (1 - \sigma)/(1 + \sigma)^2$ and $g(0) = 0$, it follows that $g(\sigma) \geq 0$ for $\sigma \in [0, 1]$, and hence $H$ is increasing on the interval $(0, 1)$. Observing also that $H(1) = 2$ implies that

$$\alpha + P \leq \frac{\varepsilon^6}{4} + \frac{\varepsilon^4}{2} \leq \varepsilon^4.$$

Since $A \geq 2$, it follows that

$$N \geq 1 - \frac{1}{2}\varepsilon^4.$$

We are now in a position to compute $\beta \cdot D\eta$ on $S$. We have

$$\begin{aligned}\beta \cdot D\eta &= 2\beta^n \frac{(x_1^n - x^n)}{(A\omega_0 r)^2} - (1+\varepsilon)\beta' \cdot \frac{x'}{r^2}(P^2 + \alpha_1^2)^{(\varepsilon-1)/2} \\ &\geq 2\frac{\beta^n}{A\omega_0 r}(N - \frac{1}{2}(1-\varepsilon^2)AP^\varepsilon) \\ &\geq 2\frac{\beta^n}{A\omega_0 r}(1 - \frac{1}{2}\varepsilon^4 - 1 + \varepsilon^4) \\ &= \varepsilon^4 \frac{\beta^n}{A\omega_0 r}.\end{aligned}$$

□

Although the explicit form of this estimate for $\beta \cdot D\eta$ may be useful, we shall typically just use the form

$$\beta \cdot D\eta \geq C(\varepsilon, \omega_0)\frac{\beta^n}{r}. \tag{1.35}$$

We now discuss some geometric aspects of this lemma. Note that, if $r$ is sufficiently small, the set $S$ may be empty. In this situation, the computation of $\beta \cdot D\eta$ is irrelevant and, in our applications, this function $\eta$ can be replaced by a simpler one, namely $\eta(x) = 1 - |x - x_1|^2/r^2$. The reason for introducing the constant $\alpha_1$ is a technical one; we want $\eta$ to be $C^2$ and this smoothness is true if and only if $\alpha_1 \in (0,1)$. Of course, if $\alpha_1 \geq 1$, the function $G$ is still defined, but then $E(x_1, r)$ is always empty. In practice, the choice of $\alpha_1$ will not be arbitrary; we shall impose a suitable smallness condition, which will be made explicit at the appropriate time.

The main application of this lemma is to prove a more sophisticated minimum principle. Roughly speaking, we show that if $u$ is nonnegative in an open set and greater than some positive constant in some small ball contained in the set, then we can obtain a quantitative lower bound on $u$. To obtain a form of this bound which leads to useful results, we need to choose our sets rather carefully. We also introduce some notation that will be used throughout this book (although often in a somewhat modified format). Assuming that condition (1.30) holds, we write $\Omega[R]$ and $\partial\Omega[R]$ for the intersection of $\Omega$ and $\partial\Omega$, respectively, with the ball $B(0, R)$.

**Lemma 1.18.** *Suppose conditions* (1.30) *are satisfied and that there are positive constants* $\mu$, $\mu_1$, $\mu_2$, *and* $\mu_3$ *and a positive function* $\lambda$ *such that conditions* (1.20) *hold and*

$$\beta^0 \geq -\mu_3 \beta^n. \tag{1.36}$$

*Let $\alpha_1 < 1$ and $r$ be positive constants, and let $x_1 = (0, x_1^n)$ be a point in $\mathbb{R}^n$ such that $E(x_1, r) \subset B(0, R)$ and (1.31) holds. Then, for any constants $\theta_1$, $\theta_2$ in $(0,1)$ with $\theta_1 < \theta_2$, there are positive constants $C_1$ and $\zeta$, determined only by $\alpha_1$, $n$, $\varepsilon$, $\omega_0$, $\mu$, $\mu_1 r$, $\mu_2 r^2$, and $\mu_3 r$, such that, if $u \in C^2(\Omega[R]) \cap C(\overline{\Omega[R]})$ satisfies*

$$Lu \le \lambda f \text{ in } \Omega[R], \quad Mu \le \beta^n G \text{ on } \partial\Omega[R] \tag{1.37}$$

*for some nonnegative function $f \in L^n(\Omega[R])$, with $\|f\|_n = F$, and nonnegative constant $G$, then, for any nonnegative constant $h$ such that*

$$|\{x \in E(\theta_1 r) : \bar{u}(x) < h\}| \le \zeta^n r^n, \tag{1.38}$$

*we have*

$$C_1 \bar{u} \ge h \text{ in } E(\theta_2 r), \tag{1.39}$$

*where $E(s) = E(x_1, s) \cap \Omega$ and $\bar{u} = u + (F + G)r$.*

***Proof.*** As in Lemma 1.13, we may assume that $c$ and $\beta^0$ are nonpositive. Let $\eta$ be the function from Lemma 1.17. Then, for any $q > 1$, we have

$$L(\eta^q) = q\eta^{q-2}[\eta a^{ij} D_{ij}\eta + (q-1)a^{ij} D_i\eta D_j\eta + \eta b^i D_i\eta + \frac{c}{q}\eta^2].$$

From the explicit form of $D\eta$, it follows that there are positive constants $C_2(\omega_0, n, \varepsilon, \theta_1)$ and $C_3(\alpha_1, \omega_0, n, \varepsilon, \theta_1)$ such that $|D\eta| \ge C_2/r$ on $E(r) \setminus E(\theta_1 r)$ and $|D^2\eta| \le C_3/r^2$ in $E(r)$, so there is a constant $q_0(\alpha_1, n, \mu, \mu_1 r, \mu_2 r^2, \omega_0, \varepsilon, \theta_1)$ such that

$$L(\eta^q) \ge 0$$

on $E(r) \setminus E(\theta_1 r)$ if $q \ge q_0$. In addition, we have

$$M(\eta^q) = q\eta^{q-1}[\beta \cdot D\eta + \eta\frac{\beta^0}{q}] \ge 0$$

on $E(r) \cap \partial\Omega[R]$ if $q \ge A\omega_0\mu_3 r/\varepsilon^4$.

We now fix $q = \max\{q_0, A\omega_0\mu_3 r/\varepsilon^4\}$. Then

$$L(\eta^q) \ge -C_3\lambda/r^2$$

on $E(\theta_1 r)$. Writing $B^*$ for the subset of $E(r)$ on which $u < h$ and $v = h\eta^q - u$, we conclude that

$$\left\| \frac{(Lv)^-}{\mathcal{D}^{1/n}} \right\|_{L^n(B^*)} \le C_4 h\zeta/r + F$$

and $Mv \ge -G\beta^n$ on $\partial B^* \cap \partial\Omega[R]$. In addition $v \le 0$ on $\partial B^* \setminus \partial\Omega[R]$, so Theorem 1.11 implies that

$$v \le C_5 r(F + G) + C_4 h\zeta$$

in $B^*$. The proof is completed in the same way as for Lemma 1.13. □

Although this result can be used to prove many interesting estimates, it will be convenient to show that a lower bound can be obtained in a much simpler geometry. Given a function $\omega$ (as in conditions (1.3)), we introduce the *Lipschitz cube* $K(y_0, r)$ as the set of all points $x$ such that the corresponding $y$ coordinates satisfy the inequalities:

$$|y^i - y_0^i| < r \text{ if } i < n, \tag{1.40a}$$

$$|y^n - y_0^n + \omega(y_0') - \omega(y')| < r. \tag{1.40b}$$

Note that the change of variables $z' = y'$, $z^n = y^n - \omega(y')$ converts $K(y_0, r)$ into the ordinary cube

$$\{z \in \mathbb{R}^n : |z^i - z_0^i| < r \text{ for } i - 1, \dots, n\}.$$

We shall take advantage of this fact in the next section. For our preliminary results, we write $K^*(y_0, r)$ for the subset of $K(y_0, r)$ on which $y^n > \omega(y')$; this set is just the intersection of $K(y_0, r)$ and $\Omega$.

**Lemma 1.19.** *Suppose conditions* (1.3) *are satisfied and that there are positive constants* $\mu$, $\mu_1$, $\mu_2$, *and* $\mu_3$ *and a positive function* $\lambda$ *such that* (1.20) *and* (1.36) *hold. Let* $r \in (0, \infty)$ *and* $x_0$ *satisfy* $K^*(y_0, 3(1 + A\omega_0)r) \subset \Omega[R]$. *Then there are positive constants* $C_2$ *and* $\zeta_1$, *determined only by* $n$, $\varepsilon$, $\omega_0$, $\mu$, $\mu_1 r$, $\mu_2 r^2$, *and* $\mu_3 r$, *such that, if* $u \in C^2(\Omega[R]) \cap C(\overline{\Omega[R]})$ *satisfies* (1.37) *for some nonnegative function* $f \in L^n(\Omega[R])$, *with* $\|f\|_n = F$ *and some nonnegative constant* $G$, *then, for any nonnegative constant* $h$ *such that*

$$|\{x \in K^*(y_0, r) : \bar{u}(x) < h\}| \le \zeta_1^n r^n, \tag{1.41}$$

*then*

$$C_2 \bar{u} \ge h \tag{1.42}$$

*in* $K^*(y_0, 3r)$.

***Proof.*** We start by proving a weaker result, namely, that there are constants $C_3$ and $\delta_1$ (determined by the same quantities as $C_2$ and $\eta_1$) so that (1.41) implies

$$C_3 \bar{u} \ge h \text{ in } K^*(y_0, (1 + \delta_1) r). \tag{1.43}$$

This result will follow from Lemma 1.18 via a suitable covering of Lipschitz cubes by sets of the form $E(y_1, \tilde{r})$.

To create the covering, we take $\alpha_1$ so that

$$(1 - \alpha_1^{1+\varepsilon})^{1/2} = 1 - \frac{\varepsilon^6}{32A}.$$

In addition, we work exclusively in $y$-coordinates now. For $y'_* \in \mathbb{R}^{n-1}$, we set

$$y^n_* = \omega(y'_*) + \min\{0, y^n_0 - \omega(y'_0) - r\}. \tag{1.44}$$

With $\tilde{r}$ to be chosen, we set

$$y'_1 = y'_*, \quad y^n_1 = y^n_* + \left(1 - \frac{\varepsilon^6}{16A}\right) A\omega_0\tilde{r}.$$

The key observation here is that, if $y' = y'_*$, then the point $(y', y^n)$ is in $E(y_1, \tilde{r})$ if and only if $|y^n - y^n_*| < (1 - \alpha_1^{1+\varepsilon})^{1/2} A\omega_0 r$. This inequality can be rewritten as

$$-\frac{\varepsilon^6}{32A} A\omega_0\tilde{r} < y^n - y^n_* < \left(2 - \frac{3\varepsilon^6}{32A}\right) A\omega_0\tilde{r}.$$

If $y \in K^*(y_0, r)$ and $y' = y'_*$, we have $0 < y^n - y^n_* < 2r$, so we want to choose $\tilde{r}$ such that

$$\left(2 - \frac{3\varepsilon^6}{32A}\right) A\omega_0\tilde{r} > 2r.$$

For the purpose of this proof, the exact choice of $\tilde{r}$ is not important, but we shall take

$$\tilde{r} = \frac{r}{(1 - \frac{\varepsilon^6}{16A})A\omega_0}.$$

With this choice for $\tilde{r}$, there is a positive constant $\delta_0$, determined only by $\varepsilon$, $n$, and $\omega_0$, such that

$$\{y : y' = y'_*, -\delta_0 < y^n - y^n_* < (2 + \delta_0)r\} \subset E(y_1, \tilde{r}).$$

Hence, there is a positive constant $\theta_0$, determined only by $\varepsilon$, $n$, and $\omega_0$, such that

$$\{y : |y' - y'_*| < \theta_0, -\tfrac{\delta_0}{2} < y^n - y^n_* < (2 + \tfrac{\delta_0}{2})r\} \subset E(y_1, \tilde{r}).$$

Therefore there are positive constants $\delta_1$ and $\varepsilon_1$, determined only by $\varepsilon$, $n$, and $\omega_0$, such that, for any $y \in K^*(y_0, (1 + \delta_1)r)$, there is a $y'_* \in \mathbb{R}^{n-1}$ with

$$|y' - y'_*| < \theta_0 r, \quad \max_{i<n} |y^i_* - y^i_0| < (1 - \varepsilon)r. \tag{1.45}$$

We therefore conclude that there are positive constants $\theta_1$ and $\theta_2$, determined only by $\varepsilon$, $n$ and $\omega_0$ with $\theta_1 < \theta_2 < 1$ such that, for every $y \in K^*(y_0, (1 + \delta_1)r)$, there is a $y'_*$ satisfying (1.45) so that $y \in E(y_1, \theta_2\tilde{r})$ and $E(y_1, \theta_1\tilde{r}) \subset K^*(y_0, r)$. With this choice for $\alpha_1$, $\theta_1$, and $\theta_2$, we use Lemma 1.18 to obtain constants $C_1$ and $\zeta$. We now choose $C_3 = C_1$ and $\zeta_1 = \zeta r/\tilde{r}$ to infer that (1.41) implies $C_2\bar{u}(y) \geq h$ for any $y \in K^*(y_0, (1 + \delta_1)r)$.

We now note that

$$|\{y \in K(y_0,(1+\delta_1)r) : \bar{u}(y) < h/C_3\}| = 0,$$

so our previous argument gives $C_3\bar{u} \geq h/C_3$ in $K^*(y_0,(1+\delta_1)^2 r)$. A simple induction argument then yields

$$C_3^m \bar{u} \geq h \text{ in } K^*(y_0,(1+\delta_1)^m r)$$

as long as $(1+\delta_1)^m \leq 3$. Since we can decrease $\delta_1$ without changing $C_3$, we may take $\delta_1$ so that $(1+\delta_1)^m r = 3$ for some positive integer $m$, and the proof is completed by taking $C_2 = C_3^m$. □

Of course, there is nothing special about the number 3 in this proof; it could be replaced by any number $N > 1$, and then the constant $C_1$ would also depend on $N$. The number 3 will be very useful in our proof of the weak Harnack inequality which we now present.

**Theorem 1.20.** *Let $u \in C^2(\Omega[R]) \cap C(\overline{\Omega[R]})$, and suppose $u \geq 0$ in $\Omega[R]$. Suppose that conditions* (1.3), (1.20a,b), *and*

$$|c| \leq \mu_2 \lambda, \tag{1.46a}$$

$$|\beta^0| \leq \mu_3 \beta^n \tag{1.46b}$$

*are satisfied. If there are a nonnegative function $f \in L^n(\Omega[R])$ and a nonnegative constant $G$ such that* (1.37) *holds, then there are positive constants $C$ and $p$, determined by $\varepsilon$, $\mu$, $\mu_1 R$, $\mu_2 R^2$, $\mu_3 R$, and $\omega_0$ such that*

$$\left(R^{-n}\int_{\Omega[R/2]} u^p\,dx\right)^{1/p} \leq C(\inf_{\Omega[R/2]} u + (F+G)R), \tag{1.47}$$

*where $F = \|f\|_n$.*

***Proof.*** We set

$$\bar{u} = u + (F+G)R,$$

we use the change of variables $z' = y'$, $z^n = y^n - \omega(y')$ to define $w(z) = -\ln \bar{u}(x)$ for $z^n \geq 0$, and we extend $w$ as an even function of $z^n$ for $z^n < 0$. By using $k = -\ln h$, we conclude that there is a positive $\delta$ such that (1.26) implies (1.41) for the constant $\zeta_1$ from Lemma 1.19. Converting the result of that lemma into the language of this lemma, we see that (1.25) is satisfied with $C_0 = -\ln C_1$.

From this point, the argument is the same as for Theorem 1.16. □

## 1.8 The strong maximum principle and uniqueness

From Lemma 1.18, we can derive several useful consequences about the behavior of solutions of oblique derivative problems. The first assertion is a strong maximum principle which asserts that if a subsolution has a nonnegative maximum, then that subsolution must be a constant. We proceed in several steps. The first step is to prove the strong maximum principle if the nonnegative maximum occurs in the domain.

**Corollary 1.21.** *Let $u \in C^2(\Omega)$ and suppose $u$ has a nonnegative maximum value at some point in $\Omega$. Suppose also that the coefficients of $L$ satisfy* (1.20) *for some positive constants $\lambda$, $\mu_1$, and $\mu_2$ and that $c \leq 0$ in $\Omega$. If $Lu \geq 0$ in $\Omega$, then $u$ is constant in $\Omega$.*

***Proof.*** We argue by contradiction. If $u$ is nonconstant and it has an interior maximum value $M$, then there is a point $x_1 \in \Omega$ on the boundary of the set on which $u = M$. Since $x_1 \in \Omega$, there is a positive constant $r_1$ such that the ball $B(x_1, r_1) \subset \Omega$ and a point $x_2 \in B(x_1, r_1/4)$ such that $u(x_2) < M$. By continuity, $u(x) < M$ in some ball centered at $x_2$ with radius $\theta_1 r_1/4$. We now use Lemma 1.13 with $r = r_1/2$ and $u$ replaced by $M - u$ to see that $M - u > 0$ in $B(x_2, r/2)$. But $x_1$ is in this ball, which contradicts our assumption that $u(x_1) = M$. Hence $u$ can't have an interior maximum if it is nonconstant. □

Next, we show that, if the nonnegative maximum occurs on the boundary and if a suitable oblique condition is satisfied near that maximum, then $u$ is constant.

**Corollary 1.22.** *Let $u \in C^2(\Omega[R]) \cap C(\overline{\Omega}[R])$ and suppose $u$ has a nonnegative maximum value at 0. Suppose also that conditions* (1.20), (1.30) *and* (1.36) *are satisfied. If $c$ and $\beta^0$ are nonpositive and if*

$$Lu \geq 0 \text{ in } \Omega[R], \quad Mu \geq 0 \text{ on } \partial\Omega[R],$$

*then $u$ is constant in $\Omega[R]$.*

***Proof.*** Again we argue by contradiction. With $r$ sufficiently small, we can find an $x_1$ with $x_1' = 0$ and $x_1^n = (A - \varepsilon^6/4)\omega_0 r$ such that $\eta(0) < 1$ and a constant $\theta_1 \in (0,1)$ such that $0 \notin \overline{E(\theta_1)}$ by choosing $\alpha_1$ sufficiently small. From Corollary 1.21, it follows that $u < u(0)$ in $E(\theta_1 r)$. Lemma 1.18 then implies that $u < u(0)$ in $E(\theta_2 r)$ for any $\theta_2 \in (\theta_1, 1)$. In particular, we have $u(0) < u(0)$. This contradiction proves the result. □

The strong maximum principle then produces a simple uniqueness result.

**Corollary 1.23.** *Let $\beta$ be uniformly oblique on $\partial\Omega$ and suppose that $|a^{ij}/\lambda|$, $|b/\lambda|$, and $|c/\lambda|$ are uniformly bounded in $\Omega$. Suppose also that $|\beta^0|/|\beta|$ is uniformly bounded on $\partial\Omega$. If $c \le 0$ in $\Omega$ and $\beta^0 \le 0$ on $\partial\Omega$, then any two solutions of*

$$Lu = f \text{ in } \Omega, \quad Mu = g \text{ on } \partial\Omega \tag{1.48}$$

*differ by a constant. Moreover, if $c$ is somewhere nonzero or if $\beta^0$ is somewhere nonzero, then* (1.48) *has at most one solution.*

***Proof.*** Let $u_1$ and $u_2$ be two solutions of (1.48) and suppose $v = u_1 - u_2$ is nonconstant. Since $v$ satisfies

$$Lv = 0 \text{ in } \Omega, \quad Mv = 0 \text{ on } \partial\Omega,$$

the strong maximum principle implies that $v$ can't have a nonnegative maximum. In other words, $v \le 0$. On the other hand, $-v$ satisfies the same differential equation and boundary condition as $v$, so it can't have a nonnegative maximum, so $-v \le 0$. It follows that $v = 0$ and hence $u_1 = u_2$.

If $v$ is a constant and $c(x_0) \ne 0$, then the differential equation implies that $c(x_0)v(x_0) = 0$ and hence $v \equiv 0$. A similar argument applies if $\beta^0(x_0) \ne 0$. □

A related result is known as the boundary point lemma. Although its statement uses a stronger smoothness assumption about $\partial\Omega$, the conclusion is stronger.

**Lemma 1.24.** *Suppose $|a^{ij}|/\lambda$, $|b|/\lambda$, and $|c|/\lambda$ are uniformly bounded in $\Omega$ and suppose $Lu \ge 0$ in $\Omega$. Suppose also that there is a point $x_0 \in \partial\Omega$ such that $u$ is continuous at $x_0$, $u(x_0) > u(x)$ for all $x \in \Omega$, and $\partial\Omega$ satisfies an interior sphere condition at $x_0$. If $c \le 0$ in $\Omega$ and if $u(x_0) \ge 0$, or if $u(x_0) = 0$, then*

$$\beta \cdot Du(x_0) < 0$$

*for any vector $\beta$ which is oblique to the interior sphere.*

***Proof.*** We first suppose that $c \le 0$. Let $B(y, R)$ be a ball which lies in $\Omega$ with $x_0 \in \partial B(y, R)$. Choose $\rho \in (0, R)$, set $A = B(y, R) \setminus \overline{B(y, \rho)}$ and define $v$ in $A$ by

$$v(x) = \exp(-\alpha|x - y|^2) - \exp(-\alpha R^2)$$

for a positive constant $\alpha$ at our disposal. It is not hard to see that $\alpha$ can be chosen so that $Lv \geq 0$ in $A$.

On $\partial B(y,\rho)$, there is a constant $\varepsilon$ so that $u - u(x_0) + \varepsilon v \leq 0$, and the definition of $v$ guarantees that this inequality is also true on $\partial B(y,R)$. Hence $u - u(x_0) + \varepsilon v \leq 0$ on $\partial A$ and $L(u - u(x_0) + \varepsilon v) \geq 0$ in $A$. The maximum principle implies that $u - u(x_0) + \varepsilon v \leq 0$ in $A$ and hence

$$\beta \cdot Du \leq \varepsilon \beta \cdot Dv < 0.$$

If $u(x_0) = 0$, then we can repeat the preceding argument with $L_0$, defined by $L_0 w = a^{ij} D_{ij} w + b^i D_i w$, in place of $L$. □

Note that Lemma 1.24 implies Corollary 1.23 under the stronger assumption that $\Omega$ satisfies an interior sphere condition at every point of $\partial\Omega$. Hence, for our purposes, this lemma is not very important; however, it has a number of important consequences. We mention here that Lemma 1.24 is the basis for studying many symmetry properties of solutions of elliptic equations. (See for example, [62].)

## 1.9 Hölder continuity

We now prove a result that is fundamental in the theory of nonlinear elliptic (and parabolic) equations. Such a result was first proved (although only for weak solutions of equations in divergence form, and only away from the boundary) by De Giorgi [35] and Nash [146]. A key step is the following iteration result. Although the proof is identical to that of, for example, Lemma 4.6 from [114], we include it to emphasize that it is true under slightly different hypotheses from the usual ones.

**Lemma 1.25.** *Let $w$ and $\sigma$ be nonnegative functions defined on some interval $[0, R_0]$, and suppose that there are positive constants $\alpha$, $\delta$, and $\tau$ with $\tau < 1$ and $\delta < \alpha$ such that*

$$\frac{\sigma(r)}{r^\delta} \leq \frac{\sigma(s)}{s^\delta} \tag{1.49}$$

*if $0 < s \leq r \leq R_0$ and*

$$w(\tau^k R_0) \leq \tau^\alpha w(\tau^{k-1} R_0) + \sigma(\tau^{k-1} R_0) \tag{1.50}$$

*if $0 < r \leq R_0$. If $\sigma$ is increasing on $(0, R_0]$ and $w$ is increasing on each interval of the form $(\tau^{k+1} R_0, \tau^k R_0]$, then there is a constant $C$ determined only by $\alpha$, $\delta$, and $\tau$ such that*

$$w(r) \leq C \left[ \left( \frac{r}{R_0} \right)^\alpha w(R_0) + \sigma(r) \right] \tag{1.51}$$

*for all* $r \in (0, R_0]$.

***Proof.*** If $r > \tau R_0$, we have

$$w(r) \le w(R_0) \le \tau^{-\alpha}\left(\frac{r}{R_0}\right)^{\alpha} w(R_0).$$

If $r \in (\tau^{k+1}R_0, \tau^k R_0]$ for some positive integer $k$, then induction gives

$$w(r) \le w(\tau^k R_0) \le \tau^{k\alpha} w(R_0) + \tau^{-\delta}\sigma(\tau^k R_0)\sum_{j=0}^{k-1}\tau^{(\alpha-\delta)j}.$$

The sum is a finite geometric series with positive terms, so it can be estimated from above by its (finite) limit $1/(1-\tau^{\delta-\alpha})$, and, because $r > \tau^{k+1}R_0$, we have $\tau^k R_0 < r/\tau$, so

$$\sigma(\tau^k R_0) \le \sigma(r/\tau) \le \tau^{-\delta}\sigma(r).$$

Combining these three inequalities yields (1.51) with $C = \max\{\tau^{-\alpha}, 1/(1-\tau^{\delta-\alpha})\}$. □

We shall use this lemma repeatedly in this book. Typically, $w$ will be assumed increasing (and this is the case for Lemma 4.6 in [114]), but we shall have several occasions to use this stronger result. It is important to point out that (1.51) provides a rate at which $w(r)$ goes to zero with $r$ if $\sigma(r) \to 0$ as $r \to 0$. In most applications, we shall even assume that $\sigma(r) = Ar^\delta$, but sometimes a general $\sigma$ will be useful.

**Theorem 1.26.** *Let* $L$ *be as in Theorem* 1.20. *If* $Lu = f$ *in* $\Omega[R]$ *and* $Mu = g$ *on* $\Sigma[R]$, *then there are constants* $C$, $\alpha$ *and* $\delta$ *determined only by* $\varepsilon$, $\mu$, $\mu_1 R$, $\mu_2 R^2$, $\mu_3 R$, *and* $\omega_0$ *such that*

$$\operatorname*{osc}_{\Omega[r]} u \le C\left(\frac{r^\alpha}{R^\alpha}\sup_{\Omega[R]} u + (F+G)r^\delta R^{1-\delta}\right), \tag{1.52}$$

*where* $F = \|f/\lambda\|_n$ *and* $G = \sup|g/\beta^n|$.

***Proof.*** If $r \ge R/4$, then the result is immediate. If $r < R/4$, we write

$$M_i = \sup_{\Omega[ir]} u, \quad m_i = \inf_{\Omega[ir]} u$$

for $i = 1, 4$. Then $L(M_4 - u) = f$ and $L(u - m_4) = f$ in $\Omega[4r]$, and $M(M_4 - u) = -g$ and $M(u - m_4) = g$ on $\Sigma[4r]$. We may therefore apply the weak Harnack inequality to these functions to obtain

$$(r^{-n}\int_{\Omega[2r]}(M_4 - u)^p\,dx)^{1/p} \le C(M_4 - M_1 + \sigma(r)),$$

$$(r^{-n}\int_{\Omega[2r]}(u - m_4)^p\,dx)^{1/p} \le C(m_1 - m_4 + \sigma(r)),$$

where $\sigma$ is defined by

$$\sigma(s) = (F + G)s.$$

We now add these two inequalities and take advantage of the appropriate version of Minkowski's inequality:

$$\left(\int (f+g)^p\right)^{1/p} \le C(p)\left[\left(\int f^p\right)^{1/p} + \left(\int g^p\right)^{1/p}\right]$$

which is valid for all $p > 0$; we recall that $C(p) = 1$ if $p \ge 1$. It follows that

$$M_4 - m_4 \le C(M_4 - m_4 + m_1 - M_1 + \sigma(r)).$$

We now set $w(s) = \operatorname{osc}_{\Omega[s]} u$ to obtain

$$w(r) \le \gamma w(4r) + \sigma(r)$$

with $\gamma = 1 - C^{-1}$, so $\gamma \in (0,1)$. The desired result now follows from this inequality via Lemma 1.25 with $\alpha$ and $\delta$ chosen to satisfy $4^{-\alpha} \ge \gamma$ and $\delta < \alpha < 1$. □

Although this result can be taken without further explanation, it is useful to give a brief description here of the Hölder spaces because they will play a crucial role in the remainder of this book.

Let $\alpha \in (0,1]$. We say that a real-valued function $f$, defined on some set $S$ is *Hölder continuous with exponent $\alpha$ on $S$* if there is a constant $M$ such that

$$|f(x) - f(y)| \le M|x - y|^\alpha$$

for all $x$ and $y$ in $S$. The smallest such $M$ for which this inequality holds is called the *Hölder seminorm of $f$ on $S$* and is written as

$$[f]_{\alpha:S}.$$

We write $H_\alpha(S)$ for the set of all functions $f$ which are bounded and Hölder continuous on $S$ and we use the norm

$$|f|_{\alpha;S} = [f]_{\alpha;S} + \sup_S |f|.$$

We also frequently abbreviate $\sup_S |f|$ to $|f|_{0;S}$. Sometimes we write $C^\alpha(S)$ in place of $H_\alpha(S)$ when $S$ is closed.

A related concept is that of local Hölder continuity. We say that $f$ is *locally Hölder continuous* on an open set $\Omega$ if $f$ is Hölder continuous on any compact subset $S$ of $\Omega$, and we use $C^\alpha(\Omega)$ to denote the set of all locally Hölder continuous functions on an open set $\Omega$. In this language, we can restate Theorem 1.26 to say that solutions of oblique derivative problems in Lipschitz domains are Hölder continuous.

The concept of Hölder continuity is paramount in our later study of solutions. For that study, we need to define additional function spaces in terms of Hölder continuity of derivatives, but we save the details for Chapter 4.

## 1.10 The local maximum principle

Occasionally, one wishes to estimate the maximum value of a solution in a neighborhood of a point in terms of local data. Since it is not generally possible to obtain such a bound without some information on the solution (for example, any constant is a solution of the equation $A^{ij}D_{ij}u = 0$ but the maximum value in a neighborhood of a given point is exactly the unknown constant), we look for weak information on the solution which implies such a bound. In this section, the additional information is a bound on some $L^p$ norm of the solution.

**Theorem 1.27.** *Let $u \in C^2(\Omega[R]) \cap C(\overline{\Omega[R]})$, and suppose $u \geq 0$ in $\Omega[R]$. Suppose that conditions (1.3), (1.20a,b), and (1.46) are satisfied. If there are a nonnegative function $f \in L^n(\Omega[R])$ and a nonnegative constant $g$ such that*

$$Lu \geq -\lambda f \text{ in } \Omega[R], \quad Mu \geq -\beta^n g \text{ on } \partial\Omega[R], \tag{1.53}$$

*then, for any $p > 0$, there is a constant $C$, determined only by $\mu$, $\omega_0$, $\varepsilon$, $\mu_1 r$, $\mu_2 r^2$, $\mu_3 r$, and $p$ such that*

$$\sup_{\Omega[R/2]} u \leq C\left[\left(R^{-n}\int_{\Omega[R]}(u^+)^p\,dx\right)^{1/p} + (F+g)R\right], \tag{1.54}$$

*where $F = \|f\|_n$.*

***Proof.*** Since the conclusion is clear if $\sup_{\Omega[R/2]} u \leq 0$, we shall assume that this supremum is positive. Working in $y$-coordinates, we let $y_0$ be a point in $\Omega[R/2]$ such that $u(y_0) \geq \frac{1}{2}\sup_{\Omega[R/2]} u$. We note that there is a constant $\tilde{A} \in (0, \frac{1}{2})$, determined only by $\omega_0$ and $\varepsilon$, so that, if $y_1' = y_0'$, $y_1^n = y_0^n + \tilde{A}R$, then

$$y_0 \in E(y_1, \tilde{A}R) \subset B((y_0', \omega(y_0')), R).$$

With $\eta$ from Lemma 1.17 corresponding to $r = \tilde{A}R$ and $q \geq 2$ to be chosen, we set $v = \eta^q u$ and we define the operators $L_0$ and $M_0$ by

$$L_0 w = a^{ij}D_{ij}w, \quad M_0 w = \beta^i D_i w.$$

Then

$$\begin{aligned} L_0 v &= q\eta^{q-2}(\eta a^{ij}D_{ij}\eta + (q-1)a^{ij}D_i\eta D_j\eta - c\eta^2)u \\ &\quad + \eta^{q-1}[2qa^{ij}D_i\eta - b^i]D_j u - \eta^q f \\ &\geq -CR^{-2}\eta^{q-2}u - CR^{-1}\eta^{q-1}|Du| - f. \end{aligned}$$

To proceed, we shall invoke Theorem 1.11, and we choose the cone $K$ so that any vector in $K$ with vertex in $E(y_1, r)$ must intersect $\partial E(y_1, r)$ inside $\Omega[R]$. (Such will be the case if every vector $\xi \in K$ satisfies $\xi_0|v'| \le \xi^n$.) It then follows that, on $G$, we have $|Dv| \le Cv/\eta$. Since $Dv = \eta^q Du + q\eta^{q-1}uD\eta$, it follows that

$$L_0 v \ge -CR^{-2}\eta^{q-2}u - f$$

on $G$. In addition,

$$M_0 v = \eta^{q-1}(q\beta^i D_i\eta - \eta\beta^0)u - \eta^q g \ge -\beta^n g$$

if $q$ is sufficiently large, say $q \ge q_0$. From Theorem 1.11, we therefore infer that

$$\sup v \le CR^{-1}\|\eta^{q-2}u\|_n + C(F+g)R.$$

We now observe that

$$\|\eta^{q-2}u\|_n \le (\sup v)^{1-(2/q)}\|u\|_{2n/q}^{2/q},$$

and hence, for any $\zeta > 0$, there is a constant $C(\zeta, n, q)$ such that

$$\|\eta^{q-2}u\|_n \le \zeta \sup v + C(\zeta, n, q)\|u\|_{2n/q}.$$

If $p \le \min\{n, 2n/q_0\}$, then we take $q = 2n/p$ and then $\zeta$ sufficiently small to conclude (1.54). The case $p > \min\{n, 2n/q_0\}$ is then obtained via Hölder's inequality. □

## 1.11 Pointwise estimates for solutions of mixed boundary value problems

For our later work, we wish to have estimates for solutions of the mixed boundary value problem

$$Lu = f \text{ in } \Omega, \quad Mu = g \text{ on } \Sigma, \quad u = \psi \text{ on } \sigma$$

when $\Sigma$ and $\sigma$ are disjoint subsets of $\partial\Omega$ whose union is all of $\partial\Omega$. Without any restrictions on $\Sigma$ and $\sigma$, this problem can be extremely complicated and a complete description is beyond the scope of this book. Here, we make some assumptions on $\Sigma$ and $\sigma$ which are fairly general and easy to state. The model configuration is commonly called a wedge or dihedral angle, which means that, in cylindrical coordinates $(r, \theta, z)$, we have

$$\Sigma = \{r > 0, \theta = \theta_0\}, \quad \sigma = \{r \ge 0, \theta = \theta_1\},$$
$$\Omega = \{r > 0, \theta_0 < \theta < \theta_1\}$$

for some numbers $\theta_0$ and $\theta_1$ satisfying $\theta_0 < \theta_1 < \theta_0 + \pi$. (In fact, our results apply if we only assume that $\theta_0 < \theta_1 < \theta_0 + 2\pi$, but this extra step requires a more complicated analysis to handle the general setting, and it will not be needed in this book.)

More generally, we shall always assume that there is a Lipschitz function $\omega$, defined for all $x'$ with $|x'| < R$ (for some positive $R$), such that $\omega(0) = 0$ and $|\omega(x') - \omega(y')| \le \omega_0 |x' - y'|$. Moreover, we write

$$\Omega[R] = \{x \in \Omega : |x| < R\},$$

and we assume that $\partial\Omega[R]$ consists of two disjoint pieces: $\Sigma[R]$ which is an open subset of the surface $\{x^n = \omega(x'), |x'| < R\}$ and $\sigma[R]$. We further assume that $y^n > \omega(y')$ on $\sigma[R]$. (As already mentioned, this further assumption is not necessary for our analysis but it simplifies the proof of results and it will suffice for our applications.) For simplicity, we refer to this set of hypotheses as condition $(\omega)$.

Our first step is to obtain a weak Harnack inequality under these hypotheses and the ones previously stated for $L$ and $M$.

**Theorem 1.28.** *Let $u \in C^2(\Omega[R]) \cap C(\overline{\Omega[R]})$ and suppose $u \ge 0$ in $\Omega[R]$. Suppose conditions $(\omega)$, (1.3c), (1.20), and (1.36) are satisfied. Suppose further that there are constants $G \ge 0$ and $m > 0$, and a function $f \in L^n(\Omega[R])$ such that*

$$Lu \ge \lambda f \text{ in } \Omega[R], \quad Mu \ge \beta^n G \text{ on } \Sigma[R], \quad u \ge m \text{ on } \sigma[R]. \tag{1.55}$$

*Then there is a constant $C$ such that*

$$\left(R^{-n} \int_{\tilde{B}(R/2)} (u_m^-)^\kappa \, dx\right)^{1/\kappa} \le C\left(\inf_{\Omega[R/2]} u + R(F+G)\right), \tag{1.56}$$

*where*

$$u_m^-(x) = \begin{cases} \min\{u(x), m\} & \text{if } x \in \Omega[R] \cap \Sigma[R], \\ m & \text{otherwise,} \end{cases}$$

*$F = \|f\|_n$, and $\tilde{B}(R/2)$ is the subset of $B(R/2)$ on which $y^n > \omega(y')$.*

***Proof.*** We first observe that Lemmata 1.18 and 1.19 are still true in this case (with the same proofs) under the additional assumption that $h \le m$. The key point in Lemma 1.19 is that $K^*(y_0, r)$ is defined to be $K(y_0, r) \cap \Omega$.

It follows that (1.25) holds for $w$ defined by

$$w(z) = -\ln(u_m^-(y) + (F+G)r)$$

as long as $k \ge -\ln m$. We then infer (1.27) for $t \in (0, m)$, and then (1.56) follows as before because $\mu(t) = 0$ if $t > m$. □

This weak Harnack inequality leads to a modulus of continuity estimate if we strengthen condition $(\omega)$ just a little.

**Corollary 1.29.** *Let $u \in C^2(\Omega[R]) \cap C(\overline{\Omega[R]})$, and suppose conditions $(\omega)$, (1.3c), (1.36a,b), and (1.46) are satisfied. Suppose also that there is a constant $A_0 \in (0,1)$ such that*

$$|\tilde{B}(r) \cap \Omega| < A_0 |\tilde{B}(r)| \tag{1.57}$$

*for all $r \in (0, R)$. If there is a concave function $\sigma_1$ such that $\operatorname{osc}_{\sigma[r]} u \le \sigma_1(r)$, then there are constants $C$, $\alpha$ and $\delta$ (determined by the same quantities as in Theorem 1.26) such that $\delta \in (0,1)$ and*

$$\operatorname*{osc}_{\Omega[r]} u \le C \left[ \frac{r^\alpha}{R_0^\alpha} \operatorname*{osc}_{\Omega[R]} u + (F+G) r^\delta R_0^{1-\delta} + \sigma_1(r^\delta R_0^{1-\delta}) \right].$$

***Proof.*** With $M_i$ and $m_i$ as in the proof of Theorem 1.26, we apply Theorem 1.28 to $M_4 - u$ and obtain

$$\left( R^{-n} \int_{\tilde{B}(2R)} \left( (M_4 - u)^-_{M_4 - \tilde{M}_4} \right)^p dx \right)^{1/p} \le C(M_4 - M_1 + (F+G)R),$$

where $\tilde{M}_4 = \sup_{\sigma[4r]} u$. The definition of $u^-_m$ implies that

$$\begin{aligned} \left( r^{-n} \int_{\tilde{B}(2r)} \left( (M_4 - u)^-_{M_4 - \tilde{M}_4} \right)^p dx \right)^{1/p} &\ge (r^{-n} |\tilde{B}(2r) \setminus \Omega|)^{1/p} (M_4 - \tilde{M}_4) \\ &\ge C(n, \omega_0, A_0)^{1/p} (M_4 - \tilde{M}_4) \end{aligned}$$

using also condition (1.57), and hence

$$M_4 - \tilde{M}_4 \le C(M_4 - M_1 + (F+G)r).$$

A similar argument with $u - m_4$ yields

$$\tilde{m}_4 - m_4 \le C(m_1 - m_4 + (F+G)r),$$

using the notation $\tilde{m}_4 = \inf_{\sigma[4r]} u$.

Adding these results and rearranging yields

$$\operatorname*{osc}_{\Omega[r]} u \le \gamma \operatorname*{osc}_{\Omega[4r]} u + (F+G)r + \sigma_1(r),$$

and the proof is completed by applying Lemma 1.25. □

For our study of problems with discontinuous $\beta$, we shall also use the local maximum principle, which takes the following simple form. Since the proof is identical to that of Theorem 1.27, we omit it.

**Theorem 1.30.** *Let $u \in C^2(\Omega[R]) \cap C(\overline{\Omega[R]})$ and suppose $u \geq 0$ in $\Omega[R]$. Suppose conditions $(\omega)$, (1.3c), (1.20), and (1.36) are satisfied. Suppose further that there are constants $G \geq 0$ and $m > 0$, and a function $f \in L^n(\Omega[R])$ such that*

$$Lu \leq \lambda f \text{ in } \Omega[R], \quad Mu \leq \beta^n G \text{ on } \Sigma[R], \quad u \leq m \text{ on } \sigma[R]. \tag{1.58}$$

*Then, for any $p > 0$, there is a constant $C$ such that*

$$\sup_{\Omega[R/2]} u \leq C \left[ \left( R^{-n} \int_{\Omega[R]} (u^+)^\kappa \, dx \right)^{1/\kappa} + R(F + G) \right], \tag{1.59}$$

*where $F = \|f\|_n$.*

## 1.12 Derivative bounds for solutions of elliptic equations

The maximum principle can also be used to derive bounds on the derivatives of solutions of constant coefficient equations. There are actually several different ways to obtain these bounds, but we present here a straightforward approach that doesn't make undue assumptions about the solution itself. Most of our estimates are in balls and half-balls, and the estimates in more general domains will derived from those in the simpler domains.

We begin with a simple pointwise estimate for solutions of general elliptic equations in a half-ball with zero data on the flat portion of the boundary.

**Lemma 1.31.** *Let $[a^{ij}]$ be a positive-definite matrix-valued function defined on $B^+(R)$ for some $R > 0$, and suppose that $\lambda$ and $\mu$ are positive constants such that the eigenvalues of $[a^{ij}]$ are in the interval $[\lambda, \mu\lambda]$. If $v \in C^2(B^+(R)) \cap C(\overline{B^+(R)})$ satisfies*

$$a^{ij} D_{ij} v = f \text{ in } B^+(R), \quad v = 0 \text{ on } B^0(R),$$

*then*

$$|v(x)| \leq 16n\mu \left( \sup_{B^+(R)} |v| \right) \frac{x^n}{R} + \left( \sup_{B^+(R)} |f| \right) Rx^n \tag{1.60}$$

*for all $x \in B^+(7R/8)$.*

***Proof.*** Let $x \in B^+(7R/8)$. If $x^n \geq R/8$, then (1.60) is immediate.

On the other hand, if $x^n < R/8$, we set $x_0 = (x', 0)$, $V = \sup_{B^+(R)} |v|$, $F = \sup_{B^+(R)} |f|$, and $r = R - |x_0|$, and we define the function $w$ by

$$w(y) = V\left(\frac{|y - x_0|^2}{r^2} + n\mu\left(\frac{y^n}{r} - \frac{(y^n)^2}{r^2}\right)\right) + F(ry^n - (y^n)^2).$$

Then

$$a^{ij} D_{ij} w \leq f \text{ in } B^+(x_0, r), \quad w \geq v \text{ on } \partial B^+(x_0, r),$$

so the maximum principle implies that $v \leq w$ in $B^+(x_0, r)$. Similarly $v \geq -w$ in $B^+(x_0, r)$ and hence $|v| \leq w$ in $B^+(x_0, r)$. In particular, this inequality is true at $x$, so

$$|v(x)| \leq V\left(\frac{(x^n)^2}{r^2} + n\mu\left(\frac{x^n}{r} - \frac{(x^n)^2}{r^2}\right)\right) + F(rx^n - (x^n)^2)$$

The proof is completed by using the inequalities $0 < x^n < r$ and $R/8 \leq r \leq R$. □

Our next step is to obtain an interior gradient bound. We do so by applying our preceding estimate in a suitable way.

**Lemma 1.32.** *Let $[A^{ij}]$ be a constant positive-definite matrix and suppose $\lambda \leq \Lambda$ are positive constants satisfying the inequality*

$$\lambda|\xi|^2 \leq A^{ij}\xi_i\xi_j \leq \Lambda|\xi|^2. \tag{1.61}$$

*If $u \in C^2(B(0, r))$ satisfies the equation $A^{ij} D_{ij} u = f$ in $B(0, r)$ for some bounded function $f$, then there is a constant $C$, determined only by $n$ and $\Lambda/\lambda$ such that*

$$|Du(0)| \leq C\left(\frac{\sup|u|}{r} + r\sup|f|\right). \tag{1.62}$$

***Proof.*** Note that the eigenvectors of $[A^{ij}]$ are orthogonal. Hence it suffices to prove that this bound is true for the directional derivative of $u$ in the direction of any eigenvector. Let us fix an eigenvector, which we may take to be a unit vector. Since the hypotheses are invariant under rotation of axes, we may assume this vector is $(0, \ldots, 0, 1)$. We then define $\varphi$ and $\psi$ in $B^+(r)$ by

$$\varphi(x', x^n) = \frac{1}{2}[u(x', x^n) - u(x', -x^n)],$$
$$\psi(x', x^n) = \frac{1}{2}[f(x', x^n) - f(x', -x^n)],$$

so

$$A^{ij}D_{ij}\varphi = \psi \text{ in } B^+(r), \quad \varphi = 0 \text{ on } B^0(r).$$

It then follows from Lemma 1.31 and the definitions of $\varphi$ and $\psi$ that

$$\begin{aligned}|\varphi(0,x^n)| &\le 16n\mu\left(\sup_{B^+(r)}|\varphi|\right)\frac{x^n}{r} + \left(\sup_{B^+(r)}|\psi|\right)rx^n \\ &\le 16n\mu\left(\sup_{B^+(r)}|v|\right)\frac{x^n}{r} + \left(\sup_{B^+(r)}|f|\right)rx^n\end{aligned}$$

we now divide this inequality by $x^n$ and take the limit as $x^n \to 0$ to infer (1.62), recalling that

$$D_n v(0) = \lim_{x^n\to 0^+}\frac{\varphi(0,x^n)}{x^n}.$$

□

From this estimate, we obtain an interior estimate which will be very useful in several later chapters.

**Theorem 1.33.** *Let $\Omega$ be an open subset of $\mathbb{R}^n$ and let $u$ be a bounded solution of $A^{ij}D_{ij}u = f$ in $\Omega$, where $A^{ij}$ is a positive definite, constant matrix. If $\lambda$ and $\Lambda$ are positive constants satisfying* (1.61), *then there is a constant $C(n,\Lambda/\lambda)$ so that*

$$d(x)|Du(x)| \le C[\sup|u| + \sup d^2|f|) \tag{1.63}$$

*for all $x \in \Omega$.*

***Proof.*** Fix $x \in \Omega$ and set $r = d(x)$. If we define $u'$ and $f'$ by $u'(y) = u(y-x)$ and $f'(y) = f(y-x)$, then $u'$ is a solution of $A^{ij}D_{ij}u' = f'$ in $B(0,r)$, so we can apply Lemma 1.32 to conclude that $|Du'(0)| \le C[\sup|u'|/r + r\sup|f'|]$. Simple rearrangement now gives (1.63). □

We shall also take advantage of some higher order derivative estimates.

**Theorem 1.34.** *Let $\Omega$ be an open subset of $\mathbb{R}^n$ and let $u$ be a bounded solution of $A^{ij}D_{ij}u = f$ in $\Omega$, where $A^{ij}$ is a positive definite, constant matrix. If $f \in C^N(\Omega)$ for some nonnegative integer $N$, then $u \in C^{N+1}(\Omega)$. Moreover, if $\lambda$ and $\Lambda$ are positive constants satisfying* (1.61), *then there is a constant $C(n,N,\Lambda/\lambda)$ so that*

$$d(x)^N|D^N u(x)| \le C[\sup|u| + \sum_{j=0}^{N-1}\sup d^{j+2}|D^j f|] \tag{1.64}$$

*for all $x \in \Omega$.*

***Proof.*** We only prove the $C^2$ estimate since the full strength of the theorem is proved by a simple modification of this argument. We shall estimate the second derivative $D_{km}u$ at a point $x$ for $k$ and $m$ integers between 1 and $n$. We then apply Theorem 1.33 to the function $u_{k,h}$, defined (for $h < d(x)/(2\sqrt{n})$) by

$$u_{k,h} = \frac{u(x + he_k) - u(x)}{h},$$

in the ball $B(x, d(x)/2)$. It follows that

$$|D_m u_{k,h}(x)| \le C \left( \sup_{B(x,d(x)/2)} |u_{k,h}|/d(x) + d(x) \sup_{B(x,d(x)/2)} |f_{k,h}| \right).$$

But

$$\sup_{B(x,d(x)/2)} |u_{k,h}| \le \sup_{B(x,3d(x)/4)} |D_k u|$$

and

$$\sup_{B(x,d(x)/2)} |f_{k,h}| \le \sup_{B(x,3d(x)/4)} |D_k f|$$

so another application of Theorem 1.33 to $u$ gives

$$\sup_{B(x,3d(x)/4)} |D_k u| \le C \left( \frac{\sup |u|}{d(x)} + \sup d|f| \right).$$

Hence

$$|D_m u_{k,h}(x)| \le C \left( \frac{\sup |u|}{d(x)} + \sup d|f| + \sup d^2 |D_k f| \right).$$

The proof is completed by recalling that

$$D_{km}u(x) = \lim_{h \to 0} D_m u_{k,h}(x).$$

□

We shall use the corresponding boundary version of this theorem.

**Lemma 1.35.** *Let $[A^{ij}]$ be a positive definite matrix satisfying (1.61) for some positive constants $\lambda$ and $\Lambda$, and let $R > 0$. If $v \in C^2(B^+(R)) \cap C(\overline{B^+(R)})$ satisfies*

$$A^{ij} D_{ij} v = 0 \text{ in } B^+(R), \quad v = 0 \text{ on } B^0(R), \tag{1.65}$$

*then there is a constant $C$, determined only by $n$ and $\Lambda/\lambda$ such that*

$$\sup_{B^+(R/2)} |D^2 v| \le C R^{-2} V, \tag{1.66}$$

*where $V = \sup_{B^+(R)} |v|$.*

***Proof.*** Our first step is to use Lemma 1.31 to infer that

$$|v(x)| \leq 16nAV\frac{x^n}{R} \text{ in } B^+(7R/8).$$

Our next step is a gradient bound for $v$. We now fix $x_1 \in B^+(3R/4)$ and use $\Omega = B(x_1, x_1^n/8)$ in Theorem 1.33 to conclude that

$$\frac{x_1^n}{8}|Dv(x_1)| \leq C\sup_{\Omega}|v| \leq CV\frac{x_1^n}{R}.$$

It follows that

$$\sup_{B^+(3R/4)} |Dv| \leq C\frac{V}{R}. \tag{1.67}$$

Now let $k \in \{1, \ldots, n-1\}$ and apply the preceding argument to $\Delta_h v$ defined by

$$\Delta_h v(x) = \frac{v(x+he_k) - v(x)}{h}$$

for $0 < |h| < R/16$ to conclude that

$$\sup_{B^+(R/2)} |D\Delta_h v| \leq CR^{-1} \sup_{B^+(2R/3)} |\Delta_h v|.$$

Since $\Delta_h v \to D_k v$ as $h \to 0$, we conclude that

$$\sup_{B^+(R/2)} |D_{ik}v| \leq CR^{-1} \sup_{B^+(2R/3)} |Dv| \leq CR^{-2}V$$

by a direct application of (1.67) as long as $i \in \{1, \ldots, n\}$ and $k \in \{1, \ldots, n-1\}$. The proof is completed by using the differential equation to conclude that

$$|D_{nn}v| \leq C\sum_{i=1}^{n}\sum_{k=1}^{n-1}|D_{ik}v|.$$

□

In fact, these estimates can be expanded considerably. We refer the reader to the Notes of this chapter for detailed information.

## Notes

The maximum principle is one of the most important tools in the theory of elliptic differential equations. The books [156] and [158] give an introduction to this fruitful theory, and the first few results in this chapter (namely, Lemmata 1.1 and 1.2) are minor variations on well-known results. We also mention Chapter 3 in [64] for a discussion of other, related maximum principles, but most of that chapter is concerned with maximum principles for the Dirichlet problem; we mention that [156] has a thorough bibliography on the topic, current at the time (1967) that book was written.

The definition of obliqueness in Section 1.2 come from [116], but it is just a slight restatement of the definition in [142]. The condition (1.3b), universally called *Lipschitz continuity* of $\omega$, was first used by Lipschitz in [123] to study existence and uniqueness of solutions for ordinary differential equations, a topic we return to in Chapter 3.

Lemma 1.3 is just Theorem 2 from [128], and the Weierstrass approximation theorem first appeared in [200] (in the special case of one spatial dimension); our proof is based on the one in Section IV.15 of [36] (which seems to be the original proof), but the idea to use this approximation to control the modulus of continuity comes from the observation in Proposition I.1.4 from [165] that the polynomial approximation also approximates derivatives uniformly. Lemmata 1.5 and 1.6 are taken directly from [116].

The proof of the change of variables formula is a simplified version of the argument in Appendix F of [176]. That source gives a sharper version of the result presented here; it defines, for $y \in \mathbb{R}^n$, $\nu(y)$ to be the number of points $x$ in $\Omega$ such that $F(x) = y$ and then shows that the function $\nu\colon \mathbb{R} \to [0, \infty]$ is measurable with

$$\int_\Omega u \circ F(x) |\det DF(x)|\, dx = \int_{\mathbb{R}^n} \nu(y) u(y)\, dy.$$

The basic ideas for the estimates in Section 1.5 come from the work of Aleksandrov [2], but the technical details in Lemma 1.8 are based on Theorem 2.1 from [122] and our choice of the function $h$ in the proof of Theorem 1.9 comes from the calculation in the proof of Theorem 9.1 from [64]. (Both [64] and [2] use $h(t) = 1/(t^{n/(n-1)} + \mu^{n/(n-1)})^{n-1}$. In [2], a more careful estimate is made to evaluate the integral

$$\int_0^R h(r) r^{n-1}\, dr,$$

which leads to a more precise description of the way in which the maximum estimate depends on $B_1$.) Theorem 1.11 comes from [198].

The arguments in Section 1.6 are taken from Section 9.8 in [64], which adapted the discussion in [178]. That work, in turn, took the ideas of [88] for parabolic equations and modified the proofs to handle elliptic equations.

The function $G$ in Section 1.7 and the arguments in Sections 1.7 and 1.8 were first introduced in [116]; the present analysis fills in some of the details omitted from that paper. The interior strong maximum principle Corollary 1.21 is standard in the theory of elliptic equations; it was first proved, by different techniques, in [72]. The value of the present proof is its application to the oblique derivative problem. Although the strong maximum principle Corollary 1.22 was first proved (under slightly stronger hypotheses, in particular, continuity of the vector $\beta$) by Nadirashvili [142] as his Lemma on the Inner Derivative, the method of proof here is quite different. The boundary point lemma (Lemma 1.24) was first proved independently by Hopf [74] and Oleinik [150] at the same time, but, in the West, the result is commonly known as the Hopf boundary point lemma.

The material in Sections 1.7, 1.9, and 1.10 has become quite standard in the literature. Our proofs are based very strongly on those in Sections 9.7 and 9.8 from [64]. We point out that the present author's work [103] is somewhat incautious about its proof of the local maximum principle (Theorem 3.3 of that work) by not taking the definition of $\Gamma_v^+$ into account correctly. Fortunately, the use of Wang's maximum principle (our Theorem 1.11) avoids that difficulty. Earlier versions of the results were proved by Nadirashvili [142–145] under stronger hypotheses on the coefficients (for example, $a^{ij}$ is assumed continuous in [144] in order to prove a Hölder estimate for $u$ in terms of the $L^n$ norm of $f/\lambda$).

The material in Section 1.11 comes from Section 8 of [116]. The higher order derivative estimates in Section 1.12 are based on work of Brandt [14], [15]. He was able to prove the interior Schauder estimates (see Chapter 2) for elliptic and parabolic differential equations using only maximum principle arguments; in particular, we mention (see also Section 3.4 in [64]) that these arguments can be used to prove a modulus of continuity estimate for the first derivatives of solutions to the equation $A^{ij}D_{ij}u = f$ based only on a pointwise bound for $f$.

## Exercises

**1.1** Show that if the hypotheses of Lemma 1.6 are satisfied, then there is a constant $C$, determined only by $\Omega$, $C_2$, and the modulus of ellipticity of $\beta$ such that

$$\sup u \leq C\left(\sup \frac{Lu^-}{|c|} + \sup \frac{Mu^-}{|\beta|}\right).$$

**1.2** Prove the following variant of Lemma 1.25.

**Lemma 1.36.** *Let $w$ and $\sigma$ be increasing functions defined on some interval $(0, R_0]$, and suppose that there are positive constants $\alpha$ and $\tau$ with $\tau < 1$ such that (1.50) holds for all $r \in (0, R_0)$. Then, for any $\mu \in (0, 1)$, there is a positive constant $C(\tau, \alpha, \mu)$ such that*

$$\omega(R) \leq C[\left(\frac{R}{R_0}\right)^{\alpha(1-\mu)} \omega(R_0) + \sigma(R^\mu R_0^{1-\mu})].$$

(Hint: See Lemma 8.23 in [64].)

**1.3** Show that Theorem 1.28 remains true if we replace the assumption that $y^n > \omega(y')$ on $\sigma[R]$ by the assumption that there is a number $A_1 \in (0, 1)$ such that, for any $r \in (0, R)$, there is a ball $B^*(A_1 r)$ centered on the set $\{y^n = \omega(y')\}$ which is a subset of $\Omega$.

**1.4** Show that the higher order version of Lemma 1.35 is valid. In other words, show that the hypotheses of the lemma imply that, for any positive integer $N$, there is a constant $C$, determined only by $n$, $N$, and $\Lambda/\lambda$ such that

$$\sup_{B^+(R/2)} |D^N v| \leq C R^{-N} V.$$

# Chapter 2

# Classical Schauder Theory from a Modern Perspective

## Introduction

A key element in the theory of the Dirichlet problem for linear elliptic equations is known as Schauder theory. Developed by Schauder [168, 167], this theory has two elements: it provides estimates on the Hölder norm of any solution of the Dirichlet problem and of the first and second derivatives of this solution, and it gives an existence result under appropriate conditions on the coefficients of the operator. Schauder's method of proof is based on classical potential theory (much of which is an improvement of the results of Hölder [71]). We refer to Chapter V in [136] for a discussion of the contributions of Cacciopoli [20], Hopf [73], and Giraud [65].

A detailed description of Schauder's approach (with various minor adjustments for better results) appears in Chapter 6 of [64], and we recommend the interested reader to refer to that work for a better understanding of the theory. In this chapter, however, we use a number of ideas that were developed much later to prove all these classical results. In particular, the works of Safonov [163] and Caffarelli [21] form the basis of the methods used here, and they will be extremely useful in later chapters when we extend the results in this chapter to domains with less regular boundary.

## 2.1 Definitions and properties of Hölder spaces

In addition to the Hölder spaces from Section 1.9, we shall use several other Hölder spaces, most of them involving the derivatives of the functions in the spaces. From now on, we use $\Omega$ to denote a bounded open subset of $\mathbb{R}^n$ which may be further restricted in certain circumstances, but we always indicate when this is the case.

For a positive real number $a = k + \alpha$, where $k$ is a nonnegative integer and $\alpha \in (0, 1]$, we write $H_a(\Omega)$ for the set of all $k$ times continuously differentiable functions with uniformly bounded derivatives of all orders less than or equal to $k$ such that its $k$th order derivatives are Hölder continuous with exponent $\alpha$. We then define a norm on $H_a$ by

$$|f|_a = \sum_{j=0}^{k} |D^j f|_0 + [D^k f]_\alpha.$$

We use without further comment the fact that $H_a$ is a Banach space under this norm. Sometimes we use $C^{k,\alpha}(\overline{\Omega})$ to denote $H_a$.

We shall also make use of several weighted Hölder spaces. To describe them, we write (as before) $d$ for the distance from a point to the boundary of $\Omega$. For $b \in \mathbb{R}$, we then define the norm

$$|f|_0^{(b)} = \sup_{\Omega} |\min\{d^b, (\operatorname{diam}\Omega)^b\} f|,$$

and the seminorm

$$[f]_\alpha^{(b)} = \sup_{\substack{x \neq y \in \Omega \\ |x-y| < d(x)/2}} \min\{d(x)^{b+\alpha}, (\operatorname{diam}\Omega)^{b+\alpha}\} \frac{|f(x) - f(y)|}{|x-y|^\alpha}.$$

From these quantities, we define $H_a^{(b)}$ to be the set of functions with finite norm

$$|f|_a^{(b)} = \sum_{j=0}^{k} |D^j f|_0^{(b+j)} + [D^k f]_\alpha^{(b+k)}.$$

In this chapter, we shall be concerned primarily with these weighted norms with $b \geq 0$, but there will be many times at which other values of $b$ will be important. In particular, if $a > 0$, then $H_a^{(-a)} = H_a$.

Our next step is to collect some useful facts about all these norms. First, if $f$ and $g$ are in $H_a$ with $a > 0$, then $fg \in H_a$ and

$$|fg|_a \leq |f|_a |g|_a. \tag{2.1}$$

If $a \in (0, 1]$, then this inequality follows from the observation that

$$[fg]_a \leq [f]_a |g|_0 + |f|_0 [g]_a,$$

and the general case follows from this one via the product rule for derivatives. More generally, if $a$, $\alpha$ and $\beta$ satisfy

$$a + \alpha \geq 0, \quad a + \beta \geq 0, \quad a + \alpha + \beta \geq 0,$$

and if $f \in H_a^{(\alpha)}$ and $g \in H_a^{(\beta)}$, then $fg \in H_a^{(\alpha+\beta)}$ and

$$|fg|_a^{(\alpha+\beta)} \le |f|_a^{(\alpha)} |g|_a^{(\beta)}. \tag{2.2}$$

Next, we have that $H_a \subset H_b$ if $b \le a \le 1$ and

$$|f|_b \le \max\{1, (\operatorname{diam}\Omega)^{a-b}\} |f|_a. \tag{2.3}$$

When $a > 1$, we need a restriction on $\Omega$ for $H_a \subset H_b$. For example, we consider the following example from page 53 in [64]. Let $\Omega$ be the set of all $(x, y) \in \mathbb{R}^2$ such that $x^2 + y^2 < 1$ and $y < |x|^{1/2}$ and let $a \in (1, 2)$. Then the function $f$, defined in $\Omega$ by

$$f(x,y) = \begin{cases} y^a & \text{if } x, y > 0, \\ -y^a & \text{if } y > 0 > x, \\ 0 & \text{if } y \le 0, \end{cases}$$

is in $H_a$ but not in $H_b$ if $b \in (a/2, 1)$. A more careful examination of this issue is made in [56]. In this book, we usually assume that $\Omega$ is a bounded Lipschitz domain and then we have $H_a \subset H_b$ as long as $a \ge b$.

**Lemma 2.1.** *Let $a$ be a positive number and let $\Omega$ be a bounded Lipschitz domain. If $b \in (0, a)$, then $H_a \subset H_b$ and*

$$|f|_b \le C(\Omega, a) |f|_a. \tag{2.4}$$

***Proof.*** Suppose $a = k + \alpha$ and $b = j + \beta$. If $j = k$, then (2.4) follows from (2.3), so we just have to prove the result for $k > j$, and an induction argument shows that we only need to check the case $k = 1$, $j = 0$, and $\beta = 1$.

Let $f \in H_{1+\alpha}$. We now show that

$$[f]_1 \le C(\Omega)(|Df|_0 + |f|_0). \tag{2.5}$$

We first note that, if $x$ and $y$ are points in $\Omega$ such that the line segment joining $x$ and $y$ is a subset of $\Omega$, then the mean value theorem gives a point $x_1$ on that line segment such that

$$f(x) - f(y) = \frac{\partial f}{\partial \xi}(x_1)(x - y)$$

where $\xi$ is the unit vector in the direction of $x - y$. Hence for any two such points, we have

$$|f(x) - f(y)| \le |Df|_0 |x - y|.$$

In particular, this equation holds if $|x - y| \le d(x)$.

Since $\Omega$ is Lipschitz, there are positive constants $R$ and $\omega_0$ such that, for each $x_0 \in \partial\Omega$, there is a function $\omega$ and a coordinate system $Z = (z^1, \dots, z^n)$ centered at $x_0$ such that

$$\Omega \cap B(x_0, R) = \{x \in B(x_0, R) : z^n > \omega(z')\}$$

and

$$|\omega(z_1') - \omega(z_2')| < \omega_0 |z_1' - z_2'|$$

for $z_1'$ and $z_2'$ with $|z_1'|, |z_2'| < R$. We now set $k_1 = 2\sqrt{1+\omega_0^2}$ and $K = 2(1+k_1)$, and we suppose that $d(x) \le |x-y| \le R/K$. Using $x_0$, a nearest point to $x$ on $\partial\Omega$ as the center of our $Z$ coordinates and writing $\xi$ for the unit vector parallel to the positive $z^n$ axis, simple calculation shows that

$$\min\{d(x+t\xi),\ d(y+t\xi)\} \ge \frac{t}{\sqrt{1+\omega_0^2}}$$

for any $t > 0$ such that $x+t\xi$ and $y+t\xi$ are in $B(x_0, R) \cap \Omega$. In particular, $t = k_1|x-y|$ satisfies this condition, so $\bar{x} = x + k_1|x-y|\xi$ and $\bar{y} = y + k_1|x-y|\xi$ are in $B(x_0, R) \cap \Omega$ and $|\bar{x} - \bar{y}| \le d(\bar{x})/2$. In addition, the line segments joining $x$ and $\bar{x}$, $\bar{x}$ and $\bar{y}$, and $\bar{y}$ and $y$ are in $\Omega$ and hence

$$\begin{aligned} |f(x) - f(y)| &\le |f(x) - f(\bar{x})| + |f(\bar{x}) - f(\bar{y})| + |f(\bar{y}) - f(y)| \\ &\le |Df|_0(|x-\bar{x}| + |\bar{x}-\bar{y}| + |\bar{y}-y|) \\ &= (2k_1+1)|Df|_0|x-y|. \end{aligned}$$

If $|x-y| \ge R/K$, then

$$|f(x) - f(y)| \le 2|f|_0 \le \frac{2K}{R}|f|_0|x-y|.$$

Combining the three cases yields (2.5), and then we have

$$|u|_1 \le C|u|_{1+\alpha}$$

for any $\alpha \in (0,1)$ because

$$|u|_0 + |Du|_0 \le |u|_{1+\alpha}. \qquad \square$$

In fact, a similar monotonicity result is true for weighted norms.

**Lemma 2.2.** *Let $a$ and $\alpha$ be real numbers with $a > 0$ and $a + \alpha \ge 0$, and let $\Omega$ be an open subset of $\mathbb{R}^n$.*

(1) *If $b \in (0,a)$ and $b + \alpha \ge 0$, then $H_a^{(\alpha)} \subset H_b^{(\alpha)}$ with*

$$|f|_b^{(\alpha)} \le |f|_a^{(\alpha)} \tag{2.6}$$

*for all $f \in H_a^{(\alpha)}$.*

(2) *If* $\gamma \ge \alpha$, *then* $H_a^{(\gamma)} \subset H_a^{(\alpha)}$ *with*

$$|f|_a^{(\alpha)} \le (\operatorname{diam}\Omega)^{\gamma-\alpha}|f|_a^{(\gamma)} \tag{2.7}$$

*for all* $f \in H_a^{(\alpha)}$.

***Proof.*** To prove (2.6), we consider several cases. First, if $a \in (0,1]$, then we use the definition of $[f]_b^{(\alpha)}$ and $[f]_b^{(\alpha)}$ to conclude that

$$[f]_b^{(\alpha)} \le [f]_a^{(\alpha)}. \tag{2.8}$$

Next, the argument in the proof of Lemma 2.1 implies that

$$[f]_1^{(\alpha)} \le |Df|_0^{(1+\alpha)}, \tag{2.9}$$

and hence (2.8) holds for $b = 1$ and $a \in (1,2]$. An easy induction argument completes the proof.

Similarly, it's easy to check that

$$[f]_a^{(\gamma)} \le (\operatorname{diam}\Omega)^{\gamma-\alpha}[f]_a^{(\alpha)}$$

for any $a$ and $\gamma \ge \alpha \ge -a$, so (2.7) is also valid. □

Another useful property of the Hölder norms is an interpolation inequality. For the $H_a^{(\alpha)}$ norms with fixed $\alpha$, the interpolation inequality is valid for any domain. The case $\alpha = 0$ appears in Proposition 4.2 of [114] and Lemma 6.32 of [64], and we give a quick description for the case of arbitrary $\alpha$ (which is very close to the case $\alpha = 0$).

**Lemma 2.3.** *Let* $0 < a < b$ *be real numbers, let* $\alpha \ge -a$, *and let* $\sigma \in (0,1)$. *Then there is a constant* $C(b)$ *such that*

$$[u]_{\sigma a+(1-\sigma)b}^{(\alpha)} \le C(b)(|u|_a^{(\alpha)})^{\sigma}(|u|_b^{(\alpha)})^{1-\sigma}.$$

***Proof.*** We consider several cases, and, to simplify some notation, we define

$$[u]_0^{(\alpha)} = \sup\{d(x)^{\alpha}|u(x)-u(y)| : |x-y| < \frac{1}{2}d(x)\}$$

if $\alpha \ge 0$.

As a first step, we show that

$$[u]_{\sigma a+(1-\sigma)b}^{(\alpha)} \le ([u]_a^{(\alpha)})^{\sigma}([u]_b^{(\alpha)})^{1-\sigma} \tag{2.10}$$

if $0 \le a < b \le 1$. To simplify some writing, we set $U_s = [u]_s^{(\alpha)}$ for $s \in [0,1]$ and $t = \sigma a + (1-\sigma)b$. For any $x$ and $y$ in $\Omega$ with $|x-y| < \frac{1}{2}d(x)$, we have

$$\begin{aligned} |u(x)-u(y)| &\le |u(x)-u(y)|^{\sigma}|u(x)-u(y)|^{1-\sigma} \\ &\le (d(x)^{-a-\alpha}|x-y|^{a}U_a)^{\sigma}(d(x)^{-b-\alpha}|x-y|^{b}U_b)^{1-\sigma} \\ &= d^{-t-\alpha}|x-y|^{t}U_a^{\sigma}U_b^{1-\sigma}. \end{aligned}$$

Multiplying this equation by $d(x)^{t+\alpha}|x-y|^t$ and taking the supremum over all $x$ and $y$ gives (2.10).

Now suppose that $0 \le a < 1 < b \le 2$. With $\mu \in (0, \frac{1}{2})$ to be chosen and $\xi$ a unit vector, we set $x_1 = x + \mu d(x)\xi$. Then Taylor's theorem gives

$$u(x_1) = u(x) + \mu d(x) Du(x) \cdot \xi + \int_0^1 \mu d(x)[Du(x + s\mu d(x)\xi) - Du(x)] \cdot \xi \, ds,$$

and hence

$$Du(x) \cdot \xi = \frac{u(x + \mu d(x)\xi) - u(x)}{\mu d(x)} - \int_0^1 [Du(x + s\mu d(x)\xi) - Du(x)] \cdot \xi \, ds.$$

Since $|Du(x)| = \sup_{|\xi|=1} Du(x) \cdot \xi$, it follows that

$$\begin{aligned} |Du(x)| &\le U_a(\mu d(x))^{a-1} d(x)^{-a-\alpha} + U_b(\mu d(x))^{b-1} d(x)^{-b-\alpha} \\ &= d(x)^{-1-\alpha}[\mu^{a-1}U_a + \mu^{b-1}U_b], \end{aligned}$$

so

$$|Du|_0^{(\alpha+1)} \le \mu^{a-1}U_a + \mu^{b-1}U_b.$$

We now take

$$\mu = \frac{1}{2}\left(\frac{U_a}{U_b + U_a}\right)^{1/(b-a)}$$

to infer that

$$|Du|_0^{(\alpha+1)} \le (2^{1-a} + 2^{b-1}) U_a^\sigma (U_b + U_a)^{1-\sigma}.$$

From Lemma 2.2, we therefore infer that

$$[u]_1^{(\alpha)} \le 6([u]_a^{(\alpha)})^\sigma (|u|_b^{(\alpha)})^{1-\sigma}.$$

The proof of (2.10) in all other cases follows via simple algebra, similar to that in the proof of case (iv) in Lemma 6.32 from [64]. Since the form of the interpolation inequality in that lemma is different from the one here, we show here how to proceed in one such case. Let us suppose that $a \in [0, 1)$, $b \in (1, 2]$ and $c = \sigma a + (1-\sigma)b < 1$. Choose $\theta$ so that $1 = \theta a + (1-\theta)b$ and $\delta$ so that $c = \delta a + (1-\delta)$. Then

$$|u|_1^{(\alpha)} \le C(|u|_a^{(\alpha)})^\theta (|u|_b^{(\alpha)})^{1-\theta}$$

and

$$|u|_c^{(\alpha)} \le (|u|_a^{(\alpha)})^\delta (|u|_1^{(\alpha)})^{1-\delta}.$$

Inserting the estimate for $|u|_1$ in the second inequality gives the desired result. □

If we want to interpolate with norms having different superscripts, then we need to restrict the class of domains. In Lemma 6.35 of [64], an interpolation inequality is presented for $H_a$ norms in smooth domains, but it will be important here to consider Lipschitz domains, so we provide a sketch of the proof of that result.

**Lemma 2.4.** *Let $\Omega$ be a bounded Lipschitz domain and let $a$, $b$, and $\sigma$ be real numbers with $0 \le a < b$ and $0 < \sigma < 1$. Then there is constant $C(b, \Omega)$ such that*

$$|u|_{\sigma a+(1-\sigma)b} \le C(b, \Omega)(|u|_a)^{\sigma}(|u|_b)^{1-\sigma}. \tag{2.11}$$

***Proof.*** As in the proof of Lemma 6.34 in [64], we consider a number of cases. To simplify the notation, we define

$$[u]_0 = \sup\{|u(x) - u(y)|\}.$$

If $0 \le a < b \le 1$, then it is easy to check that

$$[u]_{\sigma a+(1-\sigma)b} \le [u]_a^{\sigma}[u]_b^{1-\sigma}.$$

Next, we suppose that $a \in [0, 1)$, $b \in (1, 2]$, and $\sigma$ is chosen so that $\sigma a + (1 - \sigma)b = 1$. Since $\Omega$ is a bounded Lipschitz domain, there is a cone $K$ with opening angle $\theta_0$ and height $R$ such that, for each point $x \in \Omega$, there is a rotation $T_x$ such that $x + T_x K \subset \Omega$. We now fix $x \in \Omega$ and let $\xi$ be any unit vector parallel to some vector in $T_x K$. With $\mu \in (0, 1]$ to be chosen later, we set $x_1 = x + \mu R\xi$ and use the argument in Lemma 2.3 to see that

$$Du(x) \cdot \xi = \frac{u(x_1) - u(x)}{\mu R} + \int_0^1 |Du(x + t\mu R\xi) - Du(x)|\, dt,$$

and hence

$$|Du(x) \cdot \xi| \le (\mu R)^{a-1}[u]_a + (\mu R)^{b-1}[u]_b.$$

We now observe that

$$|Du(x)| \le C(\theta_0) \sup |Du(x) \cdot \xi|,$$

where the supremum is taken over all unit vectors $\xi$ parallel to some vector in $T_x K$. (Note that the constant does not depend on the rotation $T_x$.) Writing

$$U_a = R^a[u]_a,\ U_1 = R \sup |Du|,\ U_b = R^b[u]_b + U_a,$$

we infer that

$$U_1 \le C(\theta_0)[\mu^{a-1}U_a + \mu^{b-1}U_b],$$

so the choice $\mu = (U_a/U_b)^{1/(b-a)}$ gives

$$U_1 \le 2C(\theta_0)U_a^\sigma U_b^{1-\sigma}.$$

Dividing this inequality by $R$ yields

$$\sup|Du| \le 2C(\theta_0)([u]_a)^\sigma([u]_b + [u]_a R^{a-b})^{1-\sigma},$$

and then we use (2.4) and (2.5) to infer (2.11).

The proof of (2.11) in all other cases follows from these cases as before. □

In fact, we can interpolate between $H_a^{(\alpha)}$ and $H_b^{(\beta)}$ as long as $a+\alpha \ge 0$ and $b+\beta \ge 0$.

**Lemma 2.5.** *Let $\Omega$ be an open subset of $\mathbb{R}^n$ and let $a$, $b$, $\alpha$ and $\beta$ be real numbers satisfying $0 < a \le b$, $a+\alpha \ge 0$, and $b+\beta \ge 0$ with $\max\{a+\alpha, b+\beta\} > 0$. Then*

$$|f|_{\sigma a+(1-\sigma)b}^{(\sigma\alpha+(1-\sigma)\beta)} \le C(b)(|f|_a^{(\alpha)})^\sigma(|f|_b^{(\beta)})^{1-\sigma}. \tag{2.12}$$

***Proof.*** We first note that the inequality is immediate if $a = b$ or if $b \le 1$.

Now suppose that $0 < a < 1 < b \le 2$ and $a + \beta \ge 0$. Let us fix a point $x \in \Omega$ and apply Lemma 2.4 in the ball $B(x,r)$ with $r = d(x)/5$. Then, with $\sigma$ chosen so that $1 = \sigma a + (1-\sigma)b$ and $t = \sigma\alpha + (1-\sigma)\beta$), we have

$$r^{1+t}|Df(x)| \le C(r^{a+\alpha}[f]_{a;B(x,r)})^\sigma(r^{b+\beta}[f]_{b;B(x,r)} + r^{a+\beta}[f]_{a;B(x,r)})^{1-\sigma}.$$

Now we estimate the factors on the right side of this equation. Let $y$ and $z$ be points in $B(x,r)$. Then

$$d(y) \ge d(x) - |y-z| \ge \frac{4}{5}d(x)$$

and

$$|y-z| \le \frac{2}{5}d(x) \le \frac{1}{2}d(y),$$

so

$$\begin{aligned}|f(y)-f(z)| &\le [f]_a^{(\alpha)}d(y)^{-a-\alpha}|y-z|^a\\ &\le C[f]_a^{(\alpha)}d(x)^{-a-\alpha}|y-z|^a\end{aligned}$$

and hence (taking the supremum over all $y$ and $z$ in $B(x,r)$), we have

$$r^{a+\alpha}[f]_{a;B(x,r)} \le C[f]_a^{(\alpha)}.$$

Therefore

$$|Df|_0^{(\sigma\alpha+(1-\sigma)\beta-1)} \le C([f]_a^{(\alpha)})^\sigma([f]_b^{(\beta)} + [f]_a^{(\beta)})^{1-\sigma}.$$

Using (2.6), we see that

$$[f]_1^{(\sigma\alpha+(1-\sigma)\alpha)} \le C([f]_a^{(\alpha)})^\sigma(|f|_b^{(\beta)})^{1-\sigma}. \tag{2.13}$$

Next, we remove the restriction $a+\beta \ge 0$ in this inequality by considering two cases. First, if $a+\sigma\alpha+(1-\sigma)\beta > 0$, then there is a $\theta \in (\sigma, 1)$ such that $a+\theta\alpha+(1-\theta)\beta \ge 0$. We now set $c = \theta a + (1-\theta)b$ and $\gamma = \theta\alpha + (1-\theta)\beta$. With $\delta \in (0,1)$ taken so that $1 = \delta a + (1-\delta)c$, we infer from (2.13) (with $b$ replaced by $c$) that

$$[f]_1^{(\sigma\alpha+(1-\sigma)\beta)} \le C([f]_a^{(\alpha)})^\delta(|f|_c^{(\gamma)})^{1-\delta}.$$

The case $b = 1$ now implies that

$$[Df]_{c-1}^{(\gamma-1)} \le ([Df]_0^{(\sigma\alpha+(1-\sigma)\beta-1)})^\lambda(|Df|_{b-1}^{(\beta-1)})^{1-\lambda},$$

where $\lambda$ satisfies $\lambda + (1-\lambda)b = c$. It follows that

$$[f]_c^{(\gamma)} \le ([f]_1^{(\sigma\alpha+(1-\sigma)\beta)})^\lambda(|Df|_b^{(\beta)})^{1-\lambda},$$

and combining this inequality with the one for $[f]_1^{(\sigma\alpha+(1-\sigma)\beta}$ yields (2.13) in this case.

On the other hand if $a+\sigma\alpha+(1-\sigma)\beta \le 0$, then there is a constant $\theta \in (0,\sigma)$ such that $c = \theta a + (1-\theta)b$ is less than 1 and $c+\beta \ge 0$. (Here, we note that $1+\sigma\alpha+(1-\sigma)\beta > 0$.) It follows that

$$[f]_1^{(\sigma\alpha+(1-\sigma)\beta)} \le C([f]_c^{(\gamma)})^\delta(|f|_b^{(\beta)})^{1-\delta},$$

where $\gamma = \theta\alpha+(1-\theta)\beta$, just as before, but now $\delta$ satisfies $\delta\gamma+(1-\delta)\beta = 1$. In addition, we conclude from the $b \le 1$ case that

$$[f]_c^{(\gamma)} \le ([f]_a^{(\alpha)})^\lambda([f]_1^{(\sigma\alpha+(1-\sigma)\beta)})^{1-\lambda},$$

with $\lambda a + 1 - \lambda = c$. Combining these inequalities yields (2.13) in this case as well.

From this estimate, we obtain the full interpolation inequality by arguing as in Lemma 2.3. □

Note that the case $a+\alpha = b+\beta = 0$ is covered by Lemma 2.4. In this case, $\Omega$ must be a bounded Lipschitz domain and the constant $C$ depends also on $\Omega$.

We shall also have occasion to consider weighted Hölder spaces where the weighting is with respect to distance to just a part of the boundary. It's possible to give a thorough treatment under very weak assumptions (for example, the weighting can be with respect to an arbitrary non-empty portion of the boundary and there can be different weightings with respect to distance to different parts of the boundary), but we just examine a relatively simple situation. In this way, we avoid some technical points that are not necessary for our program. We refer the reader to the Notes for a discussion of some of these issues.

Let $\Sigma$ be a non-empty subset of $\partial\Omega$ and write $d^*$ for the distance to $\Sigma$. We then define $H_a^{(b)}(\Omega;\Sigma)$ to be the set of all functions with finite norm $|\cdot|^{(b)}_{a:\Omega,\Sigma}$, where the norm is defined by replacing $d$ in the definition of $|\cdot|^{(b)}_{a;\Omega}$ by $d^*$. These spaces are still Banach spaces but some care must be taken in order to guarantee that the interpolation theorems are still valid. As before, the correct condition is a Lipschitz one. We say that $\Sigma$ is a $C^{0,1}$ subset of $\partial\Omega \in C^{0,1}$ if, for each $x_0 \in \Sigma$, there are two Lipschitz functions $\omega$ and $\omega_\sigma$ defined on some ball $B'(0,R) \subset \mathbb{R}^{n-1}$ and a coordinate system $Y$ centered at $x_0$ such that

$$\Omega \cap B(x_0,R) = \{x \in B(x_0,R) : y^n > \omega(y')\}$$

and

$$\Sigma \cap B(x_0,R) = \{x \in B(x_0,R) : y^n = \omega(y'), \omega_\sigma(y') > 0\}.$$

As a simple representative example, if $\Omega$ is the half-ball

$$\{(x,y) \in \mathbb{R}^2 : x^2+y^2 < 1, y > 0\},$$

then

$$\Sigma = \{(x,y) : x^2+y^2 < 1, y = 0\}$$

is a $C^{0,1}$ subset of $\partial\Omega$. The assumption $\partial\Omega \in C^{0,1}$ can be relaxed if we appropriately modify the definition of $C^{0,1}$ subset of $\partial\Omega$ but there is no reason to do so here.

It's then easy to check that, if we replace $d$ by $d^*$ in the proofs of (2.2), Lemma 2.2, and Lemma 2.5, then we obtain the corresponding results with these partially interior norms, so we just state the results.

**Lemma 2.6.** *Let $a$ and $\alpha$ be real numbers with $a > 0$ and $a + \alpha \ge 0$, let $\Omega$ be a bounded Lipschitz domain, and let $\Sigma$ be a $C^{0,1}$ portion of $\partial\Omega$.*

(1) *Let $\beta \ge \max\{-a, -a-\alpha\}$. If $f \in H_a^{(\alpha)}(\Omega,\Sigma)$ and $g \in H_a^{(\beta)}(\Omega,\Sigma)$, then $fg \in H_a^{(\alpha+\beta)}(\Omega,\Sigma)$ and*

$$|fg|^{(\alpha+\beta)}_{a;\Omega,\Sigma} \le |f|^{(\alpha)}_{a;\Omega,\Sigma}|g|^{(\beta)}_{a;\Omega,\Sigma}. \tag{2.14}$$

(2) *Let $b \in (0,a]$ and $\beta \ge \max\{\alpha, -b\}$. Then, for any $f \in H_a^{(\alpha)}(\Omega,\Sigma)$, we have*

$$|f|^{(\beta)}_{b;\Omega,\Sigma} \le C(\Omega,\Sigma,a)|f|^{(\alpha)}_{a;\Omega,\Sigma}. \tag{2.15}$$

(3) *Let $b \in (0,a)$, $\beta \geq -b$, and $\sigma \in (0,1)$ be real numbers. Then, for any $f \in H_a^{(\alpha)}(\Omega,\Sigma) \cap H_b^{(\beta)}(\Omega,\Sigma)$, we have*

$$|f|_{\sigma a+(1-\sigma)b}^{(\sigma\alpha+(1-\sigma)\beta)} \leq C(b,\Omega,\Sigma)(|f|_a^{(\alpha)})^\sigma(|f|_b^{(\beta)})^{1-\sigma}. \tag{2.16}$$

As before, the constant $C$ in (2.16) is independent of $\Omega$ and $\Sigma$ unless $a+\alpha = b+\beta = 0$.

We close this section with a simple rule which governs the behavior of weighted Hölder norms under a change of independent variable.

**Lemma 2.7.** *Let $a$ and $\alpha$ be real numbers with $a > 0$ and $\alpha \geq -a$. Let $\Omega$ and $\Omega'$ be open subsets of $\mathbb{R}^n$ and suppose that there is a one-to-one, onto function $\Psi\colon \Omega \to \Omega'$ such that $\Psi \in H_{\max\{1,a\}}(\Omega)$ and $\Psi^{-1} \in H_1(\Omega')$. If $f \in H_a^{(\alpha)}(\Omega')$, then $f\circ\Psi \in H_a^{(\alpha)}(\Omega)$ and there is a constant $C$, determined only by $a$, $\alpha$, $|\Psi|_{\max\{1,a\}}$, and $|\Psi^{-1}|_1$ such that*

$$|f\circ\Psi|_a^{(\alpha)} \leq C|f|_a^{(\alpha)}. \tag{2.17}$$

***Proof.*** Note that there is a constant $C_1$, determined only by the Lipschitz norms of $\Psi$ and $\Psi^{-1}$, such that

$$\frac{1}{C_1}|x-y| \leq |\Psi(x)-\Psi(y)| \leq C_1|x-y|$$

and

$$\frac{1}{C_1}d(x) \leq d(\Psi(x)) \leq C_1 d(x)$$

for all $x$ and $y$ in $\Omega$. It immediately follows that

$$|f\circ\Psi^{-1}|_0^{(\alpha)} \leq C|f|_0^{(\alpha)}.$$

In addition, there is a positive integer $N$ (determined only by the Lipschitz norms of $\Psi$ and $\Psi^{-1}$) such that, if $x$ and $y$ satisfy $|x-y| \leq \frac{1}{2}d(x)$, then there are points $x_0,\dots,x_N$ with $x_0 = x$ and $x_N = y$ such that

$$|\Psi^{-1}(x_i) - \Psi^{-1}(x_{i-1})| \leq \frac{1}{2}d(\Psi^{-1}(x_i))$$

for $i \in \{1,\dots,N\}$. This observation implies the result for $a \leq 1$, and the case $a > 1$ follows from this one by the chain rule. □

## 2.2 An alternative characterization of Hölder spaces

The definition of $H_a^{(b)}$ will be very useful in understanding the behavior of solutions of oblique derivative problems; however, our analysis of this behavior is greatly facilitated by the use of other means of showing that a function is in some Hölder space. In this section, we present a characterization that is relevant to the work in this chapter. This time we refer to Lemma 12.12 from [114] for a complete proof (in the parabolic case) and just give an outline of the proof.

**Lemma 2.8.** *Let $N$ be a nonnegative integer, let $\alpha \in (0,1]$, let $R > 0$, let $x_0 \in \mathbb{R}^n$, and let $u \in C(B(x_0,R))$. Suppose that there is a positive constant $H$ such that, for each $y \in B(x_0,R/2)$ and each $r \in (0,R/2)$, there is a polynomial $P_N(y,r)$ of degree no greater than $N$ with*

$$\sup_{B(y,r)} |u - P_N(y,r)| \le Hr^{N+\alpha}. \tag{2.18}$$

*Then $u \in H_{N+\alpha}(B(x_0,R/2))$ and*

$$[u]_{N+\alpha;B(x_0,R/2)} \le C(N,\alpha)H.$$

***Proof.*** We just prove the result for $N = 0$. Let $\varphi$ be a $C^\infty(\mathbb{R}^n)$ function with support in the unit ball $B(0,1)$ such that $\varphi \ge 0$ and

$$\int_{\mathbb{R}^n} \varphi(x)\,dx = 1.$$

Set $K = \sup|D\varphi|$ and note that $\varphi$ can be chosen so that $K$ is determined only by $N$. For $\tau \in (-R/2, R/2)$ and $x \in B(x_0,R/2)$, we define

$$\bar{u}(x,\tau) = \int_{\mathbb{R}^n} u(x-\tau z)\varphi(z)\,dz.$$

By writing this equation in the more usual form

$$\bar{u}(x,\tau) = (-\tau)^{-n}\int_{\mathbb{R}^n} u(w)\varphi\left(\frac{x-w}{\tau}\right)\,dw,$$

we see that $\bar{u}$ is differentiable with respect to $x$ and $\tau$ as long as $\tau \ne 0$. Moreover,

$$\frac{\partial \bar{u}}{\partial x}(x,\tau) = -\frac{1}{\tau}\int_{\mathbb{R}^n} u(x-\tau z)D\varphi(z)\,dz,$$
$$\frac{\partial \bar{u}}{\partial \tau}(x,\tau) = -\frac{1}{\tau}\int_{\mathbb{R}^n} u(x-\tau z)D_i(z^i\varphi(z))\,dz.$$

Hence, if we write $\nabla\bar{u}$ for the $(n+1)$-dimensional vector

$$\left(\frac{\partial\bar{u}}{\partial x}, \frac{\partial\bar{u}}{\partial\tau}\right),$$

we see after a quick calculation that

$$|\nabla\bar{u}(x,\tau)| \le C(\alpha,n)H\tau^{\alpha-1}.$$

We now fix $x$ and $y$ in $B(x_0,R/2)$ and take $\tau = |x-y|/2$. Since $u(x) = \bar{u}(x,0)$, it follows that

$$|u(x)-u(y)| \le |\bar{u}(x,0)-\bar{u}(w,\tau)| + |\bar{u}(w,\tau)-\bar{u}(y,0)|$$

for $w = (x+y)/2$. Since

$$\bar{u}(x,0)-\bar{u}(w,\tau) = \int_0^1 \nabla u(sx+(1-s)w, s\tau)\cdot(x-y,\tau)\,dx,$$

it follows by integration that

$$|\bar{u}(x,0)-\bar{u}(w,\tau)| \le CH\tau^{\alpha},$$

and a similar argument gives

$$|\bar{u}(w,\tau)-\bar{u}(y,0)| \le CH\tau^{\alpha}.$$

The proof is completed by adding these inequalities and recalling the definitions of $\tau$ and $[u]_{\alpha;B(x_0,R)}$. □

We shall also need a corresponding result near the boundary, which is proved by modifying the proof of Lemma 12.12 in [114] in the same way that the proof of Lemma 4.3 in [114] was modified to obtain Lemma 4.11 of [114], specifically, we assume that $\varphi(x) = 0$ for any $x$ with $x^n < 0$. In the statement, we write $B^+(x_0,r)$ and $B^0(x_0,r)$ for the set of all $x \in B(x_0,r)$ for which $x^n > 0$ or $x^n = 0$, respectively.

**Lemma 2.9.** *Let $N$ be a nonnegative integer, let $\alpha \in (0,1]$, let $R > 0$, let $x_0 \in \mathbb{R}^n$ with $x_0^n = 0$, and let $u \in C(B(x_0,R))$. Suppose that there is a positive constant $H$ such that, for each $y \in B^0(x_0,R/2)$ and each $r \in (0,R/2)$, there is a polynomial $P_N(y,r)$ of degree no greater than $N$ with*

$$\sup_{B^+(y,r)} |u - P_N(y,r)| \le Hr^{N+\alpha}. \tag{2.19}$$

*Then $u \in H_{N+\alpha}(B^0(x_0,R/2))$ and*

$$[u]_{N+\alpha;B^0(x_0,R/2)} \le C(N,\alpha)H.$$

*Suppose also that, for every $y \in B^+(x_0,R/2)$ and each $r \in (0,y^n)$, there is a polynomial $P_N(y,r)$ of degree no greater than $n$ with*

$$\sup_{B(y,r)} |u - P_N(y,r)| \le Hr^{N+\alpha}.$$

*Then $u \in H_{N+\alpha}(B^+(x_0,R/2))$ and*

$$[u]_{N+\alpha;B^+(x_0,R/2)} \le C(N,\alpha)H.$$

## 2.3 An existence result

In this section, we show that the Dirichlet problem for a constant coefficient equation can always be solved in a ball. More precisely, we have the following existence result.

**Theorem 2.10.** *Let $[A^{ij}]$ be a positive definite matrix, let $x_0 \in \mathbb{R}^n$, let $R > 0$, and set $B = B(x_0, R)$. Then, for any polynomials $f$ and $\psi$, there is a unique solution $u \in C(\overline{B}) \cap C^2(B)$ of*

$$A^{ij} D_{ij} u = f \text{ in } B, \quad u = \psi \text{ on } \partial B. \tag{2.20}$$

***Proof.*** First, note that $A^{ij} D_{ij} \psi$ is also a polynomial, so there is no loss of generality in assuming that $\psi = 0$. For each nonnegative integer $N$, write $P_N$ for the set of all polynomials of degree no greater than $N$, and define the operator $T$ (which maps $P_N$ into $P_N$ for any $N$) by

$$Tw(x) = A^{ij} D_{ij}((R^2 - |x - x_0|^2)w(x)).$$

If $Tw = 0$, then the maximum principle implies that $u$, defined by

$$u(x) = (R^2 - |x - x_0|^2)w(x), \tag{2.21}$$

must attain its maximum and minimum values on $\partial B$. Since $u = 0$ on $\partial B$, it follows that $u = 0$ in $B$ and hence $w = 0$, so $T$ is one-to-one. By linear algebra, we conclude that $T$ is also onto.

Now, let $f$ be a given polynomial. Then there is a unique polynomial $w$ such that $Tw = f$, and hence $u$ given by (2.21) solves the boundary value problem (2.20) (with $\psi = 0$). The maximum principle implies that this $u$ is the unique solution of this boundary value problem. □

In a later section, we shall prove that (2.20) has a unique solution for a very broad choice of functions $f$ and $\psi$. As a further preliminary to this result, we note that the Weierstrass approximation theorem implies that, for any continuous function $\psi$ defined on $\partial B$, there is a sequence of polynomials converging uniformly on $\partial B$ to $\psi$. The maximum principle then implies the following result.

**Corollary 2.11.** *Let $[A^{ij}]$ be a positive definite matrix, let $x_0 \in \mathbb{R}^n$, let $R > 0$, and set $B = B(x_0, R)$. Then, for any polynomial $f$ and any continuous function $\psi$, there is a unique solution $u \in C(\overline{B}) \cap C^2(B)$ of (2.20).*

## 2.4 Basic interior estimates

We are now ready to prove some estimates for solutions of constant coefficient equations in a ball.

As a first step, we obtain an estimate with constant $A^{ij}$ but variable $f$.

**Proposition 2.12.** *Let $\Omega$ be a non-empty, open subset of $\mathbb{R}^n$, let $[A^{ij}]$ be a positive-definite matrix, let $\alpha \in (0,1)$ and suppose $u \in C^2(\Omega) \cap L^\infty(\Omega)$ satisfies*

$$A^{ij}D_{ij}u = f \text{ in } \Omega \tag{2.22}$$

*for some $f \in H_\alpha^{(2)}$. Then $u \in H_{2+\alpha}^{(0)}$ and*

$$|u|_{2+\alpha}^{(0)} \leq C(\alpha, n, \mu)(|u|_0 + |f/\lambda|_\alpha^{(2)}), \tag{2.23}$$

*where $\lambda$ and $\mu$ are constants such that*

$$A^{ij}\xi_i\xi_j \geq \lambda|\xi|^2, \max_{i,j}|A^{ij}| \leq \mu\lambda \tag{2.24}$$

*for all $\xi \in \mathbb{R}^n$.*

***Proof.*** Fix $x_0 \in \Omega$. We shall estimate $[u]_{2+\alpha;B(x_0,d(x_0)/2)}$. (As a by-product of our estimate, we show that this quantity is finite.) To this end, we set $R = d(x_0)/2$, and we fix $y \in B(x_0, R)$. Because of Lemma 2.8, we need to examine the supremum of the difference between $u$ and some polynomials on balls centered at $y$. To this end, we obtain the polynomials as the second order Taylor polynomials of appropriate solutions of boundary value problems.

We begin by setting $u_0 = u$ and defining $v_0$ to be the solution, given by Corollary 2.11, of

$$A^{ij}D_{ij}v_0 = f(y) \text{ in } B(y,R), \quad v_0 = u_0 \text{ on } \partial B(y,R).$$

Since $f(y)$ is a constant, we can use Theorem 1.34 to infer that there is a constant $C_0(n,\mu)$ so that

$$\begin{aligned}\sup_{B(y,R/2)}|D^3v_0| &\leq C_0(R^{-3}\sup_{B(y,R)}|v_0| + R^{-1}|f(y)/\lambda|)\\ &\leq C_0(R^{-3}\sup_{B(y,R)}|u_0| + R^{-1}\sup_{B(y,R)}|f/\lambda|).\end{aligned}$$

We next define the second degree polynomial $Q_0$ by

$$Q_0(x) = v_0(y) + D_iv_0(y)(x-y)^i + \frac{1}{2}D_{ij}v_0(y)(x-y)^i(x-y)^j.$$

For any $r \in (0, R/2)$, it follows that

$$\sup_{B(y,r)} |v_0 - Q_0| \le C_0 \left(\frac{r}{R}\right)^3 \sup_{B(y,R)} |u_0| + C_0 \frac{r^3}{R} \sup_{B(y,R)} |f/\lambda|. \tag{2.25}$$

It also follows from the maximum principle that

$$\sup_{B(y,R)} |u_0 - v_0| \le C_1 R^2 \sup_{B(y,R)} |(f - f(y))/\lambda|.$$

Setting $F = |f/\lambda|_\alpha^{(2)}$ and combining these two inequalities, we find that

$$\sup_{B(y,r)} |u_0 - Q_0| \le C_0 \left(\frac{r}{R}\right)^3 \sup_{B(y,R)} |u| + (C_0 + C_1)F.$$

We now set $\theta = (1 + \alpha)/2$ and fix $\tau \in (0, 1/4)$ so that $C_0\tau^3 \le \tau^{2+\theta}$ to conclude that

$$\sup_{B(y,\tau R)} |u_0 - Q_0| \le \tau^{2+\theta} \sup_{B(y,R)} |u| + (C_0 + C_1)F. \tag{2.26}$$

We obtain a suitable sequence of polynomials by induction. The method can be described briefly by saying that we replace $u$ by $u_k$, and $R$ by $R_k = \tau^k R$, but we shall be more specific. For each positive integer $m$, we set $R_m = \tau^m R$. Suppose that we have already determined second degree polynomials $Q_0, \ldots, Q_{k-1}$, and functions $u_0, \ldots, u_{k-1}$ so that

$$\sup_{B(y,\tau R_m)} |u_m - Q_m| \le \tau^{2+\alpha} \sup_{B(y,R_m)} |u_m| + (C_0 + C_1)\frac{R_m^{2+\alpha}}{R^{2+\alpha}} F \tag{2.27}$$

and

$$A^{ij} D_{ij} Q_m = \begin{cases} f(y) & \text{if } m = 0, \\ 0 & \text{if } m > 0 \end{cases}$$

for $m = 0, \ldots, k - 1$. We then set $u_k = u_{k-1} - Q_{k-1}$ and take $v_k$ to be the solution of

$$A^{ij} D_{ij} v_k = 0 \text{ in } B(y, R), \quad v_k = u_k \text{ on } \partial B(y, R).$$

We also take $Q_k$ to be the second degree Taylor polynomial of $v_k$ centered at $y$. Our preceding argument shows that

$$\sup_{B(y,\tau R_k)} |v_k - Q_k| \le \tau^{2+\theta} \sup_{B(y,R)} |u_k|,$$

and the maximum principle gives

$$\sup_{B(y,R_k)} |u_k - v_k| \le C_1 R_k^2 \sup_{B(y,R_k)} |(f - f(y))/\lambda| \le C_1 \frac{R_k^{2+\alpha}}{R^{2+\alpha}} F$$

because $A^{ij}D_{ij}u_k = f - f(y)$. Combining these two estimates yields (2.27) for $m = k$ as well (because $C_0 \geq 0$), and clearly $A^{ij}D_{ij}Q_k = 0$.

For our next step, we define polynomials $P_2(y,r)$ for any $r \in (0,R)$ as follows: if $r \in (R_k, R_{k-1}]$, we set $P_2(y,r) = \sum_{m=0}^{k-1} Q_m$. Then $w$ defined by

$$w(r) = \sup_{B(y,r)} |u - P_2(y,r)|$$

is easily seen to be increasing on each interval $(R_k, R_{k-1}]$, so Lemma 1.25 (with $\sigma(r) = F(r/R)^{2+\alpha}$, $2+\theta$ in place of $\alpha$, and $2+\alpha$ in place of $\delta$) implies that

$$w(r) \leq C\left(\frac{r}{R}\right)^{2+\alpha} [|u|_0 + F],$$

and the proof is completed by invoking Lemma 2.8 and the definition of $[u]^{(0)}_{2+\alpha}$. □

A slight modification of the preceding argument allows us to consider variable coefficients in the second order terms and nonzero coefficients in lower order terms as well.

**Proposition 2.13.** *Let $\Omega$ be a non-empty, open subset of $\mathbb{R}^n$, let $[a^{ij}]$ be a positive-definite matrix-valued function defined on $\Omega$, let $\alpha \in (0,1)$ and suppose $u \in C^2(\Omega) \cap L^\infty(\Omega)$ satisfies*

$$a^{ij}D_{ij}u + b^iD_iu + cu = f \tag{2.28}$$

*for some $f \in H^{(2)}_\alpha$. If $a^{ij} \in H^{(0)}_\alpha$, $b^i \in H^{(1)}_\alpha$, and $c \in H^{(2)}_\alpha$, and if the minimum eigenvalue of $[a^{ij}]$ is bounded away from zero and infinity, then $u \in H^{(0)}_{2+\alpha}$. Moreover, if $A$, $\lambda$, and $\mu$ are positive constants such that*

$$a^{ij}\xi_i\xi_j \geq \lambda|\xi|^2, \tag{2.29a}$$
$$a^{ij}\xi_i\xi_j \leq \mu\lambda|\xi|^2 \tag{2.29b}$$

*for all $\xi \in \mathbb{R}^n$,*

$$|a^{ij}|^{(0)}_\alpha \leq \lambda A, \tag{2.29c}$$
$$|b^i|^{(1)}_\alpha \leq \lambda A, \tag{2.29d}$$
$$|c|^{(2)}_\alpha \leq \lambda A, \tag{2.29e}$$

*then*

$$|u|^{(0)}_{2+\alpha} \leq C(A,\alpha,n,\mu)(|u|_0 + |f/\lambda|^{(2)}_\alpha). \tag{2.30}$$

***Proof.*** First, we suppose that $b^i$ and $c$ are zero and that $|u|_2^{(0)}$ is finite. As in the proof of Proposition 2.12, we fix $x_0 \in \Omega$, set $R = d(x_0)/2$, and let $y \in B(x_0, R)$. We then set $u_0 = u$ and define $v_0$ to be the solution of

$$a^{ij}(y)D_{ij}v_0 = f(y) \text{ in } B(y,R), \quad v_0 = u_0 \text{ on } \partial B(y,R).$$

Just as before, there are a polynomial $Q_0$ and a constant $C_0(n,\mu)$ so that (2.25) is satisfied. We now compute

$$a^{ij}(y)D_{ij}(u_0 - v_0) = F_0,$$

where

$$F_0 = (a^{ij}(y) - a^{ij})D_{ij}u_0 + f - f(y).$$

Since

$$|F_0| \le C(n)R^{-2}|u|_2^{(0)}A\lambda + [f]_\alpha^{(2)}R^{-2},$$

we infer from the maximum principle that

$$|u_0 - v_0| \le C(n, A, \mu)F.$$

where now

$$F = |u|_2^{(0)} + |f/\lambda|_\alpha^{(2)}. \tag{2.31}$$

With $\tau$ chosen as in the proof of Proposition 2.12, we infer (2.26). A similar argument produces (2.27) (but $F$ given by (2.31)) and hence

$$|u|_{2+\alpha}^{(0)} \le C(|u|_0 + |u|_2^{(0)} + |f/\lambda|_\alpha^{(2)}). \tag{2.32}$$

The desired result follows from this inequality via Lemma 2.3.

To remove the restriction that $b^i$ and $c$ are zero, we set $\tilde{f} = f - b^iD_iu - cu$. Then (2.2) and (2.6) imply that

$$|\tilde{f}|_\alpha^{(2)} \le |f|_\alpha^{(2)} + C(n, A, \alpha)\lambda|u|_2^{(0)}.$$

The preceding argument shows that

$$|u|_{2+\alpha}^{(0)} \le C(|u|_0 + |u|_2^{(0)} + |\tilde{f}/\lambda|_\alpha^{(2)}),$$

and hence (2.32) holds in this case as well.

To remove the restriction that $|u|_2^{(0)}$ is finite, we first write $\Omega_k$ (for $k$ a positive integer) for the set of all $x \in \Omega$ satisfying $|x| < k$ and $d(x) > 1/k$. For $k$ sufficiently large, $\Omega_k$ is nonempty, and the assumption $u \in C^2(\Omega)$ implies that $u \in H_2^{(0)}(\Omega_k)$. Hence

$$|u|_{2+\alpha;\Omega_k}^{(0)} \le C(|u|_{0;\Omega_k} + |f/\lambda|_{\alpha;\Omega_k}^{(2)}).$$

We now observe that

$$|u|_{0;\Omega_k} + |f/\lambda|_{\alpha;\Omega_k}^{(2)} \le |u|_0 + |f/\lambda|_\alpha^{(2)}$$

and that

$$|u|_{2+\alpha}^{(0)} = \lim_{k\to\infty} |u|_{2+\alpha;\Omega_k}^{(0)}$$

to complete the proof. □

For our study of oblique derivative problems, we shall need a similar estimate for a slightly different set of functions on the right hand side.

**Lemma 2.14.** *Let $\Omega$ be an open subset of $\mathbb{R}^n$ and let $u \in C^{2,\alpha}(\Omega)$ satisfy the equation $Lu = f$ in $\Omega$ for some $\alpha \in (0,1)$. Suppose the coefficients of $L$ satisfy* (2.29). *Let $b \in \mathbb{R}$. Then*

$$|u|^{(b)}_{2+\alpha} \leq C(n, \alpha, \mu, A)(\sup_{\Omega} |d^b u| + |f/\lambda|^{(2+b)}_{\alpha}). \tag{2.33}$$

***Proof.*** Let $x \neq y$ be points in $\Omega$ with $|x-y| \leq \frac{1}{2}d(x)$.

If $|x - y| \leq \frac{1}{4}d(x)$, we apply Proposition 2.13 in the ball $B = B(x, d(x)/2)$ to conclude that

$$\begin{aligned} d(x)^{2+\alpha+b}\frac{|D^2u(x) - D^2u(y)|}{|x-y|^{\alpha}} &\leq Cd(x)^b(|u|_{0;B} + |f/\lambda|^{(2)}_{\alpha;B}) \\ &\leq C(\sup_{\Omega}|d^b u| + |f/\lambda|^{(2+b)}_{\alpha}). \end{aligned}$$

If $|x-y| \in (\frac{1}{4}d(x), \frac{1}{2}d(x))$, we have

$$d(x)^{2+\alpha+b}\frac{|D^2u(x) - D^2u(y)|}{|x-y|^{\alpha}} \leq Cd(x)^{2+b}(|D^2u(x)| + |D^2u(y)|).$$

It follows that

$$[u]^{(b)}_{2+\alpha} \leq C(\sup |d^b u| + |f/\lambda|^{(2+b)}_{\alpha} + |D^2u|^{(2+b)}_0).$$

Since $|D^2u|^{(2)}_{0;B} \leq C|u|^{(0)}_{2+\alpha;B}$, we can use a similar argument to show that

$$|D^2u|^{(2+b)}_{0;B} \leq C(\sup |d^b u| + |f/\lambda|^{(2+b)}_{\alpha}),$$

and the proof is completed by combining these inequalities. □

Of course, when $b = 0$, this result is identical to Proposition 2.13. If $b < 0$, it's clear that $\sup |d^b u|$ can only be finite if $u = 0$ on $\partial\Omega$.

## 2.5 The Perron process for the Dirichlet problem

There are many ways in which we can use the estimates from the preceding section to show that the Dirichlet problem has a classical solution under weak conditions on the domain and on the coefficients of the differential equation. This section is devoted to a way that will be used also to prove the existence of solutions of oblique derivative problems via the maximum principle. It is based on Perron's method of subharmonic functions [152] with some crucial modifications from [129]. We present this method in an axiomatic format; a more relaxed description is presented in many sources (for example, Section 2.8 in [64]), but we want to make the transition to the oblique derivative problem as simple as possible. In this section, $[A^{ij}]$ denotes a constant positive definite matrix. We use $\lambda$ and $\mu$ to denote positive constants so that conditions (2.24) hold.

To begin, we say that the differential equation $A^{ij}D_{ij}u = f$ is *locally solvable in* $\Omega$ if, for every $x_0 \in \Omega$, there is an open set $N$ with $x_0 \in N$ and $\overline{N} \subset \Omega$ such that the Dirichlet problem

$$A^{ij}D_{ij}w = f \text{ in } N, \quad w = g \text{ on } \partial N \tag{2.34}$$

has a solution $w$ for every continuous function $g$. We emphasize that the open set $N$ (for each $x_0$) and the function $f$ are considered fixed here. (At present, we only know that the differential equation $A^{ij}D_{ij}u = f$ is locally solvable if $f$ is a polynomial, but this situation will change soon.) We then say that $v$ is a *Perron supersolution of* $A^{ij}D_{ij}u = f$ *in* $\Omega$ if, for every $x_0 \in \Omega$ and every continuous function $g$ such that $v \geq g$ on $\partial N$ (the open set from the definition of local solvability), we have that the solution $w$ of (2.34) is no greater than $v$ in $N$. Perron subsolutions are defined similarly.

To solve the Dirichlet problem

$$A^{ij}D_{ij}u = f \text{ in } \Omega, \quad u = 0 \text{ on } \partial\Omega, \tag{2.35}$$

we assume certain properties about $f$ and $\Omega$. In particular, we shall assume that $f \in C^\alpha(\Omega)$ for some $\alpha \in (0, 1)$. Solvability for nonzero boundary data will be discussed later.

We now assume that $A^{ij}D_{ij}u = f$ in $\Omega$ is locally solvable, we suppose that there is a Perron subsolution $v^-$ of $A^{ij}D_{ij}u = f$ in $\Omega$ such that $v^- \in C(\overline{\Omega})$ and $v = 0$ on $\partial\Omega$, and we assume that $S^+$, the set of all bounded Perron supersolutions $v$ of $Lu = f$ in $\Omega$ such that

$$\lim_{\substack{x \to y \\ x \in \Omega}} v(x) = 0$$

for all $y \in \partial\Omega$ and $v \geq v^-$ in $\Omega$, is nonempty.

The solution of (2.35) will be given by

$$u(x) = \inf_{v \in S^+} v(x)$$

for all $x \in \Omega$. So we need to verify that $u$ really is the solution of (2.35).

First, we verify that $u \in S^+$. It's easy to check that $u \geq v^-$ and that

$$\lim_{\substack{x \to y \\ x \in \Omega}} u(x) = 0$$

for all $y \in \partial\Omega$. In addition, if $x_0 \in \Omega$ and $g$ is a continuous function such that $g \le u$ on $\partial N$, then $g \le v$ on $\partial N$ for any $v \in S^+$ and the definition of Perron supersolution implies that $g \le v$ in $N$ for any $v \in S^+$, and hence $g \le u$ in $N$. Proving that $u$ actually solves the differential equation requires a little more work.

We start with a modification of the *harmonic lifting* (see (2.33) in [64]). For $x_0 \in \Omega$ and $v \in S^+$, we define a function $v_N$ in three steps as follows: First, for any continuous function $g$ with $g \le v$ on $\partial N$, we write $g_N$ for the solution of (2.34). We then define $v_N$ on $N$ by

$$v_N(x) = \sup\{g_N(x) : g \text{ is continuous and } v \ge g \ge v^- \text{ on } \partial N\},$$

and we define $v_N$ on $\overline{\Omega} \setminus N$ by

$$v_N(x) = v(x).$$

We want to show that $v_N \in S^+$, that $Lv_N = f$ in $N$, and that $v_N \le v$ in $\Omega$. It's immediate that

$$\lim_{\substack{x \to y \\ x \in \Omega}} v_N(x) = 0$$

because $v_N = v$ in a neighborhood of $\partial\Omega$. Before completing the verification that $v_N \in S^+$, we show that $A^{ij}D_{ij}v_N = f$ in $N$. In particular, it will follow that $v_N \in C^2(N)$.

Now let $x_1 \in N$ and let $R$ be a positive number so that $B(x_1, R) \subset N$. Then there is a constant $G_0$ such that every $g$ with $v \ge g \ge v^-$ satisfies $|g| \le G_0$. By the interior Schauder estimate, Proposition 2.13, we have that $|D^2 g_N|_{\alpha;B(x_1,R/2)} \le C(R, G_0)$ so the Arzela-Ascoli theorem implies that we can find a sequence $(g^{(m)})$ of continuous functions satisfying $v \ge g^{(m)} \ge v^-$ and $g_N^{(m)}$ converges in $C^2(B(x_1, R/2))$ to a limit function which must be $v_N$. Hence $A^{ij}D_{ij}v_N = f$ in $N$.

Next, we show that $v_N \le v$ in $\Omega$. Since the two functions are equal on $\Omega \setminus N$, we just have to check that $v_N \le v$ in $N$, so let $g$ be a continuous function such that $v \ge g \ge v^-$ on $\partial N$. Then $g_N \le v$ in $N$ and the definition of $V_N$ implies that $v_N \le v$ in $N$.

To see that $v_N$ is a supersolution, we fix $x_1$ and let $g$ be a continuous function such that $v_N \ge g \ge v^-$ on $\partial N_1$, where $N_1$ is the open set corresponding to $x_1$. We write $w_1$ for the solution of

$$A^{ij}D_{ij}w_1 = f \text{ in } N, \quad w_1 = g \text{ on } \partial N_1.$$

Then $v \ge w_1$ and hence $v_N \ge w_1$ in $N_1 \setminus N$. To proceed, we observe that $N^* = N_1 \cap N$ is an open set and, if $N^*$ is not empty, then $v = v_N$ on $\partial N^*$.

It follows that $v_N \ge w_1$ on $\partial N^*$ and then the maximum principle implies that $v_N \ge w_1$ in $N^*$ as well, so $v_N \ge w_1$ in $N_1$ as required.

To show that $u$ is a solution of the differential equation, we need to show that $u = u_N$ for any open set $N$ corresponding to a point $x_0 \in \Omega$. But this is easy. We have just shown that $u_N$ is in $S^+$ and that $u \ge u_N$, and the definition of $u$ implies that $u \le u_N$, so $u = u_N$ and hence $u$ satisfies $A^{ij}D_{ij}u = f$ in $N$ for any such $N$.

The only remaining step in our existence program is to show that $S^+$ is non-empty. For arbitrary domains, this set may be empty, but we show now that we can solve the Dirichlet problem in a ball under general hypotheses.

**Theorem 2.15.** *Let $B$ be a ball in $\mathbb{R}^n$ and let $\alpha$ and $\delta$ be constants in $(0,1)$. Then, for any $f \in H_\alpha^{(2-\delta)}$, there is a unique $C^2(B) \cap C(\overline{B})$ solution $u$ of*

$$A^{ij}D_{ij}u = f \text{ in } B, \quad u = 0 \text{ on } \partial B. \tag{2.36}$$

*Moreover, we have*

$$|u|_{2+\alpha}^{(-\delta)} \le C(\alpha,\delta,n,\mu)|f/\lambda|_\alpha^{(2-\delta)}. \tag{2.37}$$

***Proof.*** We begin by verifying local solvability under this hypothesis. For each $x_0 \in B$, we write $N$ for a ball centered at $x_0$, the closure of which is contained in $B$. It follows from Corollary 2.11 that (2.34) has a unique solution if $f$ is a polynomial. For arbitrary $f \in H_\alpha^{(2-\delta)}$, we note that $f \in H_\alpha(N)$, so there is a sequence of polynomials $(f_m)$ which converge in $H_\alpha(N)$ to $f$. Let us now fix a continuous function $g$ and write $w_m$ for the solution of

$$A^{ij}D_{ij}w_m = f_m \text{ in } N, \quad w_m = g \text{ on } \partial N.$$

It follows from the maximum principle that $(w_m)$ is uniformly Cauchy and hence it has a limit $w$. The interior Schauder estimate, Proposition 2.13, implies that $(D^2w_m)$ converges uniformly on compact subsets of $N$ to $D^2w$ and hence $A^{ij}D_{ij}w = f$ in $N$.

Our next step is to show the existence of subsolutions and supersolutions. To this end, we write $y$ for the center of the ball and set

$$h(x) = (R^2 - |x-y|^2)^\delta.$$

Then $h \in C^2(B) \cap C(\overline{B})$ and

$$A^{ij}D_{ij}h = \delta(R^2 - |x-y|^2)^{\delta-2}((R^2 - |x-y|^2)(-2\mathfrak{T}) + 4(\delta-1)\mathcal{E}),$$

where

$$\mathcal{T} = \sum_{i=1}^{n} a^{ii}, \quad \mathcal{E} = a^{ij}(x-y)_i(x-y)_j.$$

By examining the two cases $|x-y| < \frac{R}{2}$ and $|x-y| \ge \frac{R}{2}$, we see that

$$A^{ij} D_{ij} h \le -c_0 \lambda R^{2\delta} d^{\delta-2}$$

for some positive constant $c_0$ determined by $n$, $\delta$, and $\mu$. For $F = |f/\lambda|_0^{(2-\delta)}$, it follows that

$$v^- = -\frac{F}{c_0 R^{2\delta}} h$$

is a subsolution satisfying

$$\lim_{\substack{x \to y \\ x \in B}} v^-(x) = 0$$

for all $y \in \partial B$ and that

$$v^+ = \frac{F}{c_0 R^{2\delta}} h$$

is in $S^+$.

The preceding argument shows that (2.36) has a solution $u$, and the maximum principle implies that it's unique. Since $-v^- \ge u \ge v^-$, it also follows that $|u| \le CFd^\delta$ and Lemma 2.14 implies (2.37). □

From this result, we obtain an existence theorem and estimate for arbitrary continuous data.

**Corollary 2.16.** *With $B$, $\alpha$, $\delta$, and $f$ as in Theorem 2.15, for any continuous function $\psi$, there is a unique $C^2(B) \cap C(\overline{B})$ solution $u$ of*

$$A^{ij} D_{ij} u = f \text{ in } B, \quad u = \psi \text{ on } \partial B. \tag{2.38}$$

*Moreover, if $\psi \in H_\delta$, we have*

$$|u|_{2+\alpha}^{(-\delta)} \le C(\alpha, \delta, n, \mu)(|\psi|_\delta^{(-\delta)} + |f/\lambda|_\alpha^{(2-\delta)}). \tag{2.39}$$

***Proof.*** Suppose first that $\psi \in H_\delta$. Then there is a function $G \in H_{2+\alpha}^{(-\delta)}$ with $G = \psi$ on $\partial B$ and

$$|G|_{2+\alpha}^{(-\delta)} \le C(\alpha, n, \delta)|\psi|_\delta^{(-\delta)}.$$

If $v$ is the unique solution of

$$A^{ij} D_{ij} v = f - A^{ij} D_{ij} G \text{ in } B, \quad v = 0 \text{ on } \partial B$$

given by Theorem 2.15, then $u = v + G$ solves (2.38) and

$$\begin{aligned}|u|_{2+\alpha}^{(-\delta)} &\le |v|_{2+\alpha}^{(-\delta)} + |G|_{2+\alpha}^{(-\delta)} \\ &\le C|f - A^{ij}D_{ij}G|_{\alpha}^{(2-\delta)} + |G|_{2+\alpha}^{(-\delta)} \\ &\le C(|f|_{\alpha}^{(2-\delta)} + |G|_{2+\alpha}^{(-\delta)}) \\ &\le C(|f|_{\alpha}^{(2-\delta)} + |g|_{\delta}^{(-\delta)}).\end{aligned}$$

The existence for continuous $g$ follows from this result because any continuous function is the uniform limit of a sequence of $H_\delta$ functions. □

For later applications, we also obtain an existence result in any bounded convex domain.

**Corollary 2.17.** *Let $\Omega$ be a bounded convex open subset of $\mathbb{R}^n$. Then, for any continuous function $\psi$ defined on $\partial B^+$, there is a unique $C^2(\Omega)\cap C(\overline{\Omega})$ solution $u$ of*

$$A^{ij}D_{ij}u = 0 \text{ in } \Omega, \quad u = \psi \text{ on } \partial\Omega. \tag{2.40}$$

***Proof.*** Let $(\psi_m)$ be a sequence of polynomials converging uniformly to $\psi$, and set $f_m = A^{ij}D_{ij}\psi_m$ and

$$F_m = \frac{1}{\lambda}\sup_{\Omega}|f_m|.$$

Our first step is to show that, for any $m$, there is a solution $u_m$ of

$$A^{ij}D_{ij}u_m = f_m \text{ in } \Omega, \quad u_m = 0 \text{ on } \partial\Omega. \tag{2.41}$$

Local solvability of this problem was established in Theorem 2.15, so we only need to prove the existence of a subsolution and a supersolution. To this end, we recall that at each point $y \in \partial\Omega$, there is a supporting hyperplane and, therefore, a unit vector $\xi(y)$ such that $\xi(y)\cdot(x-y) > 0$ for all $x \in \Omega$. If we set $R = \operatorname{diam}\Omega$ and define $h_y$ by

$$h_y(x) = R[\xi(y)\cdot(x-y)] - [\xi(y)\cdot(x-y)]^2,$$

then a direct calculation gives $A^{ij}D_{ij}h_y \le -\lambda$ in $\Omega$ for all $y \in \partial\Omega$. We now claim that $h$, defined by

$$h(x) = \inf_{y\in\partial\Omega} F_m h_y(x)$$

is a supersolution. To verify this statement, let $x_0 \in \Omega$ and let $N$ be the corresponding neighborhood. If $g$ is a continuous function on $\partial N$ with

$g \geq h$ on $\partial N$, then $g \geq h_y$ on $\partial N$ (for any $y \in \partial\Omega$), so, if $w$ is the solution of

$$A^{ij}D_{ij}w = f_m \text{ in } N, w = g \text{ on } N,$$

we have $A^{ij}D_{ij}h_y \leq f_m$ in $N$, so the maximum principle implies that $w \leq h_y$ in $N$. Since the set of functions $\{h_y\}$ is equicontinuous, it followss that $h$ is continuous with $h = 0$ on $\partial\Omega$, and hence $h$ is a supersolution. A similar argument shows that $-h$ is a subsolution, and hence (2.41) has a solution.

We now set $v_m = u_m + \psi_m$ and observe that $v_m$ solves

$$A^{ij}D_{ij}v_m = 0 \text{ in } \Omega, \quad v_m = \psi_m \text{ on } \partial\Omega.$$

Since $(\psi_m)$ converges uniformly to $\psi$, it follows from the maximum principle that $(v_m)$ is uniformly Cauchy in $\overline{\Omega}$, so this sequence converges uniformly to a limit function $u$ which satisfies the boundary condition $u = \psi$ on $\partial\Omega$. The interior Schauder estimate implies that $(D^2v_m)$ is also uniformly Cauchy on compact subsets of $\Omega$, so it converges uniformly on compact subsets of $\Omega$ to a limit, which must be $D^2u$, and hence $u$ also satisfies the differential equation $A^{ij}D_{ij}u = 0$. □

## 2.6 A model mixed boundary value problem

The next step in our study of Hölder estimates for oblique derivative problems is an examination of a mixed boundary value problem, that is, one in which Dirichlet data are prescribed on part of the boundary and an oblique derivative condition is prescribed on the remainder of the boundary. For simple operators and domains, there are many ways to examine this problem, and we shall return to mixed boundary value problems several times in this book. The present approach will be useful in laying the groundwork for our estimates in later chapters.

We begin by defining a standard set in which to study our problem. For a number $\kappa \in (0,1)$ to be further specified, we define the *spherical cap* $\Gamma(R)$ for $R > 0$ by

$$\Gamma(R) = \{x \in \mathbb{R}^n : |x'|^2 + |x^n + R\kappa|^2 < R^2,\ x^n > 0\}. \tag{2.42}$$

(The introduction of the spherical cap should be viewed in the same light as that for the set $E$ in Section 1.7; it will simplify certain calculations but it leads to more elaborate geometric considerations.) We then write $\Gamma^*(R)$

and $\Gamma^0(R)$ for the subsets of $\partial\Gamma(R)$ on which $|x'|^2 + |x^n + R\kappa|^2 = R^2$ and $|x'|^2 + |x^n + R\kappa|^2 < R^2$, respectively. (Here, we have to be careful that $\Gamma^*$ is closed and $\Gamma^0$ is open.) We are concerned in this section with solutions of the problem

$$a^{ij}D_{ij}u = f \text{ in } \Gamma(R), \tag{2.43a}$$

$$\beta^i D_i u = g \text{ on } \Gamma^0(R), \tag{2.43b}$$

$$u = \psi \text{ on } \Gamma^*(R) \tag{2.43c}$$

under various hypotheses on the coefficients $a^{ij}$ and $\beta$ and the functions $f$, $g$ and $\psi$. Our goal is to show that this problem has a unique solution and that the solution satisfies appropriate estimates. We shall eventually see that the problem has a solution without any restrictions on $\kappa$, but, in the present section, we shall have to make a restriction on $\kappa$ based on information about $\beta$. When we make the assumption, the reason for making it will become clear.

In addition, we shall suppress the symbols $\Gamma(R)$ and $\Gamma^*(R)$ from our weighted norms because we always consider weighted norms with respect to distance to $\Gamma^*(R)$. So we write $H_a^{(b)}$ for $H_{a;\Gamma(R),\Gamma^*(R)}^{(b)}$ and similarly we abbreviate the weighted norms.

As a first step, we prove everything for $a^{ij} = \delta^{ij}$ and $\beta^i = \delta^{in}$. This choice is dictated by our simple existence theorem for the Dirichlet problem, which we now extend to a somewhat different set, and by the nature of the mixed boundary value problem. We define

$$\hat{\Gamma}(R) = \{x \in \mathbb{R}^n : |x'|^2 + |x^n + R\kappa|^2 < R^2, |x'|^2 + |x^n - R\kappa|^2 < R^2\},$$

and then our existence result can be extended to $\hat{\Gamma}(R)$.

**Lemma 2.18.** *Let $\alpha$ and $\delta$ be constants in $(0,1)$. then, for any $F \in H_\alpha^{(2-\delta)}(\hat{\Gamma}(R))$ and $\Psi \in C(\partial\hat{\Gamma}(R))$, there is a unique $C^2(\hat{\Gamma}(R)) \cap C(\overline{\hat{\Gamma}(R)})$ solution $v$ of*

$$\Delta v = F \text{ in } \hat{\Gamma}(R), \quad v = \Psi \text{ on } \partial\hat{\Gamma}(R).$$

***Proof.*** As in Corollary 2.16, we may assume that $\Psi = 0$, and then it suffices to show that there exist a subsolution $v^-$ and a supersolution $v^+$ (both in $C(\overline{\Omega})$) such that $v^- \le v^+$ in $\hat{\Gamma}(R)$ and $v^\pm = 0$ on $\partial\hat{\Gamma}(R)$. If we set

$$h = \min\{(R^2 - |x'|^2 - |x^n - \kappa R|^2)^\delta, (R^2 - |x'|^2 - |x^n + \kappa R|^2)^\delta\},$$

then we can take

$$v^\pm = \pm Ch$$

for a suitable constant $C$, determined by $R$, $n$, and $\delta$. □

From this existence result, we can infer an existence theorem for simple mixed boundary value problems.

**Lemma 2.19.** *Let $\alpha$ and $\delta$ be constants in $(0,1)$. If $f \in H_\alpha^{(2-\delta)}$, $g \in H_{1+\alpha}^{(1-\delta)}$, and $\psi \in C(\overline{\Gamma^*(R)})$, then there is a unique $C^2(\Gamma(R)) \cap C(\overline{\Gamma(R)})$ solution of*

$$\Delta u = f \text{ in } \Gamma(R), \tag{2.44a}$$

$$\frac{\partial u}{\partial x^n} = g \text{ on } \Gamma^0(R), \tag{2.44b}$$

$$u = \psi \text{ on } \Gamma^*(R). \tag{2.44c}$$

*Moreover, if $\psi \in H_\delta$, then $u \in H_{2+\alpha}^{(-\delta)}$ and*

$$|u|_{2+\alpha}^{(-\delta)} \le C(n,\alpha,\delta,\kappa)(|f|_\alpha^{(2-\delta)} + |g|_{1+\alpha}^{(1-\delta)} + |\psi|_\delta^{(-\delta)}). \tag{2.45}$$

***Proof.*** Since $g \in H_{1+\alpha}^{(1-\delta)}$, there is an $H_{2+\alpha}^{(-\delta)}$ function $G$ such that $\partial G/\partial x^n = g$ on $\Gamma^0(R)$. We set $F = f - \Delta G$ and $\Psi = \psi - G$, and then we extend $F$ and $\Psi$ as even functions to $\hat{\Gamma}(R)$. (That is, we define

$$F(x', x^n) = F(x', -x^n), \quad \Psi(x', x^n) = \Psi(x', -x^n)$$

for $x^n < 0$.) It follows that $\Psi$ is continuous on $\partial\hat{\Gamma}(R)$ and $F$ is Hölder continuous on $\hat{\Gamma}(R)$ so there is a unique solution $v$ of the Dirichlet problem

$$\Delta v = F \text{ on } \hat{\Gamma}(R), \quad v = \Psi \text{ on } \partial\hat{\Gamma}(R).$$

Since $w$ defined by $w(x', x^n) = v(x', -x^n)$ is another solution of this problem, it follows that $v(x', x^n) = v(x', -x^n)$ and hence $\partial v/\partial x^n = 0$ on $\Gamma^0(R)$. It's now easy to verify that $u = v + G$ is the desired function.

If $\psi \in H_\delta$, then we infer (2.45) from the argument leading to (2.39). □

Our next step is to prove the corresponding result with $\Delta$ and $\partial/\partial x^n$ replaced by constant coefficient operators. In this case, it can be shown that the existence result is still valid but the regularity of the solution near points of the form $(x', 0) \in \Gamma^*(R)$ is not as good as in this theorem, that is, even for smooth data, there may not be a solution in $H_\delta$. For this reason and for computational simplicity, we put a restriction on $\kappa$ which is connected to the vector $\beta$ in the boundary condition $\beta \cdot Du = g$. To be specific, we assume now that $[A^{ij}]$ is a constant positive-definite matrix and that $B$ is a constant vector with $B^n > 0$. To state our results succinctly, we assume that there are positive constants $\lambda$, $\mu$, $\chi$, and $\mu_0$ such that

$$A^{ij}\xi_i\xi_j \ge \lambda|\xi|^2, \tag{2.46a}$$

$$A^{ij}\xi_i\xi_j \le \mu\lambda|\xi|^2, \tag{2.46b}$$

for all $\xi \in \mathbb{R}^n$,

$$B^n \geq \chi, \tag{2.46c}$$

$$|B'| \leq \mu_0 B^n. \tag{2.46d}$$

We now fix $\kappa$ so that

$$\kappa \in \left( \frac{\mu_0}{(1+\mu_0^2)^{1/2}}, 1 \right). \tag{2.47}$$

The following preliminary estimate justifies the choice for $\kappa$. If we define $h$ by

$$h(x) = (R^2 - |x'|^2 - |x^n + \kappa R|^2)^\delta,$$

then a straightforward calculation shows that there are positive constants $c_1$ (determined only by $n$ and $\mu$) and $c_2$ (determined also by $\kappa$) such that

$$A^{ij} D_{ij} h \leq -c_1 d^{\delta-2} R^\delta \text{ in } \Gamma(R),$$
$$B^i D_i h \leq -c_2 d^{\delta-1} R^\delta \text{ on } \Gamma^0(R).$$

We emphasize that this estimate on $B^i D_i h$ is only valid if $\kappa$ satisfies (2.47). In fact $c_2 \to 0$ as $\kappa \to \mu_0/(1+\mu_0^2)^{1/2}$.

Our next step is a local gradient bound for solutions of the oblique derivative problem with constant right hand sides.

**Lemma 2.20.** *Under the preceding hypotheses on $A^{ij}$ and $B$, if $v \in C^2(B^+(R)) \cap C^1(B^+(R) \cup B^0(R))$ solves*

$$A^{ij} D_{ij} v = F \text{ in } B^+(R), \quad B \cdot Dv = G \text{ on } B^0(R) \tag{2.48}$$

*for constants $F$ and $G$, then there is a constant $C$ (determined only by $n$, $\mu$, and $\mu_0$) such that*

$$|Dv(x)| \leq C \left( \frac{RF}{\lambda} + \frac{G}{\chi} + \frac{\sup |v|}{R} \right) \tag{2.49}$$

*for $|x| < R/4$.*

***Proof.*** The proof proceeds in several steps. The main step is to estimate the maximum of the gradient in terms of its $L^1$ norm. This estimate proceeds differently for different components of the gradient.

Let $k \in \{1, \ldots, n-1\}$, and set

$$w = \frac{v(x+he_k) - v(x)}{h}$$

for $h$ small and nonzero. Then $A^{ij}D_{ij}w = 0$ in $B^+(R/2)$ and $B \cdot Dw = 0$ on $B^0(R/2)$, and it follows from the local maximum principle, Theorem 1.27, that

$$\sup_{B+(R/4)} w \le CR^{-n} \int_{B^+(R/2)} w \, dx.$$

Sending $h \to 0$, we see that

$$\sup_{B^+(R/4)} |D_k v| \le CR^{-n} \int_{B^+(R/2)} |Dv| \, dx.$$

Next, let $\varepsilon > 0$ be given, and, for $h$ small and positive, we define $w$ by

$$w(x) = \frac{v(x + h\beta) - v(x)}{h} - G.$$

Then $A^{ij}D_{ij}w = 0$ in $B^+(R/2)$ and, if $h$ is sufficiently small depending on $\varepsilon$, $|w| \le \varepsilon$ on $B^0(R/2)$. It follows that

$$\sup_{B+(R/4)} w \le C[R^{-n} \int_{B^+(R/2)} w \, dx + \varepsilon]$$

as well. Sending $\varepsilon \to 0$ (and hence $h \to 0$), we obtain

$$\sup_{B^+(R/4)} |\beta \cdot Dv| \le C \left( R^{-n} \int_{B^+(R/2)} |\beta \cdot Dv| \, dx + G \right).$$

Combining this estimate with the one for $D_k v$, we see that

$$\sup_{B^+(R/4)} |Dv| \le C \left( R^{-n} \int_{B^+(R/2)} |Dv| \, dx + \frac{G}{\chi} \right).$$

To proceed, we use the interior gradient bound, Theorem 1.33, (with $u$ replaced by $v - \inf v$) to infer that

$$|Dv(x)| \le C \frac{1}{x^n} \operatorname*{osc}_{B(x,x^n/2)} v$$

for any $x \in B^+(R/2)$. Then the Hölder estimate, Theorem 1.26, for $v$ implies that

$$\operatorname*{osc}_{B(x,x^n/2)} v \le CUR^{-\delta}(x^n)^\delta$$

for

$$U = \frac{R^2 F}{\lambda} + \frac{RG}{\chi} + \sup_{B^+(R)} |v|. \tag{2.50}$$

Therefore

$$\sup_{B^+(R/4)} |Dv| \le CUR^{-n-\delta} \int_{B^+(R/2)} (x^n)^{\delta-1} \, dx = CUR^{-1},$$

which is just the estimate (2.49). □

In fact, we can obtain estimates for higher order derivatives as well.

**Corollary 2.21.** *Under the hypotheses of Lemma 2.20, there is a constant $C$, determined by the same quantities as in that lemma, such that*

$$\sup_{B^+(R/2)} |D^2 v| \le C\left(\frac{F}{\lambda} + \frac{G}{R\chi} + \frac{\sup|v|}{R^2}\right), \tag{2.51a}$$

$$\sup_{B^+(R/2)} |D^3 v| \le C\left(\frac{F}{R\lambda} + \frac{G}{R^2\chi} + \frac{\sup|v|}{R^3}\right). \tag{2.51b}$$

***Proof.*** We begin by combining the argument in the proof of Theorem 1.33 with (2.49) and Theorem 1.33 itself to infer that

$$\sup_{B^+(3R/4)} |Dv| \le CU/R,$$

where $U$ is defined by (2.50).

To prove (2.51a), we define $v_{k,h}$ with $k \in \{1, \dots, n-1\}$ and $h$ small and nonzero by

$$v_{k,h}(x) = \frac{v(x + he_k) - v(x)}{h},$$

and use the preceding estimate for $v$ and $v_{k,h}$ to infer that

$$\sup_{B(R/2)} |Dv_{k,h}| \le CU/R^2.$$

Sending $h \to 0$ yields

$$\sup_{B^+(R/2)} |D_{ik} v| \le CU/R^2$$

for $i \in \{1, \dots, n\}$ and $k \in \{1, \dots, n-1\}$, and the differential equation for $v$ gives the same estimate for $i = k = n$.

To prove (2.51b), we define $v_{jk,h}$ with $j$ and $k$ in $\{1, \dots, n-1\}$ and $h$ small and nonzero by

$$v_{jk,h} = \frac{v(x + he_j + he_k) - 2v(x + he_j) + v(x)}{h^2}.$$

As before, we obtain

$$\sup_{B^+(R/2)} |Dv_{jk,h}| \le CU/R^3,$$

and we infer from Theorem 1.34 that $v \in C^3(B^+(R/2))$, so

$$\sup_{B^+(R/2)} |D_{ijk} v| \le CU/R^3,$$

for $i \in \{1, \dots, n\}$ and $j$ and $k$ in $\{1, \dots, n-1\}$. For each $k$, we use the differential equation for $D_k v$ to obtain the corresponding result for $D_{nnk} v$ and the estimate for $D_{nnn} v$ follows from the differential equation for $D_n v$. □

A simple induction argument shows that we can estimate derivatives of $v$ of any order in the corresponding way but with $C$ depending also on the number of derivatives.

From this estimate, we can derive a Schauder estimate and an existence result for solutions of the mixed boundary value problem with constant coefficients.

**Lemma 2.22.** *Let $\alpha$ and $\delta$ be constants in $(0,1)$ and suppose $\kappa$ satisfies (2.47). Suppose conditions (2.46) are satisfied. If $f \in H_\alpha^{(2-\delta)}$, $g \in H_{1+\alpha}^{(1-\delta)}$, and $\psi \in C(\overline{\Gamma^*(R)})$, then there is a unique $C^2(\Gamma(R)) \cap C(\overline{\Gamma(R)})$ solution of*

$$A^{ij} D_{ij} u = f \text{ in } \Gamma(R), \tag{2.52a}$$

$$B \cdot Du = g \text{ on } \Gamma^0(R), \tag{2.52b}$$

$$u = \psi \text{ on } \Gamma^*(R). \tag{2.52c}$$

*Moreover, if $\psi \in H_\delta$, then $u \in H_{2+\alpha}^{(-\delta)}$ and there is a constant $C = C(n, \alpha, \delta, \omega_0, \kappa, \mu, \mu_0)$ such that*

$$|u|_{2+\alpha}^{(-\delta)} \le C(|f/\lambda|_\alpha^{(2-\delta)} + |g/\chi|_{1+\alpha}^{(1-\delta)} + |\psi|_\delta^{(-\delta)}). \tag{2.53}$$

***Proof.*** For simplicity of notation, we replace $A^{ij}$ by $A^{ij}/\lambda$ and $B$ by $B/\chi$, so there is no loss of generality in assuming that $\lambda = \chi = 1$. Moreover, as usual, we may assume that $\psi = 0$. We also write $X$ for the set of all $H_{2+\alpha}^{(-\delta)}$ functions which vanish on $\Gamma^*(R)$ and $Y = H_\alpha^{(2-\delta)} \times H_{1+\alpha}^{(1-\delta)}$.

With these simplifications, we note that the lemma has already been proved for $A^{ij} = \delta^{ij}$ and $B^i = \delta^{in}$. We then define

$$A_s^{ij} = sA^{ij} + (1-s)\delta^{ij}, B_s^i = sB^i + (1-s)\delta^{in}$$

for $s = (0,1)$, and we define the operator $J_s \colon X \to Y$ by

$$J_s = (A_s^{ij} D_{ij} u,\ B_s^i D_i u).$$

We now imitate the proof of the method of continuity to show that $J_s$ is onto if $s$ is sufficiently small. Since $J_0$ is invertible, we see that the equation $J_s u = y$ (for $y = (f,g) \in Y$) is equivalent to

$$J_0 u = y + (J_0 - J_s)u = y + s(J_0 - J_1)u,$$

or

$$u = J_0^{-1} y + sJ_0^{-1}(J_0 - J_1)u.$$

We now define the map $T$ from $X$ to itself by

$$Tu = J_0^{-1} y + sJ_0^{-1}(J_0 - J_1)u.$$

Then $T$ is a contraction if $s$ is sufficiently small and hence it has a unique fixed point, which is the unique solution of

$$A_s^{ij} D_{ij} u = f \text{ in } \Gamma(R), \tag{2.54a}$$

$$B_s \cdot Du = g \text{ on } \Gamma^0(R), \tag{2.54b}$$

$$u_s = 0 \text{ on } \Gamma^*(R). \tag{2.54c}$$

From this argument, we conclude that (2.54) has a unique solution for any $s \in [0,1]$ if we can show that there is some constant $C$, independent of $s$, such that the uniform estimate

$$\|u\|_X \le C\|J_s u\|_Y \tag{2.55}$$

is valid for all $u \in X$, whenever (2.54) is solvable for all $y = (f, g)$ in $Y$.

To prove this estimate, we proceed in several steps. With $h$ defined (as above) by

$$h(x) = (R^2 - |x'|^2 - (x^n + \kappa R)^2)^\delta,$$

it's easy to check from the maximum principle applied to

$$CR^{-\delta}\left(\frac{R^2 F}{\lambda} + \frac{RG}{\chi}\right) h \pm u$$

that there is a constant $C$ such that

$$|u| \le C\left(\frac{R^2 F}{\lambda} + \frac{RG}{\chi}\right) R^{-\delta} d^\delta.$$

The proof is completed by using this estimate along with Corollary 2.21 in the proof of Lemma 2.14. □

Once we have this general existence result, we infer the appropriate local estimates and existence for the oblique derivative problem.

**Theorem 2.23.** *Let $a^{ij} \in H_\alpha^{(0)}$ and suppose that there are positive constants $A$, $\lambda$ and $\mu$ such that conditions (2.29a,b) are satisfied. Let $\beta \in H_{1+\alpha}^{(0)}$ and suppose that there are positive constants $B_0$, $\chi$, and $\mu_0$ such that*

$$\beta^n \ge \chi, \tag{2.56a}$$

$$|\beta'| \le \mu_0 \beta^n, \tag{2.56b}$$

$$|\beta|_{1+\alpha}^{(0)} \le B_0 \chi. \tag{2.56c}$$

*Then for every $f \in H_\alpha^{(2-\delta)}$, $g \in H_{1+\alpha}^{(1-\delta)}$ and $\psi \in C^0$, there is a unique solution $u$ of (2.43). Moreover, if $\psi \in H_\delta$, then $u \in H_{2+\alpha}^{(-\delta)}$ and there is a constant $C$, determined only by $A$, $B_0$, $n$, $\alpha$, $\delta$, $\mu$, and $\alpha$ such that*

$$|u|_{2+\alpha}^{(-\delta)} \le C(|f/\lambda|_\alpha^{(2-\delta)} + |g/\chi|_{1+\alpha}^{(-\delta)} + |\psi|_\delta^{(-\delta)}). \tag{2.57}$$

***Proof.*** As usual, we prove the result for $\psi \equiv 0$. From Lemma 2.22 and the proof of Proposition 2.13, we see that

$$|u|^{(-\delta)}_{2+\alpha} \le C(|a^{ij}D_{ij}u/\lambda|^{(2-\delta)}_{\alpha} + |\beta^i D_i u/\chi|^{(-\delta)}_{1+\alpha})$$

for all $u \in H^{(-\delta)}_{2+\alpha}$ which vanish on $\Gamma^*(R)$. We then use the method of continuity with $J_s$ defined by

$$J_s u = ([sa^{ij} + (1-s)\delta^{ij}]D_{ij}u, [s\beta^i + (1-s)\delta^{in}]D_i u)$$

to infer the existence result for $\psi = 0$. □

For our existence program for the oblique derivative problem, an existence result and estimate for operators with lower order terms will be very useful. Note that an additional condition is placed on $b$ beyond (2.29d). We refer to the Notes for a brief discussion of the significance of this addition, and only point out here that it is used to prove one of the intermediate estimates. For simplicity of notation, we introduce the operators $L$ and $M$ given by

$$Lu = a^{ij}D_{ij}u + b^i u + cu, \quad Mu = \beta^i D_i u + \beta^0 u. \tag{2.58}$$

**Theorem 2.24.** *Let $a^{ij} \in H^{(0)}_{\alpha}$, let $b^i \in H^{(1)}_{\alpha}$, let $c \in H^{(2)}_{\alpha}$, and suppose that there are positive constants $A$, $\lambda$ and $\mu$ such that conditions (2.29) are satisfied. Let $\beta \in H^{(0)}_{1+\alpha}$, let $\beta^0 \in H^{(1)}_{1+\alpha}$, and suppose that there are positive constants $B_0$, $\chi$, and $\mu_0$ such that (2.56) holds and*

$$|\beta^0|^{(1)}_{1+\alpha} \le B_0\chi. \tag{2.59}$$

*Suppose also that $c \le 0$, $\beta^0 \le 0$, and there are nonnegative constants $B_1$ and $B_2$ with $B_1 < 2(1-\delta)\kappa/(1+\kappa)$ such that*

$$b(x)\cdot\bar{x} \ge -\left(\frac{B_1}{d(x)} + \frac{B_2}{R}\right)|\bar{x}| \tag{2.60}$$

*for every $x \in \Gamma(R)$, where $\bar{x} = (x', x^n + \kappa R)$. Then for every $f \in H^{(2-\delta)}_{\alpha}$, $g \in H^{(1-\delta)}_{1+\alpha}$ and $\psi \in C^0$, there is a unique solution $u$ of*

$$Lu = f \text{ in } \Gamma(R), \quad Mu = g \text{ on } \Gamma^0(R), \quad u = \psi \text{ on } \Gamma^*(R). \tag{2.61}$$

*Moreover, if $\psi \in H_\delta$, then $u \in H^{(-\delta)}_{2+\alpha}$ and there is a constant $C$, determined only by $A$, $B_0$, $B_1$, $B_2$, $n$, $\alpha$, $\delta$, $\mu$, $\mu_0$, and $\kappa$ such that (2.57) holds.*

***Proof.*** Again, we may assume that $\psi = 0$, and we now apply the method of continuity with $J_s$ defined by

$$J_s u = (a^{ij}D_{ij}u + s(b^i D_i u + cu), \beta^i D_i u + s\beta^0 u).$$

The only step in the proof of Theorem 2.23 that needs modification is the estimate of $d^{-\delta}u$. To this end, we define $w$ by

$$w(x) = R^2 - |\bar{x}|^2,$$

and we set

$$h = \int_0^w s^{\delta-1} \exp(-B_2 s/(\kappa R^2))\, ds.$$

It's easy to check that $Mh \le -\chi C R^\delta d^{\delta-1}$ on $\Gamma^0(R)$ and that $h = 0$ on $\Gamma^*(R)$. Next, we set $\mathcal{T} = \sum_i a^{ii}$ and $\mathcal{E} = a^{ij}\bar{x}_i\bar{x}_j$, and

$$L_1 = -2\mathcal{T} - 2b^i\bar{x}_i + 4(\delta - 1)\frac{1}{w}\mathcal{E} - 4\frac{B_2}{\kappa R^2}\mathcal{E}.$$

Then

$$Lh \le a^{ij}D_{ij}h + b^i D_i h = \exp(-B_2|\bar{x}|/(R\kappa))w^{\delta-1}L_1.$$

Because $|\bar{x}| \ge \kappa R$, $\mathcal{T} \ge n\lambda$ and $\mathcal{E} \ge \lambda|\bar{x}|^2$ in $\Gamma(R)$, we have that

$$\begin{aligned} Lh &\le B_0\lambda w^{\delta-2}|\bar{x}|/d \\ &\le -C_0\lambda R^\delta d^{\delta-2}, \end{aligned}$$

where $B_0 = [4(\delta - 1)\kappa/(1 + \kappa) + 2B_1]\exp(-B_2)$ and $C_0 = -B_0\kappa(1 + \kappa)^{\delta-1}$. (Note that $B_0 < 0 < C_0$). Applying the maximum principle to $C_0(|f/\lambda|_0^{(2-\delta} + |g/\chi|_0^{(1-\delta)})R^{-\delta}h \pm u$, we see that $|u| \le C_0(|f/\lambda|_0^{(2-\delta)} + |g/\chi|_0^{(1-\delta)})d^\delta$. From this bound on $d^{-\delta}|u|$, the proof proceeds as before. □

The proof of the preceding theorem uses condition (2.60), but a weaker result is still true if we remove this assumption.

**Corollary 2.25.** *Let $a^{ij} \in H_\alpha^{(0)}$, let $b^i \in H_\alpha^{(1)}$, let $c \in H_\alpha^{(2)}$, and suppose that there are positive constants $A$, $\lambda$ and $\mu$ such that conditions (2.29) are satisfied. Let $\beta \in H_{1+\alpha}^{(0)}$, let $\beta^0 \in H_{1+\alpha}^{(1)}$, and suppose that there are positive constants $B_0$, $\chi$, and $\mu_0$ such that (2.56) and (2.59) hold. Then there is a constant $C$, determined only by $A$, $B_0$, $n$, $\alpha$, $\delta$, $\mu$, $\mu_0$, and $\kappa$, such that, if $u \in H_{2+\alpha}^{(0)}$, then*

$$|u|_{2+\alpha}^{(0)} \le C(|u|_0 + |Lu/\lambda|_\alpha^{(2)} + |Mu/\chi|_{1+\alpha}^{(1)}).$$

## 2.7 Domains with curved boundary

We are now ready to study problems in more general domains. Our first step is an appropriate (and obvious, based on our earlier discussion of Lipschitz domains,) definition of $H_a$ domains. For $a \geq 1$, we say that a bounded domain $\Omega$ is an $H_a$ *domain* or that $\partial\Omega \in H_a$ if there is a number $R > 0$, and for each $x_0 \in \partial\Omega$, there are a coordinate system $Y$ centered at $x_0$ and an $H_a$ function $\omega$ such that

$$\Omega \cap B(x_0, R) = \{x \in \mathbb{R}^n : |x - x_0| < R, y^n \geq \omega(y')\}.$$

Since $\Omega$ is bounded, we can cover $\partial\Omega$ by finitely many balls of radius $R$, so we may assume that the functions $\omega$ have uniformly bounded $H_a$ norm. Again, we refer to [56] for details.

Because of the invariance of the norms under appropriate changes of variables, we obtain the following result immediately. To simplify writing, we define the operators $L$ and $M$ by (2.58).

**Theorem 2.26.** *Let $a^{ij}$, $b^i$, and $c$ be in $H_\alpha$ for some $\alpha \in (0,1)$ and suppose that $\beta$ and $\beta^i$ are in $H_{1+\alpha}$ and that $\partial\Omega \in H_{2+\alpha}$. Suppose $A$, $B_0$, $\lambda$, $\chi$, $\mu$, and $\mu_0$ are constants such that conditions* (2.29a,b) *hold and such that*

$$\beta \cdot \gamma \geq \chi, \tag{2.62a}$$

$$|\beta| \leq \mu_0 \chi, \tag{2.62b}$$

$$|a^{ij}|_\alpha + |b|_\alpha + |c|_\alpha \leq A\lambda, \tag{2.62c}$$

$$|\beta|_{1+\alpha} + |\beta^0|_{1+\alpha} \leq B_0 \chi. \tag{2.62d}$$

*Then there is a constant $C$, determined only by $A$, $B_0$, $n$, $\alpha$, $\mu$, $\mu_0$, $\Omega$ such that*

$$|u|_{2+\alpha} \leq C(|Lu/\lambda|_\alpha + |Mu/\chi|_{1+\alpha} + |u|_0) \tag{2.63}$$

*for any $u \in H_{2+\alpha}$.*

***Proof.*** By virtue of the interpolation inequality (2.11), we may assume that $b^i$, $c$, and $\beta^0$ are all zero. Using Theorem 2.23 with $\psi = u$ at each boundary point along with the change of variables theorem for Hölder norms, we find that

$$|u|_{2+\alpha} \leq C(|u|_{2+\alpha;\Omega'} + |f/\lambda|_\alpha + |g/\chi|_{1+\alpha} + |u|_\delta)$$

for some open set $\Omega'$ such that $d(\Omega', \partial\Omega) > 0$. Since

$$|u|_{2+\alpha;\Omega'} \leq C(\Omega', \Omega)|u|^{(0)}_{2+\alpha},$$

It then follows from the interior Schauder estimate that

$$|u|_{2+\alpha} \le C(|u|_\delta + |f/\lambda|_\alpha + |g/\chi|_{1+\alpha}),$$

and another application of (2.11) completes the proof. □

In order to obtain further information, we now examine the existence theory. Our approach is a variant of the Perron process from Section 2.5.

**Theorem 2.27.** *Let $a^{ij}$, $b^i$, and $c$ be in $H_\alpha$ for some $\alpha \in (0,1)$ and suppose that $\beta$ and $\beta^i$ are in $H_{1+\alpha}$ and that $\partial\Omega \in H_{2+\alpha}$. If $c \le 0$ in $\Omega$ and $\beta^0 < 0$ on $\partial\Omega$, then*

$$Lu = f \text{ in } \Omega, \quad Mu = g \text{ on } \partial\Omega \tag{2.64}$$

*has a unique solution in $C^2(\Omega) \cap C(\overline{\Omega})$. Moreover $u \in H_{2+\alpha}$ and*

$$|u|_{2+\alpha} \le C(\Omega, \alpha, A, B_0)(|f/\lambda|_\alpha + |g/\chi|_{1+\alpha}). \tag{2.65}$$

***Proof.*** The main point is to consider the combination of differential equation and boundary condition in the same light that was used for just the differential equation in Section 2.5. Here we say that (2.64) is *locally solvable* if for each $x_0 \in \overline{\Omega}$, there is a non-empty, relatively open subset $N$ of $\overline{\Omega}$ such that, for any continuous function $\psi$ (defined on $\partial N \cap \Omega$), there is a unique solution of

$$Lu = f \text{ in } N \cap \Omega, \quad Mu = g \text{ on } N \cap \partial\Omega, \quad u = \psi \text{ on } \partial N \cap \Omega. \tag{2.66}$$

For $x_0 \in \Omega$, we use the theory of the Dirichlet problem to infer that this problem is solvable provided we take $N$ to be a ball $B$ centered at $x_0$ such that $\overline{B} \subset \Omega$. For $x_0 \in \partial\Omega$, we need to be more subtle. First, by the definition of $\partial\Omega \in C^{2,\alpha}$, we know that there is a ball $B$ centered at $x_0$ such that, after a suitable rotation, we can write

$$\Omega \cap B = \{x \in \mathbb{R}^n : |x - x_0| < R,\ x^n \ge \omega(x')\}$$

for some $H_{2+\alpha}$ function $\omega$ and some $R > 0$.

We then make the change of variables

$$y' = x' - x_0', \quad y^n = x^n - \omega(x').$$

In the $y$-coordinates, we see that, for some $r > 0$,

$$\{x \in \mathbb{R}^n : |y| < r,\ y^n > 0\}$$

is a subset of $\Omega$,

$$\{x \in \mathbb{R}^n, |y| < r,\ y^n\}$$

is a subset of $\partial\Omega$, and $y = 0$ corresponds to $x_0$. If we write the differential equation and boundary condition in $y$-coordinates, we see that local solvability is achieved via Theorem 2.24.

The calculations in the proof of Lemma 1.2 show that the function $\varphi$, defined by $\varphi(x) = \exp(-(1 + A)x^1)$, satisfies $L\varphi > 0$ in $\Omega$. Hence, there are positive constants $K_1$ and $K_2$ such that $L(K_1\varphi - K_2) \geq |f|$ in $\Omega$ and $M(K_1\varphi - K_2) \geq |g|$ on $\partial\Omega$. It follows that $K_1\varphi - K_2$ is a subsolution and that $-K_1\varphi + K_2$ is a a supersolution. Just as in the case of the Dirichlet problem, it follows that (2.64) has a solution. □

**Theorem 2.28.** *Let $a^{ij}$, $b^i$, and $c$ be in $H_\alpha$ for some $\alpha \in (0, 1)$ and suppose that $\beta$ and $\beta^i$ are in $H_{1+\alpha}$ and that $\partial\Omega \in H_{2+\alpha}$. If $u \in C^2(\Omega) \cap C(\overline{\Omega})$ is a solution of (2.64) with $f \in H_\alpha$ and $g \in H_{1+\alpha}$ is in $H_{2+\alpha}$ and there is a constant $C$, determined only by $\Omega$, $A$, $B_0$, $\alpha$, $\mu$, and $\mu_0$ such that*

$$|u|_{2+\alpha} \leq C(|f/\lambda|_\alpha + |g/\chi|_{1+\alpha} + |u|_0). \tag{2.67}$$

***Proof.*** As a first step, we note that Lemma 1.26 implies that $u \in H_\delta$ for some $\delta > 0$. Then we can use Theorem 2.27 to infer that there is a unique solution $v \in H_{2+\delta}$ of

$$\bar{L}v = -cu + f \text{ in } \Omega, \quad \bar{M}v = -(\beta^0 + 1)u + g \text{ on } \partial\Omega, \tag{2.68}$$

where $\bar{L}v = a^{ij}D_{ij}v + b^iD_iv$ and $\bar{M}v = \beta^iD_iv - v$. It then follows that $u = v$ so $u \in H_{2+\delta}$. Therefore, $v \in H_{2+\alpha}$ and hence $u \in H_{2+\alpha}$. □

## 2.8 Fredholm-Riesz-Schauder theory

Finally, we conclude that $(L, M)$ is a Fredholm operator of index zero. To explain this terminology and to prove the result, we give a brief account of Fredholm-Riesz-Schauder theory. In what follows, $X$ always denotes a Banach space.

The dual $X^*$ of $X$ is the set of all bounded, linear, real-valued functions defined on $X$. Note that $X^*$ is a Banach space. In addition, if $T$ is any bounded linear function defined on $X$ with values in $X$, we define its adjoint $T^*$ by defining $T^*y^*$, for any $y^* \in X^*$, to be the element of $X^*$ given by

$$(T^*y^*)x = y^*(Tx).$$

(Although we shall not need the additional generality, we observe that this expression allows us to define $T^*y^*$ as a real-valued function on $X$ assuming only that $T$ is a function defined on $X$ with values in $X$.) Also, for any subset $W$ of $X$, we define its orthogonal complement $W^\perp$ to be the set of all $y^* \in X^*$ such that $y^*(w) = 0$ for all $w \in W$. Similarly, if $W^* \subset X^*$, its orthogonal complement $(W^*)^\perp$ is the set of $x \in X$ such that $w^*(x) = 0$ for all $w^* \in W^*$.

We say that $T$ is a *compact linear operator on* $X$ if $T$ is a linear function defined on $X$ with values in $X$ and if $T$ maps bounded sets into compact sets. It can be shown that, for any compact linear operator $T$ on $X$, $T^*$ is a compact linear operator on $X^*$. For a given compact linear operator $T$ on $X$ and $\mu \in \mathbb{R} \setminus \{0\}$, we define the following sets:

$$N_\mu = \{x \in X : x = \mu T x\}, \quad N_\mu^* = \{x^* \in X^* : x^* = \mu T^* x^*\}.$$

We also write $R_\mu$ for the set of all $y \in X$ such that $y = x - \mu T x$ for some $x \in X$ and $R_\mu^*$ or the set of all $y^* \in X^*$ such that $y^* = x^* - \mu T^* x^*$ for some $x^* \in X^*$.

With these definitions, we can state the basic results of the Fredholm-Riesz-Schauder theory. We refer to Theorems 5.2.10 and 5.3.2 in [57] for details.

**Theorem 2.29.** *Let $X$ be a Banach space and let $T$ be a compact linear operator on $X$. Then, for any nonzero $\mu$, the spaces $N_\mu$ and $N_\mu^*$ have the same finite dimension. In addition $R_\mu = (N_\mu^*)^\perp$ and $R_\mu^* = (N_\mu)^\perp$. Finally, the set of all numbers $\mu$ for which $N_\mu$ has nonzero dimension is countable and it has no finite limit points.*

Note the numbers $\mu$ for which $N_\mu$ has nonzero dimension are the reciprocals of the eigenvalues of $T$; however, this choice will be convenient for our application.

**Theorem 2.30.** *Let $a^{ij}$, $b^i$, and $c$ be in $H_\alpha(\Omega)$ and let $\beta^0$ and $\beta$ be in $H_{1+\alpha}(\partial\Omega)$ for some domain $\Omega$ with $\partial\Omega \in H_{2+\alpha}$. Set $X = H_\alpha(\Omega) \times H_{1+\alpha}(\partial\Omega)$. Suppose also that the minimum eigenvalue of the matrix $[a^{ij}]$ is bounded from below by a positive constant and that $\beta \cdot \gamma$ is bounded from below by a positive constant. Then there is a countable set $\Sigma$ of real numbers with no finite limit points such that the problem*

$$(L + \sigma)u = f \text{ in } \Omega, \quad (M + \sigma)u = g \text{ on } \partial\Omega \tag{2.69}$$

*has a unique solution for all $(f, g) \in X$ if and only if $\mu \notin \Sigma$. In addition, for each $\sigma \in S$, there is a positive integer $I_\sigma$ such that the dimension of the solution set of the problem*

$$(L + \sigma)v = 0 \text{ in } \Omega, \quad (M + \sigma)u = 0 \text{ on } \partial\Omega \tag{2.70}$$

*is $I_\sigma$, and there is an $I_\sigma$-dimensional subspace $W_\sigma^*$ of $X^*$ such that* (2.69) *has a solution if and only if $(f, g) \in (W_\sigma^*)^\perp$.*

***Proof.*** Let $K$ be a constant such that $K > \max\{|c|_0, |\beta^0|_0\}$. Then the problem

$$(L-K)u = f \text{ in } \Omega, \quad (M-K)u = g \text{ on } \partial\Omega$$

has a unique solution $u \in H_{2+\alpha}$ for any $(f, g) \in X$. We now define the operator $T \colon X \to X$ by $T(f, g) = (u, u_{|\partial\Omega})$, where $u$ is the solution of this problem. Since $T$ maps $X$ into $H_{2+\alpha}(\Omega) \times H_{2+\alpha}(\partial\Omega)$, it follows that $T$ is a compact linear operator on $X$.

Suppose first that $\mu$ is a nonzero number such that $N_\mu$ has dimension zero. Then, for any $(f, g) \in X$, there is $(f_1, g_1) \in X$ such that

$$(I - \mu T)(f_1, g_1) = (f, g).$$

Let $u$ be the solution of

$$(L-K)u = f_1 \text{ in } \Omega, \quad (M-K)u = g_1 \text{ on } \partial\Omega.$$

Then it's easy to check that

$$(L-K-\mu)u = f \text{ in } \Omega, \quad (M-K-\mu)u = g \text{ on } \partial\Omega,$$

and $u$ is the only solution of this boundary value problem. Hence if $\sigma = -K - \mu$ for some $\mu \neq 0$ with $N_\mu$ having dimension zero (or if $\sigma = -K$), then (2.69) has a unique solution for any $(f, g) \in X$. We define $\Sigma$ to be the set of all numbers $\sigma$ of the form $\sigma = -K - \mu$ with $\mu$ nonzero and with $N_\mu$ having nonzero dimension.

If $\sigma \in \Sigma$, we set $\mu = -K - \sigma$ and note that $N_\mu$ has nonzero dimension, which we denote by $I_\sigma$. For simplicity of notation, if $u \in H_{2+\alpha}(\Omega)$, we write $\bar{u} = (u, u_{|\partial\Omega})$. It's easy to see that $u$ solves (2.70) if and only if $\bar{u} = (K+\mu)\bar{u}$, so the solution set of (2.70) has dimension $I_\sigma$. (In particular, if $(f, g) = \mu T(f, g)$ for any constant $\mu$, then $g$ must be the restriction of $f$ to $\partial\Omega$.) In addition, $u$ solves (2.69) if and only if $\bar{u} - \mu T\bar{u} = (f, g)$, so (2.69) has a solution if and only if $(f, g) \in (W_\sigma^*)^\perp$ provided $W_\sigma^*$ is just $N_\mu^*$ (that is, the set of all $x^* \in X^*$ such that $x^* = \mu T^* x^*$). □

In the language of linear operators on vector spaces, the second part of this theorem just says that the codimension of the range of $L + \lambda_k$ is equal to the dimension of its kernel.

## Notes

As previously mentioned, the Hölder condition was introduced by Hölder [71] to study potentials. The weighted norms are based on those in Chapter 6 from [64]; we have modified the definition there (which comes from Section 2 in [39], and that source suggested that the norms were well-known in the 1950's). In fact, the norm in [64] is equivalent to the one given here (see Exercise 2.1). The weighted Hölder spaces $H_a^{(b)}$ with $b < 0$ were used by Michael [129, 130] to study the Dirichlet problem for linear elliptic equations. The results in Section 2.1 are well-known (see, for example, [63]) and the proofs given here are easy modifications of the ones in Section 6.8 in [64].

The alternative characterization of Hölder norms in Section 2.2 was first explicitly noted by Safanov (see Section 2 in [163]); he ascribes the characterization to Campanato [25], who uses the $L^p$ norm (with $p$ finite) in place of the supremum norm. Our proof is different from those in [25, 163, 21] and follows instead the one in Section 3 of [180]. Note that the function $\bar{u}$ in the proof of Lemma 2.8 is commonly called a mollification of $u$ with $\varphi$ known as a mollifier; this terminology comes from Friedrich's paper [58], and the story of the origin of this name is recounted by Peter Lax in his commentary on [58] on page 117 of [59].

The basic existence result (Theorem 2.10) comes from Theorem 2.7.2 in [87]. If we assume that $A^{ij} = \delta^{ij}$, that $f \in H_\alpha(B)$, and that $\psi \in H_{2+\alpha}$, then the conclusion of Theorem 2.10 is also true; a proof of this fact without using Theorem 2.10 is given in, for example, Theorem 4.3 of [64]. In fact, that reference proves the result via an integral representation for solutions of the differential equation $\Delta v = f$.

Interior Schauder estimates were proved by Schauder in [168] (for constant coefficient equations) and in [167] (for general equations). His primary tools were integral representations of solutions of solutions of differential equations and appropriate functional analysis. We refer to Chapter 6 in [64] for more details on this approach. In particular, our Lemma 2.14 is just Lemma 6.20 from [64] (which, in turn, is essentially Theorem 4.3 of [129]), and its derivation from Proposition 2.13 is taken from [64].

The Perron process described in Section 2.5 is essentially the same as in Section 6.3 of [64] (see also Section 6.5 of [64], which is based on Section 4 of [129]). In particular, Theorem 2.15 is a simple modification of Theorem 4.9 in [64] (which comes from Theorem 4.5 in [129]).

Section 2.6 comes from Section 2 of [100] with one important exception:

the Schauder-type estimates (Lemma 2.20, Corollary 2.21, and (2.53)) were proved in [100] by using integral representations (more precisely, by pointing out that the techniques in [64] could be applied to that situation). We present independent proofs here for the first time; however, the method of proof used here is straightforward. Condition (2.60) in Theorem 2.24 is not surprising for the existence result. The theory of Fichera [44] (see also [118]) shows that the Dirichlet problem for the differential equation

$$a^{ij}D_{ij}u + \frac{1}{d}(b^i D_i u - u) = 0 \text{ in } \Omega,$$

with $a^{ij}$ and $b$ continuously differentiable, has a solution for arbitrary boundary data if and only if

$$b \cdot \gamma < a^{ij}\gamma_i\gamma_j$$

on $\partial\Omega$, and (2.60) is just a restatement of this condition.

The Fredholm-Riesz-Schauder theory is taken from Sections 5.2 and 5.3 in [57]; this theory is an extension by Schauder [166] of earlier work by Riesz [161]. Schauder's main advance was to introduce the idea of adjoint operators, thus allowing the theory to be applied in general Banach spaces.

## Exercises

**2.1** Define the seminorm $M_\alpha^{(b)}$ (for $\alpha \in (0,1]$ and $b \geq -\alpha$) by

$$M_\alpha^{(b)}(f) = \sup_{x \neq y \in \Omega} \min\{d(x), d(y)\}^{b+\alpha} \frac{|f(x) - f(y)|}{|x-y|^\alpha},$$

and define the norm $N_a^{(b)}$ (for $a = k + \alpha$) by

$$N_a^{(b)}(f) = \sum_{j=0}^{k} |D^j f|_0^{(b+j)} + M_\alpha^{(b+k)}(f).$$

Show that there is a positive constant $C$ such that

$$\frac{1}{C}|f|_a^{(b)} \leq N_a^{(b)}(f) \leq C|f|_a^{(b)}.$$

**2.2** For $g \in H_\delta(B)$, show that there is a function $G \in H_{2+\alpha}^{(-\delta)}(B)$ with $G = g$ on $\partial B$ and

$$|G|_{2+\alpha}^{(-\delta)} \leq C(n, \alpha, \delta)|g|_\delta^{(-\delta)}.$$

**2.3** Suppose that $\partial\Omega \in H_{k+2+\alpha}$ with $k$ a positive integer. Suppose also that $a^{ij}$, $b^i$ and $c$ are in $H_{k+\alpha}$ and that $\beta$ and $\beta^0$ are in $H_{k+1+\alpha}$. Prove that

$$|u|_{k+2+\alpha} \le C(|Lu|_{k+\alpha} + |Mu|_{k+1+\alpha} + |u|_0).$$

**2.4** Prove Corollary 2.25.

**2.5** Suppose that $\partial\Omega \in H_{k+2+\alpha}$ with $k$ a positive integer. Suppose also that $a^{ij}$, $b^i$ and $c$ are in $H_{k+\alpha}$, and that there is a function $\varphi \in H_{k+2+\alpha}$ such that $u = \varphi$ on $\partial\Omega$. Prove that

$$|u|_{k+2+\alpha} \le C(|Lu|_{k+\alpha} + |\varphi|_{k+2+\alpha} + |u|_0).$$

(Hint: Use Exercise 1.4.)

# Chapter 3

# The Miller Barrier and Some Supersolutions for Oblique Derivative Problems

## Introduction

In this chapter, we construct various supersolutions for both the Dirichlet problem and the oblique derivative problem in Lipschitz domains. This study is based on the construction of a suitable comparison function in a cone, due to Miller, and this construction is based the solution of a singular ordinary differential equation. We start with a discussion of solutions for ordinary differential equations in Section 3.1; some of the results are quite standard but others are not, so we give brief proofs of the results that we use. (The Notes at the end of the chapter give further references for all these results.) Section 3.2 uses the results for ordinary differential equations to construct the barriers.

## 3.1 Theory of ordinary differential equations

In this section, we present a few basic existence, uniqueness and continuous dependence results for some special second order differential equations. Although many of the results are true under much more general hypotheses, we shall give simple proofs only for the cases of interest in this chapter.

Our first step is a basic existence and uniqueness result.

**Lemma 3.1.** *Let $a < b$ be real numbers and let $H : [a, b] \times \mathbb{R} \times \mathbb{R} \to \mathbb{R}$ be a continuous function. Suppose $H$ is uniformly Lipschitz with respect to its second and third arguments. Then, for any $s_0 \in [a, b]$ and any real numbers $f_0$ and $f_1$, there is a unique solution $f \in C^2[a, b]$ of*

$$f'' + H(s, f(s), f'(s)) = 0 \text{ for } s \in (a, b), \tag{3.1a}$$

$$f(s_0) = f_0, \quad f'(s_0) = f_1. \tag{3.1b}$$

***Proof.*** Suppose first that $s_0 < b$. Write $K$ for the Lipschitz constant of $H$ with respect to its second and third arguments and set $h = \min\{b - s_0, 1, 1/(2K)\}$. We first show that the initial value problem

$$f'' + H(s, f(s), f'(s)) = 0 \text{ for } s \in (s_0, s_0 + h), \tag{3.2a}$$

$$f(s_0) = f_0, \quad f'(s_0) = f_1 \tag{3.2b}$$

has a unique solution. To this end, let $X$ denote the set of all $C^1[s_0, s_0+h]$ functions $g$ with $g(s_0) = f_0$ and $g'(s_0) = f_1$ with the norm

$$|g|_X = \sup |g(s)| + \sup |g'(s)|;$$

Note that $X$ is a complete metric space. We then define the operator $T$ on $X$ by

$$Tg(s) = f_0 + f_1(s - s_0) - \int_{s_0}^{s} \left( \int_{s_0}^{\tau} H(\sigma, g(\sigma), g'(\sigma))\, d\sigma \right) d\tau.$$

It's easy to check that $Tg \in X$ and that

$$(Tg)''(s) + H(s, g(s), g'(s)) = 0.$$

A simple calculation shows that

$$|Tg_1 - Tg_2|_X \le K[h + \frac{1}{2}h^2]|g_1 - g_2|_X.$$

Our choice of $h$ implies that $K[h + \frac{1}{2}h^2] < 1$, so $T$ is a contraction. Hence it has a unique fixed point $f$, which will be the unique solution of (3.2).

We then repeat this process a finite number of times to obtain a solution on $[s_0, b]$; a similar argument gives a solution on $[a, b]$. □

When $H$ is singular, then the existence of a unique solution is no longer guaranteed for arbitrary initial data. As a simple example, consider the equation $f'' + f'/s = 0$. It's easy to check that the general solution of this equation is $f(s) = c_1 + c_2 \ln s$. Hence any solution which is bounded at 0 must have zero derivative there. This is the situation which we wish to analyze more closely for nonlinear problems. For our applications, we now specialize the interval of existence to $[0, \pi)$.

**Lemma 3.2.** *Let $G : [0, \pi) \times \mathbb{R} \times \mathbb{R}$ be continuous, Lipschitz with respect to its second and third arguments, and suppose that there is a constant $s_0 \in (0, \pi)$ such that $G(s, z, p_1) \le G(s, z, p_2)$ for $s \in [0, s_0]$, $z \in \mathbb{R}$ and $p_1 \le p_2$ in $\mathbb{R}$. Then, for any real number $f_0$, there is a unique solution $f \in C^2[0, \pi)$ of the initial value problem*

$$f'' + G(s, f(s), f'(s)/s) = 0 \text{ for } s \in (0, \pi), \tag{3.3a}$$

$$f(0) = f_0, \quad f'(0) = 0. \tag{3.3b}$$

***Proof.*** We first note that, by continuity, if $f \in C^2[0,\pi)$ is a solution of (3.3), then $f''(0) = h$, the unique fixed point of the equation

$$h = -G(0, f_0, h).$$

We now define two functions $\hat{f}$ and $\theta$ by

$$\hat{f}(s) = f_0 + \frac{h}{2}s^2, \quad \theta(s) = h + G(s, \hat{f}(s), h).$$

For each $\varepsilon > 0$, there is a constant $\eta(\varepsilon) \in (0, \min\{\varepsilon, \pi/2\})$ such that $|\theta(s)| \leq \varepsilon$ if $0 \leq s < \eta(\varepsilon)$. We now define $\hat{\theta}$ by

$$\hat{\theta}(s;\varepsilon) = \begin{cases} \theta(s) & \text{if } 0 \leq s < \eta(\varepsilon), \\ \theta(2s - \eta(\varepsilon)) & \text{if } \eta(\varepsilon) \leq s < 2\eta(\varepsilon), \\ 0 & \text{if } 2\eta(\varepsilon) \leq s < \pi, \end{cases}$$

and we write $\hat{f}_\varepsilon$ for the unique solution of

$$\hat{f}_\varepsilon'' + G(s, \hat{f}_\varepsilon, \hat{f}_\varepsilon'/s) + \hat{\theta}(s;\varepsilon) = 0 \text{ for } s \in (\eta(\varepsilon), \pi),$$
$$\hat{f}_\varepsilon(\eta(\varepsilon)) = f_0 + \frac{1}{2}h\eta(\varepsilon)^2, \quad \hat{f}_\varepsilon'(\eta(\varepsilon)) = h\eta(\varepsilon)$$

given by Lemma 3.1. If we now define $f_\varepsilon$ by

$$f_\varepsilon(s) = \begin{cases} \hat{f}(s) & \text{if } 0 \leq s < \eta(\varepsilon), \\ \hat{f}_\varepsilon(s) & \text{if } s \geq \eta(\varepsilon), \end{cases}$$

then $f_\varepsilon \in C^2[0,\pi)$ with

$$|f_\varepsilon'' + G(s, f_\varepsilon, f_\varepsilon'/s)| < \varepsilon \text{ for } s \in (0,\pi), \quad f_\varepsilon(0) = f_0, \quad f_\varepsilon'(0) = 0.$$

To complete the proof, we must show that the family of functions $(f_\varepsilon)$ converges in $C^2$ to a limit function which must be our solution. This convergence is proved by proving a suitable *a priori* bound for $f_\varepsilon - f_\delta$ and its derivatives for $\varepsilon > \delta > 0$.

To prove the estimate, fix $\varepsilon > \delta > 0$, set $y = f_\varepsilon - f_\delta$, and write $K$ for the Lipschitz constant of $G$ with respect to the second argument. If $y'(s) > 0$, then $|y'|$ is differentiable at $s$ with $|y'|'(s) = y''(s)$ and hence

$$|y'|'(s) \leq -G(s, f_\varepsilon(s), f_\varepsilon'(s)/s) + G(s, f_\delta(s), f_\delta'(s)/s + 2\varepsilon)$$
$$\leq K|y| + 2\varepsilon$$

wherever $y'(s) > 0$ and $s \in (0, s_0)$. A similar argument (but noting that $|y'|'(s) = -y''(s)$ wherever $y'(s) < 0$ and $s \in (0, s_0)$) shows that

$$|y'|'(s) \leq K|y| + 2\varepsilon$$

wherever $y'(s) \neq 0$. For $\zeta \in (0, s_0)$, we now set

$$m(\zeta) = \sup\{|y'|'(s) : 0 \leq s < \zeta,\ y'(s) \neq 0\}.$$

If $s \in [0, \zeta)$, then $|y'(s)| \leq m(\zeta)s$ and $|y(s)| \leq \frac{1}{2}m(\zeta)s^2$, so

$$m(\zeta) \leq \frac{1}{2}K\zeta^2 m(\zeta) + 2\varepsilon.$$

It follows that, for $\zeta$ sufficiently small (so that $K\zeta^2 \leq 1$ and $\zeta \leq s_0$), we have

$$m(\zeta) \leq 4\varepsilon. \tag{3.4}$$

With this $\zeta$ fixed, it follows that

$$|f_\varepsilon(s) - f_\delta(s)| \leq 2\varepsilon s^2, |f'_\varepsilon(s) - f'_\delta(s) \leq 4\varepsilon s. \tag{3.5}$$

Since $f_\varepsilon$ and $f_\delta$ are $C^2$, it follows that the limit

$$\tilde{f} = \lim_{\varepsilon \to 0} f_\varepsilon$$

exists on $[0, \zeta]$ and (3.4) and (3.5) imply that the convergence is in $C^2[0, \zeta]$. To extend $\tilde{f}$ to $[0, \pi)$, we just take $F$ to be the solution of

$$F'' + G(s, F, F'/s) \text{ for } s \in (\zeta, \pi), \quad F(\zeta) = \tilde{f}(\zeta), \quad F'(\zeta) = \tilde{f}'(\zeta)$$

given by Lemma 3.1 and define $f$ by

$$f(s) = \begin{cases} \tilde{f}(s) & \text{if } 0 \leq s < \zeta, \\ F(s) & \text{if } \zeta \leq s < \pi. \end{cases}$$ □

In fact, the solution in this lemma depends continuously on the function $G$.

**Corollary 3.3.** *Let $G_1$ and $G_2$ be two continuous functions defined on $[0, \pi) \times \mathbb{R} \times \mathbb{R}$ which are uniformly Lipschitz with respect to their second and third arguments, and suppose that there is a constant $s_0 \in (0, \pi)$ such that $G_j(s, z, p_1) \leq G_j(s, z, p_2)$ for $j = 1, 2$, $s \in [0, s_0]$, $z \in \mathbb{R}$ and $p_1 \leq p_2$ in $\mathbb{R}$. For $j = 1, 2$, let $f_j$ be the solution of*

$$f_j'' + G_j(s, f_j, f_j/s) \text{ for } s \in [0, \pi), \quad f_j(0) = 1, \quad f_j'(0) = 0.$$

*Let $K$ be a Lipschitz constant for $G_1$ and $G_2$ with respect to the second argument and suppose $Kh^2 \leq \frac{1}{2}$ for some $h \in (0, s_0)$. Set*

$$\varepsilon = \sup\{G_1(s, f_1, f_1'/s) - G_2(s, f_1, f_1'/s)| : s \in [0, h]\}.$$

*Then $y = f_1 - f_2$ satisfies the inequalities*

$$|y(s)| \leq 2\varepsilon s^2, \quad |y'(s)| \leq 4\varepsilon s, \quad |y''(s)| \leq 4\varepsilon \tag{3.6}$$

*for all $s \in [0, h]$.*

***Proof.*** Just as in the proof of Lemma 3.2, we set

$$m = \sup\{|y'|'(s) : y'(s) \neq 0, 0 \leq s \leq h\}.$$

At a point where $y'(s) > 0$, we have

$$\begin{aligned}|y'|'(s) = y''(s) &= G_2(s, f_2, f_2'/s) - G_1(s, f_1, f_1'/s) \\ &\leq \varepsilon + G_2(s, f_2, f_2'/s) - G_2(s, f_1, f_1'/s) \\ &\leq \varepsilon + K|y(s)|.\end{aligned}$$

Arguing similarly where $y'(s) < 0$, we obtain

$$|y'|'(s) \leq \varepsilon + K|y(s)|$$

for all $s$ at which $y'(s) \neq 0$. Again, $|y(s)| \leq \frac{1}{2}ms^2$, so

$$m \leq 4\varepsilon,$$

and this inequality easily implies (3.6). □

We shall also use a result concerning the existence and structure of solutions of linear degenerate ordinary differential equations.

**Lemma 3.4.** *Let $a$ and $b$ be continuous functions on $[0, \pi)$ and suppose that there are positive constants $h < \pi$ and $\alpha$ such that $a(s) > 1 - \alpha s$ for $0 \leq s \leq h$. Then there are two functions $H \in C^2[0, \pi)$ and $K \in C^2(0, \pi)$ such that*

$$H'' + \frac{a(s)}{s}H' + b(s)H = 0 \text{ for } 0 < s < \pi, \quad H(0) = 1, \quad H'(0) = 0 \quad (3.7)$$

*and*

$$K'' + \frac{a(s)}{s}K' + b(s)K = 0 \text{ for } 0 < s < \pi, \quad \lim_{s \to 0+} K(s) = \infty. \quad (3.8)$$

*Moreover, for any solution $u \in C^2(0, \pi)$ of*

$$u'' + \frac{a(s)}{s}u' + b(s)u = 0 \text{ for } 0 < s < \pi, \quad (3.9)$$

*there are unique constants $c_1$ and $c_2$ such that $u = c_1H + c_2K$.*

***Proof.*** The existence of $H$ satisfying (3.7) follows immediately from Lemma 3.2.

To obtain $K$, we use the method of reduction of order. Specifically, we seek a nonconstant function $w$ such that $K = Hw$ satisfies the correct differential equation. By differentiating $Hw$ and using the differential equation for $H$, we see that

$$w'' + \left(\frac{a(s)}{s} + \frac{2H'}{H}\right)w' = 0.$$

Of course, this equation only makes sense where $H \neq 0$, so we now work with this equation only for $0 < s < \delta$, where $\delta \in (0, h)$ is so small that $H(s) \geq \frac{1}{2}$ for $0 \leq s \leq \delta$.

Dividing our equation by $w'$ and integrating yields

$$\ln |w'| = \int_s^\delta \frac{a(\sigma)}{\sigma}\, d\sigma - \ln(H^2) + k$$

for some constant $k$. Hence

$$\begin{aligned} |w'| &= \frac{e^k}{H^2} \exp\left(\int_s^\delta \frac{a(\sigma)}{\sigma}\, d\sigma\right) \\ &\geq \frac{e^k}{H^2} \exp\left(\int_s^\delta \frac{1 - \alpha\sigma}{\sigma}\, d\sigma\right) \\ &= \frac{e^k}{H^2} e^{\alpha(s-\delta)} \delta \frac{1}{s}. \end{aligned}$$

It follows that $|w'| \geq C/s$ for some positive constant $C$, so $w'$ is not integrable at 0, which means (if we choose $w$ to be positive) that $w(s) \to \infty$ as $s \to 0$. Since $H \geq \frac{1}{2}$ for $s < \delta$, it follows that the $K$ we have constructed tends to infinity at 0 as well and that $K(\delta)$ and $K'(\delta)$ are finite. Just as in Lemma 3.2, we can extend $K$ to a solution of the differential equation on $(0, \pi)$.

Next, we observe that $H$ and $K$ are linearly independent. If $x_0 \in (0, \pi)$ and $b_1$ and $b_2$ are constants such that

$$b_1 H(x_0) + b_2 K(x_0) = 0, \quad b_1 H'(x_0) + b_2 K'(x_0) = 0,$$

then, by linearity, the function $v = b_1 H + b_2 K$ solves the initial value problem

$$v'' + \frac{a(s)}{s} v' + b(s) v = 0 \text{ for } 0 < s < \pi, \quad v(x_0) = 0, \quad v'(x_0) = 0,$$

and hence $v \equiv 0$ by Lemma 3.1. It follows that $b_1 = b_2 = 0$. So, if $u$ is a solution of (3.9), we can choose $c_1$ and $c_2$ so that

$$u(1) = c_1 H(1) + c_2 K(1), \quad u'(1) = c_1 H'(1) + c_2 K'(1).$$

Lemma 3.1 then implies that $u = c_1 H + c_2 K$. □

## 3.2 The Miller barrier construction

In this section, we construct an optimal supersolution for a class of linear Dirichlet problems in an infinite cone. We assume that the axis of the cone is the positive $x^n$ axis and we introduce polar coordinates $r$ and $\theta$ via the equations

$$r = |x|, \quad r\cos\theta = x^n.$$

(Strictly speaking, these are two of the $n$ spherical coordinates, but we follow Miller's terminology here.) Our cones will then be defined by the inequalities $r > 0$ and $\theta \in [0, \theta_0)$ for some $\theta_0 \in (0, \pi)$, and our supersolutions will have the form $w(x) = r^\alpha f(\theta)$ for some positive constant $\alpha$ and a $C^2[0, \pi)$ function $f$. Of course, $f$ need only be defined for $\theta \in [0, \theta_0)$ but it will be very convenient to have $f$ defined on the larger interval.

For simplicity of notation, we also introduce $\mathcal{L}_\mu$, the set of all matrix-valued functions with eigenvalues in the interval $[1, \mu]$. We begin our analysis with a polar representation.

**Lemma 3.5.** *Let $\mu \geq 1$. Then for any $[a^{ij}] \in \mathcal{L}_\mu$, there are scalar functions $a$, $b$, $c$, and $d$ such that*

$$s^2 + t^2 \leq as^2 + 2bst + ct^2 \leq \mu(s^2 + t^2) \tag{3.10a}$$

*for all real numbers $s$ and $t$,*

$$1 \leq d \leq \mu, \tag{3.10b}$$

*and*

$$a^{ij}D_{ij}w = r^{\alpha-2}[a\alpha(\alpha-1)f + 2b(\alpha-1)f' + c(f'' + \alpha f) + d(n-2)(\alpha f + \cot\theta f')]. \tag{3.11}$$

*Conversely, given $a$, $b$, $c$, and $d$ satisfying* (3.10), *there is $[a^{ij}] \in \mathcal{L}_\mu$ such that* (3.11) *holds.*

***Proof.*** To begin, we define two vectors $\xi$ and $\eta$ by

$$\xi_i = \cot\theta\frac{x^i}{r} - \csc\theta\delta^{in}, \quad \eta_i = \frac{x^i}{r}.$$

It is easy to check that $\xi$ and $\eta$ are orthogonal unit vectors and that

$$\frac{\partial r}{\partial x} = \eta, \quad \frac{\partial\theta}{\partial x} = \frac{\xi}{r}.$$

Hence, if $w = r^\alpha f(\theta)$, then

$$Dw = r^{\alpha-1}(\alpha f\eta + f'\xi)$$

and

$$D_{ij}w = r^{\alpha-2}(\alpha(\alpha-1)f\eta_i\eta_j + \alpha f'\eta_i\xi_j + \alpha r f\frac{\partial\eta_i}{\partial x^j} + (\alpha-1)f'\xi_i\eta_j + f''\xi_i\xi_j + rf'\frac{\partial\xi_i}{\partial x^j}.$$

Computing the partial derivatives of $\eta$ and $\xi$ yields

$$\frac{\partial\eta_i}{\partial x^j} = \frac{1}{r}(\delta_{ij} - \eta_i\eta_j)$$

and

$$\frac{\partial\xi_i}{\partial x^j} = (-\csc^2\theta\frac{x^i}{r^2} + \csc\theta\cot\theta\frac{1}{r}\delta_{in})\xi_j + \cot\theta\frac{1}{r}(\delta_{ij} - \eta_i\eta_j).$$

Using the trigonometric identity $\csc^2\theta = \cot^2\theta + 1$ and the definition of $\xi$, we see that

$$\frac{\partial\xi_i}{\partial x^j} = -\frac{1}{r}\cot\theta\xi_i\xi_j - \frac{1}{r}\xi_i\eta_j + \cot\theta\frac{1}{r}(\delta_{ij} - \eta_i\eta_j).$$

From these calculations of partial derivatives, we find that

$$\begin{aligned} r^{2-\alpha}D_{ij}w &= \alpha(\alpha-1)f\eta_i\eta_j \\ &\quad + (\alpha-1)f'(\xi_i\eta_j + \xi_j\eta_j) \\ &\quad + (f'' + \alpha f)\xi_i\xi_j \\ &\quad + (\alpha f + \cot\theta f)(\delta_{ij} - \xi_i\xi_j - \eta_i\eta_j). \end{aligned}$$

Our polar representation now follows with

$$\begin{aligned} a &= a^{ij}\eta_i\eta_j, \\ b &= a^{ij}\xi_i\eta_j, \\ c &= a^{ij}\xi_i\xi_j, \\ d &= \frac{1}{n-2}a^{ij}(\delta_{ij} - \xi_i\xi_j - \eta_i\eta_j). \end{aligned}$$

Straightforward algebra now gives (3.10). Specifically, if $v = s\eta + t\xi$, then $|v|^2 = s^2 + t^2$ and

$$a^{ij}v_iv_j = as^2 + 2bst + ct^2.$$

Since $|v|^2 \le a^{ij}v_iv_j \le \mu|v|^2$, we immediately obtain (3.10a). If $v^{(k)}$ ($k = 1, \ldots, n-2$) are orthonormal vectors all perpendicular to $\xi$ and $\eta$, then

$$(n-2)d = a^{ij}v_i^{(k)}v_j^{(k)},$$

so (3.10b) follows.

Conversely, given $a$, $b$, $c$, and $d$ satisfying (3.10), we take $v^{(k)}$ as above, and choose $[a^{ij}]$ to be a symmetric matrix with eigenvalue $d$ corresponding to each eigenvector $v^{(k)}$ and then (because the other two eigenvectors must be perpendicular to all $v^{(k)}$) we can take the other two eigenvalues with corresponding eigenvectors in the $\xi\eta$-plane as follows: We write $(s_i, t_i)$ ($i = 1, 2$) for the eigenvectors of the matrix

$$\begin{bmatrix} a & b \\ b & c \end{bmatrix}$$

with corresponding eigenvalues $\lambda_i$. The other two eigenvectors of $[a^{ij}]$ are

$$v^{(1)} = s_1\eta + t_1\xi,$$
$$v^{(2)} = s_2\eta + t_2\xi,$$

with corresponding eigenvalues of $\lambda_1$ and $\lambda_2$. □

For $\mu \geq 1$, we now define the operator $\mathcal{M}_\mu$ on $C^2$ by

$$\mathcal{M}_\mu h(x) = \sup\{a^{ij}(x)D_{ij}h(x) : [a^{ij}] \in \mathcal{L}_\mu\}.$$

As we shall see, $\mathcal{M}_\mu$ is a fully nonlinear operator and a complete existence and uniqueness theory is known for this operator. For now, we make a few simple but crucial observations. First, if we write $\mathcal{L}_\mu^*$ for the set of all constant matrix-valued functions in $\mathcal{L}_\mu$, then we have

$$\mathcal{M}_\mu h(x) = \sup\{a^{ij}(x)D_{ij}h(x) : [a^{ij}] \in \mathcal{L}_\mu^*\}.$$

Second, for a given $h \in C^2(\Omega)$, there is a matrix-valued function $[a_h^{ij}] \in \mathcal{L}_\mu$ such that $\mathcal{M}_\mu h(x) = a_h^{ij}(x)D_{ij}h(x)$. To prove the second observation, we fix $x \in \Omega$ and let $[a_k^{ij}]$ be a sequence of matrices such that

$$\lim_{k\to\infty} a_k^{ij} D_{ij}h(x) = \mathcal{M}_\mu h(x).$$

Since $\mathcal{L}_\mu$ is a compact subset of the set of all symmetric matrices, it follows that there is a convergent subsequence of $[a_k^{ij}]$; we take $[a_h^{ij}(x)]$ to be the limit of this convergent subsequence.

Our next goal is to obtain solutions of $\mathcal{M}_\mu w = 0$ when $w = r^\alpha f(\theta)$, and we do so by solving a suitable initial value problem of the type discussed in the previous section.

We first define the function $g$ by

$$\begin{aligned} g(\alpha, a, b, c, d; \theta, z, p) = {} & \frac{a}{c}\alpha(\alpha-1)z + 2\frac{b}{c}(\alpha-1)\theta p + \alpha z \\ & + \frac{d}{c}(n-2)(\alpha z + \theta(\cot\theta)p), \end{aligned}$$

and then we define $G$ by

$$G(\alpha;\theta,z,p)=\sup\{g(\alpha,a,b,c,d;\theta,z,p):a,b,c,d\text{ satisfy (3.10)}\}.$$

Although $g$ and $G$ clearly depend on the parameter $\mu$, we suppress this dependence from the notation because it is not as important for our purposes as the dependence on $\alpha$. It then follows that $\mathcal{M}_\mu w=0$ (and hence $a^{ij}D_{ij}w\le 0$ for all $[a^{ij}]\in\mathcal{L}_\mu$) if $f$ is a solution of the initial value problem

$$f''+G(\alpha;\theta,f,f'/\theta)=0\text{ for }\theta\in(0,\pi),\quad f(0)=1,\quad f'(0)=0. \tag{3.12}$$

We saw in Lemma 3.2 that this initial value problem has a unique $C^2[0,\pi)$ solution, which we denote by $F_\alpha$. In the next lemma, we collect some important properties of the function $F_\alpha$, especially in terms of its dependence on $\alpha$.

**Lemma 3.6.** *Suppose $n\ge 3$. Then $F_\alpha$ has a first zero, which we denote by $\Psi(\alpha)$, in $(0,\pi)$. Moreover, the function $\Psi$ is continuous and strictly decreasing, $\Psi$ maps $(0,\infty)$ onto $(0,\pi)$, and $\Psi(1)=\pi/2$.*

***Proof.*** To show that $\Psi$ is defined, let $\Phi_\alpha$ be the solution of

$$\Phi_\alpha''+(n-2)\cot\theta\Phi_\alpha'+[\alpha^2+(n-2)\alpha]\Phi_\alpha=0\text{ for }\theta\in(0,\pi),$$
$$\Phi_\alpha(0)=1,\quad\Phi_\alpha'(0)=0.$$

(In other words, $\Phi_\alpha$ is just $F_\alpha$ corresponding to $\mu=1$.)

Our first step is to show that $\Phi_\alpha$ is zero somewhere in $(0,\pi)$. As a preliminary to this result, we show that $\Phi_\alpha$ is decreasing as long as it remains positive. To this end, suppose $\Phi_\alpha>0$ on $[0,h]$ for some $h\in(0,\pi)$. If $\Phi_\alpha'(\theta^*)=0$ for some $\theta^*\in(0,h)$, then $\Phi_\alpha''(\theta)<0$ from the differential equation, so $\Phi_\alpha'$ can't change from negative to positive as $\theta$ increases across $\theta^*$. Hence $\Phi_\alpha'$ is negative in $(0,h)$. It therefore follows that, if $\Phi_\alpha>0$ on $(0,\pi)$, then $\Phi_\alpha$ is bounded on $(0,\pi)$. Using the change of variables $s=\pi-\theta$, we see from Lemma 3.4 that $\Phi_\alpha\in C^2(0,\pi]$ with $\Phi_\alpha(\pi)\ne 0$ and $\Phi_\alpha'(\pi)=0$. Applying the preceding argument, we see that $\Phi_\alpha'$ must be positive on $(0,\pi)$. Since we have shown that $\Phi_\alpha'$ is also negative on this interval, we conclude that $\Phi_\alpha$ must vanish somewhere in the interval $(0,\pi)$.

Since $\Phi_\alpha(0)=1$ and $\Phi_\alpha$ is continuous on $[0,\pi)$, it follows that $\Phi_\alpha$ has a first zero in $(0,\pi)$. We denote this zero by $h_\alpha$.

If $F_\alpha>0$ on $[0,h_\alpha]$, then there is a $\beta>\alpha$ such that $F_\beta>0$ on $[0,h_\alpha]$ by continuous dependence. We now set $v=r^\alpha[\Phi_\alpha(\theta)-\frac{1}{2}F_\alpha(\theta)]$ and, for any $\varepsilon>0$, we set $v_\varepsilon=v-\varepsilon r^\beta F_\beta(\theta)$. For $\varepsilon>0$, there is a positive constant $R_\varepsilon$ such that $v_\varepsilon\le 0$ if $r\ge R_\varepsilon$. In the cone

$$K_\varepsilon=\{x:0<r<R_\varepsilon,0\le\theta<h_\alpha\},$$

we have that

$$\begin{aligned}\Delta v_\varepsilon &= -\frac{1}{2}\Delta(r^\alpha F_\alpha(\theta)) - \Delta(r^\beta F_\beta(\theta)) \\ &\geq -\frac{1}{2}\mathcal{M}_\mu(r^\alpha F_\alpha(\theta)) - \mathcal{M}_\mu(r^\beta F_\beta(\theta)) \\ &= 0.\end{aligned}$$

Since $v_\varepsilon \leq 0$ on $\partial K_\varepsilon$, it follow from the maximum principle that $v_\varepsilon \leq 0$ in $K_\varepsilon$. Therefore $v_\varepsilon \leq 0$ in

$$K = \{x : 0 < r, 0 \leq \theta < h_\alpha\}.$$

Sending $\varepsilon \to 0$, we conclude that $v \leq 0$ in $K$. But $v = \frac{1}{2}$ when $r = 1$ and $\theta = 0$. This contradiction shows that $F_\alpha$ must be zero somewhere in $[0, h_\alpha]$ and hence $F_\alpha$ has a first zero in $(0, \pi)$.

The continuity of $\Psi$ follows from the inverse function theorem after noting that $F'_\alpha(\Psi(\alpha)) \neq 0$.

To show that $\Psi$ is decreasing, let $\alpha > \beta$ and set $h = \Psi(\beta)$.

If $\Psi(\alpha) = h$, then, by the uniqueness part of Lemma 3.1, we know that $F'_\alpha(h) \neq 0$. Hence, there is a positive constant $k$ such that $F_\beta \leq kF_\alpha$ on $[0, h]$. On the other hand, if $\Psi(\alpha) > h$, then such a constant exists because $F_\alpha$ is positive on $[0, h]$.

Now we write $K$ for the cone with $\theta < h$, and set $w_1 = r^\alpha F_\alpha(\theta)$ and $w_2 = r^\beta F_\beta(\theta)$. For any $\varepsilon > 0$, there is a positive number $R_\varepsilon$ such that $\varepsilon w_1 - w_2 > 0$ for $r \geq R_\varepsilon$, and we now write $\Omega(\varepsilon)$ for the subset of $K$ on which $r < R_\varepsilon$. Then $\varepsilon w_1 - w_2 \geq 0$ on $\partial\Omega(\varepsilon)$. We now choose $[a^{ij}] \in L_\mu$ so that $a^{ij}D_{ij}w_2 = 0$. Since $a^{ij}D_{ij}w_1 \leq 0$, it follows that $a^{ij}D_{ij}(\varepsilon w_1 - w_2) \leq 0$ in $\Omega(\varepsilon)$, and then the maximum principle implies that $\varepsilon w_1 - w_2 \geq 0$ in $\Omega(\varepsilon)$. From our choice of $\Omega(\varepsilon)$, we therefore conclude that $\varepsilon w_1 - w_2 \geq 0$ in $K$. Sending $\varepsilon \to 0$, we conclude that $-w_2 \geq 0$ in $K$, which contradicts our construction of $w_2$ (which is positive in $K$). Hence we must have $\Psi(\alpha) < \Psi(\beta)$ and hence $\Psi$ is strictly decreasing.

Since

$$G(0; \theta, z, p) = \sup\{\frac{2b}{c}\theta p + (n-2)\frac{d}{c}\theta(\cot\theta)p\},$$

it follows that $F_0 \equiv 1$. By continuity, $F_\alpha(\theta) \to 1$ as $\alpha \to 0$ for any $\theta \in [0, \pi)$ and hence $\Psi(\alpha) \to \pi$ as $\alpha \to 0$.

To show that $\Psi(\alpha) \to 0$ as $\alpha \to \infty$, we show that $\Phi_\alpha$ must have a first zero in the interval $(0, \pi/(2\alpha))$. To this end, set $\varphi_\alpha(\theta) = \cos\alpha\theta$. Then

$$\Delta(r^\alpha\varphi_\alpha(\theta)) = (n-2)\alpha\sin\alpha\theta[\cot\alpha\theta - \cot\theta]$$

as long as $\theta \le \pi/(2\alpha)$. Since cot is decreasing on the interval $(0, \pi/2)$, we conclude that $\Delta(r^\alpha \varphi_\alpha(\theta)) \le 0$ in

$$K = \{x : 0 < r, 0 \le \theta < \pi/(2\alpha)\}.$$

If $\Phi_\alpha > 0$ in $[0, \pi/(2\alpha)]$, then our previous argument would show that $\Phi_\alpha \le \varphi_\alpha$ on $[0, \pi/(2\alpha)]$, which contradicts the assumption $\Phi_\alpha > 0$. It follows that $\Phi_\alpha$ must have its first zero in the interval $[0, \pi/(2\alpha)]$ and then the proof of Lemma 3.6 shows that $\Psi_\alpha$ is in this interval as well. Therefore $\Psi_\alpha \to 0$ as $\alpha \to \infty$. Since $\Psi_\alpha$ is a continuous decreasing function of $\alpha$, we conclude that $\Psi$ maps $(0, \pi)$ onto $(0, \infty)$.

Finally, we have

$$G(1; \theta, z, p) = z + \sup\{(n-2)\frac{d}{c}(z + \theta(\cot\theta)p)\},$$

so $F_1(\theta) = \cos\theta$ and hence $\Psi(1) = \pi/2$. □

In the case of two space dimensions, the arguments are generally easier because the ordinary differential equation is nondegenerate; however, because $\theta$ is allowed to be greater than $\pi$, some of the arguments need a little modification. Using the language of Riemann surfaces would allow us to use the same arguments as before, but we shall not do so here.

**Lemma 3.7.** *Suppose $n = 2$. Then $F_\alpha$ has a first zero, which we denote by $\Psi(\alpha)$, in $(0, \infty)$. In addition $\Psi$ is a continuous strictly decreasing function which maps $(0, \infty)$ onto $(0, \infty)$, and $\Psi(1) = \pi/2$.*

***Proof.*** To show that $F_\alpha$ has a first zero in $(0, \infty)$, we note that $F''_\alpha + \alpha^2 F \le 0$, so $F''_\alpha < 0$ wherever $F_\alpha > 0$. It follows that $F'_\alpha$ is decreasing and, because $F'_\alpha(0) = 0$, $F'_\alpha < 0$ as long as $F_\alpha > 0$. Hence $F_\alpha$ can't be positive for all $\theta > 0$.

Continuity of $\Psi$ again follows from the Inverse Function Theorem.

To show that $\Psi$ is strictly decreasing, we again let $\alpha > \beta$ and set $h = \Psi(\beta)$. If $\Psi(\alpha) \ge h$, then the proof of Lemma 3.6 implies that there is a positive constant $k$ such that $F_\beta \le kF_\alpha$ on $[0, h]$. We now write $\tilde{K}$ for the set of all $(r, \theta)$ such that $0 < r$ and $\theta \in (-h, h)$ and we set

$$\tilde{w}_1(r, \theta) = r^\alpha F_\alpha(\theta), \tilde{w}_2(r, \theta) = r^\beta F_\beta(\theta).$$

For any $\varepsilon > 0$, there is a positive number $R_\varepsilon$ such that $\varepsilon w_1 - w_2 > 0$ for $r \ge R_\varepsilon$, and we now write $\tilde{\Omega}(\varepsilon)$ for the subset of $\tilde{K}$ on which $r < R_\varepsilon$. Then $\varepsilon\tilde{w}_1 - \tilde{w}_2 \ge 0$ on $\partial\tilde{\Omega}(\varepsilon)$. If $\varepsilon\tilde{w}_1 - \tilde{w}_2$ is ever negative, then it attains a negative minimum at some point $(r_0,\theta_0) \in \tilde{\Omega}(\varepsilon)$. We now introduce coordinates

$$x^1 = r\cos\theta, \quad x^2 = r\sin\theta$$

in the neighborhood

$$\tilde{N} = \{(r,\theta) : 0 < r < R - \varepsilon, |\theta - \theta_0| < \min\{\frac{\pi}{2}, h - |\theta|\}\}$$

of $(r_0,\theta_0)$ and we abbreviate $x = (x^1, x^2)$. (The neighborhood is chosen so small that it is contained in $\tilde{\Omega}(\varepsilon)$ and so that distinct points in $N$ have distinct $x$ coordinates. Hence it is mapped to an $x$-neighborhood of the point $x_0$ corresponding to $(r_0,\theta_0)$.) If we define $w_j$ $(j = 1,2)$ by

$$w_j(x) = \hat{w}_j(r,\theta),$$

then it follows that there is a matrix $[a^{ij}] \in L_\mu$ such that $a^{ij}D_{ij}w_2 = 0$ in $N$ and that $a^{ij}D_{ij}w_1 \le 0$ in $N$. The strong maximum principle now implies that $\varepsilon w_1 - w_2$ is equal to its minimum in all of $N$ and hence $\varepsilon\tilde{w}_1 - \tilde{w}_2$ is equal to its minimum in all of $\tilde{N}$. Hence the set on which $\varepsilon\tilde{w}_1 - \tilde{w}_2$ attains its minimum is an open subset of $\tilde{\Omega}(\varepsilon)$ and this set is also closed by the continuity of $\tilde{w}_1$ and $\tilde{w}_2$. Hence $\varepsilon\tilde{w}_1 - \tilde{w}_2$ must be a negative constant in $\tilde{\Omega}(\varepsilon)$, in contradiction to the assumption that $\varepsilon\tilde{w}_1 - \tilde{w}_2$ is nonnegative on the boundary. Hence $\varepsilon\tilde{w}_1 - \tilde{w}_2 \ge 0$ in $\tilde{\Omega}(\varepsilon)$ and therefore in $\tilde{K}$. Arguing as in Lemma 3.6, we conclude that $\Psi$ is strictly decreasing.

As in Lemma 3.6, we obtain $F_0 \equiv 1$, so $\Psi(\alpha) \to \infty$ as $\alpha \to 0$. The proof that $\Psi(\alpha) \to 0$ as $\alpha \to \infty$ is similar to that in the lemma with the observation that, now, $\Phi_\alpha(\theta) = \cos\alpha\theta$.

Finally, the equation $\Psi(1) = \frac{\pi}{2}$ follows from the same argument as in Lemma 3.6. □

From these lemmata, we obtain the following result on positive solutions of linear elliptic equations.

**Theorem 3.8.** *Let $K$ be an infinite cone with opening angle $\theta_0$ and let $\mu \ge 1$. Then there is a positive constant $\alpha_0(\theta_0,\mu,n)$ with the following property: For any $\alpha \in (0,\alpha_0)$, there is a function $w_\alpha \in C^2(K) \cap C(\overline{K})$ with $w_\alpha \ge |x|^\alpha$ and $Lw_\alpha \le 0$ in $K$ for all $L \in \mathcal{L}_\mu$. In addition, there is a constant $C(\alpha,\mu)$ such that $|Dw| \le C(\alpha,\mu)|x|^{\alpha-1}$.*

***Proof.*** Without loss of generality, the axis of the cone is the positive $x^n$-axis. Then Lemma 3.6 implies that there is a unique $\alpha_0 \in (0,\pi)$ such that $\Psi(\alpha_0,\mu) = \theta_0$. In addition, if $\alpha \in (0,\alpha_0)$, then $F_\alpha > 0$ on $[0,\theta_0]$ and $F_\alpha$ is $C^2$ on this interval. The proof is completed by taking $w = r^\alpha F_\alpha(\theta)/F_\alpha(\theta_0)$. □

It will also be useful to obtain a supersolution with a stronger condition on $Lw$.

**Corollary 3.9.** *Let $K$ be an infinite cone with opening angle $\theta_0$ and let $\mu \ge 1$. Then there is a positive constant $\alpha_0(\theta_0, \mu, n)$ with the following property: For any $\alpha \in (0, \alpha_0)$, there is a function $w_\alpha \in C^2(K) \cap C(\overline{K})$ with $w_\alpha \ge |x|^\alpha$ and $Lw_\alpha \le -|x|^{\alpha-2}$ in $K$ for all $L \in \mathcal{L}_\mu$. In addition, there is a constant $C(\alpha, \mu)$ such that $|Dw| \le C(\alpha, \mu)|x|^{\alpha-1}$.*

***Proof.*** With the same $\alpha_0$ as in Theorem 3.8, for $\varepsilon \in \mathbb{R}$, we define $H_\varepsilon$ to be the solution of the initial-value problem

$$H_\varepsilon'' + G(\alpha; \theta, H_\varepsilon, H_\varepsilon'/\theta) + \varepsilon = 0 \text{ for } \theta \in (0, \pi),$$
$$f(0) = 1, \quad f'(0) = 0.$$

Then Corollary 3.3 implies that $H_\varepsilon > 0$ on $[0, \theta_0]$ for some $\varepsilon > 0$. For this particular $\varepsilon$, we take

$$w_\alpha = \frac{1}{\min\{\varepsilon, H_\varepsilon(\theta_0)\}} r^\alpha H_\varepsilon(\theta).$$

□

## 3.3 Construction of supersolutions for Dirichlet data

In this section, we are concerned with obtaining supersolutions of the equation

$$\mathcal{M}_\mu u = -d^{\alpha-2}$$

which vanish on $\partial\Omega$ for suitable positive $\alpha$. There are several ways to obtain the supersolution, and we include one here that is simple but it requires a rather severe limitation on $\alpha$. It is relatively elementary to obtain the same result without the strong restriction on $\alpha$ provided we are only interested in obtaining a generalized supersolution (for example, a Perron supersolution), and we discuss this situation in the Notes.

**Lemma 3.10.** *Let $\omega \colon \mathbb{R}^{n-1} \to \mathbb{R}$ be a Lipschitz function with $\omega(0) = 0$ and*

$$|\omega(x') - \omega(y')| \le \omega_0 |x' - y'| \tag{3.13}$$

*for some nonnegative constant $\omega_0$. Set*

$$\Omega = \{x = (x', x^n) : x^n > \omega(x')\}, \tag{3.14}$$

*and let $\mu \geq 1$. Then, for any $\sigma \in (0, \alpha_0(\arctan(-\omega_0), \mu, n)/(n+1))$, there are a constant $C_0(\mu, \sigma, n)$ and a function $w\colon \Omega \to \mathbb{R}$ such that*

$$d^\sigma \leq w \leq C_0 d^\sigma, \tag{3.15a}$$

$$|Dw| \leq C_0 d^{\sigma-1}, \tag{3.15b}$$

$$\mathcal{M}_\mu w \leq -d^{\sigma-2} \tag{3.15c}$$

*in $\Omega$.*

***Proof.*** From Theorem 3.8 with $\alpha = (n+1)\sigma$, there are a positive constant $C_1$ and a function $w_0$ such that

$$|x|^\alpha \leq w_0 \leq C_1|x|^\alpha,$$
$$|Dw_0| \leq C_1|x|^{\alpha-1},$$
$$Lw_0 \leq -|x|^{\alpha-2}$$

in the infinite cone $K$ with vertex at the origin, axis along the positive $x^n$ axis, and opening angle $\arctan(-\omega_0)$. We now set $F = \mathbb{R}^n \setminus \Omega$, and we define $W$ on $\Omega$ by

$$W(x) = \left(\int_F w_0(x-y)^{-1/\sigma}\,dy\right)^{-\sigma}.$$

To analyze this function, we also define, for each $x \in \Omega$, the set

$$H(x) = \{y \in F : |y-x| < 2d(x)\}.$$

We also observe that, for any number $P > n$, we have

$$\int_F |x-y|^{-P}\,dy \leq \int_{\mathbb{R}^n\setminus B(x,d(x))} |x-y|^{-P}\,dy = \frac{\omega_n}{P-n} d(x)^{n-P}, \tag{3.16a}$$

$$\int_F |x-y|^{-P}\,dy \geq \int_{H(x)} |x-y|^{-P}\,dy \geq c(\omega_0, n, P)d(x)^{n-P} \tag{3.16b}$$

because $|x-y| \leq 2d(x)$ in $H(x)$ and there is a ball of radius $C(\omega_0)d(x)$ in $H(x)$. It follows that

$$\frac{c_1}{d(x)} \leq \int_F w_0(x-y)^{-1/\sigma}\,dy \leq \frac{c_2}{d(x)}$$

for positive constants $c_1$ and $c_2$, and hence

$$\frac{1}{c_2^\sigma} d(x) \leq W(x) \leq \frac{1}{c_1^\sigma} d(x). \tag{3.17}$$

Next, we write

$$D_i W^{-1/\sigma} = -\frac{1}{\sigma}\int_F w_0^{-1-1/\sigma}(x-y) D_i w_0(x-y)\,dy,$$

so

$$D_i W(x) = W^{1+1/\sigma}(x) \int_F w_0^{-1-1/\sigma}(x-y) D_i w_0(x-y)\, dy. \tag{3.18}$$

Using (3.16) with $P = 1 + 1/\sigma$ and (3.17), we conclude that

$$|DW(x)| \le c_3 d(x)^{\sigma-1}. \tag{3.19}$$

To estimate the second derivatives of $W$, we first fix a constant matrix $[a^{ij}]$ in $\mathcal{L}_\mu$, differentiate (3.18) with respect to $x^j$ and then multiply by $a^{ij}$ and sum on $i$ and $j$. It follows that

$$a^{ij} D_{ij} W(x) = W^{1+1/\sigma} \left[ W(x)^{1/\sigma} \left(1 + \frac{1}{\sigma}\right) I_1 - \left(1 + \frac{1}{\sigma}\right) I_2 + I_3 \right],$$

where

$$I_1 = \int_{F\times F} w_0(x-y)^{-1-1/\sigma} w_0(x-z)^{-1-1/\sigma} a^{ij} D_i w_0(x-y) D_j w_0(x-z)\, dy\, dz,$$
$$I_2 = \int_F w_0(x-y)^{-2-1/\sigma} a^{ij} D_i w_0(x-y) D_j w_0(x-y)\, dy,$$
$$I_3 = \int_F w_0(x-y)^{-1-1/\sigma} a^{ij} D_{ij} w_0(x-y)\, dy.$$

From Cauchy's inequality for quadratic forms, we have

$$a^{ij} D_i w_0(x-y) D_j w_0(x-z) \le \left(a^{ij} D_i w_0(x-y) D_j w_0(x-y)\right)^{1/2} \left(a^{ij} D_i w_0(x-z) D_j w_0(x-z)\right)^{1/2},$$

so

$$I_1 \le \left( \int_F w_0(x-y)^{-1-1/\sigma} \left(a^{ij} D_i w_0(x-y) D_j w_0(x-y)\right)^{1/2} dy \right)^2$$

and another application of Cauchy's inequality (this time for integrals) gives

$$\int_F w_0(z-y)^{-1-1/\sigma} \left(a^{ij} D_i w_0(x-y) D_j w_0(x-y)\right)^{1/2} dy \le I_2^{1/2} W(x)^{-1/(2\sigma)}.$$

It follows that

$$W(x)^{1/\sigma} \left(1 + \frac{1}{\sigma}\right) I_1 - \left(1 + \frac{1}{\sigma}\right) I_2 \le 0$$

and hence

$$a^{ij} D_{ij} W(x) \le W^{1+1/\sigma} I_3 \le -c_4 d(x)^{\sigma-2}. \tag{3.20}$$

The proof is completed by taking $w = W/\min\{c_1, c_4\}$ and using (3.17), (3.19) and (3.20). □

## 3.4 Construction of a supersolution for oblique derivative problems

Our next step is to obtain a supersolution for the oblique derivative problem. The basic idea of the proof is to reverse the process used to derive the Hölder gradient estimate for the oblique derivative problem: We obtain our supersolution by integrating a modified form of the supersolution from Lemma 3.10.

**Lemma 3.11.** *Under the hypotheses of Lemma* 3.10, *for each* $r > 0$ *and each* $\alpha \in (0,1)$, *there are a nonnegative function* $\hat{w}$ *defined in*

$$\Omega[r] = \{x : x^n > \omega(x'), |x| < r\}$$

*and a constant* $C(n, \alpha, \mu)$ *such that*

$$\hat{w} \le C r^{1+\alpha}, \tag{3.21a}$$

$$|D\hat{w}| \le C r^{\alpha}, \tag{3.21b}$$

$$\mathcal{M}_\mu \hat{w} \le -d^{\alpha-1} \tag{3.21c}$$

*in* $\Omega[r]$.

***Proof.*** Let

$$\sigma = \frac{1}{2}\min\{\alpha, \alpha_0(\arctan(-\omega_0), \mu, n)/(n+1)\}.$$

With $w$ from Lemma 3.10 and $\kappa = (1+\omega_0^2)^{1/2}$, we set

$$w_1(x) = \int_{x^n}^{2\kappa r} w(x', s)\, ds.$$

Since $0 \in \partial\Omega$, it follows that

$$d(x', s) \le (|x'|^2 + s^2)^{1/2} \le (1+4\kappa^2)^{1/2} r,$$

and hence

$$0 \le w_1(x) \le C_0 2\kappa (1+4\kappa^2)^{\sigma/2} r^{1+\sigma},$$

where $C_0$ is the constant from Lemma 3.10.

Next,

$$D_i w_1(x) = \begin{cases} \int_{x^n}^{2\kappa r} D_i w(x', s)\, ds & \text{if } i < n, \\ -w(x) & \text{if } i = n. \end{cases}$$

Since $d(x',s) \ge (s-x^n)/\kappa$, it follows that

$$|D'w_1| \le \int_{x^n}^{2\kappa r} |D'w|\,ds \le C_0\kappa^{1-\sigma} \int_{x^n}^{2\kappa} (s-x^n)^{\sigma-1}\,ds$$
$$= C_0\kappa^{1-\sigma}\frac{(2\kappa r - x^n)^\sigma}{\sigma} \le 2C_0\frac{\kappa}{\sigma}r^\sigma.$$

Since $|D_n w_1| \le C_0 r^\sigma$, we conclude that

$$|Dw_1| \le C_0\frac{2\kappa+\sigma}{\sigma}r^\sigma.$$

To estimate $a^{ij}D_{ij}w_1$ for $[a^{ij}] \in \mathcal{L}_\mu$, we first rewrite

$$D_n w_1(x) = -w(x',x^n) = -w(x',2\kappa r) + \int_{x^n}^{2\kappa r} D_n w(x',s)\,ds,$$

so, for $i = 1,\dots,n$, we have

$$D_i w_1(x) = -\delta_{in} w(x',2\kappa r) + \int_{x^n}^{2\kappa r} D_i w(x',s)\,ds.$$

We therefore have that

$$D_{ij}w_1(x) = \int_{x^n}^{2\kappa} D_{ij}w(x',s)\,ds - \begin{cases} D_j w(x',2\kappa r) & \text{if } i=n, j<n, \\ D_i w(x',2\kappa r) & \text{if } j=n, \\ 0 & \text{otherwise,} \end{cases}$$

so

$$a^{ij}D_{ij}w_1(x) = \int_{x^n}^{2\kappa r} a^{ij}D_{ij}w(x',s)\,ds - 2a^{in}D_i w(x',2\kappa r).$$

Then

$$\int_{x^n}^{2\kappa r} a^{ij}D_{ij}w(x',s)\,ds \le -\int_{x^n}^{2\kappa r} d(x',s)^{\sigma-2}\,ds.$$

Because $d(x',s) \le d(x) + s - x^n$, it follows that

$$\int_{x^n}^{2\kappa r} a^{ij}D_{ij}w(x',s)\,ds \le -\int_{x^n}^{2\kappa r} (d(x)+s-x^n)^{\sigma-2}\,ds$$
$$= \frac{1}{1-\sigma}(d(x)+2\kappa r - x^n)^{\sigma-1} - \frac{1}{1-\sigma}d(x)^{\sigma-1}.$$

We now observe that $2\kappa r - x^n \ge 2\kappa r - r \ge r \ge d(x)$ to conclude that

$$\int_{x^n}^{2\kappa r} a^{ij}D_{ij}w(x',s)\,ds \le \frac{2^{\sigma-1}-1}{1-\sigma}d^{\sigma-1} \le -\frac{1}{2}d^{\sigma-1}.$$

In addition, it's easy to check that the graph of $\omega$ does not intersect the ball $B((x', 2\kappa r), r)$, so $d(x', 2\kappa r) \geq r$ and hence there is a constant $c_1$, determined only by $n$, $\alpha$, and $\mu$, such that

$$-2a^{in}D_i w(x', 2\kappa r) \leq c_1 r^{\sigma-1}.$$

Combining these estimates with our expression for $a^{ij}D_{ij}w_1$, we find that

$$\mathcal{M}_\mu w_1 \leq -\frac{1}{2}d^{\sigma-1} + c_1 r^{\sigma-1}.$$

The proof is completed by taking

$$\hat{w} = 2r^{\alpha-\sigma}w_1 + c_1 r^{\alpha-1}(r^2 - |x|^2)$$

and noting that $r^{\alpha-\sigma}d^{\sigma-1} \geq d^{\alpha-1}$ because $\sigma \leq \alpha$. □

It is interesting to observe that this estimate can be proved for any $\alpha < 1$ even though $\sigma$ in the previous supersolution construction had to be much smaller than 1.

From this construction, we can prove a weak Harnack inequality when $Lu$ behaves like $d^{\alpha-1}$ near $\Sigma[r]$. More precisely, we have the following estimate.

**Theorem 3.12.** *Let $R > 0$, and let $\omega\colon \mathbb{B}_{n-1}(R) \to \mathbb{R}$ be a Lipschitz function with $\omega(0) = 0$ and satisfying* (3.13) *(for all $x$ and $y$ in $B_{n-1}(R)$) for some nonnegative constant $\omega_0$. Set*

$$\Omega[R] = \{x = (x', x^n) : |x| < R,\ x^n > \omega(x')\}, \tag{3.22a}$$

$$\Sigma[R] = \{x = (x', x^n) : |x| < R,\ x^n = \omega(x')\} \tag{3.22b}$$

*and let $\mu \geq 1$. Suppose that there are positive constants $\lambda$, $\mu_1$, and $\mu_2$ such that*

$$\lambda|\xi|^2 \leq a^{ij}\xi_i\xi_j \leq \mu\lambda|\xi|^2 \tag{3.23a}$$

*for all $\xi \in \mathbb{R}^n$,*

$$|b| \leq \mu_1\lambda, \tag{3.23b}$$

$$c \geq -\mu_2\lambda. \tag{3.23c}$$

*Suppose also that there are positive constants $\varepsilon$ and $\mu_3$ such that*

$$\omega_0|\beta'| \leq (1-\varepsilon)\beta^n, \tag{3.24a}$$

$$\beta^0 \geq -\mu_3\beta^n. \tag{3.24b}$$

*If $u \in C^2(\Omega[R]) \cap C(\overline{\Omega[R]})$ is nonnegative and satisfies*

$$a^{ij}D_{ij}u + b^iD_iu + cu \le \lambda(Fd^{\alpha-1}) \text{ in } \Omega[R], \tag{3.25a}$$

$$\beta^iD_iu + \beta^0u \le G\beta^n \tag{3.25b}$$

*for some nonnegative constants $F$, $G$ and $\alpha \in (0,1)$, then there are positive constants $p$ and $C$, determined only by $\varepsilon$, $\mu$, $\mu_1R$, $\mu_2R^2$, $\mu_3R$, $\omega_0$, and $\alpha$ such that*

$$\left(R^{-n}\int_{\Omega[R/2]} u^p\,dx\right)^{1/p} \le C(\inf_{\Omega[R/2]} u + FR^{1+\alpha} + GR). \tag{3.26}$$

***Proof.*** With $\hat{w}$ from Lemma 3.11, we set $\hat{u} = u + F_2\hat{w}$. With $L$ defined by

$$Lh = a^{ij}D_{ij}h + b^iD_ih + cu$$

and $M$ defined by

$$Mh = \beta^iD_ih + \beta^0h,$$

we see that

$$L\hat{u} \le \lambda F_1, \quad M\hat{u} \le \beta^n(G + CFR^\alpha)$$

for some $C$ determined only by $\alpha$, $\mu$, $\mu_1R$, $\mu_2R^2$, and $\mu_3R$. It then follows from the weak Harnack inequality Theorem 1.20 that

$$\left(R^{-n}\int_{\Omega[R/2]} \hat{u}^p\,dx\right)^{1/p} \le C(\inf_{\Omega[R/2]} \hat{u} + FR^{1+\alpha} + GR),$$

and (3.26) follows from this inequality because $0 \le \hat{w} \le CR^{1+\alpha}$. □

Note that conditions (3.23) are the same as (1.20) from Lemma 1.20, and conditions (3.24) are the same as (1.3c) and (1.36).

The argument of Theorem 1.26 gives the corresponding Hölder estimate.

**Corollary 3.13.** *Let $\omega$, $L$, and $M$ be as in Theorem 3.12. Then there are positive constants $C$ and $\zeta$ determined only by such that*

$$\operatorname*{osc}_{\Omega[r]} u \le C\frac{r^\zeta}{R^\zeta}(\sup_{\Omega[R]} u + FR^{\alpha+1} + GR). \tag{3.27}$$

## 3.5 The strong maximum principle, revisited

In the next chapter, we study problems with $a^{ij} \in \mathcal{M}_\mu$ with $b$ and $c$ possibly growing near $\partial\Omega$ like $d^{\alpha-1}$ for some $\alpha \in (0,1)$. It is possible to use the function from Lemma 3.11 to extend the theory developed in Chapter 1 for equations with bounded coefficients to one for equations with coefficients of this type, but we shall not do so here, referring the reader to [115] for a details. Instead, we just derive a strong maximum principle along the lines of Corollary 1.22. For the reader's convenience, we recall the notation from Chapter 1.

First, let $\Omega$ be an open subset of $\mathbb{R}^n$, and suppose that there are positive constants $R$ and $\omega_0$ such that

$$\{x : |x| < R,\ x^n > \omega_0|x'|\} \subset \Omega.$$

Let $\varepsilon \in (0,1)$ and suppose that $\beta$ is a nonzero vector field defined on $\partial\Omega \cap B(0,R)$ such that

$$\omega_0|\beta'| \le (1-\varepsilon)\beta^n.$$

With $\alpha_1 \in (0,1)$ to be chosen and $A = 2^{1+2\varepsilon}\varepsilon^{-4\varepsilon}$, we define the function $G$ on $\mathbb{R}^n \times (0,\infty)$ by

$$G(y,r) = \left(\frac{|y'|^2}{r^2} + \alpha_1^2\right)^{(1+\varepsilon)/2} + \frac{(y^n)^2}{(A\omega_0 r)^2},$$

and for $x_1 \in \mathbb{R}^n$ and $r > 0$, we write $E(x_1,r)$ for the set of all $x \in \mathbb{R}^n$ with $G(x-x_1,r) < 1$. We also abbreviate $\Omega\cap B(0,R)$ to $\Omega[R]$ and $\partial\Omega \cap B(0,R)$ to $\partial\Omega[R]$.

A major step is the following analog of Lemmata 1.13 and 1.18.

**Lemma 3.14.** *With $\Omega$, $\beta$, $\omega_0$, $\varepsilon$, and $R$ as above, suppose $a^{ij} \in \mathcal{M}_\mu(\Omega[R])$ for some $\mu \ge 1$, and suppose that $b$ and $c$ are defined on $\Omega[R]$ with*

$$|b| \le \mu_1 d^{\alpha-1}, \quad c \ge -\mu_2 d^{\alpha-1} \tag{3.28}$$

*for some nonnegative constants $\mu_1$ and $\mu_2$. Let $\beta^0$ be defined on $\partial\Omega[R]$ and suppose that there is a nonnegative constant $\mu_3$ such that $\beta^0 \ge -\mu_3\beta^n$. Finally, let $\alpha_1 \in (0,1)$ and $r$ be positive constants and let $x_1 = (0, x_1^n)$ be a point in $\mathbb{R}^n$ with*

$$x_1^n \ge \left(A - \frac{\varepsilon^6}{4}\right)\omega_0 r,$$

*$E(x_1,r) \subset B(0,R)$, and $E(x_1,r) \cap \partial\Omega[R] \ne \emptyset$. Then, for any constants $\theta_1 < \theta_2$ and $\alpha$ in $(0,1)$, there are positive constants $C_1$, determined only*

*by $\alpha$, $\alpha_1$, $n$, $\varepsilon$, $\omega_0$, and $\mu_3 R$, and $r_0$, determined only by $\alpha$, $\omega_0$, $n$, $\varepsilon$, $\theta_1$, $\theta_2$, $\mu_1$, $\mu_2$, and $\mu_3 R$ such that, if $r \le r_0$ and $u \in C^2(\Omega[R]) \cap C(\overline{\Omega[R]})$ is nonnegative with*

$$Lu \le 0 \text{ in } \Omega[R], \quad Mu \le 0,$$

*and if $h$ is a nonnegative constant such that*

$$u \ge h \text{ in } E(x_1, \theta_1 r) \cap \Omega, \tag{3.29}$$

*then*

$$C_1 u \ge h \text{ in } E(x_1, \theta_2 r) \cap \Omega. \tag{3.30}$$

***Proof.*** For simplicity, we write $E(s)$ for $E(x_1, s) \cap \Omega$ and $E^0(s)$ for $E(x_1, s) \cap \partial\Omega$. As before, we may assume that $c \le 0$ and $\beta^0 \le 0$.

With $\eta$ defined by $\eta(x) = 1 - G(x - x_1, r)$, we conclude (as in Lemma 1.18) that there is a constant $q \ge 2$ determined only by $\mu$, $n$, $\theta_1$, and $\mu_3 R$ such that

$$L(\eta^q) \ge -q\eta^{q-1}\mu_1 d^{\alpha-1}|D\eta| - \mu_2 \eta^q d^{\alpha-1}$$

in $E(r)\setminus E(\theta_1 r)$ and $M(\eta^q) \ge 0$ on $E^0(r)$. It follows that there is a constant $k_1$, determined only by $\omega_0$ and $\varepsilon$ so that

$$L(\eta^q) \ge -k_1 q \left(\frac{\mu_1}{r} + \mu_2\right) d^{\alpha-1}.$$

Next, we note that there is a constant $k_2$, determined only by $\omega_0$ and $\varepsilon$, such that $E(r) \subset B(0, k_2 r)$. With $\hat{w}$ taken from Lemma 3.11 corresponding to $R = k_2 r$ and $K$ a positive constant to be determined, we now set

$$w_2 = \hat{w} + K r^\alpha (k_2 r - x^n).$$

It follows that there is a constant $k_3$, determined only by $\alpha$, $\mu$, $\omega_0$, and $\varepsilon$, such that

$$Lw_2 \le d^{\alpha-1}[-1 + (K + k_3\mu_1) r^\alpha]$$

in $E(r)$, and

$$Mw_2 \le \beta^n [k_3 - K] r^\alpha.$$

We now choose $K = k_3$ and suppose that $r_0 \le (k_3(2\mu_1 + 2))^{-1/\alpha}$ to conclude that

$$Lw_2 \le -\frac{1}{2} d^{\alpha-1} \text{ in } E(r), \quad Mw_2 \le 0 \text{ on } E^0(r).$$

We now apply the maximum principle to

$$h\eta^q - u - k_1 q \left(\frac{\mu_1}{r} + \mu_2\right) h w_2$$

in $E(r) \setminus E(\theta_1 r)$ to infer that

$$h\eta^q \le u + hk_1k_4(\mu_1 r^\alpha + \mu_2 r^{\alpha+1})$$

for some further constant $k_4$, determined only by $\alpha$, $\mu$, $n$, and $k_3$. We now choose $r_0$ sufficiently small that

$$k_1k_4(\mu_1 r_0^\alpha + \mu_2 r^{\alpha+1}) \le \frac{1}{2}\eta_0,$$

where $\eta_0$ is the (nonzero) minimum of $\eta^q$ on $E(\theta_2 r)$. Then (3.30) holds with $C_1 = \eta_0/2$. $\square$

Just as in Corollary 1.22, we have a strong maximum principle for oblique derivative problems with appropriate unbounded coefficients as well.

**Corollary 3.15.** *Let $u \in C^2(\Omega[R]) \cap C(\overline{\Omega}[R])$ and suppose $u$ has a nonnegative maximum value at 0. Suppose also that the hypotheses of Lemma 3.14 are satisfied for $L$, $M$, and $\Omega$. If $c$ and $\beta^0$ are nonpositive and if*

$$Lu \ge 0 \text{ in } \Omega[R], \quad Mu \ge 0 \text{ on } \partial\Omega[R],$$

*then $u$ is constant in $\Omega[R]$.*

## 3.6 A Miller barrier for mixed boundary value problems

In order to study the oblique derivative problem in a domain with an edge when the vector $\beta$ changes discontinuously across the edge, we need to examine the behavior of a mixed boundary value problem. Although this problem is interesting in its own right, we shall confine ourselves to the points that are applicable to the oblique derivative problem.

Our first step is a variant of the Miller barrier in two dimensions. The notation will be slightly modified from what was used earlier in the chapter to conform with standard polar coordinates. In other words, we define $r$ and $\theta$ by

$$x^1 = r\cos\theta, \quad x^2 = r\sin\theta.$$

A simple calculation shows that, for $w = r^\alpha f(\theta)$, we have

$$D_1 w = r^{\alpha-1}[\alpha\cos\theta f(\theta) - \sin\theta f'(\theta)],$$
$$D_2 w = r^{\alpha-1}[\alpha\sin\theta f(\theta) + \cos\theta f'(\theta)].$$

It follows that, on the line $\theta = 0$, we have

$$\beta \cdot Dw = r^{\alpha-1}[\beta^1(r,0)\alpha f(0) + \beta^2(r,0)f'(0)].$$

Our polar representation is essentially the same as before: If $[a^{ij}] \in \mathcal{L}_\mu$ for some $\mu \geq 1$, then there are scalar functions $a$, $b$, and $c$ satisfying (3.10a) such that

$$a^{ij}D_{ij}w = r^{\alpha-2}(a\alpha(\alpha-1)f + 2b(\alpha-1)f' + c(f'' + \alpha f)).$$

Conversely, given $a$, $b$, and $c$ satisfying (3.10a), there is a matrix $[a^{ij}] \in \mathcal{L}_\mu$ such that this representation is valid.

We now define the functions $g$ and $G$ by

$$g(\alpha, a, b, c; z, p) = \frac{a}{c}\alpha(\alpha-1)z + 2\frac{b}{c}(\alpha-1)p + \alpha z,$$
$$G(\alpha; z, p) = \sup\{g(\alpha, a, b, c; z, p) : a, b, c \text{ satisfy (3.10a)}\}.$$

Assuming that $\beta^1 \leq \mu_0\beta^2$ for some constant $\mu_0$, we then focus on the initial value problem

$$F_\alpha'' + G(\alpha; F_\alpha, F_\alpha') = 0 \text{ on } (0, \infty),$$
$$F_\alpha(0) = 1,\ F_\alpha'(0) = -\mu_0\alpha.$$

As before, if $F_\alpha$ is positive on the interval $(0, \theta_0)$, then $w$ satisfies the conditions

$$\mathcal{M}_\mu w \leq 0 \text{ if } 0 < \theta < \theta_0,$$
$$\beta \cdot Dw \leq 0 \text{ if } \theta = 0,$$
$$w \geq 0 \text{ if } \theta = \theta_0,$$

so we are again interested in determining $\Psi(\alpha)$, the first zero of $F_\alpha$, as a function of $\alpha$.

The proof of Lemma 3.7 shows immediately that $\Psi$ is continuous and strictly decreasing and that $\Psi$ maps $(0, \infty)$ onto itself. Moreover, because $G(1; z, p) = z$, we see that $F_1(\theta) = \cos\theta - \mu_0 \sin\theta$ and hence $\Psi(1) = \operatorname{arccot}\mu_0$. (As before, $\Psi(1)$ is independent of $\mu$.)

We therefore obtain (just as for Corollary 3.9) the following supersolution result.

**Proposition 3.16.** *Let $K$ be the sector $0 < \theta < \theta_0$ for some $\theta_0 > 0$, and let $\mu \geq 1$ and $\mu_0$ be constants. Then there is a positive constant $\alpha_1(\theta_0, n, \mu, \mu_0)$ with the following properties: For any $\alpha \in (0, \alpha_1)$, there is a function $w_\alpha \in C^2(K) \cap C(\overline{K})$ with $w_\alpha \geq |x|^\alpha$ and $Lw_\alpha \leq -|x|^{\alpha-2}$ in $K$ for all $L \in \mathcal{L}_\mu$. In addition, $\beta \cdot Dw_\alpha \leq -|x|^{\alpha-1}$ if $\theta = 0$ for all vectors $\beta$ satisfying $\beta^1 \leq \mu_0\beta^2$ and $\beta^2 > 0$. Finally, there is a constant $C(\alpha, \mu, \mu_0)$ such that $|Dw| \leq C|x|^{\alpha-1}$ in $K$.*

Our study of nonlinear problems will be based on a slight variation of the barrier constructed in this proposition. We begin by recalling the geometric setting from Section 2.6. With $\kappa \in (0,1)$ fixed, we define the spherical cap $\Gamma(R)$ for $R > 0$ by

$$\Gamma(R) = \{x \in \mathbb{R}^n : (x')^2 + (x^n + \kappa R)^2 < R^2,\ x^n > 0\}$$

and we write $\Gamma^*(R)$ and $\Gamma^0(R)$ for the subsets of $\partial\Gamma(R)$ on which $(x')^2 + (x^n + \kappa R)^2 = R^2$ and $(x')^2 + (x^n + \kappa R)^2 < R^2$, respectively. We also set

$$\Gamma^{**}(R) = \Gamma^*(R) \cap \overline{\Gamma^0(R)}$$

and we write $d^{**}(x)$ for the distance from $x$ to $\Gamma^{**}(R)$. Our barrier result takes the following form.

**Proposition 3.17.** *Let $\mu \geq 1$ and $\mu_0 \geq 0$ be constants, and set*

$$\alpha_1 = \alpha_1(\arctan \kappa, n, \mu_0, \mu).$$

*Then, for any $\alpha \in (0, \alpha_1)$, there are a positive constant $\varepsilon(n)$ and a function $\tilde{w}_\alpha \in C^2(\Gamma(R)) \cap C(\overline{\Gamma(R)})$ such that*

$$\tilde{w}_\alpha \geq (d^{**})^\alpha, \quad L\tilde{w}_\alpha \leq -(d^{**})^{\alpha-2} \tag{3.31a}$$

*on the subset of $\Gamma(R)$ where $d^{**} < \varepsilon(n)R$,*

$$\beta \cdot D\tilde{w}_\alpha \leq -(d^{**})^{\alpha-1} \tag{3.31b}$$

*on the subset of $\Gamma^0(R)$ where $d^{**} < \varepsilon(n)R$. Moreover, there is a constant $C_1(\alpha, \mu, \mu_0, n)$ such that*

$$\tilde{w}_\alpha \leq C_1(d^{**})^\alpha, \quad |D\tilde{w}_\alpha| \leq C_1(d^{**})^{\alpha-1}$$

*on the subset of $\Gamma(R)$ where $d^{**} < \varepsilon(n)R$.*

***Proof.*** First, we note that we can cover $\Gamma^{**}(R)$ by a finite number of open sets $U_1, \ldots, U_N$ such that, for each $k \in \{1, \ldots, N\}$, there is a $C^2$ function $\Phi_k \colon U_k \to \mathbb{R}^n$ which has a $C^2$ inverse (defined on $\Phi_k(U_k)$) such that $x^n = x^{n-1} = 0$ for all $x \in \Phi_k(U_k \cap \Gamma^{**}(R))$, $x^n = 0$ for all $x \in \Phi_k(U_k \cap \Gamma^0(R))$, $x^n = \kappa x^{n-1}$ for all $x \in \Phi(U_k \cap \Gamma^*(R))$, and $0 < x^n < \kappa x^{n-1}$ for all $x \in \Phi_k(U_k \cap \Gamma(R))$. Moreover, we can arrange that the number of sets $N$ and an upper bound for $C^2$ norms of all $\Phi_k$s and their inverses are completely determined by $n$.

With $\tilde{c}(n)$ a constant at our disposal, we now set

$$W_k = \tilde{c}(n)\Phi_k^{-1}(w_\alpha),$$

where $w_\alpha$ is the function from Proposition 3.16. By choosing $\tilde{c}(n)$ appropriately, we can arrange that

$$W_k \geq (d^{**})^\alpha, \quad Lw_k \leq -(d^{**})^{\alpha-1}$$

in $U_k \cap \Gamma(R)$, and

$$\beta \cdot DW_k \leq -(d^{**})^{\alpha-1}$$

on $U_k \cap \Gamma^0(R)$. Moreover there is a constant $C$, determined by the same quantities as for $C_1$ such that

$$W_k \leq C(d^{**})^\alpha, \quad |DW_k| \leq C(d^{**})^{\alpha-1}$$

in $U_k \cap \Gamma(R)$.

If $(\eta_k)$ is a partition of unity on $\Gamma^{**}(R)$ subordinate to the cover $(U_k)$, it follows that

$$\tilde{w}_\alpha = \sum_{k=1}^{N} \eta_k w_k$$

satisfies the conditions

$$\tilde{w}_\alpha \geq (d^{**})^\alpha,$$
$$L\tilde{w}_\alpha \leq (d^{**})^{\alpha-2}\left[-2 + C_2\frac{d^{**}}{R} + C_2\left(\frac{d^{**}}{R}\right)^2\right]$$

wherever $\sum \eta_k = 1$ in $\Gamma(R)$ and

$$\beta \cdot D\tilde{w}_\alpha \leq (d^{**})^{\alpha-1}\left[-2 + C_2\frac{d^{**}}{R}\right]$$

wherever $\sum \eta_k = 1$ on $\Gamma^0(R)$ for some constant $C_2$. By choosing $\varepsilon(n)$ sufficiently small, we infer that $\sum \eta_k = 1$. Further decreasing $\varepsilon$ gives the desired result. □

## Notes

The basic material in Section 3.1 on ordinary differential equations is standard and can be found in any text on the subject; the first proof of the result in this form is in [123]. Lemma 3.2 and Corollary 3.3 are simplified versions of the results in [134]. The polar representation Lemma 3.5 is proved in [133] and the barrier construction comes from [135] (see Section 2 of that work for the case $n \geq 3$ and Section 3 for $n = 2$). Our proofs of these results differ from Miller's only in some minor technical details. We mention here that Oddson [149] derived an explicit formula for $\alpha_0$ in case $n = 2$.

The construction of a supersolution for the Dirichlet problem (Lemma 3.10) is a simplified version of the construction in Theorem 3.7 of [130]. In fact, the existence of a viscosity supersolution is fairly straightforward, given the theory of viscosity solutions, and we shall present that theorem in Chapter 7. The existence of a Perron supersolution was first presented in Theorem 4.2 of [118]; that theorem actually shows that there is a classical supersolution by first showing that a Perron supersolution exists and then using the standard theory of fully nonlinear equations, some of which we present in Chapter 12. The construction of a supersolution for the oblique derivative problem with singular coefficients comes from [115]. Our Lemma 3.11 is essentially Lemma 2.1 of [115], and our Theorem 3.12 and Corollary 3.13 are essentially Theorems 4.2 and 4.3 from [115]. In fact the results in [115] allow more general unbounded lower coefficients; we present a simplified version of some of these results in the Exercises 3.1, 3.2, and 3.3.

The material in Section 3.6 is taken from [110] although, as is emphasized here, it is only marginally different from Miller's work.

We point out that most of the results in this chapter can be proved via the arguments in [133] if we are not interested in obtaining the optimal Hölder exponent of our function $w$. Our main interest in this sharp result is to show that the Hölder exponent depends on simple, sharp geometric quantities, and that the exponent can be made arbitrarily close to 1 by taking the opening angle of the exterior cone sufficiently close to $\pi/2$.

## Exercises

**3.1** Let $B_1$ be a positive constant. Adapt the proof of Lemma 3.11 to construct a function $\hat{w}_1$ such that

$$\hat{w}_1 \le Cr^{1+\alpha},$$
$$|D\hat{w}_1| \le Cr^{\alpha},$$
$$\mathcal{M}_\mu \hat{w}_1 + B_1\lambda|D\hat{w}_1| \le -d^{\alpha-1}$$

in $\Omega[r]$.

**3.2** Prove the corresponding maximum principle to Theorem 1.9 with condition (1.16) replaced by the condition $|b| \le \lambda B_1 d^{\alpha-1}$. Specifically, suppose that the eigenvalues of the $[a^{ij}]$ are in the range $[\lambda, \mu\lambda]$ for some positive function $\lambda$ and a positive constant $\mu$, let $\beta$ be an inward pointing vector and suppose that there are nonnegative constants $B_0$ and $B_1$ such that

$$|\beta| \le -B_0\beta^0 \text{ on } \partial\Omega,$$
$$|b| \le \lambda B_1(\operatorname{diam}\Omega)^\alpha d^{\alpha-1} \text{ in } \Omega.$$

Show that, if $u \in C^2(\Omega) \cap C(\overline{\Omega})$ satisfies

$$Lu \ge -f \text{ in } \Omega, \quad Mu \ge g\beta^0 \text{ on } \partial\Omega$$

for some nonnegative function $f$ and some nonnegative constant $g$, then

$$\sup u \le g + C(\alpha, n, B_1)(B_0 + \operatorname{diam}\Omega)(\operatorname{diam}\Omega)^\alpha \sup_\Omega f d^{1-\alpha}.$$

**3.3** Prove the weak Harnack inequality for supersolutions under hypotheses similar to those in the preceding exercise. In other words, show that Theorem 3.12 is still true if we replace conditions (3.23bc) by

$$|b| \le \mu_1\lambda d^{\alpha-1}, c \ge -\mu_2\lambda d^{\alpha-1}$$

and if $p$ and $C$ depend on $\mu_1 R^\alpha$ and $\mu_2 R^{\alpha+1}$.

**3.4** Prove that solutions of $Lu = f$ in $\Omega$, $Mu = g$ on $\partial\Omega$ are Hölder continuous under the hypotheses in Exercise 3.3.

# Chapter 4

# Hölder Estimates for First and Second Derivatives

## Introduction

This chapter provides Hölder estimates for the first and second derivatives of solutions of linear oblique derivative problems. As we shall see, several different methods are used to cover various situations. In particular, we show that solutions of linear oblique derivative problems are typically smoother than the domains in which they are posed. For example, solutions in Lipschitz domains are $C^{1,\alpha}$ (for some $\alpha > 0$ determined by the domain and the differential equation but not the boundary condition) while solutions in $C^{1,\alpha}$ domains are in $C^{2,\alpha}$ (with $\alpha \in (0,1)$ arbitrary).

Sections 4.1 and 4.5 give $C^{1,\alpha}$ estimates for linear oblique derivative problems in two special circumstances. The $C^{2,\alpha}$ estimates are given in Section 4.7 as a straightforward corollary of our $C^{1,\alpha}$ estimates.

## 4.1 $C^{1,\alpha}$ estimates for continuous $\beta$

We start with a model problem which will be used to study the general problem. The most important features of the model problem are that the differential equation and boundary condition have constant coefficients (and no lower order terms) and that we do not make any change of variables, so our estimates are derived in the original domain $\Omega$. To avoid later complications, we fix $R_0 > 0$ and a Lipschitz function $\omega$ (defined only for $|x'| < R_0$) with Lipschitz constant no greater than a positive constant $\omega_0$. For $|x_0| < R_0$ and $R \in (0, R_0 - |x_0|)$, we introduce the sets

$$\Omega[x_0, R] = \{x = (x', x^n) : |x - x_0| < R,\ x^n > \omega(x')\}, \tag{4.1a}$$

$$\Sigma[x_0, R] = \{x = (x', x^n) : |x - x_0| < R,\ x^n = \omega(x')\}, \tag{4.1b}$$

$$\sigma[x_0, R] = \{x = (x', x^n) : |x - x_0| = R,\ x^n \geq \omega(x')\}. \tag{4.1c}$$

Our first step is to prove a Hölder gradient estimate for solutions of constant coefficient differential equations with constant coefficient oblique boundary condition on $\Sigma[x_0, R]$. This estimate naturally comes in two parts, which are most easily described in the special case that the oblique boundary condition is $D_n u = 0$ on $\Sigma[x_0, R]$. In this case, we first obtain a Hölder estimate on $D_n u$ by observing that it satisfies the same differential equation as $u$ and that it satisfies a Dirichlet boundary condition on $\Sigma[x_0, R]$; then we obtain a Hölder estimate on the derivatives $D_1 u, \dots D_{n-1} u$. The exact formulation of our estimates is actually somewhat more subtle, as we shall see, but this description is the motivation for our estimates. The Hölder estimate for $D_n u$ (or more generally $\beta \cdot Du$) is proved exactly as described and it requires no restriction on the direction of the vector $\beta$.

**Lemma 4.1.** *Let $u \in C^2(\Omega[x_0, R]) \cap C^1(\Omega[x_0, R] \cup \Sigma[x_0, R])$ be a solution of*

$$A^{ij} D_{ij} u = 0 \text{ in } \Omega[x_0, R], \tag{4.2a}$$

$$\beta_0^i D_i u = 0 \text{ on } \Sigma[x_0, R] \tag{4.2b}$$

*for some positive definite (constant) matrix $[A^{ij}]$ and some nonzero (constant) vector $\beta_0$. Suppose also that $|\beta_0^i D_i u|$ is bounded in $\Omega[x_0, R]$ and that there are positive constants $\lambda$ and $\mu$ such that*

$$\lambda |\xi|^2 \leq A^{ij} \xi_i \xi_j \leq \mu \lambda |\xi|^2 \tag{4.3}$$

*for all $\xi \in \mathbb{R}^n$. Suppose also that there is a cone $K$ with opening $\theta_0$ with vertex at $x_0$ which contains $\Omega[x_0, R]$. If $\alpha \in (0, \alpha_0(\theta_0, \mu, n))$ (the constant from Theorem 3.8), then there is a constant $C(n, \alpha, \theta_0, \mu)$ such that*

$$|\beta_0^i D_i u(x)| \leq C \sup |\beta_0^i D_i u| \frac{|x - x_0|^\alpha}{R^\alpha}. \tag{4.4}$$

***Proof.*** From Lemma 2.1, we know that $u \in C^3(\Omega[x_0, R])$, so $v = \beta_0^i D_i u \in C^2(\Omega[x_0, R]) \cap C^0(\Omega[x_0, R] \cup \Sigma[x_0, R])$. For simplicity of notation we set $H = \sup |\beta_0^i D_i u|$. If $w$ is the function from Theorem 3.8 and if we write $K'$ for the intersection of $K$ and $\Omega[x_0, R]$, then we see that $L(Hw \pm v) \leq 0$ in $K'$ and $Hw \pm v \geq 0$ on $\partial K'$. It follows from the maximum principle that $Hw \pm v \geq 0$ in $K'$ and this inequality, in conjunction with the upper bound for $w$ in Theorem 3.8, immediately yields (4.4). □

The remainder of the Hölder gradient estimate is much more delicate. To begin with, we need to make some quantitative hypotheses on the vector $\beta_0$ which measure its obliqueness. Furthermore (as will be apparent from the proof), we do not estimate the Hölder constant of the gradient directly. Instead, we obtain a pointwise estimate on $u$ which relates to the results in Section 2.2. To state our condition on the Hölder exponent more succinctly in its statement, we define the *Miller angle* for $\Omega[x_0, R]$ to be the largest number $\theta$ such that, for each $x \in \Sigma[x_0, R]$, there is a cone $K$ with vertex $x$ and opening angle $\theta$ which is exterior to $\Omega[x_0, R]$. We also use $\alpha_0$ to denote the function from Theorem 3.8.

**Lemma 4.2.** *Suppose, in addition to the hypotheses of Lemma 4.1, that there are positive constants $\chi$ and $\varepsilon$ with $\varepsilon < 1$ such that*

$$|\beta_0'| \le \frac{1-\varepsilon}{\omega_0}\beta_0^n, \quad \beta_0^n \ge \chi. \tag{4.5}$$

*If*

$$H_0 = \sup_{\Omega[x_0,R]} \left|\frac{\beta_0^i D_i u}{\chi}\right|,$$

*then there are constants $C_0(n, \alpha, \alpha_0(\theta_0, n, \mu), \varepsilon, \omega_0) \ge 0$, where $\theta_0$ is the Miller angle for $\Omega[x_0, R]$, and $\kappa(\varepsilon, \omega_0) \in (0, 1)$ such that, for any $r \in (0, R)$, there is a linear function $P_1(x_0, r)$ with*

$$\beta_0^i D_i P_1(x_0, r) = 0, \tag{4.6a}$$

$$\sup_{\Omega[x_0,\tau r]} |u - P_1(x_0, r)| \le C_0(\tau^2 \sup_{\Omega[x_0,r]} |u| + H_0\frac{r^{1+\alpha}}{R^\alpha}) \tag{4.6b}$$

*for all $\tau \in (0, \kappa)$.*

***Proof.*** We set $\xi = \beta_0/|\beta_0|$ and

$$\bar{x}_0 = x_0 + \frac{1}{2}r\xi.$$

Then there is a positive constant $\kappa(\varepsilon, \omega_0) \in (0, 1)$ such that

$$B(\bar{x}_0, \kappa r) \subset \Omega[x_0, r].$$

With $\sigma \in (0, 1)$ to be further specified, we consider $x_1 \in \Omega[x_0, \sigma\kappa r]$ and set

$$\bar{x}_1 = x_1 + \frac{1}{2}r\xi.$$

We also set

$$p_0 = Du(\bar{x}_0) - \xi \cdot Du(\bar{x}_0)\xi,$$

and we define

$$P_1(x_0,r)(x) = p_0 \cdot (x - x_0) + u(x_0)$$

and

$$v(y) = \frac{\beta_0}{|\beta_0|} \cdot Du(y).$$

Then

$$\begin{aligned} u(x_1) - P_1(x_0,r)(x_1) = u(\bar{x}_1) - Du(\bar{x}_0)\cdot(\bar{x}_1 - \bar{x}_0) - u(\bar{x}_0) \\ + \xi \cdot Du(\bar{x}_0)\xi \cdot (x_1 - x_2) \\ + \int_0^{r/2} v(x_1 + s\xi) - v(x_0 + s\xi)\, ds. \end{aligned}$$

The second derivative estimate for $u$ then implies that

$$|u(\bar{x}_1) - Du(\bar{x}_0)\cdot(\bar{x}_1 - \bar{x}_0) - u(\bar{x}_0)| \le C(n,\mu)\sigma^2 \sup_{B(\bar{x}_0,\kappa r)} |u|$$

and Lemma 4.1 gives

$$|\xi \cdot Du(\bar{x}_0)\xi \cdot (x_1 - x_2)| \le CH_0 \frac{r^{1+\alpha}}{R^\alpha}$$

and

$$\begin{aligned} \int_0^{r/2} v(x_1 + s\xi) - v(x_0 + s\xi)\, ds &\le CH_0 \int_0^{r/2} \frac{s^\alpha}{R^\alpha}\, ds \\ &= CH_0 \frac{r^{1+\alpha}}{R^\alpha}. \end{aligned}$$

The proof is completed by choosing $\sigma = \frac{\tau}{\kappa}$, taking the supremum over all $x_1 \in \Omega[x_0, \tau r]$, and combining these three inequalities with the observation that

$$\sup_{B(\bar{x}_0,\kappa r)} |u| \le \sup_{\Omega[x_0,r]} |u|.$$

□

We next prove a Hölder gradient estimate for constant coefficient problems. Although it is possible to prove the corresponding estimate for variable coefficient problems without this intermediate result, we include it for several reasons. First, the result is of independent interest; second, the techniques involved in the variable coefficient case are more complicated versions of those for the constant coefficient case; and, third, we will use the same method to prove the variable coefficient estimate.

**Proposition 4.3.** *Under the hypotheses of Lemma 4.2, we have $u \in H_{1+\alpha}(\Omega[x_0, R/2])$ and there is a constant $C$ such that*

$$|u|_{1+\alpha;\Omega[x_0;R/2]} \le CR^{-1-\alpha}|u|_0. \tag{4.7}$$

***Proof.*** Our goal is to obtain an estimate of the form (2.19) for each $y \in \Omega[x_0, R/2] \cup \Sigma[x_0, R/2]$ and appropriate values of $r$ (as in Lemma 2.9). To this end, we consider separately the cases when $y \in \Sigma[x_0, R/2]$ and $y \in \Omega[x_0, R/2]$.

If $y \in \Sigma[x_0, R/2]$, we take $\kappa$ and $C_0$ from Lemma 4.2 and set $\tau = \min\{\kappa, (2C_0)^{1/(\alpha-1)}\}$. Our construction of $P_1$ is obtained via an iteration argument very similar to the one used in proving Proposition 2.12. With $u_0 = u$, $r = R/2$ and $R$ replaced by $R/2$, Lemma 4.2 gives a linear function $L_0$ such that

$$\sup_{\Omega[y,\tau R/2]} |u_0 - L_0| \le C_0(\tau^2 \sup_{\Omega[y,R/2]} |u_0| + H_0 R),$$

where here

$$H_0 = \sup_{\Omega[y,R/2]} \left| \frac{\beta_0^i D_i u}{\chi} \right|.$$

Repeated applications of Lemma 4.2 provide a sequence of linear functions $(L_m)$ such that the sequence of functions $(u_m)$ defined by $u_m = u_{m-1} - L_{m-1}$ satisfies the estimate

$$\sup_{\Omega[y,\tau^{m+1}R/2]} |u_{m+1}| \le C_0(\tau^2 \sup_{\Omega[x_1,\tau^m R/2]} |u_m| + H_0 R \tau^{m(1+\alpha)})$$

(with the same constant $C_0$). For $r \le R/2$, we take $k$ to be the integer such that $\tau^k R/2 < r \le \tau^{k-1} R/2$ and define

$$P_1(y,r) = \sum_{m=0}^{k} L_m.$$

It follows from Lemma 1.25 that

$$\sup_{\Omega[y,r]} |u - P_1(y,r)| \le C r^{1+\alpha}(R^{-1-\alpha}|u|_0 + H_0 R^{-\alpha}). \tag{4.8}$$

The proof of (4.8) when $y \in \Omega[x_0, R/2]$ is more subtle, but easier. First, recalling that $d$ denotes distance to $\partial\Omega$, we set $r_0 = \frac{1}{2}d(y)$ and we choose $y_1 \in \partial\Omega$ so that $|y_1 - y| = d(y)$. We also observe that $\Omega[y, r_0] \subset \Omega[y_1, 3r_0]$. It follows that

$$\begin{aligned}\sup_{\Omega[y,r_0]} |u - P_1(y_1, 3r_0)| &\le \sup_{\Omega[y_1,3r_0]} |u - P_1(y_1, 3r_0)| \\ &\le C r_0^{1+\alpha}(R^{-1-\alpha}|u|_0 + H_0 R^{-\alpha}).\end{aligned}$$

So we take $P_1(y, r_0) = P_1(y_1, 3r_0)$, and we abbreviate $L_0 = P_1(y_1, 3r_0)$. We now set $u_1 = u - L_0$ and then we define $L_1$ to be the first order

Taylor polynomial of $u_1$ centered at $y$. The second derivative estimate for $u_2 = u_1 - L_1$ then implies that

$$\sup_{\Omega[y,r]} |u_2| \le C\left(\frac{r}{r_0}\right)^{1+\alpha} \sup_{\Omega[y,r_0]} |u_1|.$$

We now take

$$P_1(y,r) = \begin{cases} L_0 + L_1 & \text{if } 0 < r \le r_0, \\ P_1(y_1, r_0) & \text{if } r_0 < r \le 3r_0, \\ P_1(y_1, r) & \text{if } r > 3r_0 \end{cases}$$

to infer (4.8) for $r \in (0, R]$, and therefore

$$|u|_{1+\alpha;\Omega[x_0;R/2]} \le C(R^{-1-\alpha}|u|_0 + R^{-\alpha}H_0).$$

To complete the proof, we need to eliminate $H_0$ from this estimate. This elimination follows from the estimate

$$H_0 \le C|u|_0/R, \tag{4.9}$$

which is an immediate consequence of Lemma 2.20. □

Corresponding estimates for variable coefficient problems with inhomogeneous differential equation and boundary condition follow the same basic scheme but several new technical details arise. Most of these details are algebraic in nature (for example, the estimate corresponding to (4.8) must take the inhomogeneity of the problem into account, leading to a more complicated right hand side) but a key point that requires more concern is related to the perturbation argument that we use here. Since the perturbation is based on an existence theorem, we need to take advantage of the rather complicated estimates from Chapter 2.

We begin with the existence theorem.

**Lemma 4.4.** *Let $x_0$, $R$, $A^{ij}$ and $\beta_0$ be as in Lemma 4.2. Then, for any $g \in C(\sigma[x_0,R])$, there is a unique solution $u \in C^2(\Omega[x_0,R]) \cap C^0(\overline{\Omega[x_0,R]})$ of*

$$A^{ij}D_{ij}u = 0 \text{ in } \Omega[x_0,R], \tag{4.10a}$$

$$\beta_0^i D_i u = 0 \text{ on } \Sigma[x_0,R], \tag{4.10b}$$

$$u = g \text{ on } \sigma[x_0,R]. \tag{4.10c}$$

***Proof.*** Let $(\omega_m)$ be a sequence of $C^{2,\alpha}$ functions which converge uniformly to $\omega$ and which satisfy

$$|\omega_m(x') - \omega_m(y')| \le \omega_0|x' - y'|$$

for all $x'$ and $y'$ in $\mathbb{R}^{n-1}$ with $|x'|, |y'| < R$. (The existence of this sequence follows from Lemma 1.4.) Without loss of generality, $\Omega_m[x_0, R]$ is nonempty for all $m$. We then obtain a solution $u_m$ to

$$\begin{aligned} A^{ij} D_{ij} u_m &= 0 \text{ in } \Omega_m[x_0, R], \\ \beta_0^i D_i u_m &= 0 \text{ on } \Sigma_m[x_0, R], \\ u_m &= g \text{ on } \sigma_m[x_0, R]. \end{aligned}$$

The maximum principle implies that the sequence $(u_m)$ is uniformly bounded and then Corollary 1.29 implies that the sequence is equicontinuous, so the Arzela-Ascoli theorem provides a uniformly convergent subsequence. Proposition 4.3 shows that $(Du_m)$ is uniformly bounded in $H_\alpha(K)$ for any compact subset $K$ of $\Omega[x_0, R] \cup \Sigma[x_0, R]$ and the interior second derivative estimates imply that $(D^2 u_m)$ is uniformly bounded in $H_\alpha(K')$ for any compact subset $K'$ of $\Omega[x_0, R]$. Hence $(Du_m)$ converges uniformly to $Du$ on compact subsets of $\Omega[x_0, R] \cup \Sigma[x_0, R]$ and $(D^2 u_m)$ converges uniformly to $D^2 u$ on compact subsets of $\Omega[x_0, R]$. It follows that $u$ is a solution of (4.10), and another application of the maximum principle shows that $u$ is unique. □

We are now ready to prove our Schauder estimate for variable coefficient equations.

**Theorem 4.5.** *Let $\theta$ be a continuous, increasing function on $[0, 1/2]$ with $\theta(0) = 0$. Let $R_0$ be a positive number and let $\omega$ be a Lipschitz function. Suppose $a^{ij}$ satisfies* (4.3) *and*

$$|a^{ij}(x) - a^{ij}(y)| \le \theta(|x - y|/d^*(x)) \tag{4.11}$$

*for all $x$ and $y$ in $\Omega[R_0]$ with $|x - y| < \frac{1}{2} d^*(x)$. Suppose also that $\beta$ satisfies* (4.5) *and*

$$|\beta|_\alpha^{(0)} \le B_0 \tag{4.12}$$

*for some $\alpha \in (0, \alpha_0(\theta_0, \mu, n))$, where $\theta_0$ is the Miller angle for $\Omega[R_0]$. Let $u \in C^2(\Omega[R_0]) \cap H_1^{(0)}(\Omega[R_0])$ with $Mu \in H_\alpha^{(1)}$, and suppose that there is a nonnegative constant $F$ such that*

$$|Lu| \le \lambda F (d^*)^{-\alpha-1} d^{\alpha-1} \tag{4.13}$$

*in $\Omega[R_0]$, where $L = a^{ij} D_{ij}$ and $M = \beta^i D_i$. Then $u \in H_{1+\alpha}^{(0)}$ and*

$$|u|_{1+\alpha}^{(0)} \le C(\theta, \mu, B_0, \omega_0, \varepsilon)(F + \left|\frac{Mu}{\chi}\right|_\alpha^{(1)} + |u|_0). \tag{4.14}$$

***Proof.*** It suffices to obtain a Hölder estimate for $Du$. To this end, we fix a point $y \in \Omega[R_0] \cup \Sigma[R_0]$ and prove a suitable estimate for the difference between $u$ and some linear polynomial. To simplify writing, we set $g = Mu/\chi$ and $G = [g]_{\alpha}^{(1)}$.

We first assume that $y \in \Sigma[R_0]$. We set $\sigma = \frac{1}{2}(\alpha + \alpha_0)$, and, with $\delta_1 \in (0, 1/2]$ to be further specified, we set $R = \delta_1 d^*(y)$. Further we define $u_0$ by

$$u_0(x) = u(x) + \frac{\beta(y)}{|\beta(y)|^2} \cdot (y - x)Mu(y),$$

and we note that

$$|\beta(x) \cdot Du_0(x)| \leq \chi G \frac{|x - y|^\alpha}{R^{1+\alpha}} \tag{4.15}$$

for $x \in \Sigma[R_0]$. Our next step is to define quantities $r_k$, $U_k$, $L_k$, $u_k$, and $v_k$ inductively as follows. Specifically, $r_k$ and $U_k$ are numbers, $u_k$ and $v_k$ are functions, and $L_k$ is a linear polynomial with $\beta(y) \cdot DL_k = 0$. First, with $\tau \in (0, 1/2)$ to be chosen, we set $r_k = \tau^k R/2$ and

$$U_k = \sup_{\Omega[y, r_k]} |u_k|,$$

we write $L_k$ for the first order Taylor polynomial of $v_k$ centered at $y$, and we set $u_{k+1} = u_k - L_k$. Assuming that we have already $u_k$, we infer from Lemma 4.4 that there is a solution $v_k$ of

$$a^{ij}(y) D_{ij} v_k = 0 \qquad \text{in } \Omega[y, r_k], \tag{4.16a}$$

$$\beta(y) \cdot Dv_k = 0 \qquad \text{on } \Sigma[y, r_k], \tag{4.16b}$$

$$v_k = u_k \qquad \text{on } \sigma[y, r_k]. \tag{4.16c}$$

This construction guarantees that $\beta(y) \cdot DL_k = 0$.

Proposition 4.3 then implies that there is a constant $C_0$ (determined only by $\omega_0$, $\sigma$, $n$, $\varepsilon$, and $\mu$) such that

$$\sup_{\Omega[y, r_{k+1}]} |v_k - L_k| \leq C_0 \tau^{1+\sigma} \sup_{\Omega[y, r_k]} |v_k|,$$

and the maximum principle implies that

$$\sup_{\Omega[y, r_k]} |v_k| \leq U_k.$$

and hence

$$\sup_{\Omega[y, r_{k+1}]} |v_k - L_k| \leq C_0 \tau^{1+\sigma} U_k. \tag{4.17}$$

To continue, we infer from Lemma 2.20 that

$$|Dv_k(y)| \le C\frac{\sup|v_k|}{r_k},$$

and hence

$$|\beta \cdot DL_k| \le C\chi \frac{U_k}{r_k}\frac{|x-y|^\alpha}{R^\alpha}.$$

It follows that

$$|Mu_k(x)| \le |Mu_{k-1}(x)| + C\delta_1^\alpha \chi \frac{U_{k-1}}{r_{k-1}}.$$

Combining this inequality with (4.15) yields

$$|Mu_k| \le C\chi(G + \delta_1^\alpha \sum_{j=0}^{k-1}\frac{U_j}{\tau^j})\frac{r_k^\alpha}{R^{1+\alpha}}$$

on $\Sigma[y, r_k]$. We now abbreviate

$$G_k = G + \delta_1^\alpha \sum_{j=0}^{k-1}\frac{U_j}{\tau^j}.$$

It follows from Corollary 3.13 that there are positive constants $C$ and $\zeta$ such that

$$|u_k(x) - u_k(x_1)| \le C(U_k + [F + G_k]\frac{r_k^{1+\alpha}}{R^{1+\alpha}})\left(\frac{|x-x_1|}{R}\right)^\zeta$$

for all $x$ and $x_1$ in $\Omega[y, r_k]$. In addition, Corollary 1.29 implies that

$$|v_k(x) - v_k(x_1)| \le C(U_k + [F + G_k]\frac{r_k^{1+\alpha}}{R^{1+\alpha}})\left(\frac{|x-x_1|}{R}\right)^\zeta$$

for all $x \in \Omega[y, r_k]$ and all $x_1 \in \sigma[y, r_k]$. It follows that, for any $\delta \in (0,1)$, we have

$$|u_k(x) - v_k(x)| \le C\delta^\zeta U_0 + C[F + G_k]\frac{r_k^{1+\alpha}}{R^{1+\alpha}}$$

if $(1-\delta)r_k < |x-y| < r_k$ (since $u_k(x_1) = v_k(x_1)$ where $x_1$ is the closest point on $\sigma[y, r_k]$ to $x$). The Hölder gradient estimate (Proposition 4.3) and the second derivative estimate (Theorem 1.34) for $v_k$ then imply that

$$|D^2 v_k| \le C(\delta) r_k^{-1-\alpha} d^{\alpha-1} U_k$$

in $\Omega[y, (1-\delta)r_k]$. It follows that

$$|Lv_k| \le C(\delta)\theta(\delta_1)\lambda r_k^{-1-\alpha} d^{\alpha-1} U_k$$

and hence

$$|L(u_k - v_k)| \le A\lambda r_k^{-1-\alpha} d^{\alpha-1}$$

in $\Omega[y,(1-\delta)r_k]$ for

$$A = C(\delta)\theta(\delta_1)U_k + F\frac{r_k^{1+\alpha}}{R^{1+\alpha}}.$$

Using the function $\hat{w}$ from Lemma 3.11 with $r_k$ in place of $r$, we infer that $w^\pm = Ar^{-1-\alpha}\hat{w} \pm (u_k - v_k)$ satisfy

$$\begin{aligned}
L(w^\pm) &\le 0 && \text{in } \Omega[y,(1-\delta)r_k],\\
M(w^\pm) &\le \chi\left[C\frac{A}{r_k} + G_k\frac{r_k^\alpha}{R^{1+\alpha}}\right] && \text{on } \Sigma[y,(1-\delta)r_k],\\
w^\pm &\ge -C\delta^\zeta U_0 - C[F+G_k]\frac{r_k^{1+\alpha}}{R^{1+\alpha}} && \text{on } \sigma[y,(1-\delta)r_k].
\end{aligned}$$

The maximum principle then implies that

$$w^\pm \ge -C\delta^\zeta U_k - C[F+G_k]\frac{r_k^{1+\alpha}}{R^{1+\alpha}} - CA$$

in $\Omega[y,(1-\delta)r_k]$, and it follows from the inequality $0 \le \hat{w} \le Cr_k^{1+\alpha}$ that

$$|u_k - v_k| \le [C(\delta)\theta(\delta_1) + C\delta^\zeta]U_0 + C(F+G_k)\frac{r_k^{1+\alpha}}{R^{1+\alpha}}$$

in $\Omega[y,(1-\delta)r_k]$, and hence this inequality is true in $\Omega[y,r_k]$. Combining this inequality with (4.17), we conclude that

$$\begin{aligned}
U_{k+1} \le [C(\delta)\theta(\delta_1) &+ C\delta^\zeta + C_0\tau^{1+\sigma}]U_k\\
&+ C(F+G_k)\frac{r_k^{1+\alpha}}{R^{1+\alpha}}.
\end{aligned}$$

We now set $\sigma' = (\alpha+\sigma)/2$ and choose $\tau$ so that $C_0\tau^{1+\sigma} \le \frac{1}{3}\tau^{1+\sigma'}$. Then we choose $\delta$ so that $C\delta^\zeta \le \frac{1}{3}\tau^{1+\sigma'}$, and finally $\delta_1$ so that $C(\delta)\theta(\delta_1) \le \frac{1}{3}\tau^{1+\sigma'}$. We then see that

$$U_{k+1} \le \tau^{1+\sigma'}U_k + C(F+G)\tau^{k(1+\alpha)} + C\delta_1^\alpha\tau^{(1+\alpha)k}\sum_{j=0}^{k-1}\frac{U_j}{\tau^j}$$

for all nonnegative integers $k$. If we set $H = C(F+G)$, this inequality takes the form

$$U_{k+1} \le \tau^{1+\sigma'}U_k + H\tau^{k(1+\alpha)} + C\delta_1^\alpha\tau^{(1+\alpha)k}\sum_{j=0}^{k-1}\frac{U_j}{\tau^j}, \tag{4.18}$$

and we now claim that

$$U_k \le \tau^{(1+\alpha)k} U_0 + 2H\tau^{(1+\alpha)(k-1)} \tag{4.19}$$

provided we take $\tau$ also small enough that $\tau^{\sigma'-\alpha} \le 1/4$ and $\delta_1$ small enough that

$$C\delta_1^\alpha \sum_{j=0}^{\infty} \tau^{\alpha j} \le \tau^{1+\alpha}/2.$$

To verify this claim, we first note that it is true for $k = 0$. Assuming that we have verified (4.19) for all $k$ less than or equal to some positive integer $m$, we infer from (4.18) that

$$\begin{aligned} U_{m+1} \le U_0 \tau^{(1+\alpha)(m+1)} [\tau^{\sigma'-\alpha} + C\delta_1^\alpha \tau^{-1-\alpha} \sum_{j=0}^{m-1} \tau^{\alpha j}] \\ + H\tau^{(1+\alpha)m} [2\tau^{\sigma'-\alpha} + 1 + C\delta_1^\alpha \tau^{-1-\alpha} \sum_{j=0}^{m-1} \tau^{\alpha j}]. \end{aligned}$$

Our smallness assumptions on $\tau$ and $\delta_1$ then imply that (4.19) is also valid for $k = m + 1$, and hence, by induction, this estimate is true for all $k$.

Rewriting (4.19) in terms of the norms of $u$ and $g$ and observing that

$$\sup_{\Omega[y,R]} |u_0| \le \sup |u| + C|g|_\alpha^{(1)}$$

now gives, for each $r \in (0, \frac{1}{2}d^*(y))$, a linear polynomial $P_1(y, r)$ such that

$$\sup_{\Omega[y,r]} |\bar{u} - P_1(y,r)| \le C\frac{r^{1+\alpha}}{R^{1+\alpha}} (\sup |u| + F + |g|_{1+\alpha}^{(1)}).$$

In particular $P_1(y, r) = 0$ if $r \ge \delta_1 d^*(y)$.

Arguing as in Proposition 4.3 gives the analogous estimate for $y \in \Omega[x_0, R]$ as well and hence we obtain (4.14). □

Once we have this local $C^{1,\alpha}$ estimate, it is relatively straightforward to obtain estimates for the second derivatives. Unfortunately, an attempt to prove local estimates leads to a consideration of weighted Hölder norms involving two weights, and this is a technical complication that we wish to avoid. For this reason, we proceed to a global estimate.

**Theorem 4.6.** *Let $\Omega$ be a bounded Lipschitz domain, and let $\beta$ be a uniformly oblique vector field defined on $\partial\Omega$ (with constants $R$, $\omega$, and $\varepsilon$). Let $[a^{ij}]$ be a matrix-valued function defined on $\Omega$ and suppose (4.3) is satisfied. Let $\alpha_0$ be the Miller exponent for $\mu$ and $\Omega$, and let $\alpha \in (0, \alpha_0)$. Suppose*

*$a^{ij} \in H_\delta^{(0)}(\Omega) \cap C(\overline{\Omega})$, $b^i \in H_\delta^{(1-\alpha)}$, $c \in H_\delta^{(1-\alpha)}$. Suppose further that $\beta$ and $\beta^0$ are in $H_\alpha(\partial\Omega)$. If $u \in H_{2+\delta}^{(-1-\alpha)}(\Omega)$ and if $\theta$ is an increasing continuous function with $\theta(0) = 0$ such that*

$$|a^{ij}(x) - a^{ij}(y)| \le \theta(|x-y|), \tag{4.20}$$

*and if $A$ and $B_0$ are nonnegative constants with*

$$|a^{ij}|_\delta^{(0)} \le \lambda A, \tag{4.21a}$$

$$|b^i|_\delta^{(1-\alpha)} \le \lambda A, \tag{4.21b}$$

$$|c|_\delta^{(1-\alpha)} \le \lambda A, \tag{4.21c}$$

$$|\beta|_\alpha \le \chi B_0, \tag{4.21d}$$

$$|\beta^0|_\alpha \le \chi B_0, \tag{4.21e}$$

*then there is a constant $C$, determined only by $A$, $B_0$, $\alpha$, $\varepsilon$, $\omega_0$, $R$, $\theta$, and $\mu$, such that*

$$|u|_{2+\delta}^{(-1-\alpha)} \le C(|Lu|_\delta^{(1-\alpha)} + |Mu|_\alpha + |u|_0). \tag{4.22}$$

***Proof.*** We first define $\hat{L}$ and $\hat{M}$ by

$$\hat{L}w = a^{ij}D_{ij}w, \quad \hat{M}w = \beta^i D_i w.$$

It follows from Theorem 4.5 (and a partition of unity argument) that

$$|u|_{1+\alpha} \le C(|\hat{L}u|_0^{(1-\alpha)} + |\hat{M}u|_\alpha + |u|_0).$$

In conjunction with Lemma 2.14, this inequality implies that

$$|u|_{2+\delta}^{(-1-\alpha)} \le C(|\hat{L}u|_\delta^{(1-\alpha)} + |\hat{M}u|_\alpha + |u|_0).$$

Since $Lu = \hat{L}u + b^i D_i u + cu$, it follows that, for any $\zeta > 0$, there is a constant $C(\zeta)$ such that

$$|Lu - \hat{L}u|_\delta^{(1-\alpha)} \le C|u|_{1+\delta}^{(0)} \le \zeta |u|_{2+\delta}^{(-1-\alpha)} + C(\zeta)|u|_0.$$

Similarly, we have

$$|Mu - \hat{M}u| \le \zeta |u|_{2+\delta}^{(-1-\alpha)} + C(\zeta)|u|_0.$$

Hence there is a constant $C_1$ such that

$$|u|_{2+\delta}^{(-1-\alpha)} \le C(|Lu|_\delta^{(1-\alpha)} + |Mu|_\alpha) + C_1\zeta|u|_{2+\delta}^{(-1-\alpha)} + C(\zeta)|u|_0.$$

The desired result follows from this inequality with $\zeta = 1/(2C_1)$. □

We now recall from Proposition 2.13 that $C^2$ solutions of $Lu = f$ are in $C^{2,\alpha}$ if $f$ and the coefficients of $L$ are in $C^\alpha$. Theorem 4.5 shows that, near a Lipschitz boundary portion $T$, $u \in C^{1,\alpha}(\Omega \cup T)$.

## 4.2 Regularized distance

To continue our study of boundary value problems in domains with boundaries that are not necesarily $C^2$, we introduce a modification of the distance function. We say that $\rho$ is a *regularized distance* for a domain $\Omega$ if $\rho \in C^2(\Omega)$ and if there are positive constants $C_1$ and $C_2$ such that $C_1\rho \le d \le C_2\rho$ in $\Omega$. There are many ways to construct such a function, and the one of most use here is a local construction.

**Lemma 4.7.** *Let $R > 0$, let $\omega \in H_1(B_{n-1}(0,R))$, and write $\Omega$ for the set of all $x \in \mathbb{R}^n$ with $x^n > \omega(x')$ and $|x| < R$. Then there is a function $\rho \in H_1(\Omega) \cap C^3(\Omega)$ such that*

$$C(\sup[\omega]_1)\rho \le d \le \frac{3}{2}\rho, \tag{4.23a}$$

$$|D\rho| \le C(n)[\omega]_1, \tag{4.23b}$$

$$|D^2\rho| \le C(n)[\omega]_1 d^{-1}. \tag{4.23c}$$

*Furthermore, for any $\varepsilon > 0$, the function $\rho$ can be taken so that*

$$D_n\rho \ge 1-\varepsilon, \quad |D'\rho| \le [\omega]_1(1+\varepsilon), \tag{4.24}$$

*in which case the constant in (4.23c) is determined also by $\varepsilon$.*

***Proof.*** First, we observe that $\omega$ can be extended to an $H_1(B_{n-1}(0,4R))$ function $\tilde{\omega}$ with

$$[\tilde{\omega}]_1 \le [\omega]_1.$$

Let $\varphi$ be a nonnegative $C^3(\mathbb{R}^{n-1})$ function with $\int_{\mathbb{R}^{n-1}} \varphi(x)\,dx = 1$ such that $\varphi(x) = 0$ if $|x| > 1$, and define

$$\bar{\omega}(x',\tau) = \int_{\mathbb{R}^{n-1}} \tilde{\omega}(x' - \frac{\tau}{L}y')\varphi(y')\,dy'$$

with $L = 2\max\{1, [\omega]_1\}$. It's easy to check that $\bar{\omega}$ is three times continuously differentiable with respect to all its arguments wherever it's defined and $\tau \ne 0$ and that $|D^2\bar{\omega}(x,\tau)| \le C[\omega]_1\tau^{-1}$. We then define $\rho(x)$ to be the unique number $\tau$ such that

$$x^n - \bar{\omega}(x',\tau) = \tau.$$

Such a number exists because

$$\left|\frac{\partial\bar{\omega}}{\partial\tau}\right| \le \frac{1}{2}.$$

We leave it to the reader to check (in Exercise 4.1) that this $\rho$ is the desired function satisfying (4.23).

If we take $L$ sufficiently large (determined only by $\varepsilon$ and $[\omega]_1$), then (4.24) is also satisfied. □

In the next section, we shall need a global version of this regularized distance which interacts well with the vector field $\beta$. Here we give the result, which follows by a simple partition of unity argument.

**Lemma 4.8.** *Let $\Omega$ be a bounded Lipschitz domain in $\mathbb{R}^n$ and suppose that $\beta$ is a continuous vector field defined in $\overline{\Omega}$ which is uniformly oblique on $\partial\Omega$. Then, for any $\alpha \in (0,1)$, there is a sequence of $H_{2+\alpha}$ domains $(\Omega_k)$ with $\Omega_k \subset \Omega$, and $\beta$ is uniformly oblique on $\Omega_k$ (with constants independent of $k$).*

***Proof.*** We first proceed locally. Fix a point $x_0 \in \partial\Omega$ and rotate axes so that $\Omega[R] = \Omega \cap B(x_0, R)$ can be represented as the set of all $x$ with $|x| < R$ and $x^n > \omega(x')$ for some Lipschitz function $\omega$ and some positive $R$. To match previous notation, we set $\omega_0 = [\omega]_1$. Since $\beta$ is uniformly oblique on $\partial\Omega$, we may assume also that there is a constant $\varepsilon \in (0,1)$ such that

$$\omega_0|\beta'| \leq (1-\varepsilon)\beta^n$$

on $\partial\Omega[R]$. From the continuity of $\beta$, it follows that there is a constant $r_1$ such that

$$\omega_0|\beta'| \leq (1-\frac{\varepsilon}{2})\beta^n$$

if $d(x) < r_1$, and hence this inequality holds on the level sets of $\rho$ close enough to $\partial\Omega[R]$; in other words there is a positive constant $r_2$ such that this inequality holds on the set $\rho = s$ if $s \in (0, r_2)$. We then take $\Omega_k[R]$ to be the level set $\rho = r_2/k$ and observe that this level set is the graph of an $H_3$ function.

The proof is completed via a straightforward partition of unity argument. □

If we assume that the function $\omega$ is more regular, then we can make stronger assertions about $\rho$. This result will be used in Section 4.4.

**Lemma 4.9.** *Let $R > 0$, let $\alpha \in (0,1]$, let $\omega \in H_{1+\alpha}(B_{n-1}(0,R))$, and write $\Omega$ for the set of all $x \in \mathbb{R}^n$ with $x^n > \omega(x')$ and $|x| < R$. Then there is a function $\rho \in H_{1+\alpha}(\Omega) \cap C^2(\Omega)$ such that*

$$C(\sup|D\omega|)\rho \leq d \leq \frac{3}{2}\rho, \tag{4.25a}$$

$$[D\rho]_\alpha \leq C(n)[D\omega]_\alpha, \tag{4.25b}$$

$$|D^2\rho| \leq C(n)[D\omega]_\alpha d^{\alpha-1}. \tag{4.25c}$$

***Proof.*** We construct $\rho$ just as in Lemma 4.7. We leave the verification of (4.25) to the reader as Exercise 4.2. □

## 4.3 Existence of solutions for continuous $\beta$

We can use the regularized distance from Lemma 4.8 to show that the problem

$$Lu = f \text{ in } \Omega, \quad Mu = g \text{ on } \partial\Omega \tag{4.26}$$

has a unique solution under the regularity assumptions of Theorem 4.6 and the additional assumptions $c \le 0$ and $\beta^0 \le 0$ but these functions are not both identically zero.

**Theorem 4.10.** *Let $\Omega$ be a bounded Lipschitz domain and let $\beta$ be a uniformly oblique vector field on $\partial\Omega$ (with constants $R$, $\omega$, and $\varepsilon$). Let $[a^{ij}]$ be a matrix-valued function defined on $\Omega$ and suppose (4.3) is satisfied. Let $\alpha_0$ be the Miller exponent for $\mu$ and $\Omega$, and let $\alpha \in (0, \alpha_0)$. Suppose $a^{ij} \in H_\delta^{(0)}(\Omega) \cap C(\overline{\Omega})$, $b^i \in H_\delta^{(1-\alpha)}$, $c \in H_\delta^{(1-\alpha)}$. Suppose further that $\beta$ and $\beta^0$ are in $H_\alpha(\partial\Omega)$. Suppose also that $c \le 0$ in $\Omega$ and $\beta^0 \le 0$ on $\partial\Omega$ but that at least one of these functions is not identically zero. Then, for all $f \in H_\delta^{(2-\alpha)}(\Omega)$ and all $g \in H_\alpha(\partial\Omega)$, there is a unique solution $u \in C^2(\Omega) \cap C(\overline{\Omega})$ of (4.26). Moreover, there is a constant $C$, determined only by $\Omega$, and the coefficients of $L$ and $M$ such that*

$$|u|_{2+\delta}^{(-1-\alpha)} \le C(|Lu|_\delta^{(1-\alpha)} + |Mu|_\alpha). \tag{4.27}$$

***Proof.*** First, let $f$ and $g$ be fixed functions, and extend $\beta$, $\beta^0$, and $g$ as $H_{1+\delta}^{(-\alpha)}$ functions so that $\beta^0 \le 0$ in $\Omega$. Let $\Omega_k$ be the sequence of domains from Lemma 4.8 and observe that there is a positive integer $k_0$ such that $c < 0$ somewhere in $\Omega_k$ for all $k \ge k_0$ if $c < 0$ somewhere in $\Omega$ and such that $\beta^0 < 0$ somewhere on $\partial\Omega_k$ if $\beta^0 < 0$ somewhere on $\partial\Omega$.

Then, for each $k \ge k_0$, the strong maximum principle implies that the only solution of

$$Lv_k = 0 \text{ in } \Omega_k, \quad Mv_k = 0 \text{ on } \partial\Omega_k$$

is $v_k = 0$, so Theorem 2.30 gives a unique solution of

$$Lu_k = f \text{ in } \Omega_k, \quad Mu_k = g \text{ on } \partial\Omega_k.$$

We now claim that the sequence $(u_k)$ is uniformly bounded. If this is not the case, then we may assume (by taking a subsequence, if necessary) that $U_k = |u_k|_0 \to \infty$. If we set $w_k = u_k/U_k$, $f_k = f/U_k$, and $g_k = g/U_k$, then $|w_k|_0 = 1$, and $|f_k|_\delta^{(1-\alpha)}$ and $|g_k|_{1+\delta}^{(-\alpha)}$ are uniformly bounded, so Theorem 4.6 and the Arzela-Ascoli theorem imply that there is a subsequence $(w_{k_m})$ which converges uniformly to an $H_{2+\delta}^{(-1-\alpha)}$ function $w$. Moreover

$$Lw = 0 \text{ in } \Omega, \quad Mw = 0 \text{ on } \partial\Omega,$$

so Corollary 3.15 implies that $w \equiv 0$, which contradicts the two conditions $|w_{k_m}|_0 = 1$ and $(w_{k_m})$ converges uniformly to $w$.

Since $(u_k)$ is bounded, there is a uniformly convergent subsequence such that its derivatives also converge uniformly on $\overline{\Omega}$ and its second derivatives converge uniformly on compact subsets of $\Omega$. Hence the limit $u$ is a solution of (4.26), and another application of Corollary 3.15 gives the uniqueness of the solution. □

## 4.4 Hölder gradient estimates for the Dirichlet problem

In this section, we prove a Hölder estimate for the gradient of the solution to the Dirichlet problem in an $H_{1+\alpha}$ domain. This estimate will be used for several different purposes. In the next section, we use it to obtain Hölder gradient estimates for problems with discontinuous $\beta$; in Section 4.7, we use it to obtain Hölder estimates on the second derivatives of solutions of suitable oblique derivative problems; in Chapter 11 to obtain both first and second derivative Hölder estimates for nonlinear oblique derivative problems. The model for the proof is that of Theorem 4.5.

Our first step is an interior $H_{1+\alpha}$ estimate, which we prove by combining ideas from the proofs of Proposition 2.13 and Theorem 4.5.

**Proposition 4.11.** *Let $\Omega$ be a non-empty, open subset of $\mathbb{R}^n$, let $[a^{ij}]$ be a positive-definite matrix-valued function defined on $\Omega$, and suppose that there are positive constants $\lambda$ and $\mu$ such that (4.3) is satisfied in $\Omega$. Suppose also that there is a continuous, increasing function $\theta$ defined on $[0, 1/2]$ with $\theta(0) = 0$ such that*

$$|a^{ij}(x) - a^{ij}(y)| \le \theta\left(\frac{|x-y|}{d(x)}\right)$$

*for all $x$ and $y$ in $\Omega$ with $|x - y| \le \frac{1}{2}d(x)$. Suppose finally that $b^i$ and $c$ are defined on $\Omega$ and that there is a positive constant $A$ such that*

$$|b^i|_0^{(1)} + |c|_0^{(2)} \le \lambda A.$$

*Then, for any $\alpha \in (0, 1)$, there is a constant $C$, determined only by $A$, $n$, $\alpha$, $\mu$, and $\theta$ such that if $u \in C^2(\Omega) \cap L^\infty(\Omega)$ satisfies*

$$a^{ij}D_{ij}u^i_b u + cu = f$$

*in $\Omega$ for some function $f \in H_0^{(2)}$, then $u \in H^{(0}_{1+\alpha}$ and*

$$|u|_{1+\alpha}^{(0)} \le C(|u|_0 + |f/\lambda|_0^{(2)}). \tag{4.28}$$

***Proof.*** Suppose first that $b^i$ and $c$ are zero. With $\delta_1 \in (0, 1/2)$ to be chosen, we fix $x_0 \in \Omega$ and set $R = \delta_1 d(x_0)$. Then set $u_0 = u$ and write $v_0$ for the solution of

$$a^{ij}(x_0) D_{ij} v_0 = 0 \text{ in } B(x_0, R), \quad v_0 = u_0 \text{ on } \partial B(x_0, R)$$

given by Corollary 2.11. From the second derivative bound Theorem 1.34, there are a linear polynomial $L_0$ and a constant $C_0$, determined only by $n$ and $\mu$ such that

$$\sup_{B(x_0,\tau R)} |v_0 - L_0| \le C_0 \tau^2 \sup_{B(x_0,R)} |v_0|$$

for all $\tau \in (0, 1)$. The argument in the proof of Theorem 4.5 then gives positive constants $C_1$ and $\zeta$ and, for any $\delta \in (0, 1)$, a constant $C(\delta)$ such that

$$\sup_{B(x_0,\tau R)} |u_0 - L_0| \le [C_0 \tau^2 + C(\delta)\theta(\delta_1) + C_1 \delta^\zeta] \sup_{B(x_0,R)} |u_0| + C_1 F,$$

where $F = |f/\lambda|_0^{(2)}$. With $\sigma = (1 + \alpha)/2$, we can find $\tau$, $\delta$ and $\delta_1$ so that $C_0 \tau^2 + C(\delta)\theta(\delta_1) + C_1 \delta^\zeta \le \tau^{1+\sigma}$, and hence

$$\sup_{B(x_0,\tau R)} |u_0 - L_0| \le \tau^{1+\sigma} \sup_{B(x_0,R)} |u_0| + C_1 F.$$

By working inductively, we obtain, for each $r \in (0, d(x_0)/2)$, a linear polynomial $P_1(x_0, r)$ such that

$$\sup_{B(x_0,r)} |u - P_1(x_0, r)| \le C \frac{r^{1+\alpha}}{R^{1+\alpha}} (|u|_0 + F),$$

and then we obtain (4.28) by invoking Lemma 2.8 and the definition of $[u]_{1+\alpha}^{(0)}$.

Now suppose that $b^i$ and $c$ are not zero. By arguing as in Proposition 2.13, we may assume that $|u|_1^{(0)}$ is finite. The preceding argument with $f - b^i D_i u - cu$ implies that

$$|u|_{1+\alpha}^{(0)} \le C(|u|_0 + |u|_1^{(0)} + |f/\lambda|_0^{(2)}),$$

and the proof is completed by using the interpolation inequality Lemma 2.3. □

We now provide some estimates in a set that is essentially a half-ball. To be specific, we say that an open subset $G$ of $\mathbb{R}^n$ is a *generalized half-ball* if there is an open subset $G'$ of $\mathbb{R}^{n-1}$ such that every point $x = (x', x^n)$ in $G$ satisfies the conditions $x' \in G'$ and $x^n > 0$. We also write $G^0 = G' \times \{0\}$. Our first estimate is a straightforward modification of Lemma 1.31.

**Lemma 4.12.** *Let $G$ be a generalized half-ball and suppose $a^{ij}$ satisfies (4.3) in $\Omega$. Suppose also that there are constants $B_0 \ge 0$ and $\alpha \in (0,1)$ such that $b$ satisfies*

$$|b(x)| \le B_0(x^n)^{\alpha-1}d^*(x)^{-\alpha} \tag{4.29}$$

*for all $x \in G$, where $d^*$ denotes the distance to $\partial G \setminus G^0$. Let $u \in C^2(G) \cap C(\overline{G})$ with $u = 0$ on $G^0$, and suppose that there is a nonnegative constant $F$ such that (4.13) holds in $G$, where $L = a^{ij}D_{ij} + b^iD_i$. Then there is a constant $C$, determined only by $n$, $\mu$, $\alpha$, and $B_0$, such that*

$$|u(x)| \le C(\sup_G |u| + F)\frac{x^n}{d^*(x)} \tag{4.30}$$

*for all $x \in G$.*

***Proof.*** If $x^n \le \frac{1}{4}d^*(x)$, we set $r = \frac{1}{3}d^*(x)$ and $x_0 = (x', 0)$. We also abbreviate $B^+(x_0, r)$ to $\Omega$ and we observe that $d^*(y) \ge r$ for all $r \in \Omega$.

Next, we define $h$ by

$$h(y) = \int_0^{y^n} \exp\left(-\frac{B_0+1}{\alpha r^\alpha}s^\alpha\right) ds$$

and abbreviate

$$E = \exp\left(-\frac{B_0+1}{\alpha r^\alpha}(x^n)^\alpha\right).$$

It follows that

$$\begin{aligned} Lh &= E(a^{nn}(B_0+1)r^{-\alpha} + b^n) \le E\lambda[-(B_0+1)r^{-\alpha} + B_0r^\alpha] \\ &= -E\lambda r^{-\alpha} \le -E_0r^{-\alpha}\lambda, \end{aligned}$$

where

$$E_0 = \exp(-(B_0+1)/\alpha).$$

If we now set $U = \sup_G |u|$ and define $w$ by

$$w(y) = U\left(\frac{|y-x_0|^2}{r^2} + \frac{2n\mu h(y)}{E_0 r}\right) + F\frac{h(y)}{E_0 r},$$

it follows that $L(w \pm u) \le 0$ in $\Omega$ and $w \pm u \ge 0$ on $\partial\Omega$, and the maximum principle implies that $w \pm u \ge 0$ in $\Omega$. Therefore

$$|u(x)| \le w(x) = U\left(\frac{(x^n)^2}{r^2} + \frac{2n\mu h(x)}{r}\right) + F\frac{h(x)}{E_0 r}.$$

Since $h(x) \in (0, x^n)$ and $x^n/r \in (0,1)$, it follows that

$$|u(x)| \le [\left(1 + \frac{2n\mu}{E_0}\right)U + \frac{F}{E_0}]\frac{x^n}{r},$$

which is just (4.30) with

$$C = 4 + \frac{8n\mu}{E_0}.$$

Since $C \ge 4$, (4.30) is also true if $x^n \ge \frac{1}{4}d^*(x)$. □

From this estimate, we obtain a regularity result with estimate for solutions of the Dirichlet problem.

**Lemma 4.13.** *Suppose, in addition to the hypotheses of Lemma* 4.12 *that there is a continuous increasing function* $\theta$ *on* $[0, 1/2]$ *with* $\theta(0) = 0$ *such that* $a^{ij}$ *satisfies* (4.11) *holds for all* $x$ *and* $y$ *in* $G$ *with* $|x - y| < \frac{1}{2}d^*(x)$. *Then* $u \in H^{(0)}_{1+\alpha}$ *and*

$$|u|^{(0)}_{1+\alpha} \le C(\theta, \mu, B_0)(F + |u|_0). \tag{4.31}$$

***Proof.*** We first suppose that $b^i = 0$. Here, we argue as in Proposition 4.11 with Corollary 2.17 in place of Corollary 2.11 and Lemma 1.35 in place of Theorem 1.34.

If $b^i$ is not necessarily zero, then Lemma 4.12 gives

$$|u(x)| \le C(F + \sup |u|)\frac{x^n}{d^*(x)}$$

and then Proposition 4.11 (applied in $B(x, \min\{d^*(x), d(x)\}/2)$ implies that

$$|Du(x)| \le C(F + \sup |u|)\frac{1}{d^*(x)}.$$

Then, if we set $f^* = f - b^i D_i u$, we conclude that

$$|f^*| \le C(F + \sup |u|)(x^n)^{\alpha-1}(d^*(x))^{-1-\alpha},$$

so we can apply the case $b^i = 0$ with $f^*$ in place of $f$ to infer (4.31) in this case as well. □

From Lemmata 4.13 and 4.9, we obtain a local $H_{1+\alpha}$ estimate for solutions of the Dirichlet problem in an $H_{1+\alpha}$ domain.

**Proposition 4.14.** *Let* $\omega$ *be an* $H_{1+\alpha}(B_{n-1}(x_0', R))$ *function for some positive numbers* $\alpha < 1$ *and* $R$ *and some* $x_0 \in \mathbb{R}^n$ *with* $\omega(x_0') = x_0^n$, *and write* $\Omega[R]$ *and* $\Sigma[R]$ *for the sets of all* $x \in B(x_0, R)$ *with* $x^n > \omega(x')$ *or* $x^n = \omega(x')$, *respectively. Suppose* $a^{ij}$ *satisfies* (4.3) *and there is a continuous increasing function* $\theta$ *on* $[0, 1/2]$ *with* $\theta(0) = 0$ *such that* (4.11) *holds for all* $x$ *and* $y$ *in* $\Omega[R]$ *with* $|x - y| < \frac{1}{2}d^*(x)$. *Suppose also that* $b$ *satisfies the condition*

$$|b| \le B_0 d^{\alpha-1}(d^*)^{-\alpha} \tag{4.32}$$

*in* $\Omega[R]$. *If*

$$a^{ij}D_{ij}u + b^i D_i u = f \text{ in } \Omega[R], \quad u = \varphi \text{ on } \Sigma[R] \tag{4.33}$$

*for some function $f$ satisfying*

$$|f| \le \lambda F_0 d^{\alpha-1}(d^*)^{-1-\alpha}$$

*and some $\varphi \in H_{1+\alpha}(\Sigma[R])$, then there is a constant $C$, determined by $n$, $\mu$, $\alpha$, $B_0$, and $|\omega|_{1+\alpha}$ such that*

$$|u|^{(0)}_{1+\alpha} \le C(|u|_0 + F_0 + |\varphi|^{(0)}_{1+\alpha}). \tag{4.34}$$

***Proof.*** As a first step, we use Exercise 4.3 to conclude that there is a function $\varphi^* \in C^2(\Omega[R])$ such that $|D^2\varphi^*| \le C|\varphi|^{(0)}_{1+\alpha} d^{\alpha-1}(d^*)^{-\alpha-1}$ in $\Omega[R]$ and $\varphi^* = \varphi$ on $\Sigma[R]$. Hence, there is no loss of generality in assuming that $\varphi = 0$.

We then note that the function $V\colon \Omega[R] \to \mathbb{R}^n$, defined by

$$V(x) = (x', \rho(x)),$$

transforms $\Omega[R]$ into a generalized half-ball, and that the hypotheses of Lemma 4.13 are all satisfied in the new variables $y = V(x)$. Changing back to the original variables gives (4.34). □

For our application in Section 4.7, we use the following result, which is proved just like Lemma 2.14.

**Corollary 4.15.** *Suppose that all the hypothesis of Proposition* 4.14 *are satisfied except that*

$$|f| \le \lambda F_0 d^{\alpha-1}(d^*)^{s-1-\alpha}$$

*for some $s > 0$. Then there is a constant $C$, determined by $n$, $\mu$, $\alpha$, $B_0$, $s$, and $|\omega|_{1+\alpha}$ such that*

$$|u|^{(s)}_{1+\alpha} \le C(|u|^{(s)}_0 + F_0 + |\varphi|^{(s)}_{1+\alpha}). \tag{4.35}$$

## 4.5 $C^{1,\alpha}$ estimates with discontinuous $\beta$ in two dimensions

Our next step is to study solutions of the oblique derivative problem in a two dimensional domain when the direction of the vector $\beta$ has a jump discontinuity at isolated points on the boundary. As mentioned in the introduction to Chapter 1, the most straightforward example of this situation is the Neumann problem for Laplace's equation in a square, but there are several important elements in the generalization of this situation that we wish to consider.

We first consider the proper interpretation of the condition $\beta\cdot Du = g$ at a point of discontinuity for $\beta$. Although we can analyze this discontinuity in great generality, we only study here a very simple case to illustrate the issues. We take $\Omega$ to be the set of all $x \in B(0,1)$ with $x^1 > 0$ and $x^2 > 0$, and we work in $\mathbb{R}^2$. We then introduce the sets

$$\Sigma_1 = \{x \in \mathbb{R}^2 : |x| < 1, x^1 > 0, x^2 = 0\},$$
$$\Sigma_1 = \{x \in \mathbb{R}^2 : |x| < 1, x^2 > 0, x^1 = 0\},$$

and we also choose $\beta$ so that $\beta = (0,1)$ on $\Sigma_1$ and $\beta = (1,0)$ on $\Sigma_2$. In this case, neither $(0,1)$ nor $(1,0)$ is oblique at $(0,0)$. Since our theory requires a boundary condition to be prescribed at $(0,0)$ and we are interested in oblique derivative problems, we note that the condition $\beta \cdot Du = g$ at $(0,0)$ will be oblique if and only if the two components of $\beta(0,0)$ are both positive. Moreover, if $u \in C^1(\overline{\Omega})$ and if $g$ is uniformly continuous on the two sets $\Sigma_1$ and $\Sigma_2$, then the gradient of $u$ is prescribed at $(0,0)$. Specifically, if we write $g_1$ and $g_2$ for the limits of $g(x)$ as $x$ approaches $(0,0)$ along $\Sigma_1$ and $\Sigma_2$, respectively, then $Du(0,0) = (g_2, g_1)$. It follows that the condition $\beta\cdot Du = g$ at $(0,0)$ can only be satisfied if $g(0,0)$ is appropriately determined from the vector $\beta(0,0)$ and the numbers $g_1$, $g_2$. In what follows, we shall use this observation explicitly when writing the boundary condition as $\beta\cdot Du = g$ (in which case, $\beta(0,0)$ will be given and $g(0,0)$ is determined from $\beta(0,0)$, $g_1$, and $g_2$) or we shall use it implicitly when writing the boundary condition as

$$\beta \cdot Du = g \text{ on } \Sigma_1 \cup \Sigma_2.$$

These two cases are also distinguished by the explicit regularity assumption that $Du$ is continuous at $(0,0)$ (or a condition that implies this assumption) when we do not explicitly prescribe a boundary condition at $(0,0)$.

For problems with variable coefficients, we use techniques that should be familiar at this stage, so the difficult part of this section is a study of the dependence of the regularity estimates on the geometry of the domain and on the interaction between the domain and the jump in the direction of the prescribed directional derivative. In fact, the analysis of this problem is relatively simple (see the proof of Lemma 4.16), and the real difficulty is that we must prove a suitable series of existence results based on a simple existence result (Lemma 4.19).

As a first step, we assume that the geometry is particularly simple, and we study the regularity estimate only for constant coefficient problems. We also emphasize that we must assume that a boundary condition is prescribed at all points of the boundary, and we examine the smoothness of the solution only near a corner point of the domain.

Moreover, we shall find it convenient to introduce polar coordinates $(r,\theta)$ satisfying

$$x^1 = r\cos\theta, \quad x^2 = r\sin\theta$$

again.

**Lemma 4.16.** *Let $R > 0$, $\theta_1$, and $\theta_2$ be constants with $\theta_1 < \theta_2 < \theta_1 + 2\pi$, and define the sets*

$$\Omega = \{x \in \mathbb{R}^2 : |x| < R,\ \theta_1 < \theta < \theta_2\}, \tag{4.36a}$$

$$\Sigma_k = \{x \in \mathbb{R}^2 : 0 < r < R,\ \theta = \theta_k\} \tag{4.36b}$$

*for $k = 1, 2$,*

$$\sigma = \{x \in \mathbb{R}^2 : r = R, \theta_1 \le \theta \le \theta_2\}. \tag{4.36c}$$

*Let $\beta_1$ and $\beta_2$ be unit vectors such that $\beta_1$ is oblique on $\Sigma_1$ and $\beta_2$ is oblique on $\Sigma_2$. Suppose also that $\beta_1$ and $\beta_2$ are not parallel. Let $[A^{ij}]$ be a positive definite matrix. Suppose more specifically that there are positive constants $\lambda$ and $\mu$ such that (4.3) is satisfied and*

$$\beta_k \cdot \gamma_k \ge 1/\mu \tag{4.37}$$

*for $k = 1, 2$, where $\gamma_k$ is the unit inner normal to $\Sigma_k$. Then there are constants $\alpha \in (0,1)$ and $C$, determined only by $\theta_2 - \theta_1$, $\mu$, and*

$$B_0 = \min\{|\beta_1 - \beta_2|, |\beta_1 + \beta_2|\},$$

*such that, if $u \in C^1(\overline{\Omega}) \cap C^2(\Omega)$ satisfies*

$$A^{ij} D_{ij} u = 0 \quad \text{in } \Omega, \tag{4.38a}$$

$$\beta_1 \cdot Du = 0 \quad \text{on } \Sigma_1, \tag{4.38b}$$

$$\beta_2 \cdot Du = 0 \quad \text{on } \Sigma_2, \tag{4.38c}$$

*then*

$$|Du(x)| \le C \sup_{\Omega} |Du| \left(\frac{|x|}{R}\right)^{\alpha}. \tag{4.39}$$

***Proof.*** First, set $B_1 = \beta_1^1\beta_2^2 - \beta_2^1\beta_1^2$, and note that $B_1$ is just the determinant of the matrix with columns $\beta_1$ and $\beta_2$. If we write $\bar\theta$ for the angle (with $0 < \bar\theta < \pi$) between $\beta^1$ and $\beta^2$, then $|B_1| = \sin\bar\theta$. When $\bar\theta \le \pi/2$, we note that $\cos\bar\theta \ge 0$, and

$$|\beta^1 - \beta^2|^2 = 2(1 - \cos\bar\theta)$$

by the law of cosines. Then

$$1-\cos\bar\theta=\frac{\sin^2\bar\theta}{1+\cos\bar\theta}\le\sin^2\bar\theta$$

and hence $|B_1|\ge B_0/\sqrt{2}$ in this case. On the other hand, if $\bar\theta>\pi/2$, then the angle between $\beta^1$ and $-\beta^2$ is $\theta'=\pi-\bar\theta$, so $\theta'\in(0,\pi/2)$ and $\sin\theta'=\sin\bar\theta$. Our previous case shows that $1-\cos\theta'\le\sin^2\theta'$ and $|\beta^1+\beta^2|^2=2(1-\cos\theta')$, so we infer that $|B_1|\ge B_0/\sqrt{2}$ in this case as well.

To continue, we note that the statement of the theorem is invariant under a rotation of axes, so we now assume that $\theta_1=0$. We now set $v_2=\beta_2\cdot Du$, and note that

$$\begin{aligned}A^{ij}D_{ij}v_2&=0\text{ in }\Omega,\\ v_2&=0\text{ on }\Sigma_2,\\ |v_2|&\le\sup|Du|\text{ on }\sigma.\end{aligned}$$

After this rotation, the vector $\beta_1$ satisfies $\beta_1^2>1/\mu$. Moreover, a simple reflection argument (like the one in Lemma 2.19) shows that $u\in C^2(\Omega\cap\Sigma_1)$. It follows that $Dv_2$ is continuous up to $\Sigma_1$ and then, with $T_1$ a constant to be chosen, we write $\bar\beta_2$ for the vector $(T_1,1)$. Then $\bar\beta_2$ is oblique on $\Sigma_1$, and

$$\bar\beta_2\cdot Dv_2=T_1[\beta_2^1D_{11}u+\beta_2^2D_{12}u]+\beta_2^1D_{12}u+\beta_2^2D_{22}u$$

on $\Sigma_1$. Also, we can differentiate the boundary condition $\beta_1\cdot Du=0$ in the direction $(0,1)$, which is tangential to $\Sigma_1$, to infer that

$$\beta_1^1D_{11}u+\beta_1^2D_{12}u=0$$

which implies that

$$D_{12}u=-\frac{\beta_1^1}{\beta_1^2}D_{11}u$$

on $\Sigma_1$. Finally, the differential equation implies that

$$\begin{aligned}D_{22}u&=-\frac{1}{A^{22}}\left(A^{11}D_{11}u+2A^{12}D_{12}u\right)\\ &=\frac{1}{A^{22}}\left(\frac{2A^{12}\beta_1^1}{\beta_1^2}-A^{11}\right)D_{11}u\end{aligned}$$

on $\Sigma_1$. It follows that

$$\bar\beta\cdot Dv_2=\frac{D_{11}u}{\beta_1^2}\left[T_1B_1+\frac{2A^{12}}{A^{22}}\beta_2^2\beta_1^2-\beta_2^1\beta_1^1-\frac{A^{11}}{A^{22}}\beta_2^2\beta_1^2\right]$$

on $\Sigma_1$, so $\bar\beta\cdot Dv_2=0$ on $\Sigma_1$ if we choose

$$T_1=\frac{-\frac{2A^{12}}{A^{22}}\beta_2^2\beta_1^2+\beta_2^1\beta_1^1+\frac{A^{11}}{A^{22}}\beta_2^2\beta_1^2}{B_1}.$$

Since $|T_1| \le C(\mu, B_0)$, Proposition 3.16 gives positive constants $\alpha$ and $K$, determined only by $B_0$, $\theta_2$, and $\mu$, along with a nonnegative function $w_\alpha$ satisfying

$$\begin{aligned} A^{ij}D_{ij}w_\alpha &\le 0 \text{ in } \Omega, \\ w_\alpha &\le KR^\alpha \text{ in } \Omega, \\ \bar\beta \cdot Dw_\alpha &\le 0 \text{ on } \Sigma_1, \\ w_\alpha &\ge R^\alpha \text{ on } \sigma. \end{aligned}$$

The maximum principle implies that $\pm v_2 \le \sup |Du| w_\alpha / R^\alpha$ in $\Omega$ and hence

$$|\beta_2 \cdot Du(x)| \le K \sup |Du| \left(\frac{|x|}{R}\right)^\alpha$$

in $\Omega$. A similar argument (but also using a reflection across the $x^1$ axis) shows that

$$|\beta_1 \cdot Du(x)| \le K \sup |Du| \left(\frac{|x|}{R}\right)^\alpha$$

in $\Omega$, with the same constants $K$ and $\alpha$, and the combination of these two estimates, along with the positivity of $B_0$, implies (4.39). □

Although this lemma is stated under fairly general geometric conditions, we shall see eventually that the assumed regularity for $u$ is only satisfied under stronger assumptions on $\beta_1$, $\beta_2$, $\Sigma_1$, and $\Sigma_2$. In particular, we shall use a stronger form of the estimate in this lemma, which is based on the following Hölder estimate.

**Lemma 4.17.** *Let $R > 0$, $\theta_0$, $\theta_1$, and $\theta_2$ be constants with $\theta_1 < \theta_0 < \theta_2 < \theta_1 + 2\pi$, and define the sets $\Omega$, $\Sigma_k$, and $\sigma$ by (4.36). Let $\theta_1'$ and $\theta_2'$ satisfy*

$$\theta_2 - \pi < \theta_2' \le \theta_0 \le \theta_1' \le \theta_1 + \pi, \tag{4.40a}$$

$$\theta_1' < \theta_0 + \pi/2, \quad \theta_2' > \theta_0 - \pi/2, \tag{4.40b}$$

*and set $\beta_k = (\cos\theta_k', \sin\theta_k')$ for $k = 1, 2$. Also, define $\beta$ on $\Sigma = \Sigma_1 \cup \Sigma_2 \cup \{0\}$ by*

$$\beta(x) = \begin{cases} \beta_1 & \text{if } x \in \Sigma_1, \\ \beta_2 & \text{if } x \in \Sigma_2, \\ (\cos\theta_0, \sin\theta_0) & \text{if } x = 0. \end{cases} \tag{4.41}$$

*Let $[a^{ij}]$ be a positive definite matrix-valued function defined on $\Omega$. Suppose more specifically that there are positive constants $\lambda$ and $\mu$ such that (4.3) is satisfied and*

$$\theta_1 + \pi - \theta_1' \ge 1/\mu, \quad \theta_2' + \pi - \theta_2 \ge 1/\mu, \tag{4.42a}$$

$$\theta_0 + \pi/2 - \theta_1' \ge 1/\mu, \quad \theta_2' + \pi/2 - \theta_0 \ge 1/\mu. \tag{4.42b}$$

*Then there are constants $\alpha \in (0,1)$ and $C$, determined only by $\theta_2 - \theta_1$ and $\mu$ such that, if $u \in C(\overline{\Omega}) \cap C^2(\Omega)$ satisfies*

$$a^{ij} D_{ij} u = 0 \text{ in } \Omega, \quad \beta \cdot Du = 0 \text{ on } \Sigma, \tag{4.43}$$

*then*

$$|u(x) - u(0)| \le C \sup_{\Omega} |u| \left( \frac{|x|}{R} \right)^{\alpha}. \tag{4.44}$$

***Proof.*** Note that $\beta$ is a unit vector field which is oblique on $\Sigma$. Moreover, after a rotation of axes, we may assume that $\theta_0 = \pi/2$, so our hypotheses imply that $\beta^1(x)x^1 \le 0$ and $\beta^2 > 0$ on $\Sigma$. Furthermore, there is a positive constant $m$ such that $x^2 \le m|x^1|$ on $\Sigma$.

We now define $G$ by

$$G(y,\rho) = \rho^{-2} \left[ (y^1)^2 + \frac{1}{16m^2}(y^2)^2 \right],$$

we set $x_1 = (0, 8m\rho)$, and we define $E(x_1,\rho)$ to be the set of all $x$ such that $G(x - x_1, \rho) < 1$. For all sufficiently small $\rho$, we see that $\eta$, defined by

$$\eta(x) = 1 - G(x - x_1, \rho)$$

satisfies

$$\beta \cdot D\eta \ge \frac{7\beta^2}{8m\rho}$$

on $\Sigma \cap E(x_1,\rho)$. By using this function in place of the function $\eta$ from Lemma 1.17, we infer the Hölder estimate by following the proofs of Lemmata 1.18 and 1.19 and Theorems 1.20 and 1.26. □

We then obtain the following pointwise version of Lemma 4.16.

**Lemma 4.18.** *Under the hypotheses of Lemma 4.17, there is a constant $C$, determined only by $\mu$ such that*

$$|u(x) - u(0)| \le C R^{-1-\alpha} |u|_0 |x|^{1+\alpha} \tag{4.45}$$

*for $|x| \le R/2$.*

***Proof.*** From Lemma 4.16 and integration, we obtain

$$|u(x) - u(0)| \le \frac{C}{R^{\alpha}} |Du|_{0,\Omega[R/2]} |x|^{1+\alpha}.$$

Then we apply the local maximum principle Theorem 1.30 to $\beta_1 \cdot Du$ and $\beta_2 \cdot Du$ to conclude that

$$|Du|_{0,\Omega[R/2]} \le C R^{-2} \int_{\Omega[3R/4]} |Du| \, dx.$$

From Lemma 4.17 and the proof of Lemma 2.20, we conclude that

$$\int_{\Omega[3R/4]} |Du| \, dx \le C R |u|_0,$$

and combining all these estimates yields (4.45). □

From this estimate, we obtain an estimate for solutions of a particular mixed boundary value problem, which will eventually lead to an existence result for this problem.

Our next step is an existence theorem for a very special mixed boundary value problem.

**Lemma 4.19.** *Let $R > 0$ and set*

$$Q(R) = \{x \in \mathbb{R}^2 : |x| < R,\ x^1 > 0,\ x^2 > 0\}, \tag{4.46a}$$
$$S_1(R) = \{x \in \mathbb{R}^2 : |x| < R,\ x^1 > 0,\ x^2 = 0\}, \tag{4.46b}$$
$$S_2(R) = \{x \in \mathbb{R}^2 : |x| < R,\ x^1 = 0,\ x^2 > 0\}, \tag{4.46c}$$
$$s(R) = \{x \in \mathbb{R}^2 : |x| = R,\ x^1 \geq 0,\ x^2 \geq 0\}. \tag{4.46d}$$

*Let $\alpha \in (0,1)$. Then, for any $f \in H_\alpha(Q(R))$, any $g_1 \in H_\alpha(S_1(R))$, any $g_2 \in H_\alpha(S_2(R))$, and any $\varphi \in C(\overline{s(R)})$, there is a unique solution $w \in C^2(Q(R)) \cap C^1(\overline{Q(R)} \setminus s(R)) \cap C(\overline{Q(R)})$ of*

$$\Delta w = f \text{ in } Q(R), \tag{4.47a}$$
$$D_2 w = g_1 \text{ on } S_1(R), \tag{4.47b}$$
$$D_1 w = g_2 \text{ on } S_2(R), \tag{4.47c}$$
$$w = \varphi \text{ on } s(R). \tag{4.47d}$$

*Moreover, there is a constant $C$, determined only by $\alpha$, such that*

$$|u|^{(0)}_{1+\alpha} \leq C(|u|_0 + |f|^{(2)}_0 + |g_1|^{(1)}_\alpha + |g_2|^{(1)}_\alpha), \tag{4.48}$$

*where the weight in these norms is the distance to $s$.*

***Proof.*** Suppose first that $g_2 \equiv 0$ and $g_1 \in H_{1+\alpha}(S_1(R))$ with $D_1 g_1(0) = 0$. Then extend $f$, $g_1$, and $\varphi$ as even functions of $x^1$ to all of $B^+(0,R)$, $B^0(0,R)$, and $B^*(0,R)$, respectively, where $B^*(0,R) = \{x \in \mathbb{R}^2 : |x| = R, x^2 \geq 0\}$. It follows from the proof of Lemma 2.19 that the problem

$$\Delta \tilde{w} = f \text{ in } B^+(0,R), \tag{4.49a}$$
$$D_2 \tilde{w} = g_1 \text{ on } B^0(0,R), \tag{4.49b}$$
$$\tilde{w} = \varphi \text{ on } B^*(0,R) \tag{4.49c}$$

has a unique solution $\tilde{w} \in C^2(B^+(0,R)) \cap C(\overline{B^+(0,R)})$. If we only have $g \in H_\alpha(S_1(R))$, then we first note that there is a sequence $(g_{1,m})$ of $H_{1+\alpha}(S_1(R))$ functions which converges uniformly to $g_1$ with $D_1 g_{1,m}(0) = 0$ and $[g_{1,m}]_\alpha \leq C(\alpha)[g_1]_\alpha$. If we write $\tilde{w}_m$ for the solution of (4.49) with $g_{1,m}$ in place of $g_1$, we see that $(\tilde{w}_m)$ is uniformly Cauchy by the maximum principle, Theorem 1.11, so it converges uniformly to a limit function $\tilde{w}$. It then follows from Proposition 2.12 that the sequence $(D^2\tilde{w}_m)$ is uniformly Cauchy on compact subsets of $B^+$ so $\Delta\tilde{w} = f$ in $Q$, and a similar argument using Theorem 4.5 shows that $D_2\tilde{w} = g_1$ on $B^0$. In other words, (4.49) has a solution in this case as well.

Now we observe that $\hat{w}$, defined by

$$\hat{w}(x^1, x^2) = \tilde{w}(-x^1, x^2),$$

is a solution of the same boundary value problem as $\tilde{w}$, so the maximum principle implies that $\hat{w} = \tilde{w}$. In other words, $\tilde{w}$ is an even function of $x^1$, so we have

$$\begin{aligned} \Delta\tilde{w} &= f \text{ in } Q(R), \\ D_2\tilde{w} &= g_1 \text{ on } S_1(R), \\ D_1\tilde{w} &= 0 \text{ on } S_2(R), \\ \tilde{w} &= \varphi \text{ on } s(R). \end{aligned}$$

To handle the case of nonzero $g_2$, we write $w_1$ for the unique solution of

$$\begin{aligned} \Delta w_1 &= 0 \text{ in } \{|x| < R, x^1 > 0\}, \\ D_1 w_1 &= g_2 \text{ on } \{|x| < R, x^1 = 0\}, \\ w_1 &= 0 \text{ on } \{|x| = R, x^1 > 0\} \end{aligned}$$

given by the previous argument and $w_2$ for the solution of

$$\begin{aligned} \Delta w_2 &= f \text{ in } Q(R), \\ D_2 w_2 &= g_1 - D_2 w_1 \text{ on } S_1(R), \\ D_1 w_2 &= 0 \text{ on } S_2(R), \\ w_2 &= \varphi \text{ on } s(R) \end{aligned}$$

that we have just derived. It follows that $w = w_1 + w_2$ is the desired solution.

Another application of Theorem 4.5 gives (4.48). □

For our further analysis, we modify the notation slightly from (4.46). For $R > 0$ and $\kappa \in (0, 1)$, we set $x_1(\kappa, R) = (0, -\kappa R)$, and we define the sets

$$\begin{aligned} Q(\kappa, R) &= \{x \in \mathbb{R}^2 : |x - x_1(\kappa, R)| < R,\ \pi/4 < \theta < 3\pi/4\}, \\ S_1(\kappa, R) &= \{x \in \mathbb{R}^2 : |x - x_1(\kappa, R)| < R,\ \theta = \pi/4\}, \\ S_2(\kappa, R) &= \{x \in \mathbb{R}^2 : |x - x_1(\kappa, R)| < R,\ \theta = 3\pi/4\}, \\ s(\kappa, R) &= \{x \in \mathbb{R}^2 : |x - x_1(\kappa, R)| = R,\ \pi/4 \le \theta \le 3\pi/4\}, \\ Q^0(\kappa, R) &= S_1(\kappa, R) \cup S_2(\kappa, R). \end{aligned}$$

For $\alpha \in (0,1)$, we write $H_\alpha(Q^0)$ for the set of all functions $g$ defined on $Q^0$ such that the restriction of $g$ to $S_1(\kappa, R)$ is in $H_\alpha(S_1(\kappa, R))$ and the restriction of $g$ to $S_2(\kappa, R)$ is in $H_\alpha(S_2(\kappa, R))$. Since $\kappa$ and $R$ will usually be fixed in what follows, we generally omit them from the notation. In what follows, $\gamma$ denotes the unit inner normal to $Q^0$.

**Lemma 4.20.** *Let $\alpha \in (0,1)$, $R > 0$, and $\kappa \in (0,1)$. Then the problem*

$$\Delta u = f \text{ in } Q, \tag{4.50a}$$

$$\gamma \cdot Du = g \text{ on } Q^0, \tag{4.50b}$$

$$u = \varphi \text{ on } s \tag{4.50c}$$

*has a unique solution $u \in C^2(Q) \cap C^1(\overline{Q} \setminus s) \cap C(\overline{Q})$ for any $f \in H_\alpha(Q)$, any $g \in H_\alpha(Q^0)$, and any $\varphi \in C(s)$.*

***Proof.*** When $\varphi = 0$, we modify the Perron process from Section 2.5 only slightly. First, we say that

$$\Delta u = f \text{ in } Q, \quad \gamma \cdot Du = g \text{ on } Q^0 \tag{4.51}$$

is *locally solvable* if, for every $x_0 \in \overline{Q} \setminus s$, there is an open set $N$ with $x_0 \in N$ and $\overline{N} \cap s = \emptyset$ such that, for any $\psi \in C(\overline{Q} \cap \partial N)$, there is a solution $\psi_N C^2(Q \cap N) \cap C^1(\overline{Q} \cap N) \cap C(\overline{Q \cap N})$ to the problem

$$\Delta \psi_N = f \text{ in } Q \cap N, \tag{4.52a}$$

$$\gamma \cdot D\psi_N = g \text{ on } Q^0 \cap N, \tag{4.52b}$$

$$\psi_N = \psi \text{ on } \overline{Q} \cap \partial N. \tag{4.52c}$$

Lemma 4.19 gives local solvability of this problem in a neighborhood of 0, Lemma 2.19 gives local solvability in a neighborhood of any point in $Q^0$, and Corollary 2.16 gives local solvability in a neighborhood of any point in $Q$. For future reference, we observe that the set $N$ corresponding to any $x_0 \neq 0$ can be taken so that $0 \notin \overline{N}$.

Next, we say that $v$ is a *Perron supersolution* of (4.51) if, for every $x_0 \in \overline{Q} \setminus s$ and every continuous function $\psi$ with $v \geq \psi$ on $Q \cap \partial N$, we have $v \geq \psi_N$ in $N$.

If we define $w$ by

$$w(x) = (R^2 - |x - x_1|^2)^{1/2},$$

then it's easy to check that $\Delta w \leq -R^2(R^2 - |x - x_1|^2)^{-3/2}$ in $Q$ and that $w = 0$ on $\sigma$. In addition, a direct calculation gives

$$\gamma \cdot Dw = -\frac{1}{\sqrt{2}}(R^2 - |x - x_1|^2)^{-1/2} \kappa R$$

on $Q^0$.

Hence,

$$\Delta w \le -C(\kappa)/R \text{ in } Q,$$
$$\gamma \cdot Dw \le -C(\kappa) \text{ on } Q^0,$$
$$w \ge 0 \text{ on } s.$$

Therefore $-Kw$ is a subsolution and $Kw$ is a supersolution if we choose $K$ sufficiently large in terms of $|f|_0$, $|g|_0$, and $R$. We now write $S^+$ for the set of Perron supersolutions $v$ which are greater than $-Kw$ and which satisfy

$$\lim_{\substack{x \to y \\ x \in Q}} v(x) = 0$$

for all $y \in s$. (Note that $S^+$ is nonempty because $Kw \in S^+$.)

We then define $u$ by

$$u(x) = \inf_{v \in S^+} v(x)$$

and note that $-Kw \le u \le Kw$, so

$$\lim_{\substack{x \to y \\ x \in Q}} u(x) = 0$$

for all $y \in s$.

Next, for each $x_0 \in \overline{Q} \setminus s$, we define the lift $v_N$ of a supersolution $v$ by

$$v_N(x) = \sup\{\psi_N(x) : \psi \text{ is continuous and } v \ge g \ge -Kw \text{ on } \overline{Q} \cap \partial N\}$$

if $x \in \overline{Q} \cap N$, and

$$v_N(x) = v(x)$$

if $x \in \overline{Q} \setminus N$.

It follows from Proposition 2.13 and the argument in Section 2.5 that $\Delta v_N = f$ in $Q \cap N$ and Lemma 2.19 implies that $\gamma \cdot Dv_N = g$ on $Q^0 \cap N$. Moreover, if $x_0 = 0$, we use (4.48) from Lemma 4.19 to infer that $v_N$ has continuous second derivatives near 0. Just as in Section 2.5, we conclude that $u$ solves (4.50).

For nonzero $\varphi$, we take $(\varphi_m)$ to be a sequence of $H_{2+\alpha}(Q)$ functions converging uniformly on $s$ to $\varphi$. Then our Perron process argument gives a solution $\bar{u}_m$ of

$$\Delta \bar{u}_m = f - \Delta\varphi_m \text{ in } Q,$$
$$\gamma \cdot \bar{D}u_m = g - \gamma \cdot D\varphi_m \text{ on } Q^0,$$
$$\bar{u}_m = 0 \text{ on } s,$$

and then $u_m = \bar{u}_m + \varphi_m$ solves

$$\Delta u_m = f \text{ in } Q,$$
$$\gamma \cdot Du_m = g \text{ on } Q^0,$$
$$u_m = \varphi_m \text{ on } s.$$

By using the maximum principle along with Proposition 2.12, Theorem 4.5, and (4.48), we conclude that $(u_m)$ converges to a solution of (4.50). □

Our next step is to obtain estimates involving less regular functions $f$. To simplify notation, we write $g = 0$ on $Q^0 \cap s$ as shorthand for the statement

$$\lim_{\substack{x \to x_0 \\ x \in Q^0}} g(x) = 0$$

for all $x_0 \in \overline{Q^0} \cap s$.

**Lemma 4.21.** *Let $\delta \in (0,1)$. For any $\alpha \in (0,1)$, there is a constant $\kappa_1 \in (0,1)$, determined only by $\alpha$ such that if $\kappa \in (\kappa_1, 1)$, then any solution $u$ of (4.50) with $\varphi = 0$ on $s$ and $g = 0$ on $Q^0 \cap s$ satisfies the estimate*

$$|u|_{2+\delta}^{(-1-\alpha)} \le C(\alpha, \delta, \kappa)[|f|_{\delta}^{(1-\alpha)} + |g|_{\alpha}^{(-\alpha)}]. \tag{4.53}$$

*Moreover, (4.50) has a unique solution for any $f \in H_{\delta}^{(1-\alpha)}$, any $g \in H_{\alpha}(Q^0)$ with $g = 0$ on $Q^0 \cap s$, and $\varphi = 0$.*

***Proof.*** We first estimate $|u|_0$. To this end, we define the function $w$ by

$$w(x) = R^{1-\alpha}(R^2 - |x - x_1|^2)^{\alpha} + R^{1+\alpha} - (d_{S_1})^{1+\alpha} + R^{1+\alpha} - (d_{S_2})^{1+\alpha},$$

where $d_{S_k}$ denotes the distance to $S_k$. A simple calculation shows that there is a constant $C(\alpha)$ such that

$$\Delta w \le -C(\alpha, \kappa) d^{\alpha - 1} \text{ in } Q,$$
$$\gamma \cdot Dw \le -C(\alpha, \kappa) R^{\alpha} \text{ on } Q^0,$$
$$w \ge 0 \text{ on } s,$$

and then the maximum principle implies that

$$\begin{aligned} |u| &\le \frac{1}{C(\alpha, \kappa)}[|f|_0^{(1-\alpha)} + |g|_0^{(-\alpha)}]w \\ &\le CR^{1+\alpha}[|f|_0^{(1-\alpha)} + |g|_0^{(-\alpha)}]. \end{aligned}$$

We now observe that the function $G$ in Section 3.6 has the special form $G(\alpha; z, p) = \alpha^2 z$ when $\mu = 1$, which will be the case for the Laplace operator. Moreover the constant $\mu_0$ in that section is zero for the normal derivative operator. It follows that the constant $\alpha_1(\pi/4, n, 1, 0)$ in Proposition

3.16 is equal to 2, and hence, for $\kappa_1$ sufficiently close to 1 (and determined just by $\alpha$), Proposition 3.16 provides a function $v_1$ such that

$$\begin{aligned}\Delta v_1 &\le -|x-y_1|^{\alpha-1} \text{ in } Q\cap N_1,\\ \gamma\cdot Dv_1 &\le -|x-y_1|^{\alpha} \text{ on } S_1\cap N_1,\\ v_1 &\ge |x-y_1|^{1+\alpha} \text{ in } Q\cap N_1,\end{aligned}$$

where $y_1$ is the point in $\sigma\cap\overline{S_1}$ and $N_1$ is the set of all $x$ such that $|x-y_1|<C_1R$ for some sufficiently small positive constant $C_1$ (determined only by $\kappa$) such that $\overline{N_1}\cap S_2=\emptyset$. Further, $C_1$ can be chosen so that $d_{S_2}\ge|x-y_1|$ in $N_1$. It follows that

$$w=2v_1-(d_{S_1})^{1+\alpha}-(d_s)^{1+\alpha}$$

satisfies

$$\begin{aligned}\Delta w &\le -d^{\alpha-1} \text{ in } Q\cap N_1,\\ \gamma\cdot Dw &\le -|x-y_1|^{\alpha} \text{ on } S_1\cap N_1,\\ w &\ge 0 \text{ on } \sigma\cap N_1,\\ w &\ge (C_1R)^{1+\alpha} \text{ in } Q\cap\partial N_1,\end{aligned}$$

so an application of the maximum principle shows that

$$|u(x)|\le C(d^{**})^{1+\alpha}[|f|_0^{(1-\alpha)}+|g|_\alpha^{(-\alpha)}],$$

where $d^{**}$ denotes distance to $\sigma\cap(S_1\cup S_2)$, if $|x-y_1|\le C_1R$. A similar argument near $S_2\cap\sigma$ shows that

$$|u(x)|\le C(d^{**})^{1+\alpha}[|f|_0^{(1-\alpha)}+|g|_\alpha^{(-\alpha)}]$$

if $|x-y_2|\le C_1R$.

Next, we replace Lemma 4.2 by Lemma 4.18 in the proof of Theorem 4.5 to conclude that

$$|u(x)-L\cdot x-u(0)|\le C|x|^{1+\alpha}[|f|_0^{(1-\alpha)}+|g|_\alpha^{(-\alpha)}]$$

where $L$ is the (unique) vector such that

$$\gamma_k\cdot L=\lim_{\substack{x\to 0\\ x\in S_k}} g(x)$$

for $k=1,2$, and $\gamma_k$ denotes the restriction of $\gamma$ to $S_k$.

Furthermore, Proposition 4.14 gives a corresponding estimate near points of $s$, so the argument in Theorem 4.5 then allows us to conclude that

$$|u|_{1+\alpha}^{(-1-\alpha)}\le C[|f|_0^{(1-\alpha)}+|g|_\alpha^{(-\alpha)}],$$

and an application of Lemma 2.14 completes the proof of (4.53).

Since (4.50) has a solution for any $f\in H_\delta$, the existence result follows by approximation and (4.53). □

From these lemmata, we now derive an existence result and estimate for problems in the standard domain but with arbitrary constant coefficients in the differential equation.

**Proposition 4.22.** *Let $\theta_1 = \theta_0 = \pi/4$ and let $\delta \in (0,1)$. Then for any $\mu \geq 1$, there is a constant $\alpha_0 \in (0,1)$, determined only by $\mu$, such that, if $\alpha \in (0,\alpha_0)$ and $[A^{ij}]$ is a constant matrix in $\mathcal{M}_\mu$, then the problem*

$$A^{ij}D_{ij}u = f \text{ in } Q, \tag{4.54a}$$

$$\gamma \cdot Du = 0 \text{ on } Q^0, \tag{4.54b}$$

$$u = 0 \text{ on } s \tag{4.54c}$$

*for any $f \in H_\delta^{(1-\alpha)}$. Moreover there is a constant $C$, determined only by $\alpha$, $\delta$, $\kappa$, and $\mu$, such that*

$$|u|_{2+\delta}^{(-1-\alpha)} \leq C|f|_\delta^{(1-\alpha)}. \tag{4.55}$$

***Proof.*** As a first step, we observe that

$$w(x) = R^{1-\alpha}\left(R^2 - |x-x_1|^2\right)^\alpha + R^{1+\alpha} - (d_{S_1})^{1+\alpha} + R^{1+\alpha} - (d_{S_2})^{1+\alpha}$$

satisfies the estimate

$$A^{ij}D_{ij}w \leq -Cd^{\alpha-1}$$

for some positive constant $C$, determined only by $\kappa$, $\alpha$, and $\mu$, so the maximum principle implies (just as in the proof of Lemma 4.21) that

$$|u|_0 \leq C(\kappa,\alpha,\mu)R^{1+\alpha}|f|_0^{(1-\alpha)}. \tag{4.56}$$

Next, we observe that the constant $\alpha_1$ from Proposition 3.16 must be greater than 1 if $\kappa > 0$, and hence the argument in Lemma 4.21 also implies that

$$|u| \leq C(\kappa,\alpha,\mu)(d^{**})^{1+\alpha}|f|_0^{(1-\alpha)}. \tag{4.57}$$

Now, for $t \in \mathbb{R}$, we define

$$A_t^{ij} = tA^{ij} + (1-t)\delta^{ij},$$

and we write $T$ for the set of all $t \geq 0$ such that the problem

$$A_t^{ij}D_{ij}u = f \text{ in } Q, \tag{4.58a}$$

$$\gamma \cdot Du = 0 \text{ on } Q^0, \tag{4.58b}$$

$$u = 0 \text{ on } s \tag{4.58c}$$

has a solution for all $f \in H_\delta^{(1-\alpha)}$. It follows from Lemma 4.21 that $0 \in T$ and we want to show that $1 \in T$. As a first step towards this end, we show that if $t \in T$, then (4.55) holds. If $t \in T$, then, for any $\rho \in (0, R)$, there is a unique solution $\psi$ of

$$A_t^{ij} D_{ij}\psi = -f \text{ in } Q(\kappa, \rho),$$
$$\gamma \cdot D\psi = 0 \text{ on } Q^0(\kappa, \rho),$$
$$\psi = u \text{ on } s(\kappa, \rho),$$

and hence $v = u + \psi$ solves

$$A_t^{ij} D_{ij} v = 0 \text{ in } Q(\kappa, \rho), \tag{4.59a}$$
$$\gamma \cdot Dv = 0 \text{ on } Q^0(\kappa, \rho), \tag{4.59b}$$
$$v = u \text{ on } s(\kappa, \rho). \tag{4.59c}$$

Using this existence result, we argue as in Lemma 4.21 to conclude (4.55) provided $\alpha_0$ is chosen to be no greater than the constant from Lemma 4.18 in addition to our previous restriction that $\alpha_0 \le \alpha_1 - 1$.

New, we fix $t \in T$ and $f \in H_\delta^{(1-\alpha)}$. With $\tau > 0$ to be chosen, we write $\mathbb{B}$ for set of all $u \in H_{2+\delta}^{(-1-\alpha)}$ satisfying

$$\gamma \cdot Du = 0 \text{ on } Q^0, \quad u = 0 \text{ on } s,$$

and we define the operator $J$ on $\mathbb{B}$ by $u = Jv$ if $u$ is the unique solution of

$$A_t^{ij} D_{ij} u = f + [A_t^{ij} - A_{t+\tau}^{ij}] D_{ij} v \text{ in } Q, \tag{4.60}$$
$$\gamma \cdot Du = 0 \text{ on } Q^0, \tag{4.61}$$
$$u = 0 \text{ on } s. \tag{4.62}$$

If $C$ is the constant from (4.55), then we obtain

$$\|Jv_1 - Jv_2\|_{\mathbb{B}} \le C\tau \|v_1 - v_2\|_{\mathbb{B}},$$

so $J$ will be a contraction if $C\tau \le 1/2$. It follows in this case that $J$ has a fixed point $u$ which solves

$$A_{t+\tau}^{ij} D_{ij} u = f \text{ in } Q,$$
$$\gamma \cdot Du = 0 \text{ on } Q^0,$$
$$u = 0 \text{ on } s.$$

Therefore $t + \tau \in T$, and a finite repetition of this argument shows that $1 \in T$, which then implies both the existence result and the estimate. □

**Corollary 4.23.** *Let $\theta_1 = \theta_0 = \pi/4$ and let $\delta \in (0, 1)$. Then for any $\mu \ge 1$, there is a constant $\alpha_0 \in (0, 1)$, determined only by $\mu$, such that, if $\alpha \in (0, \alpha_0)$ and $[A^{ij}]$ is a constant matrix in $\mathcal{M}_\mu$, then the problem*

$$A^{ij} D_{ij} u = 0 \textit{ in } Q, \tag{4.63a}$$
$$\gamma \cdot Du = g \textit{ on } Q^0, \tag{4.63b}$$
$$u = \varphi \textit{ on } s \tag{4.63c}$$

*has a unique solution* $u \in C^2(Q) \cap C^1(\bar{Q} \setminus s) \cap C(\bar{Q})$ *for any* $g \in H_\alpha(Q^0)$ *and any* $\varphi \in C(s)$.

***Proof.*** First, extend $\varphi$ to all of $B(x_1, R)$ as a continuous function. Then there is a sequence $(\varphi_m)$ of $H_{2+\delta}^{(-1-\alpha)}(B(x_1,R))$ functions converging uniformly to $\varphi$. In addition, there is an increasing sequence $(\zeta_m)$ of $C^2(\mathbb{R}^2)$ functions with $0 \le \zeta_m \le 1$ in $\mathbb{R}^2$ and $\zeta_m \equiv 1$ on $B(x_1, (1-1/m)R)$.

We infer from Lemma 4.21 that there is a unique solution $v_m \in H_{2+\delta}^{(-1-\alpha)}$ of

$$\begin{aligned} \Delta v_m &= -\Delta\varphi_m \text{ in } Q, \\ \gamma \cdot Dv_m &= (g - \gamma\cdot D\varphi_m)\zeta_m \text{ on } Q^0, \\ v_m &= \varphi_m \text{ on } s, \end{aligned}$$

and Proposition 4.22 gives a unique solution $w$ of

$$\begin{aligned} A^{ij}D_{ij}w_m &= -A^{ij}D_{ij}v_m - A^{ij}D_{ij}\varphi_m \text{ in } Q, \\ \gamma \cdot Dw_m &= 0 \text{ on } Q^0, \\ w_m &= 0 \text{ on } s. \end{aligned}$$

Then $u_m = v_m + w_m + \varphi_m$ satisfies

$$\begin{aligned} A^{ij}D_{ij}u_m &= 0 \text{ in } Q, \\ \gamma \cdot Du_m &= g_m \text{ on } Q^0, \\ u_m &= \varphi_m \text{ on } s, \end{aligned}$$

where $g_m = \zeta_m g + (1-\zeta_m)\gamma\cdot D\varphi_m$.

Since $g_m$ and $\varphi_m$ are uniformly bounded, the sequence $(u_m)$ is bounded in $C(\overline{Q})$, and the uniform continuity of $(\varphi_m)$ implies (via the arguments in Lemma 4.17 and Corollary 1.29) a uniform modulus of continuity for $(u_m)$. The Arzela-Ascoli theorem implies that $(u_m)$ converges uniformly to a function $u$, which is easily verified to be the desired solution of (4.63). □

Our analysis of mixed problems for equations with arbitrary (constant) coefficients and the normal derivative condition in this standard domain allows us to study more general boundary conditions in domains with angles other than $\pi/2$ at the corner; however, our approach needs two further steps to get to the most general results.

In the first step, we fix a constant $\theta_1 \in (0, \pi/2)$ and we set $\theta_2 = \pi - \theta_1$. Then there is a linear transformation $T$ which preserves area and orientation such that $T$ maps the region $0 < \theta < \pi/2$ onto the region $\theta_1 < \theta <$

$\theta_2$. Moreover, there is a positive constant $c_1$, determined only by $\theta_1$ (and tending to $\infty$ as $\theta_1$ goes to 0 or $\pi/2$) such that $|x|/c_1 \le |Tx| \le c_1|x|$ for all $x \in \mathbb{R}^2$. We now set

$$\tilde{E}^+(R) = T(Q(R)), \quad \tilde{E}_1(R) = T(S_1(R)),$$
$$\tilde{E}_2(R) = T(S_2(R)), \quad \tilde{e}(R) = T(s(R)),$$
$$\tilde{E}^0(R) = \tilde{E}_1(R) \cup \tilde{E}_2(R).$$

With $R$ suppressed from the notation, we now define the vector field $\beta_0$ on $\tilde{E}^0$ by setting

$$\beta_0(x) = \begin{cases} (-\cos\theta_1, \sin\theta_1) & \text{if } x \in \tilde{E}_1, \\ (\cos\theta_1, \sin\theta_1) & \text{if } x \in \tilde{E}_2. \end{cases}$$

For simplicity of notation, we use $H_\alpha(\tilde{E}^0)$ to denote the set of all functions $g$ defined on $\tilde{E}^0$ such that the restriction of $g$ to $\tilde{E}_k$ is in $H_\alpha(\tilde{E}_k)$ for $k = 1, 2$.

If $[A^{ij}]$ is a positive definite matrix, it follows from Corollary 4.23 that the problem

$$A^{ij}D_{ij}w = 0 \text{ in } \tilde{E}^+,$$
$$\beta_0 \cdot Dw = g \text{ in } \tilde{E}^0,$$
$$w = \varphi \text{ on } \tilde{e}$$

has a unique solution $w \in C^2(\tilde{E}^+) \cap C^1(\overline{\tilde{E}^+} \setminus \tilde{e}) \cap C(\overline{\tilde{E}^+})$ for all $g \in H_\alpha(\tilde{E}^0; \beta_0)$ and all $\varphi \in C(\overline{\tilde{E}^+})$.

We now set up our existence theory for problems in wedge domains with arbitrary convex angle at the corner but with a strong restriction on the connection between the two different directions for the directional derivatives and the positive $x^2$-axis. Let $R > 0$, and let $\theta_1 < \pi/2 < \theta_2$ be constants with $\theta_2 < \theta_1 + \pi$. With $m > 0$ a constant (to be further specified), we define the sets

$$E(m, R) = \{x \in \mathbb{R}^2 : (x^1)^2 + \frac{1}{16m^2}(x^2 + 8mR)^2 < R^2, \theta_1 < \theta < \theta_2\},$$
$$E_k(m, R) = \{x \in \mathbb{R}^2 : (x^1)^2 + \frac{1}{16m^2}(x^2 + 8mR)^2 < R^2, \theta = \theta_k\}$$

for $k = 1, 2$,

$$e(m, R) = \{x \in \mathbb{R}^2 : (x^1)^2 + \frac{1}{16m^2}(x^2 + 8mR)^2 = R^2, \theta_1 \le \theta \le \theta_2\},$$
$$E^0(m, R) = E_1(m, R) \cup E_2(m, R),$$

and we define $H_\alpha(E^0(m,R))$ to be the set of all functions $g$ defined on $E^0(m,R)$ such that the restriction of $g$ to $E_k(m,R)$ is in $H_\alpha(E_k(m,R))$ for $k=1,2$.

We now prove the existence of solutions for

$$A^{ij}D_{ij}u = 0 \text{ in } E(m,R), \tag{4.64a}$$
$$\beta \cdot Du = g \text{ on } E^0(m,R), \tag{4.64b}$$
$$u = \varphi \text{ on } e(m,R) \tag{4.64c}$$

under various assumptions on the constant $m$, and the functions $g$ and $\varphi$. Furthermore $\beta$ will be a vector field which is constant on $E_1(m,R)$, constant on $E_2(m,R)$ and oblique on $E^0(m,R)$, and $[A^{ij}]$ will be a fixed positive definite matrix satisfying (4.3) for positive constants $\lambda$ and $\mu$. In this context, our first step is to prove the existence of a solution for all $g$ which are sufficiently smooth on $E_1(m,R)$ and on $E_2(m,R)$, $\varphi = 0$ and $\beta = \beta_0$.

**Lemma 4.24.** *Let $[A^{ij}] \in \mathcal{M}_\mu$ for some $\mu \geq 1$. Suppose also that $\theta_1$ and $\theta_2$ satisfy*

$$\frac{1}{\mu} \leq \theta_1 \leq \frac{\pi}{2} - \frac{1}{\mu}, \qquad \theta_2 = \pi - \theta_1. \tag{4.65}$$

*Then there are constants $m_0 > 0$ and $\alpha_0 \in (0,1)$, determined only by $\mu$, such that the problem*

$$A^{ij}D_{ij}u = 0 \textit{ in } E(m,R), \tag{4.66a}$$
$$\beta_0 \cdot Du = g \textit{ on } E^0(m,R), \tag{4.66b}$$
$$u = \varphi \textit{ on } e(m,R) \tag{4.66c}$$

*has a unique solution $u \in C^2(E(m,R)) \cap C^1(\overline{E(m,R)} \setminus e(m,R)) \cap C(\overline{E(m,R)})$ for any $g \in H_\alpha(E^0(m,R))$ and any $\varphi \in C(e(m,R))$ provided $m \geq m_0$ and $\alpha \in (0,\alpha_0)$.*

***Proof.*** This time we suppress $m$ and $R$ from the notation. When $\varphi \equiv 0$, we proceed as in Lemma 4.20 via the Perron process. By invoking Corollary 4.23 (and our change of variables) in place of Lemma 4.19, we infer that

$$A^{ij}D_{ij}u = 0 \text{ in } E, \quad \beta_0 \cdot Du = g \text{ on } E^0$$

is locally solvable, that is, for every $x_0 \in \overline{E} \setminus e$, there is an open set $N$ with $x_0 \in N$ and $\overline{N} \cap s = \emptyset$ such that, for any continuous function $\psi$ defined on

$\overline{E} \cap \partial N$, there is a unique $C^2(E \cap N) \cap C^1(\overline{E} \cap N) \cap C(\overline{E \cap N})$ solution $\psi_N$ to the problem

$$\begin{aligned} A^{ij} D_{ij} \psi_N &= 0 \text{ in } E \cap N, \\ \beta_0 \cdot D\psi_N &= g \text{ on } E^0 \cap N, \\ \psi_N &= \psi \text{ on } \overline{E} \cap \partial N. \end{aligned}$$

Next, we note that $w_1$, defined by

$$w_1(x) = (R^2 - (x^1)^2 - \frac{1}{16m^2}(x^2 + 8mR)^2)^{1/2},$$

satisfies the conditions

$$\begin{aligned} A^{ij} D_{ij} w_1 &\le 0 \text{ in } E, \\ \beta_0 \cdot D w_1 &\le -C(m) \text{ on } E^0, \\ w_1 &\ge 0 \text{ on } e \end{aligned}$$

provided $m$ is sufficiently large.

By following the proof of Lemma 4.20 with $A^{ij} D_{ij}$ in place of $\Delta$, $E$ in place of $Q$, $E^0$ in place of $Q^0$, $e$ in place of $s$, and $w_1$ in place of the function $w$ in that proof, we obtain the existence result for $\varphi = 0$.

For nonzero $\varphi$, we first extend $\varphi$ as a continuous function to the set

$$e^* = \{x \in \mathbb{R}^2 : (x^1)^2 + \frac{1}{16m^2}(x^2 + 8mR)^2 = R^2\}$$

and we take $(\varphi_k)$ to be a sequence in $H_{1+\alpha}(e^*)$ converging uniformly to $\varphi$ on $e^*$. If we also define

$$E^* = \{x \in \mathbb{R}^2 : (x^1)^2 + \frac{1}{16m^2}(x^2 + 8mR)^2 < R^2\},$$

it follows from Corollary 2.17 and Proposition 4.14 that there is a solution $\bar{u}_k \in H_{1+\alpha}(E^*)$ to

$$A^{ij} D_{ij} \bar{u}_k = 0 \text{ in } E^*, \quad u_k = \varphi_k \text{ on } e^*$$

and our Perron process argument gives a solution $v_k \in H_{1+\alpha}$ to

$$\begin{aligned} A^{ij} D_{ij} v_k &= 0 \text{ in } E, \\ \beta_0 \cdot D v_k &= g - \beta_0 \cdot D\bar{u}_k \text{ on } E^0, \\ v_k &= 0 \text{ on } e. \end{aligned}$$

Hence $u_k = \bar{u}_k + v_k$ solves

$$\begin{aligned} A^{ij} D_{ij} u_k &= 0 \text{ in } E, \\ \beta_0 \cdot D u_k &= g \text{ on } E^0, \\ v_k &= \varphi_k \text{ on } e. \end{aligned}$$

As before, $(u_k)$ converges uniformly to the unique solution of (4.66). □

To continue, we introduce one further function space. For $\alpha$ and $\delta$ in $(0,1)$, we denote by $H_\alpha^{(1-\delta)}(E^0(m,R))$ the set of all functions $g$ defined on $E^0(m,R)$ such that the restriction of $g$ to $E_k(m,R)$ is in $H_\alpha^{(1-\delta)}(E_k(m,R))$ for $k=1,2$ with the weight being distance to $e(m,R)$. We then have the following existence result and estimate when $g \in H_\alpha^{(1-\delta)}$.

**Corollary 4.25.** *If $A^{ij}$, $\theta_1$, $m$ and $\alpha$ satisfy the hypotheses of Lemma 4.24 and if $\delta \in (0,1)$, then (4.66) has a unique solution in $C^2(E(m,R)) \cap C^1(\overline{E(m,R)} \setminus e(m,R)) \cap C(\overline{E(m,R)})$ for any $g \in H_\alpha^{(1-\delta)}(E^0(m,R))$ and any $\varphi \in C(e(m,R))$. Moreover, if $\varphi \equiv 0$, then $u \in H_{1+\alpha}^{(\delta)}(E(m,R))$ and*

$$|u|_{1+\alpha}^{(-\delta)} \le C(m,\alpha,\mu)|g|_\alpha^{(1-\delta)}. \tag{4.67}$$

***Proof.*** We first prove (4.67) by observing that $w_\delta$, defined by

$$w_\delta(x) = R^{-\delta}\left(R^2 - (x^1)^2 - \frac{1}{16m^2}(x^2+8mR)^2\right)^\delta,$$

satisfies the conditions

$$\begin{gathered}\Delta w_\delta \le 0 \text{ in } E,\\ \beta \cdot D w_\delta \le -C(m)\beta^2 d_e^{\delta-1} \text{ on } E^0,\\ w_\delta = 0 \text{ on } e.\end{gathered}$$

Since $w_\delta \le C(m) d_e^\delta$, the maximum principle implies that

$$|u| \le C|g|_0^{(1-\delta)} R^\delta.$$

Using our local Hölder gradient estimates then yields (4.67).

When $\varphi = 0$, we approximate $g$ by a sequence of $H_\alpha$ functions and note that the corresponding solutions to (4.66) are bounded in $H_{1+\alpha}^{(-\delta)}$, so they have a limit which is the unique solution of (4.66). When $\varphi \in H_\delta$, then we can obtain an $H_{1+\alpha}^{(-\delta)}$ extension of $\varphi$ into $E$ with $A^{ij}D_{ij}\varphi = 0$, and hence our previous case yields the existence result here. For general nonzero $\varphi$, we take a sequence of $H_\delta$ functions which converge uniformly to $\varphi$ and note that the corresponding solutions of (4.66) converge uniformly to the desired solution. □

With this existence result in hand, we can modify the proof of Proposition 4.22 to prove an existence result for a more general vector field $\beta$.

**Proposition 4.26.** *Let $[A^{ij}] \in \mathcal{M}_\mu$ for some $\mu \ge 1$ and let $\theta_1$ and $\theta_2$ satisfy (4.65). Let $\theta_1'$ and $\theta_2'$ satisfy the conditions*

$$\frac{\pi}{2} + \frac{1}{\mu} \le \theta_1' \le \pi - \frac{1}{\mu}, \tag{4.68a}$$

$$\frac{1}{\mu} \le \theta_2' \le \frac{\pi}{2} - \frac{1}{\mu} \tag{4.68b}$$

*and define $\beta$ on $E^0(m,R)$ by*

$$\beta(x) = \begin{cases} (\cos\theta_1', \sin\theta_1') & \text{if } x \in E_1(m,R), \\ (\cos\theta_2', \sin\theta_2') & \text{if } x \in E_2(m,R). \end{cases} \tag{4.69}$$

*Also, let $\delta \in (0,1)$. Then there are constants $m_1 > 0$ and $\alpha_1 \in (0,1)$, determined only by $\mu$ such that, if $m \ge m_1$ and $\alpha \in (0,\alpha_1)$, then* (4.64) *has a solution $u \in C^2(E(m,R)) \cap C^1(\overline{E(m,R)} \setminus e(m,R)) \cap C(\overline{E(m,R)})$ for any $g \in H_\alpha^{(1-\delta)}(E^0(m,R))$ and any $\varphi \in C(e(m,R))$.*

***Proof.*** For $t \in (0,1)$, we define $\beta_t$ by

$$\beta_t(x) = (\cos(t\theta_1' + (1-t)\theta_2), \sin(t\theta_1' + (1-t)\theta_2))$$

if $x \in E_1(m,R)$ and

$$\beta_t(x) = (\cos(t\theta_2' + (1-t)\theta_1), \sin(t\theta_2' + (1-t)\theta_1))$$

if $x \in E_2(m,R)$. We then define $T$ to be the set of all $t \in (0,1)$ such that the problem

$$A^{ij}D_{ij}u = 0 \text{ in } E, \tag{4.70a}$$

$$\beta_t \cdot Du = g \text{ on } E^0, \tag{4.70b}$$

$$u = 0 \text{ on } e \tag{4.70c}$$

has a solution $u \in C^2(E) \cap H_{1+\alpha}^{(-\delta)}(E)$ for any $g \in H_\alpha^{(1-\delta)}(E^0)$. We begin by noting that, if $t \in T$, then the problem

$$A^{ij}D_{ij}u = 0 \text{ in } E,$$
$$\beta_t \cdot Du = g \text{ on } E^0,$$
$$u = \varphi \text{ on } e$$

has a solution $u \in C^2(E) \cap C^1(\overline{E} \setminus e) \cap C(\overline{E})$ for any $g \in H_\alpha^{(1-\delta)}(E^0)$ and any $\varphi \in C(e)$. Corollary 4.25 implies that $0 \in T$, and the proof of (4.67) shows that there is a constant $C$ (independent of $t$) such that whenever $t \in T$, we have

$$|u|_{1+\alpha}^{(-\delta)} \le C|g|_\alpha^{(1-\delta)}$$

for any solution of (4.70). This estimate allows us to conclude that $T$ is closed. Now suppose $t \in T$, let $\tau$ be a constant at our disposal, and let $g \in H_\alpha^{(1-\delta)}$. We then define $\mathbb{B}$ to be the set of all functions $u \in H_{1+\alpha}^{(-\delta)}$ such that

$$A^{ij}D_{ij}u = 0 \text{ in } E, \quad u = 0 \text{ on } e,$$

and we define a map $J\colon \mathbb{B} \to \mathbb{B}$ by defining $Jv$ to be the solution $u$ of the problem

$$A^{ij}D_{ij}u = 0 \text{ in } E,$$
$$\beta_t \cdot Du = g - (\beta_{t+\tau} - \beta_t) \cdot Dv \text{ on } E^0,$$
$$u = 0 \text{ on } e$$

which we know to exist because $t \in T$. Since

$$\|Jv_1 - Jv_2\|_{\mathbb{B}} \le C|\beta_{t+\tau} - \beta_t|_0 |D(v_1 - v_2)|_\alpha^{(1-\delta)},$$

it follows that $J$ is a contraction mapping if $\tau$ is sufficiently small and hence $T$ is also open. Therefore $T = [0,1]$, which means that (4.64) has a solution for all $g \in H_\alpha^{(1-\delta)}(E^0)$ whenever $\varphi \equiv 0$, and the case of nonzero $\varphi$ is handled as before. □

We can now deal with the most general assumptions of interest by using our preceding results.

**Theorem 4.27.** *Let $[A^{ij}] \in \mathcal{M}_\mu$ for some $\mu \ge 1$ and let $\theta_0$, $\theta_1$, and $\theta_2$ satisfy*

$$\theta_1 \le \theta_0 - \frac{1}{\mu}, \quad \theta_0 + \frac{1}{\mu} \le \theta_2 < \theta_1 + \pi. \tag{4.71}$$

*Let $\theta_1'$ and $\theta_2'$ satisfy the conditions (4.42), and define $\beta$ on $E^0(m,R)$ by (4.69). Also, let $\delta \in (0,1)$. Then there are constants $m_1 > 0$ and $\alpha_1 \in (0,1)$, determined only by $\mu$ such that (4.64) has a solution $u \in C^2(E(m,R)) \cap C^1(\overline{E(m,R)} \setminus e(m,R)) \cap C(\overline{E(m,R)})$ for any $g \in H_\alpha^{(1-\delta)}(E^0(m,R))$ and any $\varphi \in C(e(m,R))$.*

***Proof.*** Suppose first that $\theta_2 < \theta_1 + \pi$. We set $\bar\theta = \frac{1}{2}(\theta_1 + \theta_2)$ and note that Proposition 4.26 gives the result if $\bar\theta = \pi/2$. Since the cases $\bar\theta > \pi/2$ and $\bar\theta < \pi/2$ are symmetric, we only prove the theorem when $\bar\theta > \pi/2$.

In this case, we set $\theta'' = \frac{1}{2}(\theta_2 + \bar\theta)$ and we define $\beta_t$ for $t \in [0,1]$ by

$$\beta_t(x) = (\cos(t\theta_1' + (1-t)\theta''), \sin(t\theta_1' + (1-t)\theta''))$$

if $x \in E_1$ and $\beta_t = \beta$ on $E_2$. We then write $T$ for the set of all $t \in [0,1]$ such that (4.70) has a solution in $C^2(E) \cap H_{1+\alpha}^{(-\delta)}(E)$ for any $g \in H_\alpha^{(1-\delta)}(E^0)$. It follows from Proposition 4.26 that the problem

$$A^{ij}D_{ij}u = 0 \text{ in } E, \quad \beta_0 \cdot Du = g \text{ on } E^0$$

is locally solvable at $x_0 = 0$, so the proof of Lemma 4.24 implies that $0 \in T$. The proof of Proposition 4.26 then shows that $T$ is relatively open and relatively closed in $[0,1]$, so $1 \in T$ and hence (4.64) has a solution if $\varphi = 0$, and the case of nonzero $\varphi$ is handled in exactly the same way as in that proposition.

When $\theta_2 = \theta_1 + \pi$, we note that there are sequences $(\theta_1^k)$ and $(\theta_2^k)$ satisfying all of our hypotheses (with $1/\mu$ replaced by $1/2\mu$) such that $\theta_1^k \geq \theta_1$ and $\theta_2^k \leq \theta_2$, so we obtain a sequence of solutions to the problems

$$A^{ij} D_{ij} u_k = 0 \text{ in } E(k),$$
$$\beta \cdot Du_k = g \text{ on } E^0(k),$$
$$u_k = \varphi \text{ on } e(k),$$

with $E(k)$, $E^0(k)$, and $e(k)$ defined in the same way as $E$, $E^0$, and $e$, respectively, with $\theta_j^k$ in place of $\theta_j$ for $j = 1, 2$. Our estimates are all uniform in $k$, so there is a subsequence $(u_{k(j)})$ which converges to a solution of (4.64). □

With this existence result in hand, our usual argument gives local $C^{1,\alpha}$ estimates for the oblique derivative problem with variable coefficients. We start with regularity in a wedge (that is, a set with straight sides) because of its application in later sections.

**Proposition 4.28.** *Let $\zeta$ be a continuous increasing function defined on $[0, 1/2]$ with $\zeta(0)$, and let $R_0 > 0$, $\theta_1$, and $\theta_2$ be constants with*

$$\theta_1 < \frac{\pi}{2} < \theta_2 \leq \theta_1 + \pi,$$

*and define the sets $\Omega$, $\Sigma_k$, and $\sigma$ by (4.36). Let $[a^{ij}]$ satisfy (4.3) for some constants $\lambda$ and $\mu$ and*

$$|a^{ij}(x) - a^{ij}(y)| \leq \lambda \zeta\left(\frac{|x-y|}{d^*}\right)$$

*for all $x$ and $y$ in $\Omega$ with $|x - y| \leq \frac{1}{2} d^*(x)$, where $d^*$ denotes distance to $\sigma$. Let $\theta_k'$ be an $H_{1+\alpha}(\Sigma_k)$ function for $k = 1, 2$ and some $\alpha \in (0,1)$ and suppose that conditions (4.42) are satisfied, and define $\beta_k$ for $k = 1, 2$ by*

$$\beta_k(x) = (\cos \theta_k', \sin \theta_k'). \tag{4.72}$$

*Also, define $\beta$ on $\Sigma = \Sigma_1 \cup \Sigma_2$ by setting $\beta = \beta_k$ on $\Sigma_k$. Let $u \in C^2(\Omega) \cap C^1(\Omega \cup \Sigma \cup \{0\}) \cap C(\overline{\Omega})$ and set $Lu = a^{ij} D_{ij} u$, $Mu = \beta \cdot Du$. Then there is a constant $\alpha \in (0,1)$, determined only by $\mu$ such that if $Mu \in H_\alpha^{(1)}(\Sigma_1 \cup \Sigma_2)$ and if there is a positive constant $F$ such that (4.13) holds, then $u \in H_{1+\alpha}^{(0)}(\Omega)$ and*

$$|u|_{1+\alpha}^{(0)} \leq C(\mu, \zeta, |\beta|_\alpha^{(0)})(F + |Mu|_\alpha^{(1)} + |u|_0). \tag{4.73}$$

***Proof.*** We follow the argument in the proof of Theorem 4.5 with Theorem 4.27 in place of Lemma 4.4 and Lemma 4.18 in place of Proposition 4.3 when $y = 0$. □

We now extend this result to allow curved boundary.

**Proposition 4.29.** *Let $\Omega$, $\Sigma_k$, and $\beta$ be as in Proposition 4.28 except that $\theta_l$ and $\theta_2$ are now $H_{1+\alpha}$ functions with the same $\alpha$ as in that proposition. If $u$ satisfies the hypotheses of Proposition 4.28, then $u \in H^{(0)}_{1+\alpha}(\Omega)$ and (4.73) holds with $C$ determined also by the $H_{1+\alpha}$ norms of $\theta_1$ and $\theta_2$.*

***Proof.*** By an $H^{(-1-\alpha)}_{2+\alpha}$ change of coordinates, we can transform the problem into one with $\Sigma_1$ a straight line segment. If $\theta_2 = \theta_1 + \pi$, then this change of coordinates also makes $\Sigma_2$ a straight line segment, and Proposition 4.28 leads easily to the result.

If $\theta_2 < \theta_1 + \pi$, then, after a rotation, we may assume that $\theta_1 = 0$ and that $\Sigma_2$ can be written as $x^1 = \omega(x^2)$ for some $H_{1+\alpha}$ function $\omega$. The change of variables

$$y^1 = x^1 - \omega(x^2), y^2 = x^2$$

then converts $\Sigma_2$ to a segment on the positive $y^2$ axis while $\Sigma_1$ is still a segment on the positive $y^1$ axis. A simple argument involving regularized distance shows that this change of coordinates can be made to be $H^{(-1-\alpha)}_{2+\alpha}$, and a linear change of variables then converts the problem to one with the same angle at the origin as for the original domain. We leave it to the reader to verify that the constant $\alpha$ is unaffected by the change of coordinates. □

As in Section 4.1, we can now derive a global estimate which applies to equations with lower order terms and which estimates second derivatives as well. Although the method used here provides estimates in a wide variety of geometric settings, we only consider here a representative setting. Even in this (relatively simple) case, we need a long explanation of the relationship between the geometry, the differential operator and the boundary condition in order to determine the Hölder exponent of the gradient of the solution of the boundary value problem.

**Theorem 4.30.** *Let $\Omega$ be a bounded Lipschitz domain in $\mathbb{R}^2$, let $\beta$ be an oblique vector field defined on $\partial\Omega$, and let $[a^{ij}]$ be a matrix-valued function defined on $\Omega$. Let $\delta \in (0, 1)$ and $\mu \in [1, \infty)$ be constants, and suppose $[a^{ij}]$ satisfies condition (4.3). Let $a^{ij} \in H^{(0)}_\delta \cap C(\overline{\Omega})$, $b^i \in H^{(1-\alpha)}_\delta$, $c \in H^{(1-\delta)}_\delta$. Suppose $\partial\Omega$ is the union of finitely many $H_{1+\alpha}$ curves $\Sigma_1, \ldots, \Sigma_N$ which* *meet only at their endpoints and which meet at an angle no greater than* $\pi$ *when they do meet. Suppose* $\beta$ *is discontinuous wherever any of the curves meet and that* $\beta$ *and* $\beta^0$ *are in* $H_\alpha(\Sigma_k^o)$ *for* $k = 1, \ldots, N$*, where* $\Sigma_k^o$ *denotes the interior of* $\Sigma_k$ *relative to the topology of* $\partial\Omega$*. Suppose further that, if* $\Sigma_i$ *and* $\Sigma_j$ *(with* $i \neq j$*) meet at some point* $x_{ij}$*, and if polar coordinates* $(r, \theta)$ *centered at* $x_{ij}$ *can be introduced in the intersection of* $\overline{\Omega}$ *with a neighborhood of* $x_{ij}$ *and* $i$ *and* $j$ *are relabeled (if necessary) so that* $\theta_1$ *and* $\theta_2$*, defined by*

$$\theta_1 = \lim_{x \to x_{ij} x \in \Sigma_i} \theta(x), \quad \theta_2 = \lim_{x \to x_{ij} x \in \Sigma_j} \theta(x), \tag{4.74}$$

*satisfy* $\theta_1 < \theta_2$*, and if* $\theta_1'$ *and* $\theta_2'$ *are defined by*

$$\theta_1' = \lim_{x \to x_{ij} x \in \Sigma_i} \theta(\beta(x)), \quad \theta_2' = \lim_{x \to x_{ij} x \in \Sigma_j} \theta(\beta(x)), \tag{4.75}$$

*then there is a constant* $\theta_0$ *such that conditions* (4.42) *and* (4.71) *are satisfied. Then there is a constant* $\alpha_0(\mu) \in (0, 1)$ *such if* $\alpha \in (0, \alpha_0)$*, and if* $u \in H_{2+\delta}^{(-1-\alpha)}$*, then*

$$|u|_{2+\delta}^{(-1-\alpha)} \leq C \left( \frac{1}{\lambda} |Lu|_\delta^{(1-\alpha)} + \frac{1}{\chi} |Mu|_\alpha + |u|_0 \right) \tag{4.76}$$

*for some constant* $C$*, determined only by the modulus of continuity for* $a^{ij}$ *and upper bounds for the quantities*

$$\frac{1}{\lambda}|a^{ij}|_\delta^{(0)}, \quad \frac{1}{\lambda}|b^i|_\delta^{(1-\alpha)}, \quad \frac{1}{\lambda}|b^i|_\delta^{(1-\alpha)}, \frac{1}{\chi}|\beta|_\alpha, \quad \frac{1}{\chi}|\beta^0|_\alpha,$$

*where* $\chi = \inf_{\partial\Omega} |\beta|$.

***Proof.*** We just replace Theorem 4.5 with Proposition 4.29 in the proof of Theorem 4.6. □

## 4.6 $C^{1,\alpha}$ estimates for discontinuous $\beta$ in higher dimensions

It is now possible to obtain estimates for oblique derivative problems with discontinuous $\beta$ in more than two dimensions provided we consider sufficiently simple geometric configurations. The techniques of the preceding section will be expanded to prove the corresponding estimate when the domain is piecewise smooth with smooth portions meeting only in $(n-2)$-dimensional sets. In three dimensions, this restriction means that we can consider boundary value problems in the cylinder

$$\{|x| < 1, |x^3| < 1\}$$

but not in the cube

$$\{\max |x^i| < 1\}.$$

Such a restriction seems unnatural for the estimates, but the method of proof proposed here is unable to deal with the corners of the cube.

Our scheme is essentially identical to that in Section 4.5, so we focus here on the few changes that are made.

First, we modify our notation only slightly to consider the multi-dimensional problem. We abbreviate $x'' = (x^3, \dots, x^n)$, and we introduce coordinates $(r, \theta)$ corresponding to $(x^1, x^2)$ just as before so that

$$x^1 = r\cos\theta, \quad x^2 = r\sin\theta.$$

In the higher dimensional setting, we actually need to prove the local Hölder estimate first, and this estimate is essentially the same as Lemma 4.17.

**Lemma 4.31.** *Let $R > 0$, $\theta_0$, $\theta_1$, and $\theta_2$ be constants with $\theta_1 < \theta_0 < \theta_2 \le \theta_1 + \pi$, and define the sets $\Omega$, $\Sigma_k$, and $\sigma$ by*

$$\Omega = \{x \in \mathbb{R}^n : |x| < R,\ \theta_1 < \theta < \theta_2\}, \tag{4.77a}$$

$$\Sigma_k = \{x \in \mathbb{R}^n : 0 < |x| < R,\ \theta = \theta_k\} \tag{4.77b}$$

*for $k = 1, 2$,*

$$\sigma = \{x \in \mathbb{R}^n : |x| = R, \theta_1 \le \theta \le \theta_2\}. \tag{4.77c}$$

*Let $[a^{ij}] \in \mathcal{M}_\mu$ for some $\mu \ge 1$, and Let $\theta_1'$ and $\theta_2'$ satisfy (4.42). Suppose also that there is a positive constant $m_0$ such that $|\beta''| \le m_0$ on $\Sigma$ Then there are constants $\alpha \in (0,1)$ and $C$, determined only by $\theta_2 - \theta_1$, $\mu$, and $m_0$ such that, if $u \in C(\overline{\Omega}) \cap C^2(\Omega)$ satisfies (4.43), then (4.44) is valid.*

***Proof.*** First, as in the proof of Lemma 4.17, we may assume that $\theta_0 = \pi/2$. Moreover, $\beta^1 x^1 \le 0$ and $\beta^2 > 0$ on $\Sigma$. With $m_1 = m_0/\inf\beta^2$, we then define $G$ by

$$G(y,\rho) = \rho^{-2}\left[(y^1)^2 + \frac{1}{16m^2}(y^2)^2 + \frac{1}{16m^2 m_1^2}|y''|^2\right]$$

and we set $x_1 = (0, 8m\rho, 0, \dots, 0)$. The remainder of the proof is identical to that of Lemma 4.17. □

We then have the following higher dimensional analog of Lemma 4.16.

**Lemma 4.32.** *Under the hypotheses of Lemma 4.31, there are constants $C > 0$ and $\alpha \in (0,1)$, both determined by $\mu$ and $n$, such that*

$$|Du(x) - Du(0)| \le C \sup_\Omega |Du| \left(\frac{|x|}{R}\right)^\alpha. \tag{4.78}$$

***Proof.*** We write $\Omega(\rho)$ (for $\rho \in (0, R)$) to denote the set of all $x \in \Omega$ with $|x| < \rho$.

For $m \geq 3$, we note that

$$A^{ij} D_{ij}(D_m u) = 0 \text{ in } \Omega,$$
$$\beta_k \cdot D(D_m u) = 0 \text{ on } \Sigma_k.$$

It then follows from that Theorem 1.26 that there are constants $C$ and $\theta$ such that

$$[D_m u]_{\theta,\Omega(7R/8)} \leq CR^{-\alpha} \sup_{\Omega} |D_m u|. \tag{4.79a}$$

Then the interior gradient estimate Proposition 2.13 implies that

$$|DD_m u|^{(1-\theta)}_{\theta,\Omega(3R/4)} \leq CR^{-\theta} \sup |Du|, \tag{4.79b}$$

where the norm is weighted with respect to distance to $\Sigma_1 \cup \Sigma_2$. Now fix $x_0 \in B(0, R/2)$ such that $x_0^1 = x_0^2 = 0$ and define $w$ by

$$w(x^1, x^2) = u(x^1, x^2, x_0'').$$

It follows that

$$\hat{A}^{ij} D_{ij} w = f \text{ in } \Omega^*,$$
$$\hat{\beta} \cdot Dw = g \text{ on } \Sigma^*,$$

where $\hat{A}^{ij} = A^{ij}$ for $i, j = 1, 2$, $f$ and $g$ are defined by

$$f(x^1, x^2) = -\sum_{i>2 \text{ or } j>2} A^{ij} D_{ij} u(x^1, x^2, x_0''),$$
$$g(x^1, x^2) = -\sum_{j=3}^{n} \beta^j D_j u,$$

and $\Omega^*$ and $\Sigma^*$ are defined by

$$\Omega^* = \{x \in \mathbb{R}^2 : |x| < R/4, \theta_1 < \theta < \theta_2\},$$
$$\Sigma^* = \{x \in \mathbb{R}^2 : |x| < R/4, \theta = \theta_1 \text{ or } \theta = \theta_2\}.$$

From (4.79), we conclude that

$$|f|^{(1-\theta)}_{\theta} + |g|_{\theta} \leq CR^{-\theta} \sup |Du|,$$

so Proposition 4.28 implies that

$$[Dw] \leq CR^{-\alpha} \sup |Du|$$

for some $\alpha \in (0, \theta)$. Another application of Proposition 2.13 yields

$$|D_{ij} u| \leq CR^{-\alpha} \sup |Du| d^{\alpha-1}$$

for $i, j = 1, 2$, and (4.79) implies that this inequality is also true for all other choices of $i$ and $j$. Simple integration now gives (4.39). □

Our next step is the higher-dimensional analog of Lemma 4.18. The main new element here is that, in general, we cannot expect to have $Du(0) = 0$.

**Lemma 4.33.** *Under the hypotheses of Lemma 4.31, there are constants $C > 0$ and $\alpha \in (0,1)$, both determined only by $\mu$ and $n$, such that*

$$|u(x) - u(0) - Du(0) \cdot x| \le CR^{-1-\alpha}|u|_0|x|^{1+\alpha} \tag{4.80}$$

*for $|x| \le R/2$.*

***Proof.*** This time, we use Lemma 4.32 and integration to infer that

$$|u(x) - u(0) - Du(0) \cdot x| \le CR^{-\alpha}|Du|_{0,\Omega[R/2]}|x|^{1+\alpha}.$$

Then we apply the local maximum principle Theorem 1.30 to $\beta_1 \cdot Du$ and $\beta_2 \cdot Du$ and the local maximum principle Theorem 1.27 to each component of $D''u$ to conclude that

$$|Du|_{0,\Omega[R/2]} \le CR^{-n} \int_{\Omega[3R/4]} |Du|\, dx.$$

From Lemma 4.31 and the proof of Lemma 2.20, we conclude that

$$\int_{\Omega[3R/4]} |Du|\, dx \le CR^{n-1}|u|_0,$$

and combining all these estimates yields (4.45). □

At this point, the differences between the two-dimensional case and the higher-dimensional case become relatively minor for a while. In fact, the statements and proofs of the higher-dimensional versions of Lemmata 4.19 and 4.20 differ from the two-dimensional ones only in notation (and reference to the appropriate, previously proved results), so we just recall the notation and statement of Lemma 4.20.

For $R > 0$ and $\kappa \in (0,1)$, we set $x_1(\kappa, R) = (0, -\kappa R, 0, \ldots, 0)$, and we define the sets

$$\begin{aligned}
Q(\kappa, R) &= \{x \in \mathbb{R}^n : |x - x_1(\kappa, R)| < R,\ \pi/4 < \theta < 3\pi/4\},\\
S_1(\kappa, R) &= \{x \in \mathbb{R}^n : |x - x_1(\kappa, R)| < R,\ \theta = \pi/4\},\\
S_2(\kappa, R) &= \{x \in \mathbb{R}^n : |x - x_1(\kappa, R)| < R,\ \theta = 3\pi/4\},\\
s(\kappa, R) &= \{x \in \mathbb{R}^n : |x - x_1(\kappa, R)| = R,\ \pi/4 \le \theta \le 3\pi/4\},\\
Q^0(\kappa, R) &= S_1(\kappa, R) \cup S_2(\kappa, R).
\end{aligned}$$

For $\alpha \in (0,1)$, we write $H_\alpha(Q^0)$ for the set of all functions $g$ defined on $Q^0$ such that the restriction of $g$ to $S_1(\kappa, R)$ is in $H_\alpha(S_1(\kappa, R))$ and the restriction of $g$ to $S_2(\kappa, R)$ is in $H_\alpha(S_2(\kappa, R))$. Since $\kappa$ and $R$ will usually be fixed in what follows, we generally omit them from the notation. In what follows, $\gamma$ denotes the unit inner normal to $Q$.

**Lemma 4.34.** *Let $\alpha \in (0,1)$, $R > 0$, and $\kappa \in (0,1)$. Then (4.50) has a unique solution $u \in C^2(Q) \cap C^1(\overline{Q} \setminus s) \cap C(\overline{Q})$ for any $f \in H_\alpha(Q)$, any $g \in H_\alpha(Q^0)$, and any $\varphi \in C(s)$.*

In our estimates for problems with less regular data, we need to take into account one additional fact, namely, there are points $x \in s$ with $x^1 = x^2 = 0$. Fortunately, this situation is easily handled, and Lemma 4.21 remains true in higher dimensions.

**Lemma 4.35.** *Let $\delta \in (0,1)$. For any $\alpha \in (0,1)$, there is a constant $\kappa_1$ in $(0,1)$, determined only by $\alpha$ such that, if $\kappa \in (\kappa_1, 1)$, then any solution $u$ of (4.50) with $\varphi = 0$ on $s$ and $g = 0$ on $Q^0 \cap s$ satisfies the estimate (4.53). Moreover, (4.50) has a unique solution for any $f \in H_\delta^{(1-\alpha)}$, any $g \in H_\alpha(Q^0)$ with $g = 0$ on $Q^0 \cap s$, and $\varphi = 0$.*

***Proof.*** The argument in Lemma 4.21 implies that

$$|u|_0 \le CR^{1+\alpha}[|f|_0^{(1-\alpha)} + |g|_0^{(-\alpha)}].$$

We now fix $y_0 \in \overline{Q^0} \cap s$ and note that there is a rotation $T$ such which maps the positive $x^2$ axis to the positive $z^2$ axis and such that the positive $z^1$ axis points in the direction of the vector $-y_0$, where we abbreviate $z = T(x - y)$. To continue, we introduce some notation, based on the proof of Proposition 3.16. We write $\hat{z} = (z^1, z^2)$, and we introduce polar coordinates $(\hat{r}, \hat{\theta})$ via the equations

$$z^1 = \hat{r}\cos\hat{\theta}, \quad z^2 = \hat{r}\sin\hat{\theta}.$$

A simple calculation shows that, if we take $W(z) = \hat{r}^{1+\alpha} f(\hat{\theta})$, then we can choose $f$ so that $\gamma_1 \cdot DW$ and $\gamma_2 \cdot DW$ are bounded from above by $-\hat{r}^\alpha$ when $\hat{\theta} = 0$ and such that $\mathcal{M}_1 W \le -\hat{r}^{\alpha-1}$ when $\theta \le \theta_0$ for some positive $\theta_0$. By continuity, there is a constant $\theta_1 \in (0, \theta_0)$ such that

$$\gamma_l \cdot DW \le -\frac{1}{2}\hat{r}^\alpha$$

if $\hat{\theta} \in (0, \theta_1)$, and hence, by choosing $\kappa_1$ sufficiently close to 1, we can arrange that $\hat{\theta} \le \theta_1$ in $\Omega$. We then use the comparison function

$$w(x) = A_1 W(z(x)) - A_2(d_{S_1})^{1+\alpha} - A_3(d_{S_2})^{1+\alpha} - A_4(d_\sigma)^{1+\alpha}$$

with suitable constants $A_1, \ldots, A_4$ to infer that

$$|u(x) - u(y_0)| \le C(|f|_0^{(1-\alpha)} + |g|_0^{(-\alpha)}]$$

for all $x \in \Omega$ with $z''(x) = 0$. In particular, this inequality is valid if we choose $x \in \Omega$ and then take $y_0$ to be the closest point in $S_0$ to $x$.

With this estimate in place, we can proceed as in the proof of Lemma 4.21. □

At this point, the proof of all the estimates is the same as in the two-dimensional case except for some minor notational issues and the use of the function $W$ from the proof of Lemma 4.35 to prove the analog of Proposition 4.22. We leave the details to the reader and jump directly to the statement of the higher-dimensional version of Theorem 4.27. To state this proposition, we introduce some notation, based on that for Lemma 4.24. For positive constants $m$, $m_1$, and $R$, we set

$$N(x) = (x^1)^2 + \frac{1}{16m^2}(x^2 + 8mR)^2 + \frac{1}{16m_1^2}|x''|^2$$

and we define the sets

$$\begin{aligned} E(m, m_1, R) &= \{x \in \mathbb{R}^n : N(x) < R^2, \theta_1 < \theta < \theta_2\}, \\ E_k(m, R) &= \{x \in \mathbb{R}^n : N(x) < R^2, \theta = \theta_k\} \end{aligned}$$

for $k = 1, 2$,

$$\begin{aligned} E_0(m, m_1, R) &= \{x \in \mathbb{R}^n : x^1 = x^2 = 0, |x''| < 4m_1 R\}, \\ e(m, m_1, R) &= \{x \in \mathbb{R}^n : N(x) = R^2, \theta_1 \le \theta \le \theta_2\}, \\ E^0(m, m_1, R) &= E_1(m, m_1, R) \cup E_2(m, m_1, R). \end{aligned}$$

We also define $H_\alpha(E^0(m, m_1, R))$ to be the set of all functions $g$ defined on $E^0(m, m_1, R)$ such that the restriction of $g$ to $E_k(m, m_1, R)$ is in $H_\alpha(E_k(m, m_1, R))$ for $k = 1, 2$.

**Theorem 4.36.** *Let $[A^{ij}] \in \mathcal{M}_\mu$ for some $\mu \ge 1$ and let $\theta_0$, $\theta_1$, and $\theta_2$ satisfy (4.71). Let $\theta_1'$ and $\theta_2'$ satisfy (4.68), let $\beta_1''$ and $\beta_2''$ be $(n-2)$-dimensional vectors with $|\beta_k''| \le 1/\mu$ (for $k = 1, 2$), and define $\beta$ on $E^0(m, m_1, R)$ by*

$$\beta(x) = \begin{cases} (\cos\theta_1', \sin\theta_1', \beta_1'') & \text{if } x \in E_1(m, m_1, R), \\ (\cos\theta_2', \sin\theta_2', \beta_2'') & \text{if } x \in E_2(m, m_1, R). \end{cases} \tag{4.81}$$

*Also, let $\delta \in (0, 1)$. Then there are constants $m_2 > 0$, $m_3 > 0$, and $\alpha_1 \in (0, 1)$, determined only by $\mu$ such that if $\alpha \in (0, \alpha_1)$, $m \ge m_2$, and*

*$m_1 \ge m_3 m$, then*

$$A^{ij}D_{ij}u = 0 \text{ in } E(m, m_1, R), \tag{4.82a}$$

$$\beta \cdot Du = g \text{ on } E^0(m, m_1, R), \tag{4.82b}$$

$$u = \varphi \text{ on } e(m, m_1, R) \tag{4.82c}$$

*has a solution $u \in C^2(E(m, m_1, R)) \cap C^1(\overline{E(m, m_1, R)} \setminus e(m, m_1, R)) \cap C(\overline{E(m, m_1, R)})$ for any $g \in H_\alpha^{(1-\delta)}(E^0(m, m_1, R))$ and any $\varphi \in C(e(m, m_1, R))$.*

We also have the higher-dimensional analog of Theorem 4.30, which we state without proof.

**Theorem 4.37.** *Let $\Omega$ be a bounded Lipschitz domain in $\mathbb{R}^n$, let $\beta$ be an oblique vector field defined on $\partial\Omega$, and let $[a^{ij}]$ be a matrix-valued function defined on $\Omega$. Let $\delta \in (0, 1)$ and $\mu \in [1, \infty)$ be constants, and suppose $[a^{ij}]$ satisfies condition (4.3). Let $a^{ij} \in H_\delta^{(0)} \cap C(\overline{\Omega})$, $b^i \in H_\delta^{(1-\alpha)}$, $c \in H_\delta^{(1-\delta)}$. Suppose $\partial\Omega$ is the union of finitely many $H_{1+\alpha}$ hypersurfaces $\Sigma_1, \dots, \Sigma_N$ which meet only at their edges and which meet at an angle no greater than $\pi$ when they do meet. Suppose $\beta$ is discontinuous wherever any of the hypersurfaces meet and that $\beta$ and $\beta^0$ are in $H_\alpha(\Sigma_k^o)$ for $k = 1, \dots, N$, where $\Sigma_k^o$ denotes the interior of $\Sigma_k$ relative to the topology of $\partial\Omega$. Suppose further that, if $\Sigma_i$ and $\Sigma_j$ (with $i \ne j$) meet at some point $x_{ij}$, and if cylindrical coordinates $(r, \theta, x'')$ centered at $x_{ij}$ with $\Sigma_i \cap \Sigma_j$ tangent to the set on which $x'' = 0$ can be introduced in the intersection of $\overline{\Omega}$ with a neighborhood of $x_{ij}$ and $i$ and $j$ are relabeled (if necessary) so that $\theta_1$ and $\theta_2$, defined by (4.74) satisfy $\theta_1 < \theta_2$, and if $\theta_1'$ and $\theta_2'$ are defined by (4.75), then there is a constant $\theta_0$ such that conditions (4.42) and (4.71) are satisfied. Then there is a constant $\alpha_0(\mu) \in (0, 1)$ such if $\alpha \in (0, \alpha_0)$, and if $u \in H_{2+\delta}^{(-1-\alpha)}$, then (4.76) holds for some constant $C$, determined only by the modulus of continuity for $a^{ij}$ and upper bounds for the quantities*

$$\frac{1}{\lambda}|a^{ij}|_\delta^{(0)}, \quad \frac{1}{\lambda}|b^i|_\delta^{(1-\alpha)}, \quad \frac{1}{\lambda}|b^i|_\delta^{(1-\alpha)}, \frac{1}{\chi}|\beta|_\alpha, \quad \frac{1}{\chi}|\beta^0|_\alpha,$$

*where $\chi = \inf_{\partial\Omega} |\beta|$.*

## 4.7 $C^{2,\alpha}$ estimates

In this section, we give the final Hölder estimate of the chapter. Note that it improves the classical estimate from Chapter 2 by reducing the regularity hypothesis on $\partial\Omega$.

As a first step, we obtain an estimate for the solution of a simpler problem in $\Omega[R]$.

**Lemma 4.38.** *With $\omega$ as in Proposition 4.14, let $[a^{ij}]$ be a constant matrix with eigenvalues in the interval $[\lambda, \mu\lambda]$, let $\beta$ be an $H_{1+\alpha}$ vector field defined on $\Sigma[R]$ with $|\beta'| \le \mu_0$ for some constant $\mu_0$ and $\beta^n \equiv 1$, and let $g \in H_{1+\alpha}(\Sigma[R])$. Suppose in addition that $R \le 1$. Let $\delta \in (0,1)$. If $v$ satisfies the conditions*

$$a^{ij} D_{ij} v = 0 \text{ in } \Omega[R], \quad \beta \cdot Dv = g \text{ on } \Sigma[R], \tag{4.83}$$

*then there is a positive constant $C$ determined only $\mu$, $n$, $|\beta|_{1+\alpha}$, $\delta$, and $|\omega|_{1+\alpha}$ such that*

$$|v|_{2+\alpha}^{(-\delta)} \le C(|g|_{1+\alpha} + |v|_\delta). \tag{4.84}$$

***Proof.*** To simplify the writing, we set $G = |g|_{1+\alpha}$, $B = |\beta|_{1+\alpha}$, and $V = |v|_\delta$. Extend $\beta$ and $g$ as $C^2(\Omega[R])$ functions so that $|D^2\beta(x)| \le C(n,\omega,\alpha)B(x^n)^{\alpha-1}$ and $|D^2 g(x)| \le C(n,\omega,\alpha)G(x^n)^{\alpha-1}$ in $\Omega[R]$. It follows that $w = \beta \cdot Dv - g$ satisfies

$$a^{ij} D_{ij} w = f \text{ in } \Omega[R], \quad w = 0 \text{ on } \Sigma[R]$$

for some function $f$ satisfying

$$|f| \le C(n,\alpha,\omega,B)(G + |v|_1^{(-\delta)})(x^n)^{\alpha-1}.$$

From Theorem 4.5 and the argument in Lemma 2.14, we infer that

$$|v|_1^{(-\delta)} \le C(G+V),$$

so

$$|f| \le C(G+V).$$

It then follows that

$$|w|_{1+\alpha}^{(1-\delta)} \le C(G+V). \tag{4.85}$$

We now fix $k \in \{1,\dots,n\}$ and set $v_k = D_k v$. From (4.85), we conclude that there is a function $g_k$ satisfying $|g_k|_\alpha^{(1-\delta)} \le C(G+V)$ such that

$$a^{ij} D_{ij} v_k = 0 \text{ in } \Omega[R], \quad \beta \cdot Dv_k = g_k \text{ on } \Sigma[R],$$

and then Theorem 4.5 and the argument in Lemma 2.14 imply that

$$|v_k|_{1+\alpha}^{(1-\delta)} \le C(G+V).$$

It follows that

$$|Dv|_{1+\alpha}^{(1-\delta)} \le C(G+V),$$

and this estimate implies (4.84). □

This result then implies a local $C^{2,\alpha}$ regularity result for solutions of the oblique derivative problem with $C^{1,\alpha}$ boundary.

**Lemma 4.39.** *Let $R_0$, $\alpha$ and $\delta$ be positive constants with $R_0 \le 1$ and $\alpha, \delta < 1$, and let $\omega$ be an $H_{1+\alpha}(B_{n-1}(x_0', R_0))$ function for some $x_0 \in \mathbb{R}^n$ with $x_0^n = \omega(x_0')$. For $R \in (0, R_0]$, write $\Omega[R]$ and $\Sigma[R]$ for the sets of all $x \in B(x_0, R)$ with $x^n > \omega(x')$ and $x^n = \omega(x')$, respectively. Suppose $a^{ij}$ satisfies (4.3) and $\beta$ satisfies $|\beta'| \le \mu_0$ and $\beta^n \equiv 1$. If $a^{ij} \in H_\alpha$ and $\beta \in H_{1+\alpha}$, then there is a constant $R \in (0, R_0]$ such that*

$$|u|^{(-\delta)}_{2+\alpha,\Omega[R]} \le C(|a^{ij}D_{ij}u/\lambda|^{(2-\delta)}_{\alpha,\Omega[R]} + |u|_{\delta,\Omega[R]} + |\beta \cdot Du|_{1+\alpha,\Omega[R]})$$

*for any $u \in H^{(-\delta)}_{2+\alpha,\Omega[R]}$.*

***Proof.*** First, we abbreviate $f = a^{ij}D_{ij}u$ and $g = \beta \cdot Du$. With $R$ at our disposal, we define $f_1 = f + [a^{ij}(x_0) - a^{ij}]D_{ij}u$ and set $F_1 = |f_1|^{(2-\delta)}_{\alpha,\Omega[R]}$. We then extend $f_1$ to $B(x_0, R)$ so that $f_1 \in H^{(2-\delta)}_\alpha(B(x_0, R))$ with

$$|f_1|^{(2-\delta)}_{\alpha,B(x_0,R)} \le CF_1,$$

so Corollary 2.16 implies that there is a unique solution $w$ of

$$a^{ij}(x_0)D_{ij}w = f_1 \text{ in } B(x_0, R), \quad w = 0 \text{ on } \partial B(x_0, R)$$

with

$$|w|^{(-\delta)}_{2+\alpha,B(x_0,R)} \le C(F_1 + |u|^{(-\delta)}_\delta).$$

Setting $g_1 = g - \beta \cdot Dw$, we conclude that $v = u - w$ satisfies (4.83) with $g_1$ in place of $g$. Then

$$|v|^{(-\delta)}_{2+\alpha,\Omega[R]} \le C(|g_1|^{(1-\delta)}_{1+\alpha,\Sigma[R]} + |v|^{(-\delta)}_\delta).$$

Since $|w|^{(-\delta)}_{2+\alpha,\Omega[R]} \le |w|^{(-\delta)}_{2+\alpha,B(x_0,R)}$, it follows that

$$|v|^{(-\delta)}_{2+\alpha,\Omega[R]} \le C(|u|^{(-\delta)}_\delta + |g|^{(1-\delta)}_{1+\alpha} + F_1).$$

Therefore

$$|u|^{(-\delta)}_{2+\alpha,\Omega[R]} \le C(|u|^{(-\delta)}_\delta + |g|^{(1-\delta)}_{1+\alpha} + F_1).$$

If we abbreviate $U = |u|^{(-\delta)}_{2+\alpha,\Omega[R]}$ and $U_0 = |u|^{(-\delta)}_\delta + |g|^{(1-\delta)}_{1+\alpha} + |f|^{(2-\delta)}_\alpha$, we conclude that

$$U \le C(U_0 + |[a^{ij}(x_0) - a^{ij}]D_{ij}u|^{(2-\delta)}_\alpha) \le CU_0 + CR^\alpha U + C[u]^{(-\delta)}_2.$$

Using the interpolation inequality Lemma 2.3, we conclude that

$$U \le CU_0 + CR^\alpha U$$

and the proof is concluded by choosing $R$ so small that $CR^\alpha \le 1/2$. □

From this local result, we infer the global estimate. We leave the details of the derivation to the reader in Exercise 4.4.

**Theorem 4.40.** *Let $\alpha \in (0,1)$ and let $\Omega$ be a bounded $H_{1+\alpha}$ domain. Suppose $a^{ij}$ satisfies (4.3) and $\beta$ is oblique on $\partial\Omega$. If $a^{ij}$,$b$ and $c$ are in $H_\alpha$ and if $\beta$ and $\beta^0$ are in $H_{1+\alpha}$, and if $u \in C^2(\Omega) \cap C(\overline{\Omega})$ with $Lu \in H_\alpha$ and $Mu \in H_{1+\alpha}$, then $u \in H_{2+\alpha}$ and*

$$|u|_{2+\alpha} \le C(|Lu/\lambda|_\alpha + |Mu/\chi|_{1+\alpha} + |u|_0). \tag{4.86}$$

As usual, this theorem has an associated existence theorem.

**Corollary 4.41.** *With $\Omega$, $L$, and $M$ as in Theorem 4.40, for any $f \in H_\alpha$ and $g \in H_{1+\alpha}$, the problem*

$$Lu = f \text{ in } \Omega, \quad Mu = g \text{ on } \partial\Omega \tag{4.87}$$

*has a unique solution if $c \le 0$, $\beta^0 \le 0$ and one of these two functions is not identically zero.*

## Notes

The basic ideas in this chapter were first presented in [104] and refined in [117] for continuous $\beta$ and in [109] for discontinuous $\beta$. The proof here for discontinuous $\beta$ differs from the one in [109] in several ways; the most important one is that [109] relies heavily on the theory of weak solutions (which we discuss in Chapter 5) while we avoid that theory altogether. See also [153] for an alternative approach to the Hölder gradient estimate for continuous $\beta$ in Lipschitz domains. Various estimates have been proved for discontinuous $\beta$ as well, but most of them are not the same as the ones in this chapter. We mention [68] and [126] in particular, but refer the reader to [85] for a more detailed examination of the problem in a more general setting.

Before comparing the results in this chapter to those in [68], we point out that conditions (4.40) and (4.71) are important not just because they allow our proofs to work. By examining functions of the form $r^\alpha \cos(\alpha\theta - \delta_0)$ with suitable constants $\alpha$ and $\delta_0$, we can show that there are solutions of the boundary value problem $\Delta u = 0$ in $E(m,R)$, $\beta \cdot Du = 0$ on $E^0(m,R)$ which do not have Hölder continuous gradient at the origin.

More specifically, in Chapter 6 of [68], Grisvard studies the problem

$$\Delta u = f \text{ in } \Omega,$$
$$\beta \cdot Du = 0 \text{ on } \mathcal{N},$$
$$u = 0 \text{ on } \mathcal{D},$$

where $\Omega$ is a polygon, $\mathcal{N}$ is the union of some of the edges of $\partial\Omega$, and $\mathcal{D}$ is the union of the remaining edges. Assuming that $\mathcal{D}$ is not empty, and some technical conditions are satisfied (primarily that $\beta$ is not a constant vector which is tangential on all of $\mathcal{D}$ and that $f \in H_\alpha$), he shows that there is a solution of this problem which is globally continuous and satisfies all boundary conditions classically. Moreover, as shown in Theorem 6.4.2.5 of [68], any such solution can be written as the sum of an $H_{2+\alpha}$ function and the sum of functions which near each vertex can be written as

$$r^a \cos(a\theta + b)$$

or (in certain exceptional cases) as

$$r^a[\ln r \cos(a\theta + b) + \theta \sin(a\theta + b)],$$

where $(r, \theta)$ denote polar coordinates centered at the vertex, and the numbers $a$ and $b$ are chosen so the boundary conditions are satisfied near the vertex and so that $a \in (0, 2 + \alpha)$. In fact, the exponent $a$ can be computed rather easily, and is equal to

$$1 + \frac{\theta_1' - \theta_2' + m\pi}{\theta_2 - \theta_1}$$

for suitable integers $m$, using our notation. In the cases we have studied, there is at most one such number $a$ in the interval $(0, 1)$ and the corresponding functions do not appear in any solution which satisfies the type of boundary condition which we prescribe at the vertex. For this reason, Grisvard can prove the existence of solutions but not their uniqueness.

Hölder gradient estimates for the Neumann problem (with the Laplace operator) in two-dimensional domains with angles were studied by Volkov, too. In [196], he proved that the solution is in $H_a$ for a given non-integer $a$ proved the angle at each corner is greater than $\pi/a$, assuming sufficient smoothness of the data. In [195], he proved that, if the angle at a corner is $\pi/m$ for some integer $m \geq 2$, then the solution is in $H_a$ provided suitable compatibility conditions are satisfied at the corner. Although we shall not write these compatibility conditions out for general $m$, we note here that the conditions corresponding to $a \in (2, 3)$ are exactly those that allow the solution to have continuous second derivatives at the corner, namely, if the domain is translated and rotated so that the corner is at the origin and the two sides lie locally on the positive $x$-axis and the positive $y$-axis, then the condition on the $x$-axis prescribes $u_y(x,0)$ while the condition on the $y$-axis prescribes $u_x(0,y)$. By differentiating the first boundary condition with respect to $x$ and the second with respect to $y$, we obtain two values for $u_{xy}(0,0)$, and the compatibility condition is that these two values must agree. Further results along these lines can be found in [202] and [203].

We also mention that the results for discontinuous $\beta$ are closely related to issues of reflected Brownian motion. In this connection, the papers [204, 177, 159] are of especial interest.

## Exercises

**4.1** Verify that the function $\rho$ in Lemma 4.7 satisfies all the properties listed in that lemma. (Hint: Look at Theorems 1.3 and 2.1 in [99].)

**4.2** Verify that the function $\rho$ in Lemma 4.9 satisfies all the properties listed in that lemma. (Hint: Look at Theorems 1.3 and 2.1 in [99].)

**4.3** Construct the function $\varphi^*$ described in the proof of Proposition 4.14. (Hint: Look at the proof of Lemma 6.38 in [64] and Lemma 4.24 in [114].)

**4.4** Derive Theorem 4.40 from Lemma 4.39.

# Chapter 5

# Weak Solutions

## Introduction

In this chapter, we study weak solutions of the conormal problem. The motivation for this problem is to provide a suitable generalization of the Neumann problem for Laplace's equation. Of course, our general oblique derivative problem is a generalization of this problem, but such solutions are still required to be $C^2$ inside $\Omega$. Weak solutions are only required to have first derivatives in some Sobolev space, and this weaker regularity leads to several technical difficulties. Fortunately, these difficulties are easily overcome by well-known methods. In fact, the main part of the theory in this chapter is a reworking of the material in Chapter 8 of [64]; we shall write a few results in somewhat different form (in particular, our approach to the Moser iteration scheme is slightly different from the one in [64], and we use a perturbation argument like the one in Chapter 2 to obtain Schauder estimates), but it is not difficult to rewrite all of our proofs to conform with those in [64]. We include the results just for completeness. In particular, we recall the definition and some basic properties of weak derivatives in Section 5.1; we use these results here.

The idea is to start with the Neumann problem for Laplace's equation

$$\Delta u = 0 \text{ in } \Omega, \quad \frac{\partial u}{\partial \gamma} = g \text{ on } \partial\Omega, \tag{5.1}$$

multiply the differential equation by a smooth function $\varphi$, and then integrate by parts. In this way, we obtain the integral equation

$$\int_\Omega Du \cdot D\varphi \, dx = -\int_{\partial\Omega} g\varphi \, ds, \tag{5.2}$$

where $ds$ denotes surface measure on $\partial\Omega$. Noting that this integral equation makes sense for $u \in W^{1,2}(\Omega)$, we take this integral identity as the definition of a weak solution of the boundary value problem. In other words, we say that $u \in W^{1,2}$ is a *weak solution* of (5.1) if the identity (5.2) is satisfied for all $\varphi \in C^1(\overline{\Omega})$. One of the many advantages of this weak formulation is a weaker regularity hypothesis on the function $g$ since (5.1) makes sense even if $g$ is only in $L^1(\partial\Omega)$. As we shall eventually see, a somewhat stronger assumption must be made on $g$; this problem is solvable (in the weak sense) for $g \in L^2(\partial\Omega)$ under the obvious compatibility condition

$$\int_{\partial\Omega} g\, ds = 0. \tag{5.3}$$

In fact, there is an important comment to be made here about the regularity results in this chapter as compared with the results in the preceding chapter. We were able to show that solutions of the oblique derivative problem are (globally) $C^{1,\alpha}$ even if the domain is Lipschitz. Here, we prove this regularity under the stronger hypothesis that the boundary is $C^{1,\alpha}$ (similarly to the situation in Chapter 2), but under weaker assumptions on the coefficients of the operator $L$. In addition, the results in this chapter require a relationship between the matrix $[a^{ij}]$ and the vector $\beta$ which was not needed in the preceding chapters. For linear equations, this connection is relatively unimportant; it is possible to write an oblique derivative problem (with suitably smooth data) in the form studied here by adapting the proof of Proposition 6.21 in the next chapter. For nonlinear equations, we shall see in Chapter 11 that this special relationship is a critical part of the theory.

We now give the formal definition of the weak solution of the oblique derivative problem. Let $\Omega$ be a bounded, Lipschitz domain with unit inner normal $\gamma$. Let $a^{ij}$, $b^i$, $c^i$, $c^0$, and $\beta^0$ be bounded functions and suppose that there is a positive constant $\lambda$ such that

$$a^{ij}(x)\xi_i\xi_j \ge \lambda|\xi|^2 \tag{5.4}$$

for almost all $x \in \Omega$ and all $\xi \in \mathbb{R}^n$. We then say that $u \in W^{1,2}(\Omega)$ is a *weak solution* of

$$D_i(a^{ij}D_ju + b^iu) + c^iD_iu + c^0u = D_if^i + g \text{ in } \Omega, \tag{5.5a}$$

$$a^{ij}\gamma_iD_ju + b\cdot\gamma u + \beta^0u = \psi + f\cdot\gamma \tag{5.5b}$$

if the integral identity

$$\int_\Omega [(a^{ij}D_ju + b^iu - f^i)D_i\varphi + (-c^iD_iu - c^0u + g)\varphi]\, dx = \int_{\partial\Omega} [\beta^0u - \psi]\varphi\, ds \tag{5.6}$$

is satisfied for all $\varphi \in C^2(\overline{\Omega})$. We also abbreviate (5.5) to

$$Lu = D_if^i + g \text{ in } \Omega, \quad Mu = \psi \text{ on } \partial\Omega. \tag{5.7}$$

## 5.1 Definitions and basic properties of weak derivatives

The key point here is the concept of weak derivatives. Let $\Omega$ be an open subset of $\mathbb{R}^n$ and let $i \in \{1, \dots, n\}$. We say that $v \in L^1_{\text{loc}}(\Omega)$ is the weak derivative $D_i u$ of $u \in L^1_{\text{loc}}(\Omega)$ if, for any $\varphi \in C^1_c(\overline{\Omega})$ (the set of all continuous functions which have uniformly continuous gradient on $\Omega$ and which have compact support in $\Omega$), we have

$$\int_\Omega v\varphi \, dx = -\int_\Omega u D_i \varphi \, dx.$$

Integration by parts shows that, if $u \in C^1(\Omega)$, then the usual partial derivative $D_i u$ is also the weak derivative. There are several basic properties of weak derivatives which we now reproduce, referring to [64] for the proofs.

**Lemma 5.1.** *Let $u$ and $v$ be locally integrable in $\Omega$. Then $v = D_i u$ weakly if and only if there is a sequence $(u_m)$ of $C^\infty(\Omega)$ functions such that $u_m \to u$ and $D_i u_m \to v$ in $L^1_{loc}$.*

For our next lemma, we write

$$u_+ = \max\{u, 0\}, \quad u_- = \max\{-u, 0\},$$

and we observe that $u = u_+ - u_-$ and $|u| = u_+ + u_-$.

**Lemma 5.2.** *Let $u \in L^1_{loc}$ and suppose that the weak derivative $D_i u$ exists. Then $u_+$, $u_-$, and $|u|$ have weak derivatives satisfying*

$$D_i u_+ = \begin{cases} D_i u & \textit{where } u > 0, \\ 0 & \textit{where } u \le 0, \end{cases}$$

$$D_i u_- = \begin{cases} D_i u & \textit{where } u < 0, \\ 0 & \textit{where } u \ge 0, \end{cases}$$

$$D_i |u| = \begin{cases} D_i u & \textit{where } u > 0, \\ 0 & \textit{where } u = 0, \\ -D_i u & \textit{where } u < 0. \end{cases}$$

We also write $W^{1,2}$ for the set of all $u \in L^2$ such that the weak derivatives $D_1 u, \dots, D_n u$ are also in $L^2$, and we write $W^{1,2}_0$ for the closure of $C^1_c(\Omega)$. It can be shown that any function in $W^{1,2}_0$ has a well-defined trace on $\partial\Omega$, at least for Lipschitz domains, which must be zero.

Many of the rules of calculus are also valid for weak derivatives. We note here that the product rule $D(uv) = uDv + vDu$ is valid as long as

$u$ and $v$ have weak derivatives with $uDv$ and $vDu$ being $L^1$ functions. In addition, if $\psi$ is a function that maps $\Omega$ onto some domain $\tilde{\Omega}$ in $\mathbb{R}^n$ with $\psi \in C^1(\Omega)$ and $\psi^{-1} \in C^1(\tilde{\Omega})$, then, for every $u$ which has weak derivatives in $\Omega$, the corresponding function $v = u \circ \psi^{-1}$ is weakly differentiable in $\tilde{\Omega}$ and

$$D_i u(x) = \frac{\partial y^j}{\partial x^i} \frac{\partial v}{\partial y^j}(y) \tag{5.8}$$

for almost all $x \in \Omega$, where $y = \psi(x)$. Furthermore, if $f$ is continuously differentiable on $\mathbb{R}$ with bounded derivative, and if $u$ is weakly differentiable in $\Omega$, then $f \circ u$ is also weakly differentiable in $\Omega$ and $D(f \circ u) = f'(u)Du$. (This version of the chain rule is the basis for proving Lemma 5.2.)

We can also define various inequalities on $W^{1,2}$ functions at the boundary. If $u$ and $v$ are in $W^{1,2}$, we say that $u \le v$ on $\partial\Omega$ if $(u - v)_+ \in W_0^{1,2}$. This concept will be very useful in later chapters as well.

We also recall the following results about the relation between derivatives and difference quotients of functions in Sobolev spaces. To state results succinctly, we introduce the notation

$$u_{i,h} = \frac{u(x + he_i) - u(x)}{h}$$

for $h \ne 0$ and $i = 1, \dots, n$. We then have the following results (see Lemmata 7.23 and 7.24 in [64]).

**Lemma 5.3.** *Let $u \in W^{1,p}(\Omega)$ for some $p \ge 1$. Then $u_{i,h} \in L^p(\Omega')$ for any $\Omega'$ such that $d(x) > |h|$ for all $x \in \Omega'$. In addition,*

$$\|u_{i,h}\|_{p;\Omega'} \le \|D_i u\|_{p;\Omega}.$$

**Lemma 5.4.** *Let $u \in L^p(\Omega)$ for some $p > 1$, and suppose that there is a constant $K$ such that $u_{i,h} \in L^p(\Omega')$ and $\|u\|_{p;\Omega'} \le K$ for any $h > 0$ and $\Omega' \subset \Omega$ with $d(x) > |h|$ for all $x \in \Omega'$. Then the weak derivative $D_i u$ exists, $D_i u \in L^p(\Omega)$, and $\|D_i u\|_p \le K$.*

## 5.2 Sobolev imbedding theorems

Several deeper results for functions in Sobolev spaces will be needed. The first one is a special case of the Sobolev imbedding theorem. We refer the reader to Theorem 7.10 of [64] or Theorem 6.7 of [114] for a proof.

**Theorem 5.5.** *For any domain $\Omega$ in $\mathbb{R}^n$, $W_0^{1,1}(\Omega) \subset L^{n/(n-1)}(\Omega)$ and there is a constant $C$, determined only by $n$, such that*

$$\|u\|_{n/(n-1)} \le C\|Du\|_1. \tag{5.9}$$

It will also be useful to have a version of the Sobolev imbedding theorem for functions with nonzero boundary values. For our applications to local estimates, we shall find an intermediate form of this inequality very useful, so we present it here with proof.

**Lemma 5.6.** *Let $R > 0$ and $\omega_0$ be positive constants, and let $\omega$ be a real-valued function defined on $B_{n-1}(R)$ with*

$$|\omega(x') - \omega(y')| \leq \omega_0 |x' - y'| \tag{5.10}$$

*for all $x'$ and $y'$ in $B_{n-1}(R)$ and $\omega(0) \in (0, R)$. Define*

$$\Omega[R] = \{x \in \mathbb{R}^n : x^n > \omega(x'), |x| < R\}, \tag{5.11a}$$

$$\sigma[R] = \{x \in \mathbb{R}^n : x^n \geq \omega(x'), |x| = R\}. \tag{5.11b}$$

*If $u \in W^{1,1}(\Omega[R])$ and $u = 0$ on $\sigma[R]$, then there is a constant $C$, determined only by $n$ and $\omega_0$, such that* (5.9) *holds.*

***Proof.*** By virtue of Lemma 1.3, we may assume that $\omega$ is defined on $\mathbb{R}^n$ and that (5.10) is satisfied for all $x'$ and $y'$. We extend $u$ to the set where $x^n > \omega(x')$ by defining $u(x) = 0$ if $x^n > \omega(x')$ and $|x| > R$, and we define new coordinates $z$ by $z' = x'$ and $z^n = x^n - \omega(z')$. Then $\bar{u}$ defined by $\bar{u}(z) = u(x)$ is in $W^{1,1}(\mathbb{R}^n_+)$ and the even extension of $\bar{u}$ is in $W^{1,1}(\mathbb{R}^n)$. It follows that

$$\|\bar{u}\|_{n/(n-1)} \leq C(n) \|D\bar{u}\|_1.$$

But

$$\|u\|_{n/(n-1)} \leq C(\omega_0, n) \|\bar{u}\|_{n/(n-1)}$$

and

$$\|D\bar{u}\|_1 = 2\|D\bar{u}\|_{1,\mathbb{R}^n_+} \leq C(\omega_0, n) \|Du\|_1.$$

Combining these inequalities completes the proof. □

From this lemma, a simple partition of unity argument gives the general Sobolev imbedding theorem with nonzero boundary values.

**Theorem 5.7.** *Let $\Omega$ be a bounded Lipschitz domain. Then $W^{1,1}(\Omega) \subset L^{n/(n-1)}(\Omega)$ and there is a constant $C$, determined only by $\Omega$ and $n$, such that*

$$\|u\|_{n/(n-1)} \leq C(\|u\|_1 + \|Du\|_1)$$

*for all $u \in W^{1,1}$.*

We also use a multiplicative version of the Sobolev imbedding theorem, which is a special case of the Gagliardo-Nirenberg inequality. We present three versions of this inequality, but first we point out a useful consequence of Hölder's inequality: For any measure space $(X,\mu)$ and any numbers $p$, $r$, and $\alpha$ such that $1 \le p < r < \infty$ and $\alpha \in (0,1)$, we have

$$\|u\|_{\alpha p+(1-\alpha)r} \le \|u\|_p^{\alpha}\|u\|_r^{1-\alpha} \tag{5.12}$$

for all $u \in L^p(\mu)\cap L^r(\mu)$.

**Theorem 5.8.** *Let $p$ and $N$ be constants such that $N > p \ge 1$ and $N \ge n$.*

*(a) Then there is a constant $C$, determined only by $n$, $N$, and $p$ such that*

$$\begin{aligned}\left(\int_\Omega |u|^{Np/(N-p)}\,dx\right)^{(N-p)/N}\\ \le C\left(\int_\Omega |u|^p\,dx\right)^{(N-n)/N}\left(\int_\Omega |Du|^p\,dx\right)^{n/N}\end{aligned} \tag{5.13}$$

*for any $u \in W^{1,p}(\mathbb{R}^n)$.*

*(b) If $R$, $\omega_0$, $\Omega[R]$, and $\sigma[R]$ are as in Lemma 5.6, then (5.13) holds with $\Omega[R]$ in place of $\Omega$ and $C$ determined also by $\omega_0$ for any $u \in W^{1,p}(\Omega[R])$ which vanishes on $\sigma[R]$.*

*(c) For any bounded Lipschitz domain $\Omega$ there is a constant $C$, determined only by $\Omega$, $N$, $n$, and $p$, such that*

$$\begin{aligned}\left(\int_\Omega |u|^{pN/(N-p)}\,dx\right)^{(N-p)/N}\\ \le C\left(\int_\Omega |u|^p\,dx\right)^{(N-n)/N}\left(\int_\Omega [|Du|^p+|u|^p]\,dx\right)^{n/N}.\end{aligned} \tag{5.14}$$

*for all $u \in W^{1,p}(\Omega)$.*

***Proof.*** Setting $r = np(N-1)/(N-p)(n-1)$, we begin by using (5.12) to see that

$$\int |u|^{Np/(N-p)}\,dx \le \left(\int |u|^r\,dx\right)^{\alpha}\left(\int |u|^p\,dx\right)^{1-\alpha},$$

where $\alpha$ satisfies the equation $\alpha r + (1-\alpha)p = Np/(N-p)$. We next infer from Theorem 5.5 that

$$\left(\int |u|^r\,dx\right)^{(n-1)/n} \le C(n)s\int |u|^{s-1}|Du|\,dx,$$

where $s = r(n-1)/n$. If $p > 1$, Hölder's inequality shows that

$$\int u^{s-1}|Du|\,dx \le \left(\int |u|^{Np/(N-p)}\,dx\right)^{(p-1)/p}\left(\int |Du|^p\,dx\right)^{1/p}.$$

since $(s-1)p/(p-1) = Np/(N-p)$, while $p = 1$ implies $s = 1$, in which case this inequality is immediate. Combining these three inequalities and using the explicit expression

$$\alpha = \frac{p(n-1)}{N+np-n-p}$$

yields (5.13).

By invoking Lemma 5.6 in place of Theorem 5.5, we obtain (5.13) with $\Omega[R]$ in place of $\Omega$, and we obtain (5.14) by using Theorem 5.7. □

Note that the case $p < n = N$ in Theorem 5.8 corresponds to the classical Sobolev imbedding theorem. If $p \ge n$, it follows that $W^{1,p}(\Omega) \subset L^q(\Omega)$ for any finite $q$, and it is known that functions in $W^{1,p}$ are Hölder continuous with exponent $1-n/p$ for $p > n$. We refer the interested reader to Theorem 7.17 in [64] or Theorem 6.8 in [114] for details. We also have the following compactness theorem, which is proved as Theorem 7.26 in [64].

**Theorem 5.9.** *Let $\Omega$ be a bounded Lipschitz domain in $\mathbb{R}^n$ and let $p \in [1,n)$. Then the closure of any bounded subset of $W^{1,p}(\Omega)$ is a compact subset of $L^q(\Omega)$ for any $q \in [1, np/(n-p))$.*

## 5.3 Poincaré's inequality

A related result is Poincaré's inequality. We include a direct proof because the usual one (found, for example, in Lemma 4.1.3 and Theorem 4.2.1 of [206]) is indirect. The first step is the appropriate inequality for a ball. We refer to (7.45) in [64] or Proposition 6.14 in [114] for a proof.

**Lemma 5.10.** *Let $B$ be a ball of radius $R > 0$ in $\mathbb{R}^n$, and let $f \in L^\infty(B)$ be nonnegative with $\int_B f(x)\,dx = 1$. Then there is a constant $C$, determined only by $n$ and $p$ such that*

$$\inf_{L\in\mathbb{R}} \int_B |u-L|^p\,dx \le CR^p \sup f \int_B |Du|^p\,dx \tag{5.15}$$

*for any $u \in W^{1,p}(B)$.*

We also use the following result relating the left hand side of (5.15) to other similar integrals. This result is just Lemma 6.12 in [114].

**Lemma 5.11.** *Let $\Omega$ be a domain in $\mathbb{R}^n$ and let $f \in L^\infty(\Omega)$ with $\int_\Omega f(x)\,dx = 1$. Let $p \geq 1$. Then, for all $u \in L^p(\Omega)$, we have*

$$\int_\Omega |u(x) - u_f|^p\,dx \leq 2^p \sup|f| \inf_{L\in\mathbb{R}} \int_\Omega |u - L|^p\,dx, \tag{5.16}$$

*where*

$$u_f = \int_\Omega u(y)f(y)\,dy.$$

From these two lemmata, the general Poincaré inequality follows by a change of variables and a chaining argument.

**Theorem 5.12.** *Let $\Omega$ be a bounded Lipschitz domain. Then, for any $p \geq 1$, there is a constant $C$, determined only by $\Omega$ and $p$ such that*

$$\inf_{L\in\mathbb{R}} \int_\Omega |u - L|^p\,dx \leq C \int_\Omega |Du|^p\,dx \tag{5.17}$$

*for all $u \in W^{1,p}(\Omega)$.*

***Proof.*** We observe first that Lemma 5.10 gives (5.17) when $\Omega$ is a ball.

Now suppose that there is a one-to-one Lipschitz function $\psi\colon \Omega \to B$ (for some ball $B$) such that $\psi^{-1}\colon B \to \Omega$ is also Lipschitz. For later reference, we say that $\Omega$ is biLipschitz equivalent to a ball if this condition is satisfied. In this case, there is a constant $C_0$, determined only by $\Omega$, such that the Jacobian determinant $J(\psi)$ satisfies $1/C_0 \leq |J(\psi)| \leq C_0$, and (5.17) follows from Lemma 5.10 and the change of variables formula for integrals.

Next suppose that $\Omega = S_1 \cup S_2$ with $S_1$ and $S_2$ both biLipschitz equivalent to the same ball $B$, and suppose that there is a ball $B_\delta$ with radius $\delta > 0$ such that $B_\delta \subset S_1 \cap S_2$. Setting

$$L_0 = \frac{1}{|B_\delta|} \int_{B_\delta} u\,dx,$$

we infer from Lemma 5.11 and our previous case that

$$\inf_{L\in\mathbb{R}} \int_{\Omega_j} |u - L|^p\,dx \leq C \int_{\Omega_j} |Du|^p\,dx$$

for $j = 1, 2$. Combining these two inequalities yields

$$\int_\Omega |u - L_0|^p\,dx \leq C \int_\Omega |Du|^p\,dx.$$

Finally, any bounded Lipschitz domain $\Omega$ can be covered by finitely many sets $S_1, \dots, S_N$, each of which is biLipschitz equivalent to a ball and there is a positive constant $\delta$ such that, if $S_i \cap S_j$ is non-empty, then this intersection contains a ball of radius $\delta$. The desired result then follows from the preceding argument via induction. □

We also have the following variation on the Gagliardo-Nirenberg inequality.

**Proposition 5.13.** *Let $\Omega$ be a bounded Lipschitz domain. Let $p \geq 1$, let $N$ be a positive number such that $N \geq n$ and $N > p$, and let $\theta \in (0,1)$. Then there is a constant $C$, determined only by $\Omega$, $n$, $N$, $p$, and $\theta$ such that* (5.13) *holds for any $u \in W^{1,p}(\Omega)$ such that $u = 0$ on a set with measure greater than $\theta|\Omega|$.*

***Proof.*** By using $f = \chi_S/|S|$ in Lemma 5.11, we see from Theorem 5.12 that

$$\|u\|_p \leq C\|Du\|_p.$$

The proof is completed by using this inequality along with (5.14). □

We close this section with a result that will be useful in our analysis of local properties of weak solutions. To state our result, for $k \in \mathbb{R}$ and $u$ a real-valued function defined on a set $S$, we define $A_k$ to be the subset of $S$ on which $u > k$.

**Lemma 5.14.** *Let $R$, $\omega_0$, and $\Omega[R]$ be as in Lemma* 5.6. *Let $u \in W^{1,1}(\Omega[R])$ and let $m$ and $k$ be real numbers with $m > k$ such that $A_k$ has positive measure. Then there is a constant $C$, determined only by $\omega_0$, $n$, and $R^{-n}|A_k|$ such that*

$$(m-k)|A_m|^{(n-1)/n} \leq C \int_{A_k \setminus A_m} |Du|\,dx.$$

***Proof.*** We first observe that, when $\Omega = \Omega[R]$ in Proposition 5.13, then $C$ depends on $\Omega$ only through $\omega_0$. Applying this proposition with $p = 1$ and $N = n$, and $u$ replaced by $w$, defined by

$$w(x) = \begin{cases} 0 & \text{if } u < k, \\ u - k & \text{if } k \leq u < m, \\ m - k & \text{if } m \leq u \end{cases}$$

we have

$$\left(\int_{\Omega[R]} w^{n/(n-1)}\,dx\right)^{(n-1)/n} \le C\int_{\Omega[R]} |Dw|\,dx,$$

with $C$ determined by $n$, $\omega_0$, and $R^{-n}|A_k|$ since $|\Omega[R]| \ge C(\omega_0,n)R^n$. Moreover

$$\int_{\Omega[R]} w^{n/(n-1)}\,dx \ge \int_{A_m} w^{n/(n-1)}\,dx = (m-k)^{n/(n-1)}|A_m|$$

and

$$\int_{\Omega[R]} |Dw|\,dx = \int_{A_k\setminus A_m} |Du|\,dx$$

because $Dw = 0$ except on $A_k \setminus A_m$, and $Dw = Du$ on $A_k \setminus A_m$. □

## 5.4 The weak maximum principle

Just as the maximum principles of Chapter 1 play a critical role in the analysis of classical solutions, a corresponding result is valid for weak solutions, and it plays a similar role. A crucial element of the maximum principles is the inequalities on $c$ and $\beta^0$. For weak solutions, the corresponding inequality is more subtle but is easily guessed due to the nature of our proof below.

**Theorem 5.15.** *Let $u \in W^{1,2}$ satisfy*

$$D_i(a^{ij}D_ju + b^iu) + c^iD_iu + c^0u \ge 0 \text{ in } \Omega, \tag{5.18a}$$

$$a^{ij}\gamma_iD_ju + b\cdot\gamma u + \beta^0u \ge 0 \text{ on } \partial\Omega. \tag{5.18b}$$

*(In other other words,*

$$\begin{aligned}\int_\Omega [(a^{ij}D_ju + b^iu)D_i\varphi + (-c^iD_iu - c^0u)\varphi]\,dx \\ \le \int_{\partial\Omega} \beta^0u\varphi\,ds\end{aligned} \tag{5.19}$$

*for all nonnegative $\varphi \in C^1(\overline{\Omega})$.) Suppose also that*

$$\int_\Omega c^0w - b^iD_iw\,dx + \int_{\partial\Omega} \beta^0w\,ds \le 0 \tag{5.20}$$

*for all nonnegative $w \in C^1(\overline{\Omega})$. Then either $u$ is a constant or else $u \le 0$ in $\Omega$.*

***Proof.*** Let us assume that $\sup u > 0$. We write (5.18) in weak form as

$$\int_\Omega a^{ij} D_j u D_i v - (b^i + c^i) v D_i u \, dx \le \int_\Omega c^0 uv - b^i D_i(uv) \, dx + \int_{\partial\Omega} \beta^0 uv \, ds$$

for all nonnegative $v \in C^1(\overline{\Omega})$. Because we have assumed that all coefficients of $L$ and $M$ are bounded, it follows that this inequality holds for all nonnegative $v \in W^{1,2}$ such that $uv \ge 0$ in $\Omega$. Since $c^0$ and $b^i$ are bounded, it follows that (5.20) is valid for all nonnegative $w \in W^{1,1}$ and, in particular, for $w = uv$. Therefore

$$\int_\Omega a^{ij} D_j u D_i v \, dx \le \int_\Omega (b^i + c^i) v D_i u \, dx.$$

From our bounds, we conclude that

$$\int_\Omega a^{ij} D_i v D_j u \, dx \le C \int_\Omega v |Du| \, dx. \tag{5.21}$$

We now let $k$ be a positive constant such that $k < \sup u_+$. We take $v = (u-k)_+$, and we write $\Gamma_k$ for the support of $Dv$. It follows that $Du = Dv$ in $\Gamma_k$ and $Dv = 0$ in $\Omega \setminus \Gamma_k$, and hence

$$\int_\Omega a^{ij} D_i v D_j v \, dx \le C \int_{\Gamma_k} v |Dv| \, dx.$$

Using Hölder's inequality and the ellipticity of $[a^{ij}]$, we find that

$$\int_\Omega |Dv|^2 \, dx \le C \int_{\Gamma_k} v^2 \, dx.$$

If we combine this estimate with the Gagliardo-Nirenberg inequality, Proposition 5.13, taking $v$ in place of $u$ and $N = n+1$, we infer that

$$\left( \int_\Omega v^q \, dx \right)^{1/q} \le C \left( \int_\Omega v^2 \, dx \right)^{(N-n)/N} \left( \int_{\Gamma_k} v^2 \, dx \right)^{n/N}$$

for $q = 2N/(N-2)$. Hölder's inequality implies that

$$\int_S v^2 \, dx \le \left( \int_S v^q \, dx \right)^{2/q} |S|^{(q-2)/q}$$

for any measurable set $S$, and hence

$$\|v\|_{q;\Omega} \le C |\Gamma_k|^{(q-2)n/(2N)} \|v\|_{q;\Omega}.$$

It follows that $|\Gamma_k|$ is bounded from below by a positive constant, independent of $k$, and hence

$$\Gamma' = \cap_{0<k<\sup u_+} \Gamma_k$$

has positive measure. Therefore, sending $k \to \sup u_+$, we see that $u$ attains its maximum on a set $\Gamma^*$ of positive measure. Hence the supremum must be finite, and, on $\Gamma^*$, we have $Du = 0$. On the other hand, $\Gamma' \subset \Gamma^*$, and $Du \ne 0$ on $\Gamma'$. This contradiction shows that $u_+$ must be a constant.

Since we have shown that $u_+$ is a constant, it follows that $Du_+ = 0$ and hence $Du_- = -Du$, and then Poincaré's inequality implies that $u$ is a constant. □

## 5.5 Trace theorems

It will be important in our analysis of the oblique derivative problem to understand the sense in which a function $u \in W^{1,2}(\Omega)$ has boundary values if $\Omega$ is a bounded Lipschitz domain. As a first step, we observe that there is a smooth function $\bar{\gamma}$ which acts likes the unit inner normal to $\partial\Omega$.

**Lemma 5.16.** *Let $\Omega$ be a bounded Lipschitz domain and let $\gamma$ be the unit inner normal to $\partial\Omega$. Then there is a $C^1(\overline{\Omega})$ vector field $\bar{\gamma}$ such that*

$$\bar{\gamma} \cdot \gamma \geq C(\Omega) \text{ on } \partial\Omega \tag{5.22a}$$

$$|\bar{\gamma}| \leq 1 \text{ in } \Omega. \tag{5.22b}$$

***Proof.*** Cover $\partial\Omega$ by a finite number of balls $B(x_i, R)$ (with the same radius) so that $\Omega[x_i, R]$ can be represented in a suitably rotated coordinate system $(Y_i)$ as the graph of a Lipschitz function with Lipschitz constant uniformly bounded by some constant $\omega_0$. We then define the vector field $\bar{\gamma}_i$ to be the unit vector pointing in the direction of the $y_i^n$ axis. With $(\psi_i)$ a partition of unity subordinate to $(\Omega[x_i, R])$, we define

$$\bar{\gamma} = \sum \psi_i \bar{\gamma}_i.$$

□

Next, we show that any function in $W^{1,1}$ has $L^1$ boundary values and that the $L^1$ norm of the boundary values can be estimated in terms of the $L^1(\Omega)$ norms of $Du$ and $u$.

**Lemma 5.17.** *Let $\Omega$ be a bounded Lipschitz domain, let $\bar{\gamma}$ be a $C^1(\overline{\Omega})$ vector field satisfying $\bar{\gamma} \cdot \gamma \geq 0$ on $\partial\Omega$ and* (5.22b)*, and let $K_0$ be a nonnegative constant such that* $\operatorname{div} \bar{\gamma} \geq -K_0$ *in $\Omega$. Then, for all $u \in W^{1,1}(\Omega)$, we have that $u \in L^1(\partial\Omega)$ and*

$$\int_{\partial\Omega} |u| \gamma \cdot \bar{\gamma} \, dx \leq \int_{\Omega} |Du| + K_0 |u| \, dx. \tag{5.23}$$

***Proof.*** If $u \in C^1(\overline{\Omega})$ is nonnegative, then the divergence theorem tells us that

$$\int_{\partial\Omega} u\gamma \cdot \bar{\gamma} \, dx = \int_{\Omega} \operatorname{div}(u\bar{\gamma}) \, dx.$$

We then write $\operatorname{div}(u\bar{\gamma}) = \bar{\gamma} \cdot Du + u \operatorname{div} \bar{\gamma}$, and use (5.22b) to conclude that

$$\operatorname{div}(u\bar{\gamma}) \leq |Du| + K_0 u.$$

We therefore have (5.23) in this case. For an arbitrary $u \in W^{1,1}$, we first observe that $|u| \in W^{1,1}$ and then approximate $|u|$ by a sequence $(u_i)$ of nonnegative $C^1(\overline{\Omega})$ functions. Since this sequence is Cauchy in $L^1(\partial\Omega)$, it follows that the limit is also in $L^1(\partial\Omega)$ with (5.23) being satisfied. □

Note that the vector field $\bar{\gamma}$ from Lemma 5.16 satisfies the hypotheses of Lemma 5.17. If $\partial\Omega \in C^2$, then we can take $\bar{\gamma} = \gamma$ on $\partial\Omega$. As a first step in proving this statement, we present a quantitative result concerning the distance function.

**Lemma 5.18.** *Let $\partial\Omega \in C^2$. Then there is a positive constant $R_0$, determined only by $\Omega$ such that $d \in C^2(\overline{\Gamma})$, where $\Gamma$ denotes the subset of $\Omega$ in which $d < R_0$, and the eigenvalues of the matrix $D^2d$ are in the interval $[-1/R_0, 1/R_0]$.*

***Proof.*** For each $y \in \partial\Omega$, we write $\kappa_1(y), \dots, \kappa_{n-1}(y)$ for the principal curvatures of $\partial\Omega$, and we set

$$\bar{\kappa} = \sup\{|\kappa_i(y)| : 1 \le i \le n-1, y \in \partial\Omega\}.$$

It follows from Lemma 14.16 in [64] that $d \in C^2(\overline{\Gamma})$ if $R_0 \le 1/\bar{\kappa}$, and Lemma 14.17 in [64] gives the required bound on the eigenvalues if $R_0 \le 1/(2\bar{\kappa})$. Specifically, (14.99) in [64] says that the eigenvalues of $D^2d(x)$ are

$$\frac{-\kappa_1(y)}{1-\kappa_1(y)d(x)}, \dots, \frac{-\kappa_{n-1}(y)}{1-\kappa_{n-1}(y)d(x)}, 0,$$

where $y$ is the closest point on $\partial\Omega$ to $x$. □

From this lemma, we introduce a globally defined regularized distance as follows. Let $g \in C^2(\mathbb{R})$ be a function such that

$$g(t) = \begin{cases} t & \text{if } t \le R_0/3, \\ R_0/2 & \text{if } t \ge R_0, \end{cases}$$

and $0 \le g'(t) \le 1$ for all $t \in \mathbb{R}$. Noting that we can take $g$ so that $|g''(t)| \le C(R_0)$, we define $\rho = g(d)$. Then $\rho \in C^2(\overline{\Omega})$ and there is a constant $\rho_2$, determined only by $R_0$ such that $|D^2\rho| \le \rho_2$. Since $|D\rho| \le 1$, it follows that $\bar{\gamma}$ also satisfies the hypotheses of Lemma 5.17. We therefore have the following estimate in $C^2$ domains.

**Lemma 5.19.** *Let $\Omega$ be a bounded domain with $\partial\Omega \in C^2$. Then there is a nonnegative constant $K_1$, determined only by $\Omega$, such that*

$$\int_{\partial\Omega} |u|\,dx \le \int_\Omega |Du| + K_1|u|\,dx \tag{5.24}$$

*for all $u \in W^{1,1}(\Omega)$.*

For the remainder of this section, we return to the setting of general Lipschitz domains. Our goal is to examine the regularity of $T(u)$ when $u \in W^{1,p}$ for arbitrary $p \geq 1$. We begin with the basic regularity result.

**Lemma 5.20.** *Let $\Omega$ be a bounded Lipschitz domain and let $p \geq 1$. Then there is a bounded linear function $T\colon W^{1,p}(\Omega) \to L^p(\partial\Omega)$ such that $T(u)$ is the restriction of $u$ to $\partial\Omega$ for any $u \in C^1(\overline{\Omega})$. Moreover, if $p < n$, then the range of $T$ is in $L^q(\partial\Omega)$ with $q = (n-1)p/(n-p)$. If $p = n$, then the range of $T$ is in $L^q(\partial\Omega)$ for any finite $q$. If $p > n$, then the range of $T$ is in $C^\alpha(\partial\Omega)$ for $\alpha = 1 - n/p$.*

***Proof.*** From Lemma 5.17, we know that there is a constant $C_1$, determined only by $\Omega$, such that

$$\int_{\partial\Omega} |u|\, ds \leq C_1 \int_\Omega |u| + |Du|\, dx \tag{5.25}$$

for any $u \in W^{1,1}(\Omega)$. Replacing $u$ by $u^p$ and applying Young's inequality, we find that

$$\|u\|_{p;\partial\Omega} \leq (C_1 p)^{1/p} \|u\|_{W^{1,p}(\Omega)}. \tag{5.26}$$

If $p = 1$, then $q = 1$, so the range of $T$ is in $L^q(\partial\Omega)$. If $1 < p < n$, then we have

$$\begin{aligned}\int_{\partial\Omega} |u|^q\, ds &\leq C_2 \int_\Omega |u|^{q-1}[q|Du| + |u|]\, dx \\ &\leq C_2 \left(\int_\Omega |u|^{(q-1)p/(p-1)}\, dx\right)^{(p-1)/p} \left(\int_\Omega (q|Du| + |u|)^p\, dx\right)^{1/p}\end{aligned}$$

for $C_2$ determined only by $\Omega$. Since $(q-1)p/(p-1) = np/(n-p)$, we infer that

$$\|u\|_{q;\partial\Omega} \leq C(C_2, n, p)\|u\|_{W^{1,p}(\Omega)}$$

from the Sobolev imbedding theorem. The cases $p = n$ and $p > n$ are left to the reader. □

More generally, we have the following versions of the Gagliardo-Nirenberg inequality concerning boundary integrals, obtained by modifying the proof of Lemma 5.20 along the lines of Theorem 5.8.

**Theorem 5.21.** *Let $p > 1$, let $N$ be a positive constant with $N > p$ and $N \geq n$, and set $q = (N-1)p/(N-p)$,*

$$\alpha = \frac{(N-n)(p-1)}{(N-p)p}, \quad \beta = \frac{N - p + np - n}{(N-p)p}.$$

*(a) Let $\Omega$ be a bounded Lipschitz domain in $\mathbb{R}^n$. Then there is a constant $C$, determined only by $N$, $n$, $p$, and $\Omega$ such that*

$$\int_{\partial\Omega} |u|^q\,ds \le C\left(\int_\Omega |u|^p\,dx\right)^\alpha \left(\int_\Omega [|Du|^p + |u|^p]\,dx\right)^\beta \tag{5.27}$$

*for all $u \in W^{1,p}(\Omega)$.*

*(b) Let $\Omega$ be a bounded Lipschitz domain in $\mathbb{R}^n$ and let $\theta \in (0,1)$. Then there is a constant $C$, determined only by $N$, $n$, $p$, $\theta$, and $\Omega$ such that*

$$\int_{\partial\Omega} |u|^q\,ds \le C\left(\int_\Omega |u|^p\,dx\right)^\alpha \left(\int_\Omega |Du|^p\,dx\right)^\beta \tag{5.28}$$

*for all $u \in W^{1,p}(\Omega)$ which vanish on a set with measure greater than $\theta|\Omega|$.*

*(c) Let $R$, $\omega_0$, $\Omega[R]$, and $\sigma[R]$ be as in Lemma 5.6, and define*

$$\Sigma[R] = \{x \in \mathbb{R}^n : |x| < R, x^n = \omega(x')\}. \tag{5.29}$$

*Then there is a constant $C$, determined only by $N$, $n$, $p$, and $\omega_0$ such that*

$$\int_{\Sigma[R]} |u|^q\,ds \le C\left(\int_{\Omega[R]} |u|^p\,dx\right)^\alpha \left(\int_{\Omega[R]} |Du|^p\,dx\right)^\beta \tag{5.30}$$

*for all $u \in W^{1,p}(\Omega[R])$ which vanish on $\sigma[R]$.*

We can also prove an interior $L^p$ estimate in terms of the $L^P$ norm of the gradient and the the $L^p$ norm of the trace.

**Lemma 5.22.** *Let $\Omega$ be a bounded Lipschitz domain and let $p \ge 1$. Then there is a constant $K_2$, determined only by $\Omega$ and $p$, such that*

$$\int_\Omega |u|^p\,dx \le K_2\left(\int_\Omega |Du|^p\,dx + \int_{\partial\Omega} |u|^p\,ds\right) \tag{5.31}$$

*for any $u \in W^{1,p}(\Omega)$.*

***Proof.*** We imitate the proof of Poincaré's inequality. Set

$$L_0 = \frac{1}{|\partial\Omega|}\int_{\partial\Omega} u\,ds, \quad L_1 = \frac{1}{|\Omega|}\int_\Omega u\,dx.$$

Then the triangle inequality implies that

$$\int_\Omega |u - L_0|^p\,dx \le p\left(\int_\Omega |u - L_1|^p\,dx + \int_\Omega |L_1 - L_0|^p\,dx\right).$$

Because $L_1 - L_0$ is a constant,

$$\int_\Omega |L_1 - L_0|^p \, dx = |\Omega||L_1 - L_0|^p,$$

and

$$|L_1 - L_0|^p = \left| \frac{1}{|\partial\Omega|} \int_{\partial\Omega} (u - L_1) \, ds \right|^p \le \frac{1}{|\partial\Omega|} \int_{\partial\Omega} |u - L_1|^p \, ds$$

by Hölder's inequality. It follows that

$$\int_\Omega |u - L_0|^p \, dx \le C \left( \int_\Omega |u - L_1|^p \, dx + \int_{\partial\Omega} |u - L_1|^p \, ds \right),$$

and (5.26) gives a constant $C$, determined only by $\Omega$ and $p$ such that

$$\int_{\partial\Omega} |u - L_1|^p \, ds \le C \left( \int_\Omega |u - L_1|^p \, dx + \int_\Omega |Du|^p \, dx \right).$$

Applying Poincaré's inequality gives

$$\int_\Omega |u - L_0|^p \, dx \le C \int_\Omega |Du|^p \, dx,$$

and the proof is completed by observing that

$$\int_\Omega |u|^p \, dx \le p \left( \int_\Omega |u - L_0|^p \, dx + \frac{|\Omega|}{|\partial\Omega|} \int_{\partial\Omega} |u|^p \, ds \right).$$

□

We also obtain a slightly improved version of the preceding estimate, which will be used later.

**Corollary 5.23.** *Let $\Omega$ be a bounded Lipschitz domain and let $p > 1$. Then there is a constant $C$, determined only by $\Omega$ and $p$, such that*

$$\int_\Omega |u|^p \, dx \le \varepsilon \int_\Omega |Du|^p \, dx + C\varepsilon^{-1/(p-1)} \int_{\partial\Omega} |u|^p \, ds \tag{5.32}$$

*for any $u \in W^{1,p}(\Omega)$ and all $\varepsilon \in (0,1)$.*

***Proof.*** By using Young's inequality in the sharper form

$$ab \le \frac{\varepsilon a^p}{p} + \frac{(p-1)\varepsilon^{-1/(p-1)} b^{p/(p-1)}}{p},$$

we obtain

$$\int_{\partial\Omega} |u|^p \, ds \le \varepsilon \int_\Omega |Du|^p \, dx + C\varepsilon^{-1/(1-p)} \int_\Omega |u|^p \, dx.$$

The proof now follows that of Lemma 5.22 with this inequality replacing (5.26). □

## 5.6 Existence of weak solutions

To prove the existence of weak solutions to the oblique derivative problem, we use an extension of the Riesz representation theorem for Hilbert spaces, due to Lax and Milgram. We refer to Theorem 5.8 in [64] for a proof of this result.

**Theorem 5.24.** *Let $\mathbb{H}$ be a Hilbert space and suppose $\mathbf{B}$ is a real-valued function defined on $\mathbb{H} \times \mathbb{H}$ such that*

$$\mathbf{B}(\alpha x_1 + \beta x_2, y) = \alpha \mathbf{B}(x_1, y) + \beta \mathbf{B}(x_2, y) \tag{5.33a}$$

$$\mathbf{B}(x, \alpha y_1 + \beta y_2) = \alpha \mathbf{B}(x, y_1) + \beta \mathbf{B}(x, y_2) \tag{5.33b}$$

*for all real $\alpha$ and $\beta$ and all $x, y, x_1, x_2, y_1, y_2$ in $\mathbb{H}$ and there is a positive constant $K$ such that*

$$|\mathbf{B}(x, y)| \le K\|x\|\|y\|, \tag{5.34a}$$

$$\mathbf{B}(x, x) \ge \frac{1}{K}\|x\|^2 \tag{5.34b}$$

*for all $x$ and $y$ in $\mathbb{H}$. Then, for any bounded linear functional $F$ defined on $\mathbb{H}$, there is a unique $f \in \mathbb{H}$ such that $\mathbf{B}(x, f) = F(x)$ for all $x \in \mathbb{H}$.*

We recall that a function $\mathbf{B}\colon \mathbb{H} \times \mathbb{H} \to \mathbb{R}$ satisfying (5.33) is called a *bilinear form*. If $\mathbf{B}$ also satisfies (5.34a), it is called *bounded*; if $\mathbf{B}$ also satisfies (5.34b), it is called *coercive*.

To apply this result to the oblique derivative problem, we define the bilinear form $\mathcal{L}$ on $\mathbb{H} = W^{1,2}$ by

$$\mathcal{L}(u, v) = \int_\Omega [(a^{ij} D_j u + b^i u) D_i v - (c^i D_i u + c^0 u) v \, dx - \int_{\partial\Omega} \beta^0 uv \, ds,$$

and we define the bounded linear functional $F$ on $\mathbb{H}$ by

$$F(v) = \int_\Omega f^i D_i v - gv \, dx - \int_{\partial\Omega} \psi v \, ds.$$

Then $u$ is a weak solution of the boundary value problem (5.5) if and only if $\mathcal{L}(u, v) = F(v)$ for all $v \in \mathbb{H}$. Note that $\mathcal{L}$ is not coercive if, for example, $b^i$, $c^i$, $c^0$ and $\beta^0$ are all zero, so it will not be possible to apply the Lax-Milgram theorem directly to our existence question; however, a coercivity-type estimate is true.

**Lemma 5.25.** *Suppose there is a positive constant $\lambda$ such that (5.4) is satisfied. Suppose further that there are constants $\mu_1 \ge 0$ and $q > n$ such that*

$$\|b - c\|_q + \|c^0\|_{q/2} + \|\beta^0\|_{q-1} \le \mu_1 \lambda. \tag{5.35}$$

*Then there is a constant $C$, determined only by $\mu_1$ and $\Omega$ such that*

$$\mathfrak{L}(u,u) \geq \frac{\lambda}{2}\int_\Omega |Du|^2\,dx - C\lambda\int_\Omega u^2\,dx. \tag{5.36}$$

***Proof.*** We write

$$\mathfrak{L}(u,u) = \int_\Omega a^{ij} D_i u D_j u\,dx + I_1 + I_2 + I_3,$$

where

$$\begin{aligned} I_1 &= \int_\Omega (b^i - c^i) D_i u u\,dx, \\ I_2 &= -\int_\Omega c^0 u^2\,dx, \\ I_3 &= -\int_{\partial\Omega} \beta^0 u^2\,ds. \end{aligned}$$

We then estimate these three integrals. First, by Hölder's inequality and (5.14) with $N = q$,

$$I_1 \geq -C\lambda \left(\int_\Omega [|Du|^2 + u^2]\,dx\right)^{(n+q)/2q} \left(\int_\Omega u^2\,dx\right)^{(q-n)/2q}$$

(with $C$ determined by $n$, $q$, $\mu_1$ and $\Omega$) and then Young's inequality yields

$$I_1 \geq -\frac{1}{6}\lambda\int_\Omega |Du|^2\,dx - C\lambda\int_\Omega u^2\,dx.$$

Similarly, we have

$$I_2 \geq -\frac{1}{6}\lambda\int_\Omega |Du|^2\,dx - C\lambda\int_\Omega u^2\,dx.$$

If we use (5.28) in place of (5.14), we also obtain

$$I_3 \geq -\frac{1}{6}\lambda\int_\Omega |Du|^2\,dx - C\lambda\int_\Omega u^2\,dx.$$

The proof is completed by using (5.4) to estimate

$$\int_\Omega a^{ij} D_i u D_j u\,dx.$$

□

From this lemma, we can prove a general existence result.

**Theorem 5.26.** *Let $\Omega$ be a bounded Lipschitz domain, and suppose that there are positive constants $\lambda$ and $\mu_1$ such that conditions* (5.4) *and* (5.35) *are satisfied. Then there is a countable set* $\Sigma$ *of real numbers with no finite limit points such that the problem*

$$(L+\sigma)u = D_i f^i + g \text{ in } \Omega, \quad (M+\sigma)u = \psi \text{ on } \partial\Omega \tag{5.37}$$

*has a unique solution for all* $(f, g, \psi) \in X = L^2(\Omega)^{n+1} \times L^2(\partial\Omega)$ *if and only if* $\sigma \notin \Sigma$. *Moreover, for each* $\sigma \in \Sigma$, *the dimension of the solution set of the problem*

$$(L+\sigma)u = 0 \text{ in } \Omega, \quad (M+\sigma)u = 0 \text{ on } \partial\Omega \tag{5.38}$$

*is a positive integer* $I_\sigma$, *and there is an* $I_\sigma$*-dimensional subspace* $W^*_\sigma$ *of* $X^*$ *such that* (5.37) *has a solution if and only if* $(f^i, g, \psi) \in (W^*_\sigma)^\perp$.

***Proof.*** We begin by setting $\mathbb{H} = W^{1,2}(\Omega)$ and noting that any $x^* \in \mathbb{H}^*$ can be represented by some $(f, g, \psi) \in X$ so that

$$x^*(v) = \int_\Omega -f^i D_i v + gv\, dx + \int_{\partial\Omega} \psi v\, ds.$$

Choosing $\sigma_0$ to be a positive number such that $\sigma_0 > C\lambda$, where $C$ is the constant from Lemma 5.25, we define the bilinear operator $\mathbf{B}$ by

$$\mathbf{B}(u,v) = \mathcal{L}(u,v) + \sigma_0 \int_\Omega uv\, dx + \sigma_0 \int_{\partial\Omega} uv\, ds.$$

It follows from Lemma 5.25 that $\mathbf{B}$ is a bounded, coercive bilinear form on $\mathbb{H}$ and hence, for any $(f, g, \psi) \in X$, there is a unique function $u \in W^{1,2}$ satisfying (5.37) with $\sigma_0$ in place of $\sigma$. We now define $L' \colon \mathbb{H}^* \to \mathbb{H}$ by $L'(x^*) = u$, the solution of (5.37) with $\sigma_0$ in place of $\sigma$ and $(f, g, \psi)$ representing $x^*$, and observe that $L'$ is a well-defined continuous, one-to-one mapping of $\mathbb{H}^*$ onto $\mathbb{H}$.

Next, we define the imbedding $I \colon \mathbb{H} \to \mathbb{H}^*$ by

$$Iu(v) = \int_\Omega uv\, dx + \int_{\partial\Omega} uv\, ds.$$

It's straightforward to check that $I$ is compact (by virtue of Theorem 5.9) and that the equations (5.7) are equivalent to

$$u + \sigma_0 L' I u = L'(f, g, \psi).$$

Then $\sigma_0 L' I$ is also compact and hence the Fredholm alternative (Theorem 2.29) implies the desired result, just as in the proof of Theorem 2.30. □

From our general theory, we obtain one simple set of hypotheses which guarantees the unique solvability of (5.5).

**Corollary 5.27.** *Let $\Omega$ be a bounded Lipschitz domain, and suppose that there are positive constants $\lambda$ and $\mu_1$ such that conditions* (5.4) *and* (5.35) *are satisfied. Suppose also that* (5.20) *holds for all nonnegative $w \in C^1(\overline{\Omega})$ and that there is some $w \in C^1(\overline{\Omega})$ such that*

$$\int_\Omega b^i D_i w - c^0 w\,dx \neq \int_{\partial\Omega} \beta^0 w\,ds. \tag{5.39}$$

*Then* (5.5) *has a unique solution for all $f^i$ and $g$ in $L^2(\Omega)$ and all $\psi \in L^2(\Omega)$.*

***Proof.*** The proof of the weak maximum principle implies that the only solutions of $Lu = 0$ in $\Omega$, $Mu = 0$ on $\partial\Omega$ are constants, and condition (5.39) implies that zero is the only constant solution of this boundary value problem. □

We also note that the case when $b^i$, $c^i$, $c^0$ and $\beta^0$ all vanish is particularly simple because then the weak maximum principle implies that the only solutions of problem (5.38) with $\sigma = 0$ are constants and it's immediate to check that any constant is a solution of this problem. Hence $I_0 = 1$, and, by using the test function $v = 1$ in the weak form of the boundary value problem (5.5), we find that this problem is solvable only if

$$\int_\Omega g\,dx + \int_{\partial\Omega} \psi\,ds = 0.$$

It follows that (5.5) is solvable if and only if this compatibility condition is satisfied. In particular, as mentioned in the introduction to this chapter, problem (5.1) is solvable if and only if (5.3) holds.

## 5.7 Higher regularity of solutions

We start with an interior regularity result, which is just Theorem 8.8 [64]. Because the method of proof will be important for our boundary estimate, we sketch the proof here.

**Theorem 5.28.** *Let $\Omega$ be a domain in $\mathbb{R}^n$, and suppose that there is a positive constant $\lambda$ such that* (5.4) *holds. Suppose also that $a^{ij}$ and $b^i$ are Lipschitz and that $c^i$ and $c^0$ are bounded. Let $u \in W^{1,2}$ be a solution of $Lu = g$ in $\Omega$ with $g \in L^2(\Omega)$. Then, for any $\Omega' \subset \Omega$ with $\operatorname{dist}(\Omega', \partial\Omega) > 0$,*

*$u \in W^{2,2}(\Omega')$ and there is a constant $C$, determined only by $n$, $\operatorname{dist}(\Omega', \partial\Omega)$ and $K$, where*

$$K = |a^{ij}/\lambda|_1 + |b^i/\lambda|_1 + \|c^i/\lambda\|_\infty + \|c^0/\lambda\|_\infty, \tag{5.40}$$

*such that*

$$\|u\|_{W^{2,2}(\Omega')} \le C(\|u\|_{W^{1,2}(\Omega)} + \|g/\lambda\|_2). \tag{5.41}$$

*Moreover*

$$a^{ij} D_{ij} u + (b^i + c^i) D_i u + (D_i b^i + c^0) u = g \tag{5.42}$$

*a.e. in $\Omega$.*

***Proof.*** First we set $g_0 = (b^i + c^i) D_i u + (D_i b^i + c^0) u - g$ and observe that $u$ is a weak solution of the equation

$$D_i(a^{ij} D_j u) = g_0 \text{ in } \Omega. \tag{5.43}$$

Now, let $\Omega'$ be given, set $d' = \operatorname{dist}(\Omega', \partial\Omega)$, and let $\Omega''$ be a subdomain of $\Omega$ such that $\operatorname{dist}(\Omega'', \partial\Omega) \ge d'/2$ and $\overline{\Omega'} \subset \Omega''$. Also we fix $k \in \{1, \ldots, n\}$ and $h \in (-d'/4, d'/4)$. Then for any function $v \in C_0^1(\Omega'')$, we take $\varphi = v_{k,h}$ in the weak form of (5.43) to conclude that

$$\int_\Omega a^{ij}(x + he_k) D_j u_{k,h} D_i v \, dx = -\int_\Omega (f^i D_i v + g_0 v) \, dx,$$

where $f_i = a^{ij}_{k,h} D_j u$. From the definitions of $f^i$ and $g_0$, we infer via Lemma 5.4 that

$$\int_\Omega a^{ij}(x + he_k) D_j u_{k,h} D_i v \, dx \le C\lambda(\|u\|_{W^{1,2}} + \|g/\lambda\|_2)\|Dv\|_2.$$

In particular, if $\eta \in C_0^1(\Omega'')$ satisfies the conditions $0 \le \eta \le 1$ and $|D\eta| \le 4/d'$ in $\Omega''$ and $\eta = 1$ on $\Omega'$, we have

$$\begin{aligned} \lambda \int_{\Omega'} |Du_{k,h}|^2 \, dx &\le \lambda \int_\Omega \eta^2 |Du_{k,h}|^2 \, dx \\ &\le \int_\Omega \eta^2 a^{ij}(x + he_k) D_i u_{k,h} D_j u_{k,h} \, dx. \end{aligned}$$

With $v = \eta^2 u_{k,h}$, we also have $\eta^2 D_j u_{k,h} = D_j v - 2\eta u_{k,h} D_j \eta$ and hence

$$\begin{aligned} \lambda \int_{\Omega'} |Du_{k,h}|^2 \, dx \le{}& C\lambda(\|u\|_{W^{1,2}} + \|g/\lambda\|_2)\|Dv\|_2 \\ &- 2\int_\Omega \eta u_{k,h} a^{ij}(x + he_k) D_i u_{h,k} D_j \eta \, dx. \end{aligned}$$

The Schwarz inequality implies that

$$- 2\int_\Omega \eta u_{k,h}a^{ij}(x+he_k)D_iu_{h,k}D_i\eta\,dx$$
$$\leq \mu\lambda\left(\int_\Omega \eta^2|D_iu_{h,k}|^2\,dx\right)^{1/2}\left(\int_\Omega |D\eta|^2|u_{h,k}|^2\,dx\right)^{1/2},$$

and it follows from simple rearrangement and Lemma 5.4 that

$$\int_{\Omega'} |Du_{k,h}|^2\,dx \leq C(\|u\|^2_{W^{1,2}} + \|f\|^2).$$

Sending $h\to 0$ completes the proof of (5.41), and (5.42) follows from (5.6). □

A slight modification of the proceeding proof gives the corresponding regularity up to the boundary.

**Theorem 5.29.** *Let $\Omega$ be a domain in $\mathbb{R}^n$ with $\partial\Omega\in C^2$, and suppose that there is a positive constant $\lambda$ such that (5.4) holds. Suppose also that $a^{ij}$, $b^i$ and $\beta^0$ are Lipschitz and that $c^i$ and $c^0$ are bounded. Let $u\in W^{1,2}$ be a solution of $Lu=g$ in $\Omega$ and $Mu=\psi$ with $g\in L^2(\Omega)$ and $\psi\in W^{1,2}(\partial\Omega)$. Then $u\in W^{2,2}(\Omega)$ and there is a constant $C$, determined only by $n$, $K$ (the constant from (5.40)), and $|\beta^0/\lambda|_1$, such that*

$$\|u\|_{W^{2,2}(\Omega)} \leq C(\|u\|_{W^{1,2}(\Omega)} + \|g/\lambda\|_2 + \|\psi/\lambda\|_{W^{1,2}(\partial\Omega)}).$$

***Proof.*** From Theorem 5.28, it suffices to prove a local $W^{2,2}$ estimate for $u$ near the boundary. Moreover, by a change of independent variable, we only need to prove a bound for $D^2u$ in the half-ball $B^+(R)$, assuming that $Lu=g$ in $B^+(2R)$ and $Mu=\psi$ on $B^0(2R)$. Using the notation from the proof of Theorem 5.28 and setting $\Psi=(\beta^0+b^n)u-\psi$, we find that

$$\int_{B^+(2R)} a^{ij}(x+he_k)D_ju_{k,h}D_iv\,dx = -\int_{B^+(2R)} (f^iD_iv + g_0v_{k,-h})\,dx + \int_{B^0(2R)} \Psi_{k,h}v\,dx'$$

if $k<n$ and $|h|<R$. This time, we use the divergence theorem to conclude that

$$\int_{B^0(2R)} \Psi_{k,h}v\,dx' = \int_{B^+(2R)} \Psi_{k,n}D_nv - D_n\Psi D_nv\,dx,$$

and then the argument of Theorem 5.28 implies that

$$\|D_{ik}u\|_{2,B^+(R)} \leq C(\|u\|_{W^{1,2}} + \|g/\lambda\|_2 + \|\Psi\|_{W^{1,2}})$$

for $i \le n$ and $k < n$. The proof is completed by noting that

$$D_{nn}u = \frac{1}{a^{nn}}\left(g - (b^i + c^i)D_i u + (D_i b^i + c^0)u + \sum_{(i,j)\neq(n,n)} a^{ij}D_{ij}u\right). \qquad \square$$

We shall see in Chapter 6 that the hypotheses of these theorems can be relaxed considerably, but the techniques involved are quite different from those of the current chapter.

## 5.8 Global boundedness of weak solutions

We now show that weak solutions of (5.5) are bounded under appropriate conditions on the data of the problem. Although the arguments we shall use are based on a nonlinear structure (see Section 8.5 in [64] for details), we prefer to write everything in linear form here; the corresponding result for nonlinear problems will be presented in a later chapter. As was the case in Chapter 1, we are concerned with subsolutions and supersolutions of (5.5). We say that $u$ is a *weak subsolution of* (5.5) if $u \in W^{1,2}$ and if

$$\begin{aligned}\int_\Omega [(a^{ij}D_j u + b^i u - f^i)D_i\varphi + (-c^i D_i u - c^0 u + g)\varphi]\,dx \\ \le \int_{\partial\Omega} [\beta^0 u - \psi]\varphi\,ds\end{aligned} \tag{5.44}$$

for all $\varphi \in C^1(\overline{\Omega})$. For brevity, we also write

$$Lu \ge D_i f^i + g \text{ in } \Omega, \quad Mu \ge \psi \text{ in } \partial\Omega.$$

Weak supersolutions are defined by reversing all these inequalities. Our first result is that subsolutions are bounded above. Before proving this result, we demonstrate an iteration argument, which is usually called Moser iteration.

**Lemma 5.30.** *Let $(X, \mu)$ be a measure space, let $w \in L^p(\mu)$ for all $p \ge 1$ be nonnegative, and suppose that there are positive constants $K$, $\kappa > 1$, $m \ge 1$, and $s_0 \ge 1$ such that*

$$\left(\int_X w^{\kappa s}\,d\mu\right)^{1/\kappa} \le K s^m \int_X w^s\,d\mu \tag{5.45}$$

*for all $s \ge s_0$. Then $w \in L^\infty(\mu)$ and there is a constant $C$, determined only by $K$, $\kappa$, $m$, and $s_0$, such that*

$$\|w\|_\infty \le C\|w\|_1. \tag{5.46}$$

***Proof.*** For brevity, we set $I(q) = \|w\|_{qs_0}$, so (5.45) says that

$$I(\kappa s) \le K_1^{1/s} s^{m_1/s} I(s)$$

for $s \ge 1$, where $K_1 = K^{1/s_0} s_0^{m/s_0}$ and $m_1 = m/s_0$. By induction, it follows that

$$I(\kappa^j) \le K_1^{a(j)} \kappa^{b(j)} I(1)$$

for any positive integer $j$, where

$$a(j) = \sum_{i=1}^{j} \kappa^{1-i}, \quad b(j) = \sum_{i=1}^{j} \kappa^{1-i}(i-1)m_1.$$

Since these are partial sums of convergent series, it follows that there is a constant $\sigma$, determined only by $\kappa$, so that $a(j) \le \sigma$ and $b(j) \le \sigma m_1$. It follows that $w \in L^\infty$ with

$$\|w\|_\infty \le (K_1 \kappa^{m_1})^\sigma \|w\|_{s_0}.$$

The desired result follows from this inequality because

$$\|w\|_{s_0} \le \|w\|_\infty^{(s_0-1)/s_0} \|w\|_1^{1/s_0}.$$

□

**Theorem 5.31.** *Let $\Omega$ be a bounded Lipschitz domain, let $L$ and $M$ satisfy (5.4) and (5.35), and suppose there is $q > n$ such that $f^i \in L^q(\Omega)$, $g \in L^{q/2}(\Omega)$, and $\psi \in L^{q-1}(\partial\Omega)$. If $u \in W^{1,2}(\Omega)$ is a weak subsolution of (5.5), then $u_+ \in L^\infty(\Omega)$. Moreover, if we set $k = (\|f\|_q + \|g\|_{q/2} + \|\psi\|_{q-1})/\lambda$, then there is a constant $C$, determined only by $n$, $q$, $\mu_1$, and $\Omega$ such that*

$$\sup_\Omega u \le C(\|u_+\|_2 + k). \tag{5.47}$$

***Proof.*** We begin by proving a basic energy inequality. Let $\chi$ be a Lipschitz function defined on $[k, \infty)$, and suppose that there is a nonnegative constant $c_\chi$ such that

$$0 \le \chi'(s)(s-k) \le c_\chi \chi(s) \tag{5.48}$$

for all $s \ge k$. For notational simplicity, we define $\chi(s) = 0$ if $s < k$. We then take $\varphi = \chi(u)(u-k)_+$ as test function and observe that

$$D\varphi = \begin{cases} (\chi'(u)(u-k) + \chi(u))Du & \text{if } u > k, \\ 0 & \text{if } u \le k. \end{cases}$$

Hence $a^{ij} D_j u D_i \varphi \ge \chi |Du|^2$ wherever $u > k$. It follows that

$$I_0 \le I_1 + I_2 + I_3 + I_4,$$

where

$$I_0 = \int_\Omega \chi |Du|^2 \, dx,$$

$$I_1 = \frac{1}{\lambda} \int_\Omega \left( \frac{f^i}{u} - b^i \right) (\chi'(u-k) + \chi) D_i u u \, dx,$$

$$I_2 = \frac{1}{\lambda} \int_\Omega \chi c^i D_i u u(u-k) \, dx,$$

$$I_3 = \frac{1}{\lambda} \int_\Omega \left( c^0 - \frac{g}{u} \right) \chi u(u-k) \, dx,$$

$$I_4 = \frac{1}{\lambda} \int_{\partial\Omega} \left( \beta^0 - \frac{\psi}{u} \right) \chi u(u-k) \, ds.$$

We first use (5.48) and the observation that

$$\int_{\{u>k\}} \left| \frac{f}{u} - b \right|^q dx \le \lambda^q (1+\mu_1)^q$$

to infer that

$$I_1 \le (1+\mu_1)(1+c_\chi) I_0^{1/2} \left( \int_\Omega [\chi u^2]^{q/(q-2)} \, dx \right)^{q-2)/2q}.$$

To estimate the integral in this inequality, we adapt the proof of Theorem 5.8 slightly. We set $r = 2n(q-1)/(q-2)(n-1)$, $A = 2(n-1)/(q+n-2)$, and $s = r(n-1)/n$. (These correspond to $N = q$ in the proof of Theorem 5.8.) We also set $B = -rA/(1-A)$. Then Hölder's inequality implies that

$$\int_\Omega [\chi u^2]^{q/(q-2)} \, dx \le \left( \int_\Omega [\chi (u-k)^2]^{r/2} \, dx \right)^A \left( \int_\Omega \chi u^2 \left( 1 - \frac{k}{u} \right)^B dx \right)^{1-A}.$$

Moreover, using the Sobolev imbedding theorem and another application of Hölder's inequality, we find that

$$\begin{aligned} &\left( \int_\Omega [\chi (u-k)^2]^{r/2} \, dx \right)^{(n-1)/n} \\ &\qquad \le C(n,\Omega) \left( \int_\Omega [\chi (u-k)^2]^{q/(q-2)} \, dx \right)^{1/2} \\ &\qquad\qquad \times \left( \int_\Omega |D(\chi^{1/2}(u-k))|^2 \, dx + \int_\Omega \chi (u-k)^2 \, dx \right)^{1/2}. \end{aligned}$$

Because of (5.48), we have $|D(\chi^{1/2}(u-k))| \le (1+c_\chi)\chi^{1/2}|Du|$ and hence (5.13) implies that

$$\begin{aligned} &\left( \int_\Omega [\chi (u-k)^2]^{q/(q-2)} \, dx \right)^{(q-2)/q} \\ &\le C(n,\Omega,q) \left( \int_\Omega \chi (u-k)^2 \, dx \right)^{(q-n)/q} \left( (1+c_\chi)^2 I_0 + \int_\Omega \chi (u-k)^2 \, dx \right)^{n/q}. \end{aligned}$$

It then follows, via Young's inequality and the simple inequality $(u-k)^2 \le u^2(1-k/u)^B$ that

$$I_1 \le \frac{1}{4} I_0 + C(1+c_\chi)^{(2q+n)/(q-n)} \int_\Omega \chi u^2 \left(1 - \frac{k}{u}\right)^B dx.$$

Similar reasoning shows that

$$I_2 + I_3 + I_4 \le \frac{1}{4} I_0 + C(1+c_\chi)^{(2q+n)/(q-n)} \int_\Omega \chi u^2 \left(1 - \frac{k}{u}\right)^B dx,$$

and hence

$$\int_\Omega \chi |Du|^2 \, dx \le C(1+c_\chi)^{(2q+n)/(q-n)} \int_\Omega \chi u^2 \left(1 - \frac{k}{u}\right)^B dx. \tag{5.49}$$

Now choose $N \in (2,q)$ so that $n \ge N$, set

$$\alpha = N - \frac{BN}{2}, \quad \beta = \frac{B(N-2)}{2} - N$$

and, with $s \ge 1$ and $Z > k$ parameters at our disposal, set

$$\chi = \left(1 - \frac{k}{u}\right)_+^{\alpha s + \beta} \min\{u, Z\}^{2s-2}.$$

It follows that (5.48) is satisfied with $c_\chi = C(n,q)s$. We now observe that

$$h = \left(1 - \frac{k}{u}\right)_+^{(\alpha s + \beta + 2)/2} \min\{u, Z\}^{s-1} u$$

satisfies the inequalities

$$|Dh|^2 \le Cs^2 \chi |Du|^2$$

and

$$h^2 \le \chi u^2 \left(1 - \frac{k}{u}\right)^B,$$

so we may apply the Gagliardo-Nirenberg inequality (5.14) along with (5.49) to obtain

$$\left(\int_\Omega h^{2\kappa} \, dx\right)^{1/\kappa} \le Cs^{(2q+n)/(q-n)} \int_\Omega \chi u^2 \left(1 - \frac{k}{u}\right)^B dx,$$

where $\kappa = N/(N-2)$. Since $\kappa(\beta+2) = \beta + B$ and $\min\{u,Z\} \le u$, it follows that

$$\left(\int_\Omega \left(1 - \frac{k}{u}\right)_+^{\alpha\kappa s + \beta + B} \min\{u, Z\}^{2\kappa s} \frac{u^2}{\min\{u,Z\}^2} \, dx\right)^{1/\kappa} \le \left(\int_\Omega h^{2\kappa} \, dx\right)^{1/\kappa}$$

and

$$\int_\Omega \chi u^2 \left(1 - \frac{k}{u}\right)^B dx = \int_\Omega \left(1 - \frac{k}{u}\right)_+^{\alpha s+\beta+B} \min\{u, Z\}^{2s} \frac{u^2}{\min\{u, Z\}^2}\, dx.$$

We now set $m = (2q+n)/(q-n)$ and

$$w = \left(1 - \frac{k}{u}\right)_+^{\alpha} \min\{u, Z\}^2,$$

and we define the measure $\mu$ by

$$d\mu = \left(1 - \frac{k}{u}\right)_+^{B-\beta} \frac{u^2}{\min\{u, Z\}^2}\, dx.$$

In this way, we conclude that

$$\left(\int_\Omega w^{\kappa s}\, d\mu\right)^{1/\kappa} \le C s^m \int_\Omega w^s\, d\mu.$$

It then follows from Lemma 5.30 that

$$\sup w \le C \int_\Omega w\, d\mu = C \int_{\Omega(k)} u^2\, dx,$$

where $\Omega(k)$ is the subset of $\Omega$ on which $u \ge k$. Since this inequality is true for all $Z > k$, it follows that

$$\sup \left(1 - \frac{k}{u}\right)_+^{\alpha} u^2 \le C \int_{\Omega(k)} u^2\, dx,$$

and the desired result is an immediate consequence of this inequality. □

Replacing $u$ by $-u$ gives the corresponding result for supersolutions.

**Corollary 5.32.** *Suppose all the hypotheses of Theorem* 5.31 *are satisfied except that* $u$ *is a supersolution of* (5.5) *rather than a subsolution. Then* $u_- \in L^\infty$ *and*

$$\sup_\Omega (-u) \le C(\|u_-\|_2 + k). \tag{5.50}$$

In general, we can't expect a global bound for $u$ (independent of the $L^2$ norm) without some restrictions on $b^i$, $c^0$, and $\beta^0$. For example, if $b^i$, $c^0$, and $\beta^0$ are all zero, then any constant is a solution of $Lu = 0$ in $\Omega$, $Mu = 0$ on $\partial\Omega$. We now provide some sufficient conditions for a global bound.

Our first set of conditions is based on those in Theorem 5.15.

**Theorem 5.33.** *Let* $\Omega$ *be a bounded Lipschitz domain, let* $L$ *and* $M$ *satisfy* (5.4) *and* (5.35), *and suppose there is* $q > n$ *such that* $f^i \in L^q(\Omega)$, $g \in$

*$L^{q/2}(\Omega)$, and $\psi \in L^{q-1}(\partial\Omega)$. Suppose also that there is a positive constant $\varepsilon$ such that*

$$\int_\Omega (c^0 + \varepsilon)w - b^i D_i w\, dx + \int_{\partial\Omega} \beta^0 w\, ds \le 0 \tag{5.51}$$

*for all nonnegative $w \in C^1(\overline{\Omega})$. Then there is a constant $C$, determined only by $\mu_1$, $\varepsilon$, $\Omega$, and $q$ such that any weak subsolution $u \in W^{1,2}(\Omega)$ of (5.5) satisfies the estimate*

$$\sup u \le C(\|f\|_q + \|g\|_{q/2} + \|\psi\|_{q-1})/\lambda. \tag{5.52}$$

***Proof.*** First, we infer from the proof of Theorem 5.15 that

$$\begin{aligned}\int_\Omega a^{ij} D_j u D_i v\, dx &+ \lambda\varepsilon \int_\Omega uv\, dx \\ &\le \int_\Omega (b^i + c^i) D_i uv\, dx + \int_\Omega f^i D_i v\, dx + \int_\Omega gv\, dx + \int_{\partial\Omega} \psi v\, ds\end{aligned}$$

for all nonnegative $v \in W^{1,2}(\Omega)$ with $uv \ge 0$ in $\Omega$. In particular, we can take $v = u_-^s$ with $s > 1$ at our disposal. We now take $k$ as in Theorem 5.31 and use that theorem to conclude that $U = \sup u_+$ is finite. The integrals in this inequality are now easily estimated. First,

$$\int_\Omega a^{ij} D_j u D_i v\, dx \ge \lambda s \int_\Omega |Du|^2 u_+^{s-1}\, dx.$$

Then

$$\begin{aligned}\int_\Omega (b^i + c^i) D_i uv\, dx &\le \frac{s}{2} \int_\Omega \lambda |Du|^2 u_+^{s-1}\, dx + C\lambda U^{s+1}, \\ \int_\Omega f^i D_i v\, dx &\le \frac{s}{2} \int_\Omega \lambda |Du|^2 u_+^{s-1}\, dx + Csk^2\lambda U^{s-1}, \\ \int_\Omega gv\, dx &\le CkU^s, \\ \int_{\partial\Omega} \psi v\, ds &\le CkU^s.\end{aligned}$$

It follows that

$$\varepsilon \int_\Omega u_+^{s+1}\, dx \le \frac{C}{s} U^{s+1} + CkU^s + Ck^2 U^{s-1}.$$

We now use Theorem 5.31 to infer that

$$U \le C(\|u_+\|_2 + k),$$

and then (5.12) implies that

$$U^{s+1} \le C\left(\int_\Omega u_+^{s+1} + k^{s+1}\right).$$

Therefore

$$\varepsilon U^{s+1} \le \frac{C}{s} U^{s+1} + CkU^s + Ck^2 U^{s-1} + Ck^{s+1}.$$

We now choose $s$ so that $C/s \le \varepsilon/2$ and then use Young's inequality to complete the proof. □

We can also obtain a global bound if we assume that $c^0$ is sufficiently negative in place of (5.51).

**Theorem 5.34.** *Let $\Omega$ be a bounded Lipschitz domain, let $L$ and $M$ satisfy (5.4) and (5.35), and suppose there is $q > n$ such that $f^i \in L^q(\Omega)$, $g \in L^{q/2}(\Omega)$, and $\psi \in L^{q-1}(\partial\Omega)$. Suppose also that there is a positive constant $\bar{\mu}$ such that*

$$\|b - c\|_2 \le \bar{\mu}\lambda, \quad \|\beta^0\|_1 \le \bar{\mu}^2\lambda. \tag{5.53}$$

*Then there are positive constants $C_0$ determined only by $\mu_1$, $\Omega$, and $q$, and $C_1$, determined also by $\bar{\mu}$, such that, if*

$$c^0 \le -C_0\bar{\mu}^2\lambda, \tag{5.54}$$

*then any weak subsolution $u \in W^{1,2}(\Omega)$ of (5.5) satisfies the estimate*

$$\sup u \le C_1(\|f\|_q + \|g\|_{q/2} + \|\psi\|_{q-1})/\lambda. \tag{5.55}$$

***Proof.*** Now we use the test function $v = u_+$ to see that

$$\begin{aligned}\int_\Omega a^{ij} D_i v D_j v\,dx + C_0\bar{\mu}^2\lambda \int_\Omega v^2\,dx \\ &\le \int_\Omega (b^i - c^i) D_i v v\,dx \\ &\quad + \int_\Omega f^i D_i v\,dx + \int_\Omega gv\,dx + \int_{\partial\Omega} \beta^0 v^2 - \psi v\,ds.\end{aligned}$$

With $U = \sup v$, we find from Hölder's inequality that

$$C_0\bar{\mu}^2 \int_\Omega v^2\,dx \le 2\bar{\mu}^2 U^2 + \frac{1}{2}\|f\|_2^2/\lambda^2 + \frac{\|\psi\|_1 + \|g_{-1}}{\lambda} U,$$

and Theorem 5.31 gives a constant $C_2$ such that

$$U^2 \le C_2 \left( \int_\Omega v^2\,dx + k^2 \right).$$

The proof is completed by taking $C_0 = 1/2C_2$. □

Finally, we present a result if $\beta^0$ is large and negative, rather than $c^0$.

**Theorem 5.35.** *Let $\Omega$ be a bounded Lipschitz domain, let $L$ and $M$ satisfy (5.4) and (5.35), and suppose there is $q > n$ such that $f^i \in L^q(\Omega)$, $g \in L^{q/2}(\Omega)$, and $\psi \in L^{q-1}(\partial\Omega)$. Suppose also that there is a positive constant $\bar{\mu}$ such that*

$$\|b - c\|_2 \leq \bar{\mu}\lambda, \quad \|c^0\|_1 \leq \bar{\mu}^2\lambda. \tag{5.56}$$

*Then there are positive constants $C_0$ determined only by $\mu_1$, $\Omega$, and $q$, and $C_1$, determined also by $\bar{\mu}$, such that, if*

$$\beta^0 \leq -C_0 \max\{\bar{\mu}^2, \bar{\mu}^4\}\lambda, \tag{5.57}$$

*then any weak subsolution $u \in W^{1,2}(\Omega)$ of (5.5) satisfies the estimate (5.55).*

***Proof.*** We again use the test function $v = u_+$, and this time we obtain

$$\begin{aligned}\int_\Omega a^{ij} D_i v D_j v\, dx + C_0 \max\{\bar{\mu}^2, \bar{\mu}^4\}\lambda \int_{\partial\Omega} v^2\, ds \\ \leq \int_\Omega (b^i - c^i) D_i v v\, dx + \int_\Omega c^0 v^2\, dx \\ + \int_\Omega f^i D_i v\, dx + \int_\Omega g v\, dx - \int_{\partial\Omega} \psi v\, ds,\end{aligned}$$

and hence

$$\frac{1}{2}\int_\Omega |Dv|^2\, dx + C_0 \max\{\bar{\mu}^2, \bar{\mu}^4\} \int_{\partial\Omega} v^2\, ds \leq 2\bar{\mu}^2 U^2 + Ck^2 + CkU.$$

Now Corollary 5.23 gives a constant $C_3 > 1$, determined only by $\Omega$, such that

$$\int_\Omega v^2\, dx \leq \theta \int_\Omega |Dv|^2\, dx + \frac{C_3}{\theta}\int_{\partial\Omega} v^2\, ds$$

for any $\theta \in (0, 1)$. We now take $C_0 \geq 4C_3$ and use $\theta = \min\{1, \bar{\mu}^{-2}\}$ to infer that

$$\frac{C_0}{2C_3}\bar{\mu}^2 \int_\Omega v^2\, dx \leq 2\bar{\mu}^2 U^2 + Ck^2 + CkU$$

and the proof is completed as before. □

## 5.9 The local maximum principle

The preceding maximum estimate also has a local analog similar to Theorem 1.27. We state our results under the hypotheses of Lemma 5.6, that is, given a real-valued Lipschitz function $\omega$ defined on $B_{n-1}(R)$ with Lipschitz constant $\omega_0$, we define $\Omega[R]$ and $\sigma[R]$ by (5.11) and $\Sigma[R]$ by (5.29). We say that $u \in W^{1,2}(\Omega[R])$ is a weak subsolution of

$$Lu = D_i f^i + g \text{ in } \Omega[R], \quad Mu = \psi \text{ on } \Sigma[R] \tag{5.58}$$

if (5.44) is satisfied (with $\Omega$ replaced by $\Omega[R]$ and $\partial\Omega$ replaced by $\Sigma[R]$) for all $\varphi \in C^1(\overline{\Omega[R]})$ which vanish on $\sigma[R]$.

We rewrite condition (5.35) in a way that will be convenient for the form of our estimates. We assume that there are nonnegative constants $q > n$, $\mu$, $\mu_2$, and $\mu_3$ such that

$$a^{ij}\xi_i\zeta_j \le \mu\lambda|\xi||\zeta| \tag{5.59a}$$

for all $\xi$ and $\zeta$ in $\mathbb{R}^n$,

$$\|b^i\|_q + \|c^i\|_q \le \mu_2\lambda, \tag{5.59b}$$

$$\|c^0\|_{q/2} \le \mu_2^2\lambda, \tag{5.59c}$$

$$\|\beta_0\|_{q-1} \le \mu_3\lambda. \tag{5.59d}$$

We also set

$$k(R) = \frac{1}{\lambda}\left(R^{1-n/q}\|f\|_q + R^{2(1-n/q)}\|g\|_{q/2} + R^{(q-n)/(q-1)}\|\psi\|_{q-1}\right).$$

**Theorem 5.36.** *Let $\Omega$ be a bounded Lipschitz domain, let $L$ and $M$ satisfy (5.4) and (5.59), and suppose that $f^i \in L^q(\Omega)$, $g \in L^{q/2}(\Omega)$, and $\psi \in L^{q-1}(\partial\Omega)$. Let $u$ be a weak subsolution of (5.58). Then, for any $p > 0$, there is a constant $C$, determined only by $\mu$, $\mu_2 R^{1-n/q}$, $\mu_3 R^{(q-n)/(q-1)}$, $n$, $p$, $q$, and $\omega_0$, such that*

$$\sup_{\Omega[R]} u \le C(R^{-n/p}\|u_+\|_{p,\Omega[2R]} + k(R)). \tag{5.60}$$

***Proof.*** We imitate the proof of Theorem 5.31 with one important change. With $\chi$ as in the proof of that theorem and $\zeta$ a nonnegative $C^1$ function which vanishes on $\sigma[R]$, we use the test function $\varphi = \chi(u)(u - k(R))_+\zeta^2$. It follows that

$$I_0 \le I_1 + I_2 + I_3 + I_4 + I_5,$$

where

$$
\begin{aligned}
I_0 &= \int_\Omega \chi |Du|^2 \zeta^2\,dx,\\
I_1 &= \frac{1}{\lambda}\int_\Omega \left(\frac{f^i}{u} - b^i\right)(\chi'(u-k)+\chi)\zeta^2 D_i u u\,dx,\\
I_2 &= \frac{1}{\lambda}\int_\Omega \chi c^i D_i u u(u-k)\zeta^2\,dx,\\
I_3 &= \frac{1}{\lambda}\int_\Omega \left(c^0 - \frac{g}{u}\right)\chi u(u-k)\zeta^2\,dx,\\
I_4 &= \frac{1}{\lambda}\int_{\partial\Omega} \left(\beta^0 - \frac{\psi}{u}\right)\chi u(u-k)\zeta^2\,ds,\\
I_5 &= \frac{2}{\lambda}\int_\Omega \left(\frac{f^i}{u} - b^i\right)\chi(u-k)_+\zeta D_i\zeta\,dx,\\
I_6 &= \frac{2}{\lambda}\int_\Omega a^{ij} D_j u D_i \zeta \zeta\chi(u-k)\,dx.
\end{aligned}
$$

These integrals are estimated in a similar fashion to that used in Theorem 5.31 except for the use of Lemma 5.6 in place of Theorem 5.5. For example, we have

$$
\begin{aligned}
&\left(\int_{\Omega[R]} [\chi(u-k)^2\zeta^2]^{q/(q-2)}\,dx\right)^{(q-2)/q}\\
&\quad \le C\left(\int_{\Omega[R]} \chi(u-k)^2\zeta^2\,dx\right)^{(q-n)/q}\left(\int_{\Omega[R]} |D(\chi^{1/2}(u-k)\zeta)|^2\,dx\right)^{n/q},
\end{aligned}
$$

and straightforward calculation shows that

$$
|D(\chi^{1/2})(u-k)\zeta| \le |\chi(u-k)||D\zeta| + (1+c_\chi)\chi^{1/2}\zeta|Du|.
$$

Our energy inequality now becomes

$$
\begin{aligned}
&\int_{\Omega[R]} \chi|Du|^2\zeta^2\,dx\\
&\qquad \le C(1+c_\chi)^{(2q+n)/(q-n)} R^{-2}\int_{\Omega[R]} \chi u^2\left(1-\frac{k}{u}\right)^B \zeta^2\,dx \qquad (5.61)\\
&\qquad + C(1+c_\chi)^{(2q+n)/(q-n)}\int_{\Omega[R]} \chi(u-k)^2|D\zeta|^2\,dx
\end{aligned}
$$

with

$$
B = -\frac{4n(q-1)}{(q-2)(q-n)}
$$

as before.

To continue, we fix a cut-off function $\eta$ with $|D\eta| \le C/R$ (for example, $\eta = (1 - |x|^2/R^2)_+$), we set

$$\alpha = \frac{(2-B)N}{p}, \quad \beta = \frac{B(N-2)}{2} - N,$$

and, with $s \ge 1$ and $Z > k$ constants at our disposal, we take

$$\chi = \left(1 - \frac{k}{u}\right)_+^{\alpha s + \beta} \min\{u, Z\}^{2s-2}, \quad \zeta = \eta^{(\alpha s+\beta)/2}.$$

Using

$$h = \left(1 - \frac{k}{u}\right)_+^{(\alpha s + \beta + 2)/2} \min\{u, Z\}^{s-1}\eta^{(\alpha s+\beta+2)/2}$$

in the Gagliardo-Nirenberg inequality, we find that

$$w = \left(1 - \frac{k}{u}\right)_+^{\alpha p} \eta^{\alpha p} \min\{u, Z\}^p$$

satisfies the inequality (5.45) with $m = (4q-n)/(q-n)$ and $\mu$ defined by

$$d\mu = R^{-n}\left(1 - \frac{k}{u}\right)_+^{\beta+B} \frac{u^2}{\min\{u, Z\}^2}\,dx$$

and $s_0$ sufficiently large, depending on $p$. Again, Moser iteration and sending $Z \to \infty$ completes the proof. □

## 5.10 The DeGiorgi class

In this section, we discuss a special subset of $W^{1,p}(\Omega[R])$ that will be very useful in the next few sections and also when we study local regularity of weak solutions of nonlinear problems. Let $S$ be a measurable subset of $\mathbb{R}^n$ with positive measure, let $u \in W^{1,p}(S)$ for some $p \ge 1$, and set

$$A(k, S) = \{x \in S : u(x) < k\}$$

for $k \in \mathbb{R}$. For nonnegative constants $\chi$ and $K$ and $\varepsilon \in (\max\{0, (p-n)/np\}, 1/n]$, we then define $DG^-(p, S, \chi, \varepsilon, K)$ to be the set of all nonnegative $u \in W^{1,p}(S)$ such that

$$\int_{A(k,S\cap B(\sigma r))} |Du|^p\,dx \le K^p\left(\sup_{S\cap B(r)} \frac{(u-k)_-}{(1-\sigma)r} + kr^{-n\varepsilon}|A(k))|^{\varepsilon - 1/n}\right)^p |A(k)| \tag{5.62}$$

for all balls $B(r)$ with center in $S$ and all $k \geq \chi r^{n\varepsilon}$, and we have abbreviated $A(k, S \cap B(r))$ to $A(k)$.

Our first step is to estimate the size of the set on which $u$ is larger than some constant in some ball if we know that $u$ is larger than a given constant on some fraction of a given, smaller ball.

**Lemma 5.37.** *Let $p \geq 1$. For any constants $\theta_1$ and $\theta_2$ in $(0,1)$, there is a positive integer $s$, determined also by $K$, $p$, and $\varepsilon$ such that, for any $u \in DG^-(p, \Omega[2R], \chi, \varepsilon, K)$ with*

$$\mu = \inf_{\Omega[2R]} u \leq 2^{-s} h, \tag{5.63a}$$

$$\chi(2R)^{n\varepsilon} \leq 2^{-s} h, \tag{5.63b}$$

$$|A(\mu + \chi(2R)^{n\varepsilon} + h, \Omega[2R])| \leq (1 - \theta_1)|\Omega[2R]| \tag{5.63c}$$

*for some $h > 0$, we have*

$$|A(\mu + \chi(2R)^{n\varepsilon} + 2^{-s} h, \Omega[R])| \leq \theta_2 |\Omega[R]|. \tag{5.64}$$

***Proof.*** With $s$ to be determined, we define

$$k_j = \mu + \chi(2R)^{n\varepsilon} + 2^{-j} h$$

for $j \in \{1, \dots, s\}$. From (5.62) with $\sigma = \frac{1}{2}$, $r = 2R$, and $k$ replaced by $k_j$, we infer that

$$\begin{aligned}
&\int_{A(k_j, \Omega[R])} |Du|^p \, dx \\
&\qquad \leq K^p \left( \sup_{\Omega[2R]} \frac{(u - k_j)_-}{R} + k_j (2R)^{-n\varepsilon} |A(k_j)|^{\varepsilon - 1/n} \right)^p |A(k_j)| \\
&\qquad \leq (c(n, \varepsilon) K)^p k_j^p R^{n-p}
\end{aligned}$$

for $j = 1, \dots, s$ since $|A(k_j)| \leq 2^n \omega_n R^n$ and $(u - k_j)_- \leq k_j$. For ease of notation, we now set $A_j = A(k_j, \Omega[R])$ and $\Delta_j = A(k_j, \Omega[R]) \setminus A(k_{j+1}, \Omega[R])$ and conclude that

$$\int_{\Delta_j} |Du| \, dx \leq |\Delta_j|^{(p-1)/p} \left( \int_{A_j} |Du|^p \, dx \right)^{1/p}$$

by Hölder's inequality. Combining these inequalities gives

$$\int_{\Delta_j} |Du| \, dx \leq |\Delta_j|^{(p-1)/p} c(n, \varepsilon) R^{(n-p)/p} K k_j,$$

and then Lemma 5.14 implies that

$$(k_j - k_{j+1}) |A_{j+1}|^{(n-1)/n} \leq c(n, \varepsilon, K, \theta_1) |\Delta_j|^{(p-1)/p} k_j R^{(n-p)/p}.$$

We now rewrite this inequality. First, it follows from (5.63a) and (5.63b) that $k_j \le 2^{-j+1}h$, while $k_j - k_{j+1} = 2^{-j-1}h$, so

$$|A_{j+1}|^{(n-1)/n} \le c(n, \varepsilon, K, \theta_1)|\Delta_j|^{(p-1)/p}R^{(n-p)/p}.$$

Next, we set $X_j = |A_j|/|\Omega[R]|$, $Y_j = |\Delta_j|/|\Omega[R]|$, and

$$\eta = \frac{np-p}{np-n}$$

to rewrite our estimate as

$$X_{j+1}^{\eta} \le CY_j.$$

Summing this inequality on $j$ from 1 to $s-1$ then yields

$$(s-1)X_s^{\eta} \le C\sum_{j=1}^{s-1} Y_y \le C$$

because the $\Delta_j$'s are disjoint subsets of $\Omega[R]$ and $A_{j+1} \subset A_j$. Noting that $\eta > 0$, we conclude that

$$X_s \le \left(\frac{C}{s-1}\right)^{1/\eta}$$

and the proof is completed by choosing $s$ so that the right hand side of this inequality is no greater than $\theta_2$. □

To proceed, we need a simple iteration lemma.

**Lemma 5.38.** *Let $(Y_j)_0^{\infty}$ be a sequence of nonnegative numbers and suppose that there are positive constants $C$, $\eta$, and $b$ with $b > 1$ such that*

$$Y_{j+1} \le Cb^jY_j^{1+\eta} \tag{5.65}$$

*for all $j$. If, in addition,*

$$Y_0 \le C^{-1/\eta}b^{-1/\eta^2}, \tag{5.66}$$

*then $\lim_{j\to\infty} Y_j = 0$.*

***Proof.*** It's easy to check by induction that

$$Y_j \le C^{-1/\eta}b^{-j/\eta}b^{-1/\eta^2},$$

so $Y_j \to 0$ as $j \to \infty$. □

Using this iteration lemma, we now show that, if $u$ is larger than a suitable positive constant on most of $\Omega[2R]$, then it is larger than a corresponding positive constant on all of $\Omega[R]$.

**Lemma 5.39.** *Let $p \geq 1$. Then there is a constant $\theta^* \in (0,1)$ such that, if $u \in DG^-(p, \Omega[2R], \chi, \varepsilon, K)$ and $h$ is some positive constant such that*

$$|A(\mu + \chi(2R)^{n\varepsilon} + h, \Omega[2R])| \leq \theta^* |\Omega[2R]|, \tag{5.67a}$$

$$\mu + \chi(2R)^{n\varepsilon} \leq h, \tag{5.67b}$$

*then $u \geq \mu + \chi(2R)^{n\varepsilon} + \frac{1}{2}h$ on $\Omega[R]$.*

***Proof.*** For each nonpositive integer $j$, we set

$$\rho_j = (1 + 2^{-j})R, \quad k_j = \mu + \chi(2R)^{n\varepsilon} + \frac{h}{2}(1 + 2^{-j}),$$

and we define the sets

$$A_j = A(k_j, \Omega[\rho_j]), \quad \Delta_j = A(k_j, \omega[\rho_{j+1}]) \setminus A_{j+1}.$$

We then apply (5.62) for each $j$ with $\sigma_j = \rho_{j+1}/\rho_j$ in place of $\sigma$, $\rho_j$ in place of $r$, and $k_j$ in place of $k$. Setting $Y_j = |A_j|/|\Omega[2R]|$ and noting that $1 - \sigma_j \geq 2^{-j-1}$, $Y_j \leq 1$, and

$$\sup_{\Omega[\rho_j]} (u - k_j)_- \leq k_j,$$

we conclude that

$$\int_{\Delta_j} |Du|^p \, dx \leq C k_j^p 2^{jp} Y_j^{1+p\varepsilon-p/n} R^{n-p}.$$

and hence, from Hölder's inequality and the inequality $|\Delta_j| \leq C Y_j R^n$, we have

$$\int_{\Delta_j} |Du| \, dx \leq C k_j 2^j Y_j^{1+\varepsilon-1/n} R^{n-1}.$$

Now we use Lemma 5.14 along with the inequality $k_j \leq 2h$ and the equality $k + j - k_{j+1} = 2^{-j-1}$ to infer that

$$2^{j+2} k_j Y_{j+1}^{(n-1)/n} R^{n-1} \leq C \int_{\Delta_j} |Du| \, dx.$$

It follows that $Y_{j+1}^{(n-1)/n} \leq C 4^j Y_j^{\varepsilon+(n-1)/n}$. We now invoke Lemma 5.38 with $b = 4^{n/(n-1)}$ and $\eta = \varepsilon n/(n-1)$ to see that there is a constant $\theta \in (0,1)$ such that $Y_j \to 0$ as $j \to \infty$ if $Y_0 \leq \theta$. We take $\theta^* = \min\{1/2, \theta\}$, so (5.67a) implies that $Y_0 \leq \theta$. The proof is completed by observing that $Y_j \to \infty$ implies that $u \geq \mu + \chi(2R)^{n\varepsilon} + h/2$ in $\Omega[R]$. □

Our final step is to prove an inequality which is the analog of Lemma 1.19. We begin by recalling some notation from Section 1.7. The *Lipschitz cube* $K(y_0, r)$ is the set of all points $x$ such that

$$|x^i - y_0^i| < r \text{ if } i < n, \tag{5.68a}$$

$$|x^n - y_0^n + \omega(y_0') - \omega(x')| < r. \tag{5.68b}$$

For our preliminary results, we write $K^*(y_0, r)$ for the subset of $K(y_0, r)$ on which $x^n > \omega(x')$.

**Lemma 5.40.** *Let $p > 1$, and let $K$, $\chi$ and $\varepsilon$ be nonnegative constants with $0 < \varepsilon \leq 1/n$. There are constants $C_2$ (determined only by $K$, $n$, $\omega_0$, $\varepsilon$) and $k_1$ (determined only by $\omega_0$ and $n$) such that if $y_0 \in \Omega[R]$ and $K(y_0, 3r) \subset B(x_0, k_1 R)$, $u \in DG^-(p, \Omega[R], \chi, \varepsilon, K)$, and*

$$|\{x \in K^*(y_0, r) : \bar{u}(x) < h\}| \leq \frac{1}{2}|K^*(y_0, r)|, \tag{5.69}$$

*then $C_2\bar{u} \geq h$ in $K^*(y_0, 3r)$.*

***Proof.*** As a first step, we observe that there is a constant $k_2$ determined only by $\omega_0$ and $n$ such that $K(y_0, r) \subset B(y_0, k_2 r)$. We then take $k_1 = k_2/12$. Under our hypotheses, we then have that $\Omega[y_0, 4k_1 r] \subset \Omega[R]$. It follows that there is a constant $\theta_1 \in (0, 1)$, determined only by $\omega_0$ and $n$ such that

$$\frac{1}{2}|K^*(y_0, r)| \geq \theta_1 |\Omega[y_0, 4k_1 r]|$$

and hence

$$|\{x \in \Omega[y_0, 4k_1 r] : \bar{u}(x) \leq h\}| \leq (1 - \theta_1)|\Omega[y_0, 4k_1 r]|.$$

We use $s$ to denote the constant from Lemma 5.37 corresponding to $\theta_2 = \theta^*$, where $\theta^*$ is the constant from Lemma 5.39. Also we set

$$\mu = \inf_{\Omega[y_0, 4k_1 r]} u.$$

If $h \leq 2^{s+1}\mu$, then it follows that $\bar{u} \geq u \geq 2^{s+1}h$ in $\Omega[y_0 k_1 r]$. If $h \leq 2^{s+1}((4k_1)^{n\varepsilon} + 1)\chi r^{n\varepsilon}$, we conclude that

$$\bar{u} \geq \frac{1}{(4k_1)^{n\varepsilon} + 1} h$$

in $\Omega[y_0, k_1 r]$. Hence we may assume that

$$h > 2^{s+1} \max\{\mu, \chi((4k_1)^{n\varepsilon} + 10r^{n\varepsilon}\}.$$

It then follows from Lemma 5.37 with $h$ replaced by $(1 - 2^{-s-1})h$ that

$$|\{x \in \Omega[y_0, 2k_1 r] : \bar{u} < 2^{-s-1}h\}| \leq \theta^*|\Omega[y_0, 2k_1 r]|.$$

and the proof is completed by invoking Lemma 5.39 with $h$ replaced by $2^{-s-1}$and $R$ replaced by $k_1 r$. □

Although this lemma, along with the arguments in Chapter 1, immediately yields results for functions in the De Giorgi class, we shall not state the results here. Instead, we apply them in the next several sections to weak solutions of the differential equations. Of course, the applications in this chapter only need the case $p = 2$, but the more general case will be very useful in studying nonlinear problems, and it requires no additional effort.

## 5.11 Membership of supersolutions in the De Giorgi class

We now show that positive weak supersolutions of (5.1) are in a suitable De Giorgi class provided the coefficients of $L$ and $M$ belong to certain Lebesgue spaces. Specifically, we have the following result.

**Theorem 5.41.** *Suppose that there are positive constants $q > n$, $\lambda$, $\mu$, $\mu_2$, and $\mu_3$ such that conditions* (5.4) *and* (5.59) *are satisfied. Suppose also that $f \in L^q$, $g \in L^{q/2}$ and $\psi \in L^{q-1}$. Then any positive supersolution of* (5.58) *is in $DG^-(p, \Omega[R/2], \chi, \varepsilon, K)$ with*

$$\chi = \|f\|_q + \|g\|_{q/2} R^{1-n/q} + \|\psi\|_{q-1} R^{(q-n)/q(q-1)},$$

*$\varepsilon = (q-n)/nq$ and $K$ determined only by $n$, $q$, $\omega_0$, $\mu_2 R^{1-n/q}$, and $\mu_3 R^{(q-n)/(q-1)}$.*

***Proof.*** Let $r > 0$, $\sigma \in (0,1)$, $x_0 \in \Omega$. We then use $\varphi = (u-k)_- \eta^2$ as test function in the weak form of the inequalities for $u$, where $\eta$ is Lipschitz, with $0 \le \eta \le 1$ in $B(x_0, r)$ and $\eta = 0$. Writing $w = (u-k)_-$, we obtain

$$I_1 \le \sum_{m=2}^{7} I_m,$$

where

$$I_1 = \int a^{ij} D_i w D_j w \eta^2 \, dx,$$

$$I_2 = -\int 2a^{ij} D_i w D_j \eta \eta w \, dx,$$

$$I_3 = \int (b^i u - f^i) D_i w \eta^2 \, dx,$$

$$I_4 = \int 2(b^i u - f^i) D_i \eta \eta w \, dx,$$

$$I_5 = \int c^i D_i w w \eta^2 \, dx,$$

$$I_6 = -\int (c^0 u - g) w\eta^2\,dx,$$
$$I_7 = \int_{\partial\Omega} (\beta^0 u - \psi) w\eta^2\,ds.$$

We now proceed to estimate these integrals. First is the easy one:

$$I_1 \geq \lambda \int |Dw|^2 \eta^2\,dx.$$

Next, from the Cauchy-Schwarz inequality, we have

$$I_2 \leq \frac{1}{8} I_1 + 8\mu^2 \int w^2 |D\eta|^2\,dx.$$

The Cauchy-Schwarz inequality along with Hölder's inequality then yields

$$I_3 \leq \frac{1}{8} I_1 + 2\lambda A^{1-2/q}[k^2\mu_1^2 + \chi^2],$$
$$I_4 \leq \lambda \int w^2 |D\eta|^2\,dx + \lambda A^{1-2/q}[k^2\mu_1^2 + \chi^2],$$
$$I_5 \leq \frac{1}{8} I_0 + 2\lambda k^2 A^{1-2/q}\mu_1^2,$$

where we use $A$ to denote the measure of the support of $\varphi$, and we have used the inequality $0 \leq u \leq k$ wherever $\varphi > 0$. In a similar vein (but only using Hölder's inequality), we also find that

$$I_6 \leq \lambda A^{1-2/q}[k^2\mu_2^2 + k\chi R^{(n-q)/q}].$$

To estimate $I_7$, we imitate the proof of Lemma 5.20. First, Hölder's inequality implies that

$$I_7 \leq [k\mu_3 + \chi R^{(n-q)/q(q-1)}] \left( \int_{\Sigma[R]} (w\eta^2)^{(q-1)/(1-2)}\,ds \right)^{(q-2)/(q-1)}.$$

and (5.30) (with $q = p = 1$) implies that

$$\int_{\Sigma[R]} (w\eta^2)^{(q-1)/(1-2)}\,ds \leq C(\omega_0, q) \int_{\Omega[R]} (w\eta^2)^{1/(q-2)} |D(w\eta^2)|\,dx.$$

We now apply the Cauchy-Schwarz inequality and note that $|D(w\eta^2)|^2 \leq 4\eta^2|Dw|^2 + 4w^2|D\eta|^2$ to infer that

$$\int_{\Omega[R]} (w\eta^2)^{1/(q-2)} |D(w\eta^2)|\,dx \leq 2(I_1 + \int_{\Omega[R]} w^2|D\eta|^2\,dx)^{1/2} k^{1/(q-2)} A^{1/2},$$

and hence

$$I_7 \leq \frac{1}{8} I_1 + C(\omega_0, q)\lambda \left( \int w^2 |D\eta|^2\,dx + (\mu_3 k + \chi R^{(n-q)/q(q-1)})^{2-2/q} k^{2/q} \right).$$

Combining all these inequalities gives us

$$\int |Dw|^2\eta^2\,dx \le C\left(\int w^2|D\eta|^2\,dx + k^2r^{-n\varepsilon}A^{1-2/q}\right)$$

with $C$ determined only by $\omega_0$, $q$, $\mu_2 R^{1-n/q}$, $\mu_3 R^{(q-n)/(q-1)}$, and the proof is completed by taking

$$\eta(x) = \begin{cases} 1 & \text{if } |x-x_0| \le \sigma r, \\ \frac{1-|x-x_0|/r}{1-\sigma} & \text{if } \sigma r < |x-x_0| \le r, \\ 0 & \text{if } |x-x_0| > r. \end{cases}$$

□

## 5.12 Consequences of the local estimates

Lemma 5.40 leads to several important consequences. We begin with a weak Harnack inequality, which follows from that lemma via the argument of Theorem 1.20.

**Theorem 5.42.** *Let $\Omega$ be a bounded Lipschitz domain, let $L$ and $M$ satisfy (5.4) and (5.59), and suppose that $f^i \in L^q(\Omega)$, $g \in L^{q/2}(\Omega)$, and $\psi \in L^{q-1}(\partial\Omega)$. Let $u$ be a nonnegative weak supersolution of*

$$Lu = D_i f^i + g \text{ in } \Omega[4R], \quad Mu = \psi \text{ on } \Sigma[4R].$$

*Then there are positive constants $C$ and $p$, determined only by $\mu$, $\mu_2 R^{1-n/q}$, $\mu_3 R^{(q-n)/(q-1)}$, $n$, and $\Omega$, such that*

$$\left(R^{-n}\int_{\Omega[2R]} |u|^p\,dx\right)^{1/p} \le C(\inf_{\Omega[R]} u + k(R)). \tag{5.70}$$

In fact, (5.70) is true for any $p \in (0, n/(n-2))$ if $n > 2$ and any $p > 0$ if $n = 2$. Since this stronger result will not be used in this book, we defer the proof to Exercise 5.1.

Next, by arguing exactly as in Corollary 1.22, we obtain the following strong maximum principle, which may also be deduced in several other ways. We sketch some of these methods in Exercise 5.2.

**Theorem 5.43.** *Let $\Omega$ be a bounded Lipschitz domain, and let $L$ and $M$ satisfy (5.4), (5.59), and (5.20). Suppose $u \in W^{1,2}(\Omega)$ satisfies $Lu \ge 0$ in $\Omega$ and $Mu \ge 0$ on $\partial\Omega$. If $\sup u \ge 0$, then $u$ is constant.*

Combining Theorems 5.36 and 5.42 gives a full Harnack inequality.

**Theorem 5.44.** *Let $\Omega$ be a bounded Lipschitz domain, let $L$ and $M$ satisfy* (5.4) *and* (5.59), *and let $q > n$. Then there is a constant $C$, determined only by $\mu$, $\mu_2 R$, $\mu_3 R^2$, $n$, $p$, and $\Omega$, such that*

$$\sup_{\Omega[R]} u \leq C(\inf_{\Omega[R]} u + k(R))$$

*for any nonnegative weak solution $u$ of* (5.5) *with $f^i \in L^q(\Omega)$, $g \in L^{q/2}(\Omega)$, and $\psi \in W^{1-1/q,q}(\partial\Omega)$.*

Also, the proof of Theorem 1.26 (with Theorem 5.42 in place of Theorem 1.20) gives the following Hölder estimate for weak solutions of (5.5).

**Theorem 5.45.** *Let $\Omega$ be a bounded Lipschitz domain, let $L$ and $M$ satisfy* (5.4) *and* (5.59), *and suppose there is $q > n$ such that $f^i \in L^q(\Omega)$, $g \in L^{q/2}(\Omega)$, and $\psi \in W^{1-1/q,q}(\partial\Omega)$. If $Lu = D_i f^i + g$ in $\Omega[R]$ and $Mu = \psi$ on $\Sigma[R]$, then there are constants $C$, $\alpha$ and $\delta$ determined only by $\mu$, $\mu_2 R$, $\mu_3 R^2$, $n$, $p$, and $\Omega$, such that*

$$\operatorname*{osc}_{\Omega[r]} u \leq C\left(\frac{r^\alpha}{R^\alpha} \sup_{\Omega[R]} u + k(R)\right). \tag{5.71}$$

For solutions of the Dirichlet problem, a similar result is true.

**Theorem 5.46.** *Let $\Omega$ be a bounded Lipschitz domain, let $L$ satisfy* (5.4), *and suppose there is $q > n$ such that $f^i \in L^q(\Omega)$ and $g \in L^{q/2}(\Omega)$. If $Lu = D_i f^i + g$ in $\Omega[R]$ and $u = \psi$ on $\Sigma[R]$, then there are constants $C$, $\alpha$ and $\delta$ determined only by $\mu$, $\mu_2 R$, $\mu_3 R^2$, $n$, $p$, and $\Omega$, such that*

$$\operatorname*{osc}_{\Omega[r]} u \leq C\left(\frac{r^\alpha}{R^\alpha} \sup_{\Omega[R]} u + k(R) + \operatorname*{osc}_{\Sigma[R]} \psi\right). \tag{5.72}$$

## 5.13 Integral characterizations of Hölder spaces

We now show that certain integral inequalities for functions imply that these functions are in certain Hölder spaces. To this end, we first recall the definition of the Campanato spaces $\mathcal{L}^{1,q}$ for $q \geq 0$. We say that $u \in \mathcal{L}^{1,q}(\Omega)$ for some open set $\Omega$ if $u \in L^1(\Omega)$ and if there is a constant $H$ such that

$$\inf_{L\in\mathbb{R}} \int_{B(y,r)\cap\Omega} |u(x) - L|\, dx \leq H r^q$$

for all $r \in (0, \operatorname{diam}\Omega)$ and all $y \in \Omega$. The smallest such constant $H$ is called the $\mathcal{L}^{1,q}$ seminorm of $u$ and is written $[u]_{\mathcal{L}^{1,q}}$. Note that adding a constant to $u$ gives another function with the same $\mathcal{L}^{1,q}$ seminorm, so we define a norm on $\mathcal{L}^{1,q}$ by

$$\|u\|_{1,q} = [u]_{\mathcal{L}^{1,q}} + \|u\|_1.$$

It's easy to check that $u$ must be a constant if $u \in \mathcal{L}^{1,q}$ with $q > n+1$ and that $C^\alpha(\overline{\Omega}) \subset \mathcal{L}^{1,n+\alpha}$ if $\alpha \in (0,1]$. In fact, for most domains of interest (and, in particular, all bounded Lipschitz domains), these spaces are the same. For our purposes, we shall prove a slightly weaker result first.

**Lemma 5.47.** *Let $f \in L^1(B(x_0, 2R))$ for some $x_0 \in \mathbb{R}$ and $R > 0$. If there are positive constants $\alpha \le 1$ and $H$ and a function $g(y,r)$ defined on $B(x_0,R) \times (0,R)$ such that*

$$\int_{B(y,r)} |f(x) - g(y,r)|\, dx \le H r^{n+\alpha} \tag{5.73}$$

*for all $r \in (0,R)$ and all $y \in B(x_0,R)$, then $f \in C^\alpha(B(x_0,R))$ and*

$$[f]_{\alpha;B(x_0,R)} \le c(n,\alpha)H.$$

***Proof.*** The proof is essentially the same as for Lemma 2.8. As in that lemma, we let $\varphi$ be a nonnegative $C^1(\mathbb{R}^n)$ function with support in $B(0,1)$ such that

$$\int_{\mathbb{R}^n} \varphi(x)\, dx = 1,$$

and set $K = \sup |D\varphi|$. For $(x,\tau) \in B(x_0,R) \times (0,R)$, we define the function $\bar{f}$ by

$$\bar{f}(x,\tau) = \int_{\mathbb{R}^n} f(x - \tau)\varphi(z)\, dz. \tag{5.74}$$

We then have that

$$\frac{\partial \bar{f}}{\partial x}(x,\tau) = \tau^{-n-1} \int_{B(x,\tau)} [f(w) - g(x,\tau)] D\varphi\left(\frac{x-w}{\tau}\right) dw,$$

and a similar expression for $\partial\bar{f}/\partial\tau$. Writing $\nabla\bar{f}$ for the $(n+1)$-dimensional vector

$$\left(\frac{\partial \bar{f}}{\partial x}, \frac{\partial \bar{f}}{\partial \tau}\right),$$

we conclude that

$$|\nabla \bar{f}(x,\tau)| \le (n+1)HK\tau^{\alpha-1},$$

and the proof is completed exactly as for Lemma 2.8. □

A similar result is true near the boundary of the domain provided a suitable assumption is made about the regularity of the boundary. Here we only consider Lipschitz boundaries, but the result is actually true under a weaker assumption on the regularity of the boundary.

**Lemma 5.48.** *Let $\Omega$ be a bounded domain in $\mathbb{R}^n$ and, for any $x_0 \in \mathbb{R}^n$ and $R > 0$, set*

$$\Omega[x_0, R] = \Omega \cap B(x_0, R).$$

*Let $x_0 \in \partial\Omega$ and $R > 0$, and suppose there is a function $\omega$ such that*

$$\Omega[x_0, 2R] = \{(x', x^n) : x^n > \omega(x'), |x| < 2R\} \tag{5.75a}$$

$$|\omega(x_1') - \omega(x_2')| \le \omega_0 |x_1' - x_2'| \tag{5.75b}$$

*for any $x_1'$ and $x_2'$ in $\mathbb{R}^{n-1}$ with $|x_0' - x_i'| < R$ for $i = 1, 2$. Let $f \in L^1(\Omega[x_0, 2R])$ and suppose that there are positive constants $\alpha \le 1$ and $H$ and a function $g$ defined on $\Sigma[x_0, R] \times (0, R)$, where $\Sigma[x_0, R] = \partial\Omega \cap B(x_0, R)$, so that*

$$\int_{\Omega[y,r]} |f(x) - g(y, r)|\, dx \le H r^{n+\alpha}. \tag{5.76}$$

*Then*

$$[f]_{\alpha;\Sigma[x_0,R]} \le c(n, \alpha, \omega_0) H.$$

***Proof.*** The proof is essentially the same as for Lemma 5.47 but more care must be taken on some technical points.

First, we let $\varphi_0$ be a nonnegative $C^1(\mathbb{R}^n)$ function supported in $B(0,1)$ and $\int_{\mathbb{R}^n} \varphi_0(x)\, dx = 1$, we set $\kappa = (1 + \omega_0^2)^{1/2}$, and we define $\varphi$ by

$$\varphi(x) = \kappa^n \varphi_0(\kappa(2x - e_n)),$$

where $e_n = (0, \dots, 0, 1)$. It follows that $\varphi$ is a nonnegative $C^1(\mathbb{R}^n)$ function supported in $B(0,1)$ and $\int_{\mathbb{R}^n} \varphi(x)\, dx = 1$. Moreover if $\varphi(z) \ne 0$, if $y \in \Sigma[x_0, R]$, and if $r \in (0, R)$, then $y + rz \in \Omega[x_0, R]$.

Because $\partial\Omega$ is only Lipschitz, it no longer makes sense to talk about $\bar{f}$ (defined by (5.74)) being continuously differentiable. Instead, we use a slightly different expression, which will be continuously differentiable with respect to $r$ but only Lipschitz with respect to $y$. To write everything conveniently, we first write $B'(x_0', R)$ for the ball in $\mathbb{R}^{n-1}$ with center $x_0'$ and radius $R$. We define $\bar{f}$ on $B'(x_0', R) \times (0, R)$ by

$$\bar{f}(x', \tau) = \int_{\mathbb{R}^n} f((x', \omega(x')) - \tau z)\varphi(z)\, dz,$$

so $\bar{f}$ is Lipschitz with respect to $x'$ and continuously differentiable with respect to $\tau$ (for $\tau \neq 0$. It follows now that

$$\frac{\partial \bar{f}}{\partial x^i}(x',\tau) = -\frac{1}{\tau}\int_{\mathbb{R}^n} f((x',\omega(x')) - \tau z)[D_i\varphi(z) + D_i\omega(x')D_n\varphi(z)]\,dz$$

(for $i = 1,\ldots,n-1$) almost everywhere with respect to $x$ because $\omega$ is almost everywhere differentiable. It follows that

$$\left|\frac{\partial \bar{f}}{\partial x^i}(y',\tau)\right| \le c(n,\omega_0)HK\tau^{\alpha-1},$$

and then the proof is completed as before. □

## 5.14 Schauder estimates

In this section, we prove some Hölder estimates for the derivatives of weak solutions of conormal problems. Clearly, we shall need stronger hypotheses on the coefficients than just boundedness. Conveniently, Hölder continuity of (some) coefficients implies Hölder continuity of the gradient of the solution with the same exponent.

We begin with an improved version of the regularity results from Section 5.7 for problems with constant coefficients.

**Lemma 5.49.** *Let $[A^{ij}]$ be a constant, positive definite matrix, let $F$ be a constant vector, and suppose $v$ is a solution of*

$$D_i(A^{ij}D_jv) = 0 \text{ in } B^+(0,R), \quad A^{nj}D_jv = 0 \text{ on } B^0(0,R). \tag{5.77}$$

*Then $v \in C^2(\overline{B^+(0,R/2)})$.*

***Proof.*** We begin by showing that $D_1v,\ldots,D_{n-1}v$ are uniformly Hölder continuous in $B^+(0,3R/4)$. To this end, we fix $k \in \{1,\ldots,n-1\}$ and let $h \in (0,R/8)$ be a parameter at our disposal. We then define $v_h$ in $\overline{B^+(7R/8)}$ by

$$v_h(x) = \frac{v(x+he_k) - v(x)}{h},$$

where $e_k$ is the vector with $k$th component equal to one and all other components equal to zero. Then $v_h$ satisfies the conditions

$$D_i(A^{ij}D_jv_h) = 0 \text{ in } B^+(0,7R/8), \quad A^{nj}D_jv_h = 0 \text{ on } B^0(0,7R/8),$$

so there are positive constants $\gamma$ (determined only by $n$ and the ratio of the maximum and minimum eigenvalues of $[A^{ij}]$) and $C$ (determined also by $R$) so that

$$[v_h]_{\gamma;B^+(0,3R/4)} \le C\|v_h\|_{2;B^+(0,7R/8)} \le C\|D_k v\|_{2;B^+(0,R)}.$$

Since $v_h \to D_k v$ in $L^2$ as $h \to 0^+$, it follows that $D_k u$ is Hölder continuous in $B^+(0,3R/4)$ with

$$[D_k v]_{\gamma;B^+(0,3R/4)} \le C.$$

The second derivative estimate for $v$ then implies that

$$|D_{ik}v(x)| \le C(x^n)^{\gamma-1}$$

for $i = 1,\ldots,n$, $k = 1,\ldots,n-1$ and any $x \in B^+(0,5R/8)$. The differential equation then implies that this inequality is also valid for $i = k = n$. It then follows from simple integration that

$$[D_n v]_{\gamma;B^+(0,9R/16)} \le C.$$

Hence $v \in C^{1,\gamma}(\overline{B^+(0,9R/16)})$.

We now repeat this argument with $v_h$ in place of $v$ to conclude that

$$[Dv_h]_{\gamma;B^+(0,R/2)} \le C,$$

and hence we get a uniform Hölder estimate for $D_{ik}v$ in $B^+(0,R/2)$ with $i = 1,\ldots,n$, $k = 1,\ldots,n-1$. The differential equation then gives a Hölder estimate for $D_{nn}v$ and hence $v \in C^2(\overline{B^+(0,R/2)})$. □

From this regularity result, we easily obtain some useful integral estimates for $v$.

**Lemma 5.50.** *Let $[A^{ij}]$ be a constant, positive definite matrix, and suppose $v$ is a solution of* (5.77). *Suppose $\lambda$ and $\mu$ are positive constants such that*

$$\lambda|\xi|^2 \le A^{ij}\xi_i\xi_j, \tag{5.78a}$$

$$A^{ij}\xi_i\eta_j \le (\mu\lambda|\xi|^2)^{1/2}(A^{ij}\eta_i\eta_j)^{1/2} \tag{5.78b}$$

*for all $\xi$ and $\eta$ in $\mathbb{R}^n$. Then there is a constant $C$, determined only by $n$ and $\mu$ so that*

$$\int_{B^+(R/2)} |Dv|^2\,dx \le CR^{-2}\int_{B^+(R)} v^2\,dx, \tag{5.79a}$$

$$\int_{B^+(r)} v^2\,dx \le C\left(\frac{r}{R}\right)^n \int_{B^+(R)} v^2\,dx, \tag{5.79b}$$

$$\int_{B^+(r)} |v - \{v\}_r|^2\,dx \le C\left(\frac{r}{R}\right)^{n+2} \int_{B^+(R)} |v - \{v\}_R|^2\,dx \tag{5.79c}$$

*for* $r \in (0, R)$, *where*

$$\{v\}_r = \frac{1}{|B^+(r)|} \int_{B^+(r)} v\,dx. \tag{5.80}$$

***Proof.*** Set $\zeta = 1 - |x|^2/R^2$. By using $\zeta^2 v$ as test function, we see that

$$\int_{B^+(R)} \zeta^2 A^{ij} D_i v D_j v\,dx = -2 \int_{B^+(R)} \zeta v A^{ij} D_j v D_i \zeta\,dx.$$

We then estimate the right hand side through Cauchy's inequality and (5.78b):

$$\begin{aligned} -2 &\int_{B^+(R)} \zeta v A^{ij} D_j v D_i \zeta\,dx \\ &\le 2 \left( \int_{B^+(R)} \zeta^2 A^{ij} D_i v D_j v\,dx \right)^{1/2} \left( \int_{B^+(R)} \mu\lambda |D\zeta|^2 v^2\,dx \right)^{1/2}. \end{aligned}$$

It follows that

$$\int_{B^+(R)} \zeta^2 A^{ij} D_i v D_j v\,dx \le 4\mu\lambda \int_{B^+(R)} |D\zeta|^2 v^2\,dx.$$

Using (5.78a) and dividing the resulting inequality by $\lambda$ yields

$$\int_{B^+(R)} \zeta^2 |Dv|^2\,dx \le 4\mu \int_{B^+(r)} |D\zeta|^2 v^2\,dx.$$

We now infer (5.79a) by observing that $|D\zeta| \le 2/R$ in $B^+(R)$ while $\zeta \ge 1/2$ in $B^+(R/2)$.

To prove (5.79b), we note that this inequality is clear for $r \ge R/2$, so we assume that $r < R/2$. It then follows that

$$\int_{B^+(r)} v^2\,dx \le C(n) r^n \left( \sup_{B^+(R/2)} |v| \right)^2.$$

We now use the local maximum principle to infer that

$$\sup_{B^+(R/2)} |v| \le C(n,\mu) \left( R^{-n} \int_{B^+(R)} v^2\,dx \right)^{1/2}.$$

Combining these two inequalities gives (5.79b).

For our final inequality, we first use Poincaré's inequality to infer that

$$\int_{B^+(r)} |v - \{v\}_r|^2\,dx \le C(n) r^{-2} \int_{B^+(r)} |Dv|^2\,dx. \tag{5.81}$$

If $r < R/2$, we then apply (5.79b) with $D_k v$ ($k = 1, \dots, n-1$) in place of $v$ to obtain

$$\int_{B^+(r)} |D_k v|^2\,dx \le C(n,\mu)\left(\frac{r}{R}\right)^n \int_{B^+(R/2)} |D_k v|^2\,dx.$$

A similar result is true with $n$ in place of $k$ but the proof requires a little more care. If we set $w = A^{nj}D_j v$, then we have

$$D_i(A^{ij}D_j w) = 0 \text{ in } B^+(R), \quad w = 0 \text{ on } B^0(R).$$

A straightforward modification of the proof of the local maximum principle as given here shows that $w$ satisfies the estimate

$$\sup_{B^+(R/4)} w \le C\left(R^{-n}\int_{B^+(R/2)} w^2\,dx\right)^{1/2}$$

and hence

$$\int_{B^+(r)} w^2\,dx \le C(n,\mu)\left(\frac{r}{R}\right)^n \int_{B^+(R/2)} w^2\,dx.$$

We now use the inequalities

$$|w| \le C\lambda(\sum_{k=1}^{n-1} |D_k v| + |D_n v|),$$

$$|D_n v| \le C(\frac{|w|}{\lambda} + \mu\sum_{k=1}^{n-1} |D_k v|)$$

to conclude that

$$\int_{B^+(r)} |D_n v|^2\,dx \le C(n,\mu)\left(\frac{r}{R}\right)^n \left(\int_{B^+(R/2)} |Dv|^2\,dx\right),$$

and hence

$$\int_{B^+(r)} |Dv|^2\,dx \le C(n,\mu)\left(\frac{r}{R}\right)^n \left(\int_{B^+(R/2)} |Dv|^2\,dx\right). \tag{5.82}$$

We now repeat the proof of (5.79a) with $v - \{v\}_R$ in place of $v$ to infer that

$$\int_{B^+(R/2)} |Dv|^2\,dx \le CR^{-2}\int_{B^+(R)} |v - \{v\}_R|^2\,dx.$$

In combination with (5.81) and (5.82), this inequality implies (5.79c) provided $r < R/4$. For $r \ge R/4$, we observe that

$$\int_{B^+(r)} |v - \{v\}_r|^2\,dx = \inf_{L\in\mathbb{R}} \int_{B^+(r)} |v - L|^2\,dx, \tag{5.83}$$

so

$$\begin{aligned}\int_{B^+(r)} |v-\{v\}_r|^2\,dx &\le \int_{B^+(r)} |v-\{v\}_R|^2\,dx\\ &\le \int_{B^+(R)} |v-\{v\}_R|^2\,dx\\ &\le 4^{n+2}\left(\frac{r}{R}\right)^{n+2}\int_{B^+(R)} |v-\{v\}_R|^2\,dx.\end{aligned}$$

□

The Schauder estimates are now proved by a simple idea: Consider the solutions of general problems as perturbations of solutions of constant coefficients. Estimates for the constant coefficient problems are already presented, so we just need to prove the appropriate perturbation estimates. These are relatively straightforward but they involve a large amount of calculation. Rather than provide one, extremely long calculation, we break the procedure down into several steps, all of which refer to the problem

$$D_i(a^{ij}D_ju+f^i)=g \text{ in } B^+(x_0,R_0), \tag{5.84a}$$

$$a^{nj}D_ju+\psi=0 \text{ on } B^0(x_0,R_0). \tag{5.84b}$$

In order to describe our results in a convenient form, we make some definitions. First, for any $y\in\mathbb{R}$ with $y^n=0$ and any $r>0$, we set

$$B^*(y,r)=\{x\in\mathbb{R}^n, |x-y|=r,\ x^n>0\}$$

and we write $d^*$ for distance to $B^*(x_0,R_0)$. For $q\in(0,n)$ and $\delta\ge 0$, we write $M^{q,\delta}$ for the set of all functions $g$ such that the norm

$$\|g\|_{q,\delta}=\sup_{\substack{y\in B^+(x_0,R_0)\\ r\in(0,d^*(y))}}\left(r^{-q}d^*(y)^{-\delta}\int_{B(y,r)\cap B^+(x_0,R_0)}|g|^2\,dx\right)^{1/2}$$

is finite. Note that $M^{q,\delta}$ is a weighted Morrey space. In [114], the corresponding space for parabolic equations was denoted $M^{(b)}_{2,q}$ with $b=(\delta+n+2-q)/2$.

We assume that there are positive constants $\lambda$ and $\mu$ so that (5.78) holds and we assume that $a^{ij}\in H^{(0)}_\alpha(B^+(x_0,R_0))$. To quantify our regularity assumption on $a^{ij}$, we take $A$ to be a constant such that

$$[a^{ij}]^{(0)}_\alpha\le A\lambda. \tag{5.85}$$

Assumptions on the regularity of $f^i$, $g$, and $\psi$ will be made below.

**Lemma 5.51.** *Suppose that conditions* (5.78) *and* (5.85) *are satisfied. If $u$ satisfies* (5.84)*, then there is a constant $C$, determined only by $n$, $\mu$, $\alpha$, and $A$ such that*

$$\int_{B^+(y,r)} |Du|^2\,dx \le C\frac{r^{n-\alpha}}{d^*(y)^{2-\alpha}}(\sup|u| + F_0)^2 \tag{5.86}$$

*for all $y \in B^0(x_0, R_0)$ and all $r \in (0, d^*(y)/2)$, where*

$$F_0 = |f/\lambda|_0^{(1)} + \|g/\lambda\|_{n-2+2\alpha,2+2\alpha} + |\psi/\lambda|_0^{(1)}. \tag{5.87}$$

***Proof.*** Fix $y \in B^0(x_0, R_0)$.

As a first step, we obtain an estimate for the $L^2$ norm of $Du$ over $B^+(y, d^*(y)/2)$. Define $\zeta$ by

$$\zeta = \left(1 - \frac{4|x-y|^2}{d^*(y)^2}\right)_+$$

and note that $0 \le \zeta \le 1$ and $|D\zeta| \le 8/d^*(y)$ in $\mathbb{R}^n$. Using $\zeta^2 u$ as test function in the weak form of the (5.84) yields

$$\begin{aligned}\int_{B^+} \zeta^2 a^{ij} D_i u D_j u\,dx = &-\int_{B^+} 2\zeta a^{ij} D_i\zeta D_j u u\,dx\\ &-\int_{B^+} \zeta^2 f^i D_i u + 2\zeta f^i D_i\zeta u\,dx\\ &-\int_{B^+} g\zeta^2 u\,dx + \int_{B^0} \psi\zeta^2 u\,ds,\end{aligned}$$

where we have abbreviated $B^+(y, d^*(y)/2)$ to $B^+$ and $B^0(y, d^*(y)/2)$ to $B^0$. From (5.78) and the Cauchy-Schwarz inequality, we infer that

$$\int_{B^+} \zeta^2 a^{ij} D_i u D_j u\,dx \le C(\mu, n)\lambda d^*(y)^{n-2}[\sup|u|^2 + F_0^2]$$

and hence (after noting that $\zeta \ge 3/4$ on $B^+(y, d^*(y)/4)$)

$$\int_{B^+(y,d^*(y)/4)} |Du|^2\,dx \le C(\mu, n) d^*(y)^{n-2}[\sup|u|^2 + F_0^2]. \tag{5.88}$$

With $\tau \in (0,1)$ and $\theta \in (0, 1/4)$ to be further specified, we set $r_0 = \theta d^*(y)$ and we define $r_k = \tau^k r_0$ for positive integers $k$. We also take $v_k$ to be the solution of

$$\begin{aligned} D_i(a^{ij}(y)D_j v_k) &= 0 \text{ in } B^+(y, r_k),\\ a^{nj}(y)D_j v_k &= 0 \text{ on } B^0(y, r_k),\\ v_k &= u \text{ on } B^*(y, r_k).\end{aligned}$$

It then follows from (5.82) that

$$\int_{B^+(y,r_{k+1})} |Dv_k|^2\,dx \le C\tau^n \int_{B^+(y,r_k)} |Dv_k|^2\,dx. \tag{5.89}$$

Moreover, if we use $w_k = v_k - u$ as test function in the weak forms of the equations for $u$ and $v_k$, we find that

$$\begin{aligned}\int_{B^+(y,r_k)} a^{ij}(y) D_i w_k D_j w_k\,dx &= \int_{B^+(y,r_k)} [a^{ij} - a^{ij}(y)] D_i u D_j w_k\,dx \\ &\quad + \int_{B^+(y,r_k)} f^i D_i w_k + g w_k\,dx \\ &\quad - \int_{B^0(y,r_k)} \psi w_k\,dx.\end{aligned}$$

We now observe that

$$|a^{ij}(x) - a^{ij}(y)| \le \lambda A|x-y|^\alpha d^*(y)^{-\alpha} \le C\lambda\theta^\alpha$$

for $y \in B^+(y,r_k)$ and hence, after some rearrangement along with the Cauchy-Schwarz inequality and the Sobolev inequality, we obtain

$$\int_{B^+(y,r_k)} |Dw|^2\,dx \le C\theta^{2\alpha} \int_{B^+(y,r_k)} |Du|^2\,dx + CF_0^2 r_k^{n-\alpha} d^*(y)^{\alpha-2}.$$

In combination with (5.89), this inequality implies that

$$\int_{B^+(y,r_{k+1})} |Du|^2\,dx \le C[\tau^n + \theta^{2\alpha}] \int_{B^+(y,r_k)} |Du|^2\,dx + CF_0^2 r_k^{n-\alpha} d^*(y)^{\alpha-2}.$$

We now choose $\tau$ so that $C\tau^n \le \tau^{n-\alpha/2}/2$ and then $\theta$ so that $C\theta^{2\alpha} \le \tau^{n-\alpha/2}/2$. Applying Lemma 1.25 and (5.88) completes the proof. □

Essentially the same idea then provides a mean oscillation estimate for $Du$ if $f$ and $\psi$ are in $H_\alpha^{(1)}$. To simplify the notation, we define (similarly to (5.80))

$$\{g\}_{y,r} = \frac{1}{|B^+(y,r)|} \int_{B^+(y,r)} g(x)\,dx$$

for any $y \in B^+(x_0,R_0)$, any $r \in (0,d^*(y))$ and $g \in L^1(B^+(y,r))$.

**Lemma 5.52.** *Assume the hypotheses of Lemma* 5.51. *Also, for* $y \in B^0(x_0,R_0)$ *and* $r \in (0,d^*(y))$, *define* $\{Du\}^*_{y,r}$ *to be the vector with components* $L_1,\dots,L_n$ *defined by*

$$L_i = \begin{cases} \{D_i u\}_{y,r} & \text{if } i < n, \\ [f(y) - \psi(y) - \sum_{j=1}^{n-1} a^{nj}(y) L_j]/a^{nn}(y) & \text{if } i = n. \end{cases}$$

*Then there is a constant $C$, determined only by $A$, $\alpha$, $n$, and $\mu$ such that*

$$\int_{B^+(y,r)} |Du - \{Du\}^*_{y,r}|^2\,dx \le C(\sup|u| + F)^2 \frac{r^\alpha}{d^*(y)^\delta} \tag{5.90}$$

*for any $y \in B^0(x_0, R_0)$ and $r \in (0, d^*(y))$, where*

$$F = |f/\lambda|^{(1)}_\alpha + \|g/\lambda\|_{n-2+2\alpha,2+2\alpha} + |\psi/\lambda|^{(0)}_\alpha. \tag{5.91}$$

***Proof.*** With $y$ fixed and $\tau \in (0,1)$ to be chosen, we define $r_0 = d^*(y)$ and $r_k = \tau^k r_0$, and we write $v_k$ for the weak solution of

$$\begin{aligned} D_i(a^{ij}(y)D_jv_k + f^i(y)) &= 0 \text{ in } B^+(y,r_k),\\ a^{nj}(y)D_jv_k + f^n(y) &= \psi(y) \text{ on } B^0(y,r_k),\\ v_k &= u \text{ on } B^*(y,r_k). \end{aligned}$$

Then (5.79c) applied to $D_iv_k$ with $i < n$ yields

$$\int_{B^+(y,r_{k+1})} |D_iv_k - \{D_iv_k\}_{y,r_{k+1}}|^2\,dx \le C\tau^{n+2}\int_{B^+(y,r_k)} |D_iv_k - \{D_iv_k\}_{y,r_k}|^2\,dx,$$

and a similar argument applied to $v_{k,n} = a^{nj}(y)D_jv_k + f^n(y) - \psi(y)$ yields

$$\int_{B^+(y,r_{k+1})} |v_{k,n}|^2\,dx \le C\tau^{n+2}\int_{B^+(y,r_k)} |v_{k,n}|^2\,dx.$$

Combining these two inequalities, we obtain

$$\int_{B^+(y,r_{k+1})} |Dv_k - \{Dv_k\}^*_{y,r_{k+1}}|^2\,dx \le C\tau^{n+2}\int_{B^+(y,r_k)} |Dv_k - \{Dv_k\}^*_{y,r_k}|^2\,dx.$$

Again, we set $w_k = v_k - u$ and use this function as test function in the weak forms of the boundary value problems for $v_k$ and $u$ to obtain

$$\begin{aligned} \int_{B^+(y,r_k)} a^{ij}(y)D_iw_kD_jw_k\,dx &= \int_{B^+(y,r_k)} [a^{ij} - a^{ij}(y)]D_iuD_jw_k\,dx\\ &\quad + \int_{B^+(y,r_k)} [f^i - f^i(y)]D_iw_k + gw_k\,dx\\ &\quad - \int_{B^0(y,r_k)} [\psi - \psi(y)]w_k\,dx. \end{aligned}$$

This time, we observe that

$$|a^{ij}(x) - a^{ij}(y)| \le \lambda\frac{r_k^\alpha}{d^*(y)^\alpha}$$

and use (5.86) to conclude that

$$\int_{B^+(y,r_k)} |Dw_k|^2\,dx \le C(\sup|u| + F)^2\frac{r_k^{n+\alpha}}{d^*(y)^{2+\alpha}}. \tag{5.92}$$

If we set

$$I(r) = \int_{B^+(y,r)} |Du - \{Du\}^*_{y,r}|^2\,dx,$$

we conclude that

$$I(r_{k+1}) \le C\tau^{n+2} I(r_k) + C(\sup|u| + F)^2 \frac{r_k^{n+\alpha}}{d^*(y)^{2+\alpha}},$$

and (5.90) follows from this inequality via Lemmata 1.25 and 5.51. □

Of course, this mean oscillation estimate only provides an $H^{(0)}_{1+\alpha/2}$ estimate for $u$, but the full $H^{(0)}_{1+\alpha}$ estimate follows by a easy extra step.

**Proposition 5.53.** *Let $\alpha \in (0,1)$, let $x_0 \in \mathbb{R}^n$ with $x_0^n = 0$, and let $R_0 > 0$. Suppose that there are positive constants $A$, $\lambda$, and $\mu$ such that conditions (5.78) and (5.85) are satisfied. If $u$ is a weak solution of (5.84) with $f \in H^{(1)}_\alpha$, $g \in M^{n-2+2\alpha,2+2\alpha}$, and $\psi \in H^{(1)}_\alpha$, then there is a constant $C$, determined only by $A$, $\mu$, $n$ such that*

$$|u|^{(0)}_{1+\alpha} \le C(|u|_0 + F) \tag{5.93}$$

*with $F$ given by (5.91).*

***Proof.*** By repeating the argument of Lemma 5.52 with $y \in B^+(x_0, R_0)$ and using Lemma 5.48, we conclude that

$$[u]^{(0)}_{1+\alpha/2} \le C(|u|_0 + F),$$

and the interpolation inequality (Lemma 2.3) implies that

$$[u]^{(0)}_1 \le C(|u|_0 + F).$$

With the notation from the proof of Lemma 5.52, this inequality implies that

$$\int_{B^+(y,r_k)} |Dw_k|^2\,dx \le C(\sup|u| + F)^2 \frac{r_k^{n+2\alpha}}{d^*(y)^{2+2\alpha}}.$$

We then argue as before but with this inequality in place of (5.92) to infer (5.93). □

Via a simple change of variables, we obtain a global Hölder gradient estimate. To state this result, for $q \in (0,n)$, we write $M^q$ for the set of all $g \in L^2(\Omega)$ such that the norm

$$\|g\|_{M^q} = \sup_{\substack{y\in\Omega \\ r\in(0,\operatorname{diam}\Omega)}} \left( r^{-q} \int_{B(y,r)\cap\Omega} |g|^2\,dx \right)^{1/2}$$

is finite. Although our notation here is somewhat nonstandard, we point out that our $M^q$ is the usual Morrey space (written as $M^{2,q}$ in [114]).

**Theorem 5.54.** *Let $\alpha \in (0,1)$, and let $\partial\Omega \in H_{1+\alpha}$. Suppose $a^{ij} \in H_\alpha(\Omega)$ satisfies (5.78). Suppose $b \in H_\alpha$, and $c$ and $c^0$ are in $L^2(\Omega)$ and that there is a constant $A$ such that*

$$[a^{ij}]_\alpha \le A\lambda, \tag{5.94a}$$

$$|b|_\alpha + \|c\|_{n-2+2\alpha} + \|c^0\|_{n-2+2\alpha} \le A\lambda. \tag{5.94b}$$

*If $u$ is a weak solution of (5.7) with $f$ and $\psi$ in $H_\alpha$ and $g \in M^{n-2+2\alpha}$, then $u \in H_{1+\alpha}$ and there is a constant $C$ determined only by $A$, $n$, $\mu$,and $\Omega$, such that*

$$|u|_{1+\alpha} \le C(|u|_0 + |f/\lambda|_\alpha + \|g/\lambda\|_{M^q} + |\psi/\lambda|_\alpha). \tag{5.95}$$

***Proof.*** Assuming initially that $u \in H_{1+\alpha}$, we see that

$$D_i(a^{ij}D_ju) = \hat{g} - D_i\hat{f}^i \text{ in } \Omega, \quad a^{ij}D_ju\gamma_i = \hat{\psi} + \hat{f}\cdot\gamma \text{ on } \partial\Omega$$

with

$$\hat{f}^i = -b^iu + f^i, \quad \hat{g} = -c^iD_iu + -c^0u + g, \quad \hat{\psi} = -\beta^0u + \psi.$$

By combining Proposition 5.53 with an $H_{1+\alpha}$ change of variable and the corresponding interior estimate, we find that

$$|u|_{1+\alpha} \le C(|u|_1 + |f/\lambda|_\alpha + \|g/\lambda\|_{M^q} + |\psi/\lambda|_\alpha),$$

and the interpolation inequality (2.11) completes the proof in this case.

The general case follows by a simple approximation argument which we omit. □

We note here that $g \in L^{n/(1-\alpha)}$ implies that $g \in M^{n-2+2\alpha}$.

## Notes

In many ways, this chapter is just a rehash of Chapters 7 and 8 of [64] and Chapter 2 of [91], and we refer to those chapters for a history of the theory of Sobolev spaces and weak solutions of elliptic equations although we mention the pioneering work [172] of Sobolev for an early look at the theory of what are now called Sobolev spaces and [173] for Sobolev's more thorough description of the theory. Further information can be found in [206, 127]. We also mention [124] as a source for reading about the ideas that led to Sobolev's theory of weak solutions as well as the deep theory of distributions.

Some of our results on Sobolev spaces do not appear in [91] or [64]. Our proof of Poincarè's inequality is based on that in [11]; a more general result using this argument was proved by Hurri-Syrjänen [75]. In addition, the various Sobolev inequalities and trace theorems on $\Omega[R]$ are not stated in [91] or [64], but the proofs are easy modifications of the ones in those sources. Specifically, Corollary 5.23 seems to be new, but it most likely appeared much earlier in a reference unknown to this author because versions of Lemma 5.22 are present in [172] and [76].

In fact, our approach combines certain elements from [91] with others from [64]. Like [91], we have used De Giorgi's method to prove Hölder continuity of weak solutions, but, in order to keep the proof close to that in Chapter 1, we have followed the argument in [37] to prove a weak Harnack inequality as an intermediate result towards proving Hölder continuity. On the other hand, we used Moser's method (see [64] for a discussion of this method which is based on [140]) to prove the maximum and local maximum principles. Our proof of the Schauder estimates differs from that in [91] and [64]. We have followed Campanato's proof (from [26]; see also [102]) of the Schauder estimates in Section 5.14. The boundary estimates in [26] are only presented for the Dirichlet problem, but the extension to conormal problems given here is a minor variant of the techniques in [26].

## Exercises

**5.1** Modify the proof of the local maximum principle to show that, if $u$ is a nonnegative supersolution of (5.58), then, for any two numbers $p_1$ and $p_2$ with $0 < p_1 < p_2$ (and $p_2 < n/(n-2)$ if $n > 2$), there is a constant $C$, determined only by $p_1$, $p_2$, $\mu$, $\mu_2 R^{1-n/q}$, $\mu_3 R^{(q-n)/(q-1)}$, $n$, and $\omega_0$, such that

$$\left(R^{-n}\int_{\Omega[R/2]}|u|^{p_2}\,dx\right)^{1/p_2} \le C\left[\left(R^{-n}\int_{\Omega[R]}|u|^{p_1}\,dx\right)^{1/p_1} + k(R)\right].$$

Use this estimate to show that (5.70) is valid for any $p > 0$ with $p < n/(n-2)$ if $n > 2$ provided the constant $C$ depends also on $p$. (See also the proof of Theorem 8.18 in [64].)

**5.2** In this exercise, we provide two alternative proofs for the strong maximum principle Theorem 5.43.

(a) Imitate the proof of Theorem 5.15 to prove the strong maximum principle.

(b) Use the weak Harnack inequality (5.70) with $p = 1$ to prove the strong maximum principle. (This is essentially the proof of Theorem 8.19 in [64].)

**5.3** Show that (5.83) holds. (Hint: Note that $\int_{B^+(r)} |v - L|^2 \, dx$ is a quadratic function of $L$.)

## Chapter 6

# Strong Solutions

### Introduction

In this chapter, we study strong solutions of the oblique derivative problem. We define a strong solution of the problem

$$Lu = f \text{ in } \Omega, \quad Mu = g \text{ on } \partial\Omega \tag{6.1}$$

to be a function $u \in W^{2,n}_{\text{loc}}(\Omega) \cap C(\overline{\Omega})$ which satisfies the equation $Lu = f$ a.e. in $\Omega$ and which satisfies the boundary condition $Mu = g$ is the usual sense.

### 6.1 Pointwise estimates for strong solutions

We begin by showing that strong solutions satisfy the Aleksandrov-Bakel′man-Pucci maximum principle. To this end, we recall some definitions from Section 1.5. For a continuous function $u$ defined on $\Omega$, we write $\Gamma(u)$ for the set of all $x \in \Omega$ for which there is a vector $p \in \mathbb{R}^n$ such that

$$u(y) \le u(x) + p \cdot (x - y) \tag{6.2}$$

for all $y \in \Omega$. In addition, we write $\chi_u$ for the *normal mapping*, that is, $\chi_u(x)$ is the set of all vectors $p$ such that (1.8) holds for all $y \in \Omega$. In particular, $\chi_u(x)$ is non-empty if and only if $x \in \Gamma(u)$. We also recall that a vector field $\beta$ is *inward pointing* if, for every $x_0 \in \partial\Omega$, there is a positive constant $h$ such that $x_0 + t\beta(x_0) \in \overline{\Omega}$ for all $t \in (0, h)$.

Our first result is that Theorem 1.9 is valid for strong solutions.

**Theorem 6.1.** *Suppose $[a^{ij}]$ is positive definite and set $\mathcal{D} = \det[a^{ij}]$. Suppose $c \le 0$ in $\Omega$. Suppose $\beta$ is inward pointing and $\beta^0 < 0$ and suppose that* *there are nonnegative constants $B_0$ and $B_1$ such that*

$$|\beta| \le -B_0\beta^0 \text{ on } \partial\Omega, \tag{6.3a}$$

$$\int_\Omega \frac{|b|^n}{\mathcal{D}}\,dx \le B_1. \tag{6.3b}$$

*If $u \in W^{2,n}_{loc}(\Omega) \cap C(\overline{\Omega})$ and if*

$$Lu \ge -f \text{ in } \Omega, \quad Mu \ge g\beta^0 \text{ on } \partial\Omega \tag{6.4}$$

*for some nonnegative function $f$ and some nonnegative constant $g$, then*

$$u \le g + C(n, B_1)(B_0 + \operatorname{diam}\Omega)\left\|\frac{f}{\mathcal{D}^{1/n}}\right\|_{L^n(\Gamma^+(u))}, \tag{6.5}$$

*where $\Gamma^+(u)$ is the subset of $\Gamma(u)$ on which $u > g$.*

***Proof.*** We proceed via approximation from the proof of Theorem 1.9.

As a first step, we let $\varepsilon > 0$ be given and we write $G_\varepsilon$ for the subset of all $x \in \Gamma(u)$ on which $u(x) \ge g + \varepsilon + B_0|p|$ for some $p$ satisfying (6.2). We want to show that there is a $\delta > 0$ such that $d(x) > 2\delta$ for all $x \in G_\varepsilon$, and we argue by contradiction. If there were no such $\delta$, then there would be a sequence $(x_m)$ in $G_\varepsilon$ with a corresponding sequence $(p^m)$ in $\mathbb{R}^n$ such that

$$u(y) \le u(x_m) + p^m \cdot (y - x_m)$$

for all $y \in \Omega$ and $u(x_m) \ge g+\varepsilon+B_0|p^m|$, and $x_m \to x_0$, a point in $\partial\Omega$. Since $(p^m)$ is bounded (assuming without loss of generality that $B_0 > 0$), we can take subsequences (also labeled $(x_m)$ and $(p^m)$) so that $(p^m)$ converges to some vector $p^0$ with

$$u(y) \le u(x_0) + p^0 \cdot (y - x_0) \tag{6.6a}$$

for all $y \in \Omega$ and

$$u(x_0) \ge g + \varepsilon + B_0|p^0|. \tag{6.6b}$$

By taking $y = x_0 + h\beta(x_0)$ for all sufficiently small $h$ in (6.6a) and then dividing the resultant inequality by $h$ and sending $h \to 0$, we conclude that

$$\beta(x_0) \cdot Du(x_0) \le \beta(x_0) \cdot p^0.$$

Combining this inequality with the boundary condition for $u$, we conclude that

$$\beta^0 g \le \beta(x_0) \cdot p^0 + \beta^0(x_0)u(x_0)$$

and then the inequality (6.6b) implies that

$$\beta(x_0) \cdot p^0 \le -\beta^0 B_0|p^0| \le -\beta^0(x_0)u(x_0) + \beta^0(x_0)[g + \varepsilon].$$

Combining these inequalities gives

$$\beta^0(x_0)g \le \beta^0(x_0)[g+\varepsilon],$$

which contradicts $\beta^0 < 0$ and $\varepsilon > 0$. Hence there is some $\delta > 0$ such that $d(x) > 2\delta$ if $x \in G_\varepsilon$.

Now let $\eta$ be a $C^2(\overline{\Omega})$ function vanishing in a neighborhood of $\partial\Omega$ such that $\eta(x) = 1$ if $d(x) > \delta$, let $(\tilde{u}_m)$ be a sequence of $C^2(\overline{\Omega})$ functions which converge on the support of $\eta$ to $u$, and set $u_m = \eta\tilde{u}_m + (1-\eta)u$. We also write $G_{\varepsilon,m}$ for the set of all $x \in \Omega$ such that there is a vector $p \in \mathbb{R}^n$ such that

$$u_m(y) \le u_m(x) + p \cdot (y-x)$$

for all $y \in \Omega$ and

$$u_m(x) \ge g + \varepsilon + B_0|p|.$$

Our next step is to show that, for any open set $U$ such that $G_\varepsilon \subset U$, there is a positive integer $m_0$ such that $G_{\varepsilon,m} \subset U$ if $m \ge m_0$. To this end, we let $(x_m)$ be a convergent sequence with $x_m \in G_{\varepsilon,m} \setminus U$ and write $p^m$ for the corresponding vector. If $x_0 = \lim_{m\to\infty} x_m$, then, without loss of generality, we may assume that $(p^m)$ converges to some $p^0$. It follows that

$$u(y) \le u(x_0) + p^0 \cdot (y - x_0)$$

for all $y \in \Omega$ and

$$u(x_0) \ge g + \varepsilon + B_0|p_0|.$$

Hence $x_0 \in G_\varepsilon$, and hence there is a positive integer $m_0$ such that $x_m \in U$ for $m \ge m_0$.

Now let $h$ be a positive $L^1_{\text{loc}}(\mathbb{R}^n)$ function, and define the function $H$ by

$$H(r) = \int_{B(0,r)} h(p)\,dp.$$

For any open set $U$ with $G_\varepsilon \subset U$, and $m \ge m_0$, we then have $\beta \cdot Du_m + \beta^0 u_m \ge \beta^0 g$ on $\partial\Omega$, so Corollary 1.10 implies that

$$\sup u_m \le g + \varepsilon + (B_0 + \operatorname{diam}\Omega)H^{-1}\left(\int_U h(Du_m)\det D^2u_m\,dx\right)$$

because $u \in C^2(G_{\varepsilon,m})$.

Let us now assume further that $h$ is bounded and then write

$$\int_U h(Du_m)\det D^2u_m - h(Du)\det D^2u_m\,dx = I_{1,m} + I_{2,m},$$

where

$$I_{1,m} = \int_U h(Du_m)(\det D^2 u_m - \det D^2 u)\, dx,$$
$$I_{2,m} = \int_U [h(Du_m) - h(Du)] \det D^2 u\, dx.$$

Since $h$ is bounded, we have

$$|I_{1,m}| \le \sup h \int_U |\det D^2 u_m - \det D^2 u|\, dx \to 0$$

as $m \to \infty$. Also, $[h(Du_m) - h(Du)] \det D^2 u$ converges in measure to 0 as $m \to \infty$ and

$$|[h(Du_m) - h(Du)] \det D^2 u| \le 2 \sup h |\det D^2 u|,$$

so the dominated convergence theorem implies that $I_{2,m} \to 0$ as $m \to \infty$. It follows that

$$\int_U h(Du_m) \det D^2 u_m\, dx \to \int_U h(Du) \det D^2 u\, dx$$

and hence

$$\sup u \le g + \varepsilon + (B_0 + \operatorname{diam} \Omega) H^{-1} \left( \int_U h(Du) \det D^2 u\, dx \right).$$

Since this inequality is true for all open $U$ with $G_\varepsilon \subset U$, it follows that

$$\begin{aligned} \sup u &\le g + \varepsilon + (B_0 + \operatorname{diam} \Omega) H^{-1} \left( \int_{G_\varepsilon} h(Du) \det D^2 u\, dx \right) \\ &\le g + \varepsilon + (B_0 + \operatorname{diam} \Omega) H^{-1} \left( \int_G h(Du) \det D^2 u\, dx \right). \end{aligned}$$

Sending $\varepsilon \to 0$, we infer that

$$\sup u \le g + (B_0 + \operatorname{diam} \Omega) H^{-1} \left( \int_G h(Du) \det D^2 u\, dx \right).$$

The proof is completed by noting that the argument giving Theorem 1.9 from Lemma 1.8 is still valid for $W^{2,n}_{\mathrm{loc}}(\Omega) \cap C(\overline{\Omega})$ functions. □

It follows immediately that all of the results in Chapter 1 are valid for strong solutions. We leave the exact statements of these results to the reader.

## 6.2 A sharp trace theorem

In Section 5.5, we showed that the trace map $T$ maps $W^{1,p}$ into $L^q(\partial\Omega)$ for $q = (n-1)p/(n-p)$ if $p < n$. A natural question to raise is: What is the range of $T$? The purpose of this section is to answer that question.

We begin with a definition. For $q > 1$ and $\alpha \in (0,1)$, we define the $W^{\alpha,q}(\partial\Omega)$ seminorm $\langle\cdot\rangle_{\alpha,q;\partial\Omega}$ by

$$\langle u\rangle_{\alpha,q;\partial\Omega} = \left(\int_{\partial\Omega}\int_{\partial\Omega} \frac{|u(x)-u(y)|^q}{|x-y|^{n-2+q\alpha}}\, ds_x\, ds_y\right)^{1/q}.$$

We then use $W^{\alpha.q}(\partial\Omega)$ to denote the set of elements of $L^q(\partial\Omega)$ with $\langle u\rangle_{\alpha,q;\partial\Omega}$ finite. It is a Banach space under the norm $\|\cdot\|_{\alpha,q;\partial\Omega}$ defined by

$$\|u\|_{\alpha,q;\partial\Omega} = \langle u\rangle_{\alpha,q;\partial\Omega} + \|u\|_{q,\partial\Omega}.$$

**Theorem 6.2.** *Let $\Omega$ be a bounded Lipschitz domain and let $p > 1$. Then there is a bounded linear function $T\colon W^{1,p}(\Omega) \to W^{1-1/p,p}(\partial\Omega)$ such that $T(u)$ is the restriction of $u$ to $\partial\Omega$ for any $u \in C^1(\overline{\Omega})$. Moreover, $T$ maps $W^{1,p}(\Omega)$ onto $W^{1-1/p,p}(\partial\Omega)$.*

The proof of this theorem is based on several basic calculus results. The first one is known as Hardy's inequality.

**Lemma 6.3.** *Let $p > 1$, let $a < b$ be real numbers, and let $f \in L^p(a,b)$. Then*

$$\int_a^b \left|\frac{1}{x-a}\int_a^x f(t)\,dt\right|^p dx \le \left(\frac{p}{p-1}\right)^p \int_a^b |f(x)|^p\,dx. \tag{6.7}$$

***Proof.*** Set

$$F(x) = \frac{1}{x-a}\int_a^x |f(t)|\,dt.$$

Integration by parts yields

$$\begin{aligned}\int_a^\infty F^p(x)\,dx &= (b-a)F^p(b) - p\int_a^\infty (x-a)F^{p-1}(x)F'(x)\,dx \\ &= (b-a)F^p(b) - p\int_a^\infty |f(x)|F^{p-1}(x)\,dx + p\int_a^\infty F^p(x)\,dx\end{aligned}$$

because

$$F'(x) = \frac{|f(x)| - F(x)}{x-a}$$

on $(a, b)$. By rearrangement, Hölder's inequality and the nonnegativity of $F$, we obtain

$$\int_a^\infty F^p(x)\,dx = \frac{p}{p-1}\int_a^\infty |f(x)|F^{p-1}(x)\,dx - \frac{b-a}{p-1}F^p(b)$$
$$\leq \frac{p}{p-1}\left(\int_a^\infty |f(x)|^p\,dx\right)^{1/p}\left(\int_a^\infty F^p(x)\,dx\right)^{(p-1)/p}.$$

One more rearrangement gives (6.7). □

Related to Hardy's inequality is an inequality due to Morrey.

**Lemma 6.4.** *Let $h \in (0, 1)$ and let $f \in L^p(0, 1)$ for some $p \geq 1$. Then*

$$\int_0^{1-h}\left|\frac{1}{h}\int_x^{x+h} f(t)\,dt\right|^p dx \leq \int_0^1 |f(x)|^p\,dx.$$

***Proof.*** We begin by setting $g = |f|^p$. Then Hölder's inequality implies that

$$\int_0^{1-h}\left|\frac{1}{h}\int_x^{x+h} f(t)\,dt\right|^p dx \leq \int_0^{1-h}\frac{1}{h}\int_x^{x+h} g(t)\,dt\,dx.$$

(So there is really no loss of generality in assuming that $p = 1$ and $f$ is nonnegative.) We then then write

$$\int_x^{x+h} g(t)\,dt = \int_0^h g(x+t)\,dt$$

and then apply Fubini's theorem to obtain

$$\int_0^{1-h}\frac{1}{h}\int_0^h g(x+t)\,dt\,dx = \int_0^h \frac{1}{h}\int_0^{1-h} g(x+t)\,dx\,dt$$
$$= \int_0^h \frac{1}{h}\int_t^{1-h+t} g(\xi)\,d\xi\,dt.$$

The proof is completed by noting that $g$ is nonnegative, so

$$\int_t^{1-h+t} g(\xi)\,d\xi \leq \int_0^1 g(x)\,dx.$$

□

Our next lemma is a simplified form of the general trace theorem for a two-dimensional region.

**Lemma 6.5.** *Let $T$ be the triangle with vertices $(0, 0)$, $(1, 0)$, and $(1, 1)$, let $p > 1$, and set $c_p = 2^{p+1}(p/(p-1))^p$. Then*

$$\int_0^1\int_0^1\left|\frac{V(t,t)-V(\tau,\tau)}{t-\tau}\right|^p dt\,d\tau \leq c_p\int_T |DV|^p\,dx \tag{6.8}$$

*for any $V \in W^{1,p}(T)$.*

***Proof.*** By symmetry we have

$$\int_0^1 \int_0^1 \left| \frac{V(t,t) - V(\tau,\tau)}{t-\tau} \right|^p dt\, d\tau = 2I,$$

where

$$I = \int_0^1 \int_0^t \left| \frac{V(t,t) - V(\tau,\tau)}{t-\tau} \right|^p dt\, d\tau.$$

Hence we only need to estimate $I$.

To this end, we first observe that, for any $t$ and $\tau$ with $0 \le \tau < t \le 1$, we have

$$\begin{aligned} V(t,t) - V(\tau,\tau) &= V(t,t) - V(t,\tau) + V(t,\tau) - V(\tau,\tau) \\ &= \int_\tau^t D_2V(t,s)\, ds + \int_\tau^t D_1V(\sigma,\tau)\, d\sigma, \end{aligned}$$

and hence

$$\begin{aligned} \left| \frac{V(t,t) - V(\tau,\tau)}{t-\tau} \right|^p &\le 2^{p-1} \left| \frac{1}{t-\tau} \int_\tau^t D_1V(\sigma,\tau)\, d\sigma \right|^p \\ &\quad + 2^{p-1} \left| \frac{1}{t-\tau} \int_\tau^t D_2V(t,s)\, ds \right|^p \end{aligned}$$

We now use Hardy's inequality to estimate the integrals of the terms on the right hand side of this inequality. Specifically,

$$\int_\tau^1 \left| \frac{1}{t-\tau} \int_\tau^t D_1V(\sigma,\tau)\, d\sigma \right|^p dt \le \left( \frac{p}{p-1} \right)^p \int_\tau^1 |D_1V(\sigma,\tau)|^p\, d\sigma$$

so

$$\int_0^1 \int_\tau^1 \left| \frac{1}{t-\tau} \int_\tau^t D_1V(\sigma,\tau)\, d\sigma \right|^p dt\, d\tau \le \left( \frac{p}{p-1} \right)^p \int_T |D_1V|^p\, dx,$$

and a similar argument shows that

$$\int_0^1 \int_0^t \left| \frac{1}{t-\tau} \int_\tau^t D_2V(t,s)\, ds \right|^p d\tau\, dt \le \left( \frac{p}{p-1} \right)^p \int_T |D_2V|^p\, dx.$$

It follows that

$$I \le 2^{p-1} \left( \frac{p}{p-1} \right)^p \int_T |D_1V|^p + |D_2V|^p\, dx \le \frac{1}{2} c_p \int_T |DV|^p\, dx. \qquad \square$$

By a simple change of variables, we obtain an actual trace theorem.

**Lemma 6.6.** *Let $S$ be the square with vertices $(0,0)$, $(1,0)$, $(1,1)$ and $(0,1)$. Then*

$$\int_0^1 \int_0^{1-h} \left| \frac{U(0,y+h) - U(0,y)}{h} \right|^p dy\, dh \le c_p \int_S |DU|^p\, dx \tag{6.9}$$

*for all $U \in W^{1,p}(S)$.*

***Proof.*** We define $V$ in $T$ by $V(x,y) = U((x-y)/2,(x+y)/2)$. It's easy to check that

$$\int_S |DU|^p\,dx \geq \int_I |DV|^p\,dx$$

and that

$$\int_0^1\int_0^{1-h}\left|\frac{U(0,y+h)-U(0,y)}{h}\right|^p dy\,dh \leq \int_0^1\int_0^1\left|\frac{V(t,t)-V(\tau,\tau)}{t-\tau}\right|^p dt\,d\tau.$$
□

With these preliminaries, we are ready to prove our theorem.

***Proof.*** [Proof of Theorem 6.2] We use the map $T$ from Section 5.5. By virtue of a partition of unity argument and the invariance of the norm with respect to a uniformly Lipschitz change of independent variable, there is no loss of generality in assuming that $\partial\Omega$ is on the hyperplane $x^n = 0$ and that $x^n > 0$ in $\Omega$. To prove that $T$ maps into $W^{1-1/p,p}(\partial\Omega)$, we may assume that $\in W^{1,p}(\Omega)$ vanishes for any $x$ such that $x_i$ is outside of the interval $(0,1/2)$ for some $i$. It is elementary to verify that, in this setting, the seminorm $\langle u\rangle_{1-1/p,p;\partial\Omega}$ is equivalent to one given by

$$\Lambda_p(u) = \sum_{j=1}^{n-1}\left(\int_0^1\int_{S_j(h)}\left|\frac{u(x'+he_j)-u(x')}{h}\right|^p dx'dh\right)^{1/p},$$

where

$$S_j(h) = \{x' : 0 \leq x^i \leq 1 \text{ if } i \neq j, 0 \leq x^j \leq 1-h\}$$

and $e_j$ is the $(n-1)$-dimensional vector with $i$th component $e_j^i = \delta_j^i$. It follows from Lemma 5.20 that $T$ maps into $L^p(\partial\Omega)$, so we just need to show that

$$\Lambda_p(u) \leq C\|Du\|_p,$$

and this inequality follows from Lemma 6.6. (If $n = 2$, the inequality is exactly (6.9), and the case $n > 2$ follows by simple integration.)

To show that $T$ is onto, we write $\Pi$ for the pyramid defined by

$$\Pi = \{x \in \mathbb{R}^n : 0 \leq x_i \leq 1 - x_n \text{ if } i < n,\ 0 \leq x^n \leq 1\},$$

and we write $S$ for the intersection of $\Pi$ with the hyperplane $x^n = 0$. For any element $\varphi$ of $W^{1-1/p,p}(S)$, we define $u$ on $\Pi$ by the expression

$$u(x) = \frac{1}{(x^n)^{n-1}}\int_{x^1}^{x^1+x^n}\cdots\int_{x^{n-1}}^{x^{n-1}+x^n}\varphi(\xi)\,d\xi.$$

(Note that, here, $\xi = (\xi^1, \dots, \xi^{n-1})$.) Because of Lemma 5.22, we just have to show that $\|Du\|_p \le C\Lambda_p(\varphi)$. If $i < n$, we use the simple differentiation expression

$$D_i u(x) = \frac{1}{(x^n)^{n-1}} \int_{\Sigma_i(x^n)} [\varphi(\bar{\xi} + x^n e_i) - \varphi(\bar{\xi})]\, d\xi_i',$$

where $\bar{\xi}$ is the vector with $\bar{\xi}^j = \xi$ if $j \neq i$ and $\bar{\xi}^i = x^i$, $\xi_i'$ is the $(n-2)$-dimensional vector consisting of all $\xi^j$ with $j < n$ and $j \neq i$, and $\Sigma_i(x^n)$ is the set on which $\xi^j$ is between $x^j$ and $x^j + x^n$. Integrating with respect to $x^1, \dots, x^n$ and using Lemma 6.4, we conclude that

$$\int_\Pi |D_i u|^p\, dx \le C \int_0^1 \int_{S_i(h)} \left| \frac{\varphi(x' + he_i) - \varphi(x')}{h} \right|^p dx'\, dh$$

for $i = 1, \dots, n-1$.

To estimate the $L^p$ norm of $D_n u$, we first observe that

$$\begin{aligned} D_n u &= \sum_{j=1}^{n-1} \frac{1}{(x^n)^n} \int_{x^1}^{x^1+x^n} \cdots \int_{x^{n-1}}^{x^{n-1}+x^n} (\xi^j - x^j) D_j \varphi(\xi)\, d\xi \\ &= \sum_{j=1}^{n-1} \frac{1}{(x^n)^n} \int_{x^1}^{x^1+x^n} \cdots \int_{x^{n-1}}^{x^{n-1}+x^n} [\varphi(\bar{\xi}_j) - \varphi(\xi)]\, d\xi, \end{aligned}$$

where $\bar{\xi}_j$ has $i$th component $\xi^i$ if $i \neq j$ and $j$th component $x^j + x^n$. Since $x^j + x^n - \xi^j \ge x^n$, we can imitate the proof of the estimate for $D_i u$ to conclude that

$$\int_\Pi |D_n u|^p\, dx \le C\Lambda_p(\varphi).$$

It follows that $u \in W^{1,p}$ as required. □

## 6.3 Results from harmonic analysis

In this section, we recall some results from harmonic analysis that will be useful in our further study of strong solutions.

The first result is a covering lemma of Wang.

**Lemma 6.7.** *Let $A$ be a bounded subset of $\mathbb{R}^n$ and let $\mathcal{C}$ be a covering of $A$ by balls of bounded radius. Then there is a sequence $(B(x_k, R_k))$ of disjoint elements of $\mathcal{C}$ such that*

$$A \subset \bigcup B(x_k, 5R_k). \tag{6.10}$$

***Proof.*** We define the sequence $(B(x_k, R_k))$ as follows, writing $B_k = B(x_k, R_k)$. First, we choose $B_1 \in \mathcal{C}$ so that

$$R_1 \geq \frac{1}{2} \sup\{R : R \text{ is the radius of some } B \in \mathcal{C}\}.$$

Once we have $B_1, \dots, B_{k-1}$, we write $\mathcal{C}_k$ for the set of all $B \in \mathcal{C}$ such that $B \cap B_i = \emptyset$ for all $i \leq k-1$, and we take $B_k \in \mathcal{C}_k$ so that

$$R_k \geq \frac{1}{2} \sup\{R : R \text{ is the radius of some } B \in \mathcal{C}_k\}.$$

To prove that (6.10) holds, we fix $x \in A$ and choose a ball $B(y, r) \in \mathcal{C}$ so that $x \in B(y, r)$, and we consider two cases. If the sequence $(B_k)$ is finite, then there is a last integer $k$ such that $R_k \geq \frac{1}{2}r$. If the sequence is infinite, then, without loss of generality, we may assume that $B \cap A$ is nonempty for all $B \in \mathcal{C}$ and hence $\sum |B_k| = |\cup B_k|$ is finite. It follows that $|B_k| \to 0$ as $k \to \infty$ and hence there is a last integer $k$ such that $R_k \geq \frac{1}{2}r$ in this case as well. Letting $k$ be this last integer, we note that the definition of $\mathcal{C}_k$ implies that $B_i \cap B(y, r) \neq \emptyset$ for some $i \leq k$. For this $i$, we write $y_i$ for any point in $B_i \cap B(y, r)$ and note that

$$\begin{aligned} |x - x_i| &\leq |x - y| + |y - y_i| + |y_i - x_i| \\ &\leq r + r + R_i \\ &< 5R_i. \end{aligned}$$

In other words, $x \in B(x_i, 5R_i)$, as required. □

We now introduce the Hardy-Littlewood maximal operator $\mathcal{M}$, defined on $L^1(\mathbb{R}^n)$ by

$$\mathcal{M}v(x) = \sup_{r>0} \frac{1}{|B(x,r)|} \int_{B(x,r)} |v(y)|\, dy,$$

and we recall the following properties of this operator, proved, for example, as Lemma 7.9 of [114].

**Lemma 6.8.** *If $v \in L^1(\mathbb{R}^n)$, then*

$$\sup_{\alpha>0} \alpha |\{x : \mathcal{M}v(x) > \alpha\}| \leq 5^n \|v\|_1. \tag{6.11}$$

*If $v \in L^p(\mathbb{R}^n)$ for some $p > 1$, then*

$$\|\mathcal{M}v\|_p \leq C(n, p) \|v\|_p.$$

As a consequence, we have the Lebesgue differentiation theorem and a key inequality between $v$ and $\mathcal{M}v$.

**Corollary 6.9.** *If $v \in L^1(\mathbb{R}^n)$, then*

$$v(x) = \lim_{r\to 0} \frac{1}{|B(x,r)|} \int_{B(x,r)} v(y)\,dy$$

*for almost all $x \in \mathbb{R}^n$. Hence $|v| \le \mathcal{M}v$ a.e.*

We also have a further variant of the Vitali covering lemma, which is proved as Lemma 7.11 of [114].

**Lemma 6.10.** *Let $\varepsilon < 1$ and $\rho$ be positive constants, let $x_0 \in \mathbb{R}^n$, and suppose that $A$ and $C$ are measurable subsets of $B(x_0,\rho)$ such that $A \subset C$ and $|A| < \varepsilon|B(x_0,r)|$. Suppose in addition that, for any $x \in B(0,\rho)$ and any $r \in (0,\rho)$ such that $|A \cap B(x,r)| \ge \varepsilon|B(x,r)|$, we have $B(x,r) \cap B(x_0,\rho) \subset C$. Then*

$$|A| \le 20^n \varepsilon |C|.$$

Finally, we have two simple geometric covering results.

**Lemma 6.11.** *For every positive integer $n$, there is a constant $K$, determined only by $n$, and, for any $r > 0$, there is a countable set of points $(x_i)$ in $\mathbb{R}^n$ with $\mathbb{R}^n = \cup_{i=1}^{\infty} B(x_i,r)$ and no point of $\mathbb{R}^n$ is in more than $K$ balls of the form $B(x_i,2r)$.*

***Proof.*** We take $(x_i)$ to be the collection of all points with each coordinate an integer multiple of $r/\sqrt{n}$. □

**Lemma 6.12.** *For every positive integer $n$, there is a constant $K$, determined only by $n$, and, for any $r > 0$, there is a countable set of points $(x_i)$ in $\overline{\mathbb{R}^n_+}$ with $x_i^n \ge r/2$ if $x_i^n > 0$, $\overline{\mathbb{R}^n_+} \subset \cup_{i=1}^{\infty} B(x_i,r_i)$ and no point of $\mathbb{R}^n$ is in more than $K$ balls of the form $B(x_i,2r_i)$, where $r_i = r$ if $x_i^n = 0$ and $r_i = r/8$ if $x_i^n > 0$.*

***Proof.*** Now we take $(x_i)$ to be the collection of all points such that each coordinate is an integer multiple of $r/n$ with two additional restrictions: First, $x_i^n \ge 0$ and $x_i^n \ge r/2$ if $x_i^n > 0$. □

## 6.4 Some further estimates for boundary value problems in a spherical cap

We recall the definitions from Section 2.6, but with a special choice of the parameter $\kappa$. Throughout this section, we consider $R$ to be a fixed constant, so we do not include it in the notation, and we assume that $[A^{ij}]$ is a positive definite matrix (not necessarily symmetric) such that there are positive constants $\lambda$ and $\mu$ with

$$\lambda|\xi|^2 \leq A^{ij}\xi_i\xi_j \leq \mu\lambda|\xi|^2 \tag{6.12}$$

for all $\xi \in \mathbb{R}^n$. We also assume that $\beta_0$ is a nonzero vector and that there is a positive constant $\mu_0$ such that

$$|\beta_0'| \leq \mu_0\beta_0^n. \tag{6.13}$$

With $\kappa \in (\mu_0/(1+\mu_0^2)^{1/2}, 1)$, we then write $\Gamma$ for the set of all $x \in \mathbb{R}^n$ with $|x'|^2+|x^n+\kappa R|^2 < R^2$ and $x^n > 0$. We write $\Gamma^0$ and $\Gamma^*$ for the subsets of $\partial\Gamma$ on which $|x'|^2+|x^n+\kappa R|^2 < R^2$ and $|x'|^2+|x^n+\kappa R|^2 = R^2$, respectively, and we write $\Gamma^{**}$ for the set of all $x$ with $x^n = 0$ and $|x'| = (1-\kappa)^{1/2}R$. In other words $\Gamma^{**}$ is the intersection of the closures of $\Gamma^0$ and $\Gamma^*$. We shall also find it convenient to look at the distance from these sets to a point in $\overline{\Gamma}$, so we write $d$ for the distance to $\partial\Gamma$, $d_0$ for the distance to $\Gamma^0$, $d_1$ for the distance to $\Gamma^*$, and $d_2$ for the distance to $\Gamma^{**}$.

Our first result is a pointwise gradient bound for solutions of the oblique derivative problem.

**Lemma 6.13.** *Let $x_0 \in \overline{\Gamma}$ and let $r \in (0, R)$. If $u$ satisfies the conditions*

$$A^{ij}D_{ij}u = 0 \text{ in } B(x_0, r)\cap\Gamma, \tag{6.14a}$$

$$\beta_0 \cdot Du = 0 \text{ on } B(x_0; r)\cap\Gamma^0, \tag{6.14b}$$

$$u = 0 \text{ on } B(x_0, r)\cap\Gamma^*, \tag{6.14c}$$

*then there is a constant $C$, determined only by $n$, $\kappa$, $\mu$, and $\mu_0$ such that*

$$\sup_{B(x_0,r/2)\cap\Gamma} |Du| \leq C\left(r^{-n}\int_{B(x_0,r)\cap\Gamma} |Du|^2\,dx\right)^{1/2}.$$

***Proof.*** To simplify the writing, we note that there is a constant $\delta \in (0, 1/2]$, determined only by $\kappa$, such that $\min\{d_0, d_1\} > \delta d_2$. Let $y \in B(x_0, r/2)\cap\Gamma$. We also set $b = \beta_0/|\beta_0|$ and note that $b$ is a unit vector with $b^n \geq 1/(1+\mu_0^2)^{1/2}$.

Suppose first that $d_2(y) \geq r/8$. If $d_1(y) < \delta r/32$, we choose $x_1 \in \Gamma^*$ so that $d_1(y) = |x_1 - y|$ and observe that $B(x_1, \delta r/16)$ doesn't intersect $\Gamma^0$. By arguing as in Lemma 1.35, we find that

$$|Du(y)| \leq \sup_{B(x_1,\delta r/32)\cap\Gamma} |Du| \leq \frac{C}{r} \sup_{B(x_1,\delta r/24)\cap\Gamma} |u|.$$

A simple variation of the local maximum principle (Theorem 1.27) then shows that

$$\sup_{B(x_1,\delta r/24)\cap\Gamma} |u| \leq C\left(r^{-n}\int_{B(x_1,\delta r/16)\cap\Gamma} |u|^2\,dx\right)^{1/2}$$

and Poincarè's inequality implies that

$$\int_{B(x_1,\delta r/16)\cap\Gamma} |u|^2\,dx \leq Cr^2\int_{B(x_1,\delta r/16)\cap\Gamma} |Du|^2\,dx.$$

Since $B(x_1, \delta r/16) \subset B(x_0, r)$, we conclude that

$$|Du(y)| \leq C\left(r^{-n}\int_{B(x_0,r)\cap\Gamma} |Du|^2\,dx\right)^{1/2}. \tag{6.15}$$

Next, if $d_1(y) \geq \delta r/32$ and $d_0(y) < \delta r/256$, we choose $x_1 \in \Gamma^0$ so that $d_0(y) = |x_1 - y|$ and note that $B(x_1, \delta r/128)$ doesn't intersect $\Gamma^*$. If we now set $w = |D'u|^2 + |b \cdot Du|^2$, then it's easy to check that

$$\begin{aligned} a^{ij}D_{ij}w &= 2\sum_{s=1}^{n-1} a^{ij}D_{is}uD_{js}u + 2a^{ij}D_{ik}ub^kD_{jm}ub^m \\ &\geq 0 \end{aligned}$$

in $B(x_1, \delta r/128) \cap \Gamma$ and $\beta_0 \cdot Dw = 0$ on $B(x_1, \delta r/128)$, and hence the local maximum principle Theorem 1.27 implies that

$$\sup_{B(x_1,\delta r/256)\cap\Gamma} w \leq Cr^{-n}\int_{B(x_1,\delta r/128)\cap\Gamma} w\,dx.$$

It's easy to check that $|Du|^2 \geq 2w$ and $|Du|^2 \leq C(\mu_0)w$, so

$$\sup_{B(x_1,\delta r/256)\cap\Gamma} |Du|^2 \leq Cr^{-n}\int_{B(x_1,\delta r/128)\cap\Gamma} |Du|^2\,dx.$$

Since $y \in B(x_1, \delta r/256) \cap \Gamma$ and $B(x_1, \delta r/128) \subset B(x_0, r)$, we again infer (6.15). Further, if $d_1(y) \geq \delta r/32$ and $d_0(y) \geq \delta r/256$, then $B(y, \delta r/256) \subset B(x_0, r) \cap \Gamma$, so we can apply the local maximum principle directly to each component of $Du$ in $B(y, \delta r/256)$ to infer (6.15).

Now suppose that $d_2(y) < r/8$ and choose $x_1 \in \Gamma^{**}$ so that $|x_1 - y| = d_2(y)$. To simplify writing, we set $\rho = d_2(y)$ and

$$U = \left( r^{-n} \int_{B(x_0,r)\cap\Gamma} |Du|^2 \, dx \right)^{1/2}.$$

As a first step here, we note that $B(x_1, r/4) \subset B(x_0, r)$ and use Poincarè's inequality along with the local maximum principle to infer that

$$\sup_{B(x_1,r/6)\cap\Gamma} |u| \le CrU.$$

Now, we note that the constant $\alpha_1$ in Proposition 3.16 corresponding to $\theta_0 = \arctan \kappa$ is greater than 1 and hence (after using a suitable rotation of axes) there is a nonnegative function $w_1$ such that

$$A^{ij} D_{ij} w_1 \le -|x - x_1|^{-1} \text{ in } B(x_1, r/6) \cap \Gamma,$$
$$b \cdot Dw_1 \le -1 \text{ on } B(x_1, r/6) \cap \Gamma^0,$$
$$w \le C|x - x_1| \text{ in } B(x_1, r/6) \cap \Gamma.$$

It follows that there is a constant $K_1$ such that $w = |x - x_1|^2/r + K_1 w_1$ satisfies

$$A^{ij} D_{ij} w \le 0 \text{ in } B(x_1, r/6) \cap \Gamma,$$
$$b \cdot Dw \le 0 \text{ on } B(x_1, r/6) \cap \Gamma^0,$$
$$w \ge 0 \text{ on } B(x_1, r/6) \cap \Gamma^*,$$
$$w \ge r \text{ on } \partial B(x_1, r/6) \cap \Gamma.$$

It follows from the maximum principle in $B(x_1, r/6) \cap \Gamma$ that $|u| \le CUw$ in $B(x_1, r/6)$ and hence that

$$\sup_{B(y,\rho/4)\cap\Gamma} |u| \le CU\rho.$$

Arguing as in the previous cases (with $\rho$ in place of $r$), we find that (6.15) holds here as well. □

We also need a result which shows that the mixed boundary value problem with a conormal boundary condition has a unique solution which satisfies a corresponding estimate.

**Lemma 6.14.** *For any $f \in L^2(\Gamma^0)$, there is a unique weak solution of*

$$D_i(A^{ij} D_j u) = D_i f^i \text{ in } \Gamma, \tag{6.16a}$$
$$A^{nj} D_j u = f^n \text{ on } \Gamma^0, \quad u = 0 \text{ on } \Gamma^*. \tag{6.16b}$$

*Moreover,*

$$\|Du\|_2 \le \|f\|_2/\lambda. \tag{6.17}$$

***Proof.*** We begin by proving (6.17). To this end, we use $u$ as test function in the weak version of the boundary value problem to obtain

$$\int_\Gamma A^{ij} D_i u D_j u\, dx = \int_\Gamma Du \cdot f\, dx.$$

Applying the lower bound $A^{ij}\xi_i\xi_j \geq \lambda|\xi|^2$ for any vector $\xi$ and the Cauchy-Schwarz inequality gives (6.17).

The existence proof is essentially that of Theorem 5.26 along with the observation that $\sigma = 0$ is not in the set $\Sigma$, and we leave the details of the proof to the reader. □

## 6.5 $L^p$ estimates for solutions of constant coefficient problems in a spherical cap

In this section, we continue the notation of the previous section with one additional condition: we write $\beta_0$ for the vector with components $\beta_0^j = A^{nj}$. Since $[A^{ij}]$ is not assumed to be symmetric, the number $\mu_0$ is not controlled by $\mu$, so we continue to use both quantities (and $\kappa$).

The main part of the proof of our $L^p$ estimates is a measure-theoretic estimate for the maximal function of the solutions of a constant coefficient problem. Specifically, we study the problem

$$D_i(A^{ij}D_j u) = D_i f^i \text{ in } \Gamma, \quad A^{nj}D_j u = f^n \text{ on } \Gamma^0, \quad u = 0 \text{ on } \Gamma^*. \tag{6.18}$$

In addition, if $S$ is any measurable subset of $\mathbb{R}^n$, we define the operator $\mathcal{M}_S$ by

$$\mathcal{M}_S v(x) = \sup_{r>0} \frac{1}{|B(x,r)|} \int_{B(x,r)\cap S} |v(y)|\, dy.$$

(In other words, if $v$ is defined on $S$ and $w$ is defined by

$$w(x) = \begin{cases} v(x) & \text{if } x \in S, \\ 0 & \text{if } x \notin S, \end{cases}$$

then $\mathcal{M}_S v = \mathcal{M} w$.)

**Proposition 6.15.** *Let $u$ be a weak solution of* (6.18) *for some $f \in L^2(\Gamma)$. Let $x_1 \in \Omega$ and $r \in (0, R/6)$. Then, for any $\varepsilon > 0$, there are constants $M(\mu,\mu_0,\kappa,n) > 0$ and $\delta(\varepsilon,\mu,\mu_0,\kappa,n) \in (0,1]$ such that, if there is a point $y \in B(x_1,r)\cap\Gamma$ with $\mathcal{M}_\Gamma|f|^2(y) \leq \delta^2\lambda^2\theta^2$ and $\mathcal{M}_\Gamma(|Du|^2)(y) \leq \theta^2$ for some $\theta > 0$, then*

$$|\{x \in B(x_1,r)\cap\Gamma : \mathcal{M}_\Gamma(|Du|^2)(x) > \theta^2 M^2\}| \leq \varepsilon|B(x_1,r)|. \tag{6.19}$$

***Proof.*** For $k > 0$, we abbreviate $B(x_1, kr) \cap \Gamma$ to $B_k$. We have

$$\frac{1}{\omega_n (6r)^n} \int_{B(y,6r)\cap\Gamma} |Du|^2 \, dx \le \theta^2$$

and

$$\frac{1}{\omega_n (6r)^n} \int_{B(y,6r)\cap\Gamma} |f|^2 \, dx \le \delta^2 \lambda^2 \theta^2.$$

Since $B_4 \subset B(y, 6r) \cap \Gamma$, it follows that

$$\int_{B_4} |Du|^2 \, dx \le 2^n \theta^2 \omega_n (4r)^n.$$

We now let $w$ be the solution of

$$\begin{aligned} D_i(A^{ij} D_j w) &= D_i f^i \text{ in } B^+(x_1, 4r), \\ A^{nj} D_j w &= f^n \text{ on } B^0(x_1, 4r), \\ w &= 0 \text{ on } B^*(x_1, 4r) \end{aligned}$$

given by Lemma 6.14 and set $h = u - w$. Then (6.17) implies that

$$\int_{B_4} |Dw|^2 \, dx \le \frac{1}{\lambda^2} \int_{B_4} |f|^2 \, dx \le 2^n \delta^2 \theta^2 \omega_n (4r)^n,$$

and hence (recalling that $\delta \in (0, 1]$)

$$\int_{B_4} |Dh|^2 \, dx \le 2 \int_{B_4} \left( |Dw|^2 + |Du|^2 \right) \, dx \le 2^{n+1} \theta^2 \omega_n (4r)^n.$$

Moreover, Lemma 6.13 implies that there is a constant $M_0$, determined only by $\mu$ and $n$ such that

$$\sup_{B_3} |Dh|^2 \le M_0 \theta^2.$$

It follows that

$$|Du|^2 \le 2|Dw|^2 + 2M_0^2 \theta^2 \text{ in } B_3, \tag{6.20}$$

and we infer from (6.11) (with $\alpha = \theta^2$) that

$$5^{-n} \theta^2 |\{x \in B_1 : \mathcal{M}_{B_3} |Dw|^2 > \theta^2\}| \le \int_{B_3} |Dw|^2 \, dx \le 2^n \delta^2 \theta^2 \omega_n (4r)^n.$$

It follows that

$$|\{x \in B_1 : \mathcal{M}_{B_3} |Dw|^2 > \theta\}| \le 40^n \delta^2 |B(x_1, r)|. \tag{6.21}$$

Now let $z \in B_1$ be a point at which $\mathcal{M}_{B_3} |Dw|^2(z) \le \theta^2$. For $\rho \in (0, 2r]$, we have $B(z, \rho) \cap \Gamma \subset B_3$, and therefore

$$\frac{1}{|B(z,\rho)|} \int_{B(z,\rho)\cap\Gamma} |Du|^2 \le 2\mathcal{M}_{B_3} |Dw|^2(z) + 2M_0 \theta^2 \le (2M_0 + 2)\theta^2.$$

On the other hand, for $\rho > 2r$, we have $B(z,\rho) \subset B(y,2\rho)$ and therefore

$$\frac{1}{|B(z,\rho)|}\int_{B(z,\rho)\cap\Gamma}|Du|^2\,dx \le \frac{1}{|B(z,\rho)|}\int_{B(y,2\rho)\cap\Gamma}|Du|^2\,dx \le 2^n\theta^2.$$

With $M = \max\{2^n/2,(2M_0+2)^{1/2}\}$, we conclude that $\mathcal{M}_\Gamma|Du|^2(z) \le M^2\theta^2$ and hence

$$\{x \in B_1 : \mathcal{M}_\Omega|Du|^2 > M^2\theta^2\} \subset \{x \in B(x_1,r); \mathcal{M}_{B_3}|Dw|^2 > \theta^2\}.$$

It follows from this inequality and (6.21) that

$$|\{x \in B_1 : \mathcal{M}_\Omega|Du|^2 > M^2\theta^2\}| \le 40^n M_0\delta^2|B(x_1,r)|,$$

and (6.19) follows if we take

$$\delta = \frac{\min\{1,\varepsilon\}}{(1+40^n M_0)^{1/2}}.$$

□

Our next step is to prove some conclusions about the measure of the set on which $\mathcal{M}_\Gamma|Du|^2$ is large than a certain number.

**Corollary 6.16.** *Let $u$ be a weak solution of* (6.18) *for some $f \in L^2(\Gamma)$. Let $\varepsilon > 0$, set $\varepsilon_1 = 20^n\varepsilon$, and let $M$ and $\delta$ be the constants from Proposition* 6.15. *If $g \in L^1(\Gamma)$ and $h \ge 0$, write $\{\mathcal{M}_\Gamma g > h\}$ for the subset of $\Gamma$ on which $\mathcal{M}_\Gamma g > h$. If $\theta$ is a positive constant such that*

$$|\{\mathcal{M}_\Gamma|Du|^2 > \theta^2M^2\}| \le \varepsilon|B(x_0,R)|, \tag{6.22}$$

*then*

$$|\{\mathcal{M}_\Gamma|Du|^2 > \theta^2M^{2j}\}| \le \varepsilon_1^j|\{\mathcal{M}_\Gamma|Du|^2 > \theta^2\}| + \sum_{i=1}^{j}\varepsilon_1^i|\{\mathcal{M}_\Gamma|f|^2 > \theta^2M^{2(j-i)}\delta^2\}| \tag{6.23}$$

*for any positive integer $j$.*

***Proof.*** We set

$$A = \{\mathcal{M}_\Gamma|Du|^2 > \theta^2M^2\},$$
$$B = \{\mathcal{M}_\Gamma|Du|^2 > \theta^2\} \cup \{\mathcal{M}_\Gamma|f|^2 > \theta^2\delta^2\}.$$

From the contrapositive of Proposition 6.15, we conclude that $|A \cap B(x,r)| \ge \varepsilon|B(x,r)|$ implies that $B(x,r)\cap B(x_0,R) \subset B$, and hence Lemma 6.10 implies (6.23) for $j = 1$.

For any $\alpha \ge 1$, we can replace $\theta$ by $\alpha\theta$ in (6.23) with $j = 1$ to conclude that

$$|\{\mathcal{M}_\Gamma|Du|^2 > \alpha^2\theta^2M^2\}| \le \varepsilon_1^j|\{\mathcal{M}_\Gamma|Du|^2 > \alpha^2\theta^2\}| + \varepsilon_1|\{\mathcal{M}_\Gamma|f|^2 > \alpha^2\theta^2M^2\delta^2\}|,$$

and (6.23) then follows from this inequality by induction, using $\alpha = 1, M, \dots, M^j$. □

From this corollary, we can now prove our $L^p$ estimate for $p > 2$.

**Proposition 6.17.** *Let $p > 2$ and let $u$ be a weak solution of (6.16) for some $f \in L^p(\Gamma)$. Then there is a constant $C$, determined only by $\mu$, $\mu_0$, $\kappa$, $p$ and $n$ such that*

$$\|Du\|_p \leq \frac{C}{\lambda}\|f\|_p. \tag{6.24}$$

***Proof.*** With $M$ the constant from Proposition 6.15, set

$$\varepsilon = \frac{1}{2}\min\{1, \frac{1}{20^n M^p}\},$$

and let $\delta$ be the corresponding constant from Proposition 6.15. We also set

$$F = \frac{1}{\lambda}\|f\|_p$$

and observe that

$$\frac{1}{\lambda}\|f\|_2 \leq F|B(x_0, R)|^{(p-2)/(2p)}.$$

It then follows from (6.17) that

$$\|Du\|_2 \leq CF|B(x_0, R)|^{(p-2)/(2p)},$$

and hence (6.11) gives a constant $c_1$ such that

$$|\{x \in \Gamma : \mathcal{M}_\Gamma |Du|^2 > \theta^2 M^2\} \leq c_1 \frac{1}{R^{2n/p}\theta^2 M^2} F^2 |B(x_0, R)|$$

for any $\theta > 0$. The choice

$$\theta = \frac{c_1^{1/2} F}{M R^{n/p} \varepsilon}$$

then gives (6.22).

Now, we use the inequality $|Du|^2 \leq \mathcal{M}_\Gamma |Du|^2$ from Corollary 6.9 and Lemma 1.14 to infer that

$$\begin{aligned}\int_\Gamma |Du|^p\, dx &\leq \int_\Gamma (\mathcal{M}_\Gamma |Du|^2)^{p/2}\, dx \\ &= p\int_0^\infty \sigma^{p-1} |\{\mathcal{M}_\Gamma |Du|^2 > \sigma^2\}|\, d\sigma.\end{aligned}$$

We then rewrite the right hand side of this inequality as

$$p\int_0^{\theta M} \sigma^{p-1} |\{\mathcal{M}_\Gamma |Du|^2 > \sigma^2\}|\, d\sigma + p\sum_{j=1}^\infty \int_{\theta M^j}^{\theta M^{j+1}} \sigma^{p-1} |\{\mathcal{M}_\Gamma |Du|^2 > \sigma^2\}|\, d\sigma.$$

Then

$$p\int_0^{\theta M}\sigma^{p-1}|\{\mathcal{M}_\Gamma|Du|^2>\sigma^2\}|\,d\sigma\le\theta^pM^p|B(x_0,R)|,$$

and

$$p\int_{\theta M^j}^{\theta M^{j+1}}\sigma^{p-1}|\{\mathcal{M}_\Gamma|Du|^2>\sigma^2\}|\,d\sigma\le\theta^pM^{(j+1)p}|\{\mathcal{M}_\Gamma|Du|^2>\theta^2M^{2j}\}|.$$

From Corollary 6.16, we then infer that

$$\int_\Gamma|Du|^p\,dx\le\theta^pM^p(|B(x_0,R)|+I_1+I_2),$$

where

$$I_1=\sum_{j=1}^\infty M^j\varepsilon_1^j|\{\mathcal{M}_\Gamma|Du|^2|>\theta^2M^2\}|,$$

$$I_2=\sum_{j=1}^\infty M^{jp}\sum_{i=1}^j\varepsilon_1^i|\{\mathcal{M}_\Gamma|f|^2>M^{2(j-i)}\delta^2\theta^2\}.$$

Since $M^p\varepsilon_1\le1/2$, it follows that

$$I_1\le|B(x_0,R)|,$$

and simple algebra shows that

$$\begin{aligned}I_2&=\sum_{i=1}^\infty\varepsilon_1^iM^{ip}\sum_{j=i}^\infty M^{(j-i)p}|\{\mathcal{M}|f|^2>M^{2(j-i)}\delta^2\theta^2\}|\\&=\sum_{i=1}^\infty\varepsilon_1^iM_{ip}\sum_{j=0}^\infty M^{jp}|\{\mathcal{M}_\Gamma|f|^2>M^{2j}\delta^2\theta^2\}|\\&\le\sum_{j=0}^\infty M^{jp}|\{\mathcal{M}_\Gamma|f|^2>M^{2j}\delta^2\theta^2\}|\\&\le|\{\mathcal{M}_\Gamma|f|^2>\delta^2\theta^2\}|+M^p\sum_{j=1}^\infty M^{(j-1)p}|\{\mathcal{M}_\Gamma|f|^2>M^{2j}\delta^2\theta^2\}|.\end{aligned}$$

Since

$$|\{\mathcal{M}_\Gamma|f|^2>\delta^2\theta^2\}|\le|B(x_0,R)|$$

and a simple calculation shows that

$$\begin{aligned}\sum_{j=1}^\infty M^{(j-1)p}|\{\mathcal{M}_\Gamma|f|^2>M^{2j}\delta^2\theta^2\}|&\le M^p\int_0^\infty\sigma^{p-1}|\{\mathcal{M}_\Gamma|f|^2>\sigma^2\delta^2\theta^2\}|\,d\sigma\\&=M^p\int_{B^+(x_0,6R)}\frac{\mathcal{M}_\Gamma|f|^p}{(\delta\theta)^p}\,dx\\&\le c(n,p)M^p\int_{B^+(x_0,6R)}\frac{|f|^p}{(\delta\theta)^p}\,dx,\end{aligned}$$

it follows that

$$I_2 \le C(n,p)|B(x_0,R)|.$$

Combining all these inequalities gives

$$\int_\Gamma |Du|^p\,dx \le C(n,\mu,p)F^p$$

which is just (6.24). □

In order to prove the $L^p$ estimates when $p \in (1,2)$, we argue via duality.

**Proposition 6.18.** *Proposition* 6.17 *is true for any* $p > 1$.

***Proof.*** When $p > 2$, this proposition is exactly Proposition 6.17, and when $p = 2$, it's just Lemma 6.14.

When $p < 2$, we set $q = p/(p-1)$, and $g = |Du|^{p-2}Du$. If $v$ is the weak solution of

$$D_j(A^{ij}D_iv) = D_jg^j \text{ in } \Gamma, \quad A^{in}D_iv = g^n \text{ on } \Gamma^0, \quad v = 0 \text{ on } \Gamma^*,$$

then Proposition 6.17 implies that

$$\|Dv\|_q \le C\|g\|_q = C\|Du\|_p^{p-1}.$$

By using $u$ as test function in the weak version of the boundary value problem for $v$ and then using $v$ as test function in the weak version of the boundary value problem for $u$, it follows that

$$\int_\Gamma D_iug^i\,dx = \int_\Gamma A^{ij}D_juD_iv\,dx = \int_\Gamma D_ivf^i\,dx.$$

But $D_iug^i = |Du|^p$, and hence

$$\|Du\|_p^p \le \|Dv\|_q\|f\|_p \le C\|Du\|_p^{p-1}\|f\|_p.$$

Dividing this inequality by $\|Du\|_p^{p-1}$ gives (6.24). □

## 6.6 Local estimates for strong solutions of constant coefficient problems

We now prove some estimates for strong solutions of constant coefficient problems which will be used, in the next section, to prove our $L^p$ estimates for variable coefficient problems. We first prove an estimate for the solution of a differential equation in a ball and then an estimate for the solution of a boundary value problem in a half-ball. To simplify notation, we use $x_1$ to denote an arbitrary point in $\mathbb{R}^n$ and $r$ to denote a fixed positive number.

We also assume that $[A_0^{ij}]$ is a symmetric matrix and that there are positive constants $\lambda$ and $\mu$ such that

$$\lambda|\xi|^2 \le A_0^{ij}\xi_i\xi_j \le \mu\lambda|\xi|^2 \tag{6.25}$$

for all $\xi \in \mathbb{R}^n$, and we define the operator $L_0$ by

$$L_0 h = A_0^{ij} D_{ij} h.$$

Our first proposition gives the estimate for strong solutions of constant coefficient equations in a ball.

**Proposition 6.19.** *For any $p \in (1,\infty)$, there is a positive constant $C$, determined only by $n$, $p$, and $\mu$ such that any $u \in W^{2,p}(B(x_0,2r))$ satisfies the estimate*

$$\|D^2u\|_{p,r} \le C\left(\frac{1}{\lambda}\|L_0u\|_{p,2r} + \frac{1}{r}\|Du\|_{p,2r} + \frac{1}{r^2}\|u\|_{p,2r}\right), \tag{6.26}$$

*where we have abbreviated $\|g\|_{p,B(x_1,\rho)}$ to $\|g\|_{p,\rho}$ for any function $g$ and any $\rho \in (0,2r)$.*

***Proof.*** We begin by noting that there is a nonnegative function $\eta \in C^2(\mathbb{R}^n)$ with support in $B(x_1,2r)$ which is identically 1 in $B(x_1,r)$ such that $r|D\eta| + r^2|D^2\eta| \le C(n)$. For $k = 1,\dots,n$, we set $v_k = D_k(\eta u)$ and define $f_k$ to be the vector-valued function with components

$$f_k^i = \delta_k^i[\eta(L_0u) + 2A_0^{mj}D_m\eta D_j u + L_0\eta u].$$

Without loss of generality, we may assume that $x_1^n = 2r$ and then we set $x_0 = (x_1', 0)$. There is a constant $\mu_0$, determined only by $n$ and $\mu$ such that

$$A_0^{nj}\xi_j \le \mu_0 A_0^{nn}|\xi|$$

for any vector $\xi \in \mathbb{R}^n$ with $\xi_n = 0$. With

$$\kappa = \frac{\mu_0^{1/2}}{(1+\mu_0^2)^{1/4}},$$

we have $\kappa \in (\mu_0/\sqrt{1+\mu_0^2}, 1)$ and we can choose $R = C(n,\mu)r$ so that $B(x_1, 2R) \subset \Gamma$. Hence $v_k$ is a weak solution of

$$D_i(A_0^{ij}D_jv_k) = D_if_k^i \text{ in } \Gamma, \tag{6.27a}$$

$$A_0^{nj}D_jv_k = f_k^n \text{ on } \Gamma^0, \tag{6.27b}$$

$$v_k = 0 \text{ on } \Gamma^*. \tag{6.27c}$$

It follows from Proposition 6.18 that

$$\|Dv_k\|_{p,\Gamma} \le \frac{C}{\lambda}\|f_k\|_{p,\Gamma},$$

and then (6.26) follows from this inequality because

$$\|D^2u\|_{p,r} \le \sum_k \|Dv_k\|_{p,\Gamma}$$

and

$$\|f_k\|_{p,\Gamma} \le C\|Lu\|_{p,2r} + C\lambda\left(\frac{1}{r}\|Du\|_{p,2r} + \frac{1}{r^2}\|u\|\right). \qquad \square$$

in order to prove our second derivative estimate for solutions of the oblique derivative problem, we use a special cut-off function.

**Lemma 6.20.** *Let $\mu$ be a nonnegative constant and let $\beta_0$ be a constant vector such that $|\beta_0'| \le \mu\beta_0^n$. Then, there is a nonnegative function $\eta \in C^2(\overline{B(0,2r)})$ with support in $\overline{B(0,3r/2)}$ such that $\eta \le 1$ and $r|D\eta| + r^2|D^2\eta| \le C(n,\mu)$ in $B(0,2r)$, $\eta \equiv 1$ in $B(0,r)$, and $\beta_0 \cdot D\eta \equiv 0$ on $B^0(0,2r)$.*

***Proof.*** Let $g \in C^2([0,1])$ with $g(0) = 0$, $g'(0) = 1$, and $0 \le 18\mu g \le 1$ on $[0,1]$. Also, let $h \in C^2([0,\infty))$ be decreasing with $h \equiv 1$ in $[0,1/3)$ and $h \equiv 0$ in $[1/2,\infty)$. We then define $G$ on $B(0,2r)$ by

$$G(x) = \frac{|x|^2}{4r^2} + 9\mu^2\left[g\left(\frac{x^n}{2r}\right)\right]^2 - g\left(\frac{x^n}{2r}\right)\frac{\beta_0'\cdot x}{\beta_0^n r},$$

and we take $\eta = h \circ G$. Cauchy's inequality implies that

$$\frac{2}{9}\frac{|x|^2}{r^2} \le G(x),$$

and hence $\eta(x) = 0$ if $|x| \ge 3r/2$, the choice of $h$ implies that $0 \le \eta \le 1$ and the estimates on the derivatives of $\eta$, and Cauchy's inequality also implies that

$$G(x) \le \frac{5}{18}\frac{|x|^2}{r^2} + 18\mu^2\left[g\left(\frac{x^n}{2r}\right)\right]^2,$$

so $\eta \equiv 1$ on $B(0,r)$. A simple calculation (using $g(0) = 0$ and $g'(0) = 1$) leads to $\beta_0 \cdot D\eta \equiv 0$ on $Q^0$. $\square$

We are now ready for our local second derivative estimate for strong solutions of the oblique derivative problem.

**Proposition 6.21.** *Let $x_0 \in \mathbb{R}^n_0$, let $\beta_0$ be a constant nonzero vector satisfying (6.13), and let $p \in (1,\infty)$. Then there is a constant $C$, determined only by $n$, $\mu$, and $\mu_0$, such that any $u \in W^{2,p}(B(x_0,2r))$ with $\beta_0 \cdot Du = 0$ on $B(x_0,2r)$ satisfies (6.26) with $\|g\|_{p,\rho}$ now standing for $\|g\|_{p,B^+(x_0,\rho)}$.*

***Proof.*** With $\kappa$ as in the proof of Proposition 6.19 (but with $\mu_0$ from the hypotheses of this proposition), we again can choose $R = C(n,\mu,\mu_0)r$ so that $B^+(x_0,2R) \subset \Gamma$. With $\eta$ the function from Lemma 6.20, we set $v_k = D_k(\eta u)$ and we define $f_k$ to be the vector-valued function with components

$$f_k^i = \delta_k^i[\eta L_0 u + 2A_0^{mj} D_m\eta D_j u + L_0\eta u].$$

If we define

$$A^{ij} = \begin{cases} A_0^{ij} & \text{for } i,j < n, \\ A_0^{nn}\beta_0^j/\beta_0^n & \text{for } i = n, j \le n, \\ A_0^{in} + A_0^{ni} - A_0^{nn}\beta^i/\beta_0^n & \text{for } i < n, j = n, \end{cases}$$

then we find that $v_k$ is a weak solution of (6.27) if $k = 1,\dots,n-1$. Hence we have

$$\|Dv_k\|_{p,\Gamma} \le \frac{C}{\lambda}\|f_k\|_{p,\Gamma}$$

and

$$\|f_k\|_{p,\Gamma} \le C\|Lu\|_{p,2r} + C\lambda\left(\frac{1}{r}\|Du\|_{p,2r} + \frac{1}{r^2}\|u\|\right)$$

for this range of $k$. The proof is completed by using the differential equation $A_0^{ij}D_{ij}u = f$ to infer that

$$\|D^2u\|_{p,r} \le C\left(\sum_{k=1}^{n-1}\|Dv_k\|_{p,\Gamma} + \frac{1}{\lambda}\|f\|_{p,2r}\right). \qquad \square$$

## 6.7 Local interior $L^p$ estimates for the second derivatives of strong solutions of differential equations

Our next step is to prove estimates for strong solutions of equations with variable coefficients. We argue via a standard perturbation argument which takes advantage of the following interpolation lemma, which is proved (as part of Theorem 9.11) on page 237 of [64].

**Lemma 6.22.** *For $k = 0,1,2$, define the weighted seminorms:*

$$[v]_k = \sup_{0<s<1}(1-s)^k r^k\|Dv\|_{k,B(x_1,sr)}. \tag{6.28}$$

*Then there is a constant $C$, determined only by $n$, such that*

$$[v]_1 \le \varepsilon[v]_2 + \frac{C}{\varepsilon}[v]_0$$

*for all $\varepsilon > 0$.*

Our local interior estimate has the following form.

**Theorem 6.23.** *Let $\Omega$ be a domain in $\mathbb{R}^n$, let $[a^{ij}]$, $b$, and $c$ be defined on $\Omega$ and define the operator $L$ by*

$$Lh = a^{ij}D_{ij}h + b^i D_i h + ch. \tag{6.29}$$

*Suppose that there are positive constants $\lambda$, $\mu$, and $B$ such that*

$$\lambda|\xi|^2 \le a^{ij}\xi_i\xi_j \le \mu\lambda|\xi|^2 \tag{6.30a}$$

*for all $\xi \in \mathbb{R}^n$,*

$$d|b| + d^2|c| \le B. \tag{6.30b}$$

*Suppose finally that there is a positive, continuous increasing function $\zeta$ defined on $[0, 1/2]$ with $\zeta(0) = 0$ such that*

$$|a^{ij}(x) - a^{ij}(y)| \le \zeta\left(\frac{|x-y|}{d(x)}\right) \tag{6.31}$$

*for all $x$ and $y$ in $\Omega$ with $|x-y| \le \frac{1}{2}d(x)$. Then, for any $\theta > 0$ and any $p > 1$, there is a constant $C$, determined only by $n$, $p$, $\mu$, $\theta$, and $\zeta$ such that*

$$\|D^2u\|_{p,\Omega'} \le C\left(\frac{1}{\lambda}\|Lu\|_{p,\Omega} + \|u\|_{p,\Omega}\right) \tag{6.32}$$

*for any $u \in W^{2,p}_{loc}(\Omega)$ and any $\Omega' \subset \Omega$ with $d(\Omega', \partial\Omega) < \theta$.*

***Proof.*** Fix $x_0 \in \Omega$, define $L_0$ by $L_0 h = a^{ij}(x_0)D_{ij}h$, and set $f = Lu$. For $r \in (0, d(x_0)/2)$ to be chosen, let $\eta \in C^2(\mathbb{R}^n)$ satisfy $\eta = 1$ in $B(x_0, r)$, $\eta = 0$ outside $B(x_0, 2r)$ and

$$|D\eta| \le \frac{2}{r}, \quad |D^2\eta| \le \frac{C(n)}{r^2}.$$

Then $v = \eta u$ satisfies

$$L_0 v = (a^{ij}(x_0) - a^{ij})D_{ij}v + (2a^{ij}D_i\eta - b^i\eta)D_i u + (a^{ij}D_{ij}\eta - c\eta)u + \eta f.$$

It then follows from Proposition 6.19 that there are constants $C_1(\mu, n)$ and $C(\mu, n, B)$ such that

$$\begin{aligned}\|D^2v\|_p &\le C_1\zeta(r/d(x_0))\|D^2v\|_p \\ &\quad + C\left(\frac{1}{\lambda}\|f\|_{p,2r} + \frac{1}{r}\|Du\|_{p,2r} + \frac{1}{r^2}\|u\|_{p,2r}\right).\end{aligned}$$

We now choose $\delta$ so small that $C_1\zeta(\delta) \le 1/2$ and assume that $r \le \delta d(x_0)/2$ to infer that

$$\|D^2v\|_p \le C\left(\frac{1}{\lambda}\|f\|_{p,2r} + \frac{1}{r}\|Du\|_{p,2r} + \frac{1}{r^2}C\|u\|_{p,2r}\right).$$

Since

$$\|D^2u\|_{p,r} \le \|D^2v\|_p,$$

we infer that

$$\|D^2u\|_{p,r} \le C\left(\frac{1}{\lambda}\|f\|_{p,2r} + \frac{1}{r}\|Du\|_{p,2r} + \frac{1}{r^2}C\|u\|_{p,2r}\right).$$

We now fix $y \in \Omega$ with $d(y) > \theta/2$ and set $R = \delta\theta/2$. For each $s \in (0,1)$, we use the previous estimate with $r = (1-s)R/4$ and $x_0$ replaced by any $x_i$ in the covering given by Lemma 6.11 which is in $B(y, sR)$. Modifying notation slightly, we now write $\|g\|_{p,\rho}$ for $\|g\|_{p,B(y,\rho)}$ and $\|g\|_{p,\rho,i}$ for $\|g\|_{p,B(x_i,\rho)}$. Then we have

$$\begin{aligned}\|D^2u\|_{p,sR} &\le C\sum_i\left(\frac{1}{\lambda}\|f\|_{p,r,i} + \frac{1}{(1-s)R}\|Du\|_{p,2r,i} + \frac{1}{(1-s)^2r^2}\|u\|_{p,2r,i}\right)\\ &\le CK\left(\frac{1}{\lambda}\|f\|_{p,R} + \frac{1}{(1-s)R}\|Du\|_{p,(1+s)R/2} + \frac{1}{(1-s)^2r^2}\|u\|_{p,R}\right)\end{aligned}$$

by Lemma 6.11. Using the notation from Lemma 6.22, we infer that

$$[u]_2 \le C\left(\frac{R^2}{\lambda}[f]_0 + [u]_1 + [u]_0\right).$$

so Lemma 6.22 then gives us

$$[u]_2 \le C\left(\frac{R^2}{\lambda}[f]_0 + [u]_0\right),$$

and, therefore, there is a constant $C_2(n, \mu, B)$ such that

$$\|D^2u\|_{p,R/2} \le C_2\left(\frac{1}{\lambda}\|f\|_{p,R} + R^2\|u\|_{p,R}\right).$$

Arguing again via Lemma 6.11 yields (6.32). □

## 6.8 Local $L^p$ second derivative estimates near the boundary

We now prove the analog of Theorem 6.23 for the oblique derivative problem. In addition to ideas already presented, we shall take advantage of two basic results about functions in $W^{1,p}$ and $W^{2,p}$.

Our first result is a variant of the trace theorem.

**Lemma 6.24.** *Let $p > 1$, let $w \in W^{1,p}(\mathbb{R}^n_+)$ and suppose $v(x) = 0$ if $x^n > R$ for some positive constant $R$. Then, for any $\alpha \in (1 - 1/p, 1)$, there is a constant $C$, determined only by $p$ and $\alpha$, such that*

$$\int_{\mathbb{R}^n_+} (x^n)^{p(\alpha-1)}|v|^p\,dx \le CR^{p\alpha}\int_{\mathbb{R}^n_+} |Dv|^p\,dx. \tag{6.33}$$

***Proof.*** Suppose first that $v \in C^1$. For fixed $x' \in \mathbb{R}^{n-1}$, define $w$ by $w(s) = v(x', s)$. Then

$$\begin{aligned}\int_0^\infty s^{p(\alpha-1)}|w(s)|\,ds &= \int_0^R s^{p(\alpha-1)}|w(s)|\,ds \\ &= \int_0^R s^{(p\alpha-1)}\left|\int_s^R w'(t)\,dt\right|^p ds \\ &\le \int_0^R s^{(p\alpha-1)}\left(\int_0^R |w'(t)|\,dt\right)^p ds.\end{aligned}$$

We now use Hölder's inequality to conclude that

$$\left(\int_0^R |w'(t)|\,dt\right)^p \le R^{p-1}\int_0^R |w'(t)|^p\,dt,$$

and hence

$$\begin{aligned}\int_0^\infty s^{p(\alpha-1)}|w(s)|\,ds &\le \int_0^R |w'(t)|^p\,dt R^{p-1}\int_0^R s^{p(\alpha-1)}\,ds \\ &= \frac{1}{p\alpha-p+1}R^{p\alpha}\int_0^\infty |w'(t)|^p\,dt.\end{aligned}$$

We now rewrite this inequality in terms of $v$ and $x$ and then integrate over $\mathbb{R}^{n-1}$ to infer (6.33).

The general case follows by approximation. □

Our next result concerns the representation of the trace of various functions in terms of other, simpler functions.

**Lemma 6.25.** *Let $p > 1$ and $\alpha \in (1-1/p, 1)$, let $v \in W^{1,p}(\mathbb{R}^n_+)$ and suppose that there are a number $r > 0$ and a point $x_0 \in \mathbb{R}^n_0$ such that $v(x) = 0$ if $|x - x_0| > r$. Then there is a constant $C$, determined only by $n$, $p$, and $\alpha$, and, for any $h \in H_\alpha(B_{n-1}(x'_0, r))$, there is a function $z \in W^{2,p}(\mathbb{R}^n_+)$ such that*

$$z = 0,\ D_n z = hv \text{ on } \mathbb{R}^n_0, \tag{6.34a}$$

$$z(x) = 0 \text{ if } |x - x_0| > r, \tag{6.34b}$$

$$\|D^2 z\|_p \le C(|h|_0 + [h]_\alpha r^\alpha)\|v\|_p. \tag{6.34c}$$

***Proof.*** First, by virtue of Lemma 1.3, we may assume that $h$ has been extended to all of $\mathbb{R}^n_0$ without increasing $|h|_0$ or $[h]_\alpha$. Then we extend $h$ to all of $\mathbb{R}^n_+$ so that

$$(x^n)^{1-\alpha}|Dh| + (x^n)^{2-\alpha}|D^2 h| \le C(n,\alpha)[h]_\alpha. \tag{6.35}$$

Finally, we define $z$ by

$$z(x) = x^n h(x) \int_{R^n} v\left(x - \frac{x^n y}{2}\right) \varphi(y)\, dy,$$

where, as usual, $\varphi \in C^2(\mathbb{R}^n)$ is nonnegative with $\|\varphi\|_{L^1} = 1$ and $\varphi(y) = 0$ if $|y| \geq 1$. Straightforward algebra verifies (6.34a) and (6.34b).

In addition, we have

$$D_{ij} z(x) = I_1 + I_2 + I_3$$

for

$$I_1 = (x^n D_{ij} h(x) + \delta_{in} D_j h(x) + \delta_{jn} D_i h(x)) \int_{\mathbb{R}^n} v\left(x - \frac{x^n y}{2}\right) \varphi(y)\, dy,$$

$$I_2 = D_i h(x) \int_{\mathbb{R}^n} v\left(x - \frac{x^n y}{2}\right) \left[\frac{\partial}{\partial y^j}\varphi(y) + \delta_{jn}\frac{\partial}{\partial y^k}(y^k \varphi(y))\right] dy,$$

$$I_3 = h(x) \int_{\mathbb{R}^n} D_i v\left(x - \frac{x^n y}{2}\right) \left[\frac{\partial}{\partial y^j}\varphi(y) + \delta_{jn}\frac{\partial}{\partial y^k}(y^k \varphi(y))\right] dy.$$

We now use (6.35) along with Tonelli's theorem and the observation that

$$\frac{1}{2}x^n \leq x^n - \frac{x^n y}{2} \leq \frac{3}{2}x^n$$

if $\varphi(y) \neq 0$ to conclude that

$$\|I_1\|_p + \|I_2\|_p \leq C[h]_\alpha \left(\int_{\mathbb{R}^n_+} (x^n)^{p(\alpha-1)} |v|^p\, dx\right)^{1/p},$$

and then Lemma 6.24 gives

$$\|I_1\|_p + \|I_2\|_p \leq C[h]_\alpha r^\alpha \|Dv\|_p.$$

A similar argument (but without Lemma 6.24) yields

$$\|I_3\|_p \leq C|h|_0 \|Dv\|_p.$$

The combination of these last two inequalities yields (6.34c). □

From these results, we can prove estimates for the oblique derivative problem near the boundary. Our first estimate is one near a flat boundary portion.

**Theorem 6.26.** *Let $[a^{ij}]$, $b$, and $c$ be defined on $B(x_0, R)$ for some $x_0 \in \mathbb{R}^n_0$ and some $R > 0$. Define the operator $L$ by (6.29), and suppose that there are positive constants $\alpha < 1$, $\lambda$, $\mu$, and $B$ such that*

$$\lambda|\xi|^2 \leq a^{ij}\xi_i\xi_j \leq \mu\lambda|\xi|^2 \tag{6.36a}$$

*for all* $\xi \in \mathbb{R}^n$,

$$R^\alpha (x^n)^{1-\alpha}|b| + R^2|c| \le B. \tag{6.36b}$$

*Suppose further that there is a positive, continuous increasing function* $\zeta$ *defined on* $[0,1/2]$ *with* $\zeta(0)=0$ *such that*

$$|a^{ij}(x) - a^{ij}(y)| \le \zeta\left(\frac{|x-y|}{R}\right) \tag{6.37}$$

*for all* $x$ *and* $y$ *in* $\Omega$ *with* $|x-y| \le R/2$. *Let* $\beta$ *and* $\beta^0$ *be defined on* $B^0(x_0,R)$ *and suppose that*

$$|\beta| \le \mu\beta^n, \quad R|\beta^0| \le B\beta^n \tag{6.38a}$$

*on* $B^0(x_0,R)$,

$$R^\alpha|\beta(x) - \beta(y)| + R^{1+\alpha}|\beta^0(x) - \beta^0(y)| \le B\beta^n(x)|x-y|^\alpha \tag{6.38b}$$

*for all* $x$ *and* $y$ *in* $B_0(x_0,R)$ *with* $|x-y| \le R/2$. *Then, for any* $p \in (1, 1/(1-\alpha))$, *there is a constant* $C$, *determined only by* $n$, $p$, $\mu$, $B$, *and* $\zeta$ *such that*

$$\|D^2u\|_{p;R/2} \le C\left(\frac{1}{\lambda}\|Lu\|_{p;R} + \|u\|_{p;R} + \left\|\frac{Mu}{\beta^n}\right\|_{1-1/p,p;R}\right) \tag{6.39}$$

*for any* $u \in W^{2,p}(B(x_0,R))$, *where* $Mu = \beta\cdot Du + \beta^0 u$ *and*

$$\|g\|_{p;\rho} = \|g\|_{p,B^+(x_0,\rho)}, \quad \|g\|_{1-1/p,p;\rho} = \|g\|_{1-1/p,p,B^0(x_0,\rho)}.$$

***Proof.*** We begin by observing that there is no loss of generality in assuming that $\beta^n \equiv 1$. Fixing $x_1 \in B^0(x_0,R/2)$ and $r \in (0,R/4)$, we take $\eta$ from Lemma 6.20 with $\beta_0 = \beta(x_1)$, we find that

$$\begin{aligned}\|L_0 v\|_p &\le C_1\left(\zeta\left(\frac{r}{R}\right) + B\left(\frac{r}{R}\right)^\alpha\right)\|D^2v\|_p \\ &\quad + C\left(\frac{1}{\lambda}\|f\|_{p,2r} + \frac{1}{r}\|Du\|_{p,2r} + \frac{1}{r^2}\|u\|_{p,2r}\right).\end{aligned}$$

The only new element is that we estimate the quantity $w = b^iD_iu\eta$ in three steps. First, we use the structure condition to see that

$$\|b^iD_iu\eta\|_p \le C(n)BR^{-\alpha}\|(x^n)^{\alpha-1}\eta|Du|\|_p.$$

Then Lemma 6.24 implies that

$$\|(x^n)^{\alpha-1}\eta|Du|\|_p \le C(n,p,\alpha)r^\alpha\|D(\eta|Du|)\|_p.$$

Finally,

$$|D(\eta|Du|)| \le C|D^2v| + \frac{C}{r}|Du| + \frac{C}{r^2}|u|.$$

We then introduce three functions from Lemma 6.25. The most straightforward function is $z_1$, the function (labeled $z$ in the lemma) corresponding to $v$ and $h = -\beta^0$. Next, for each $k \in \{1, \dots, n\}$, we write $z_{k,2}$ for the function corresponding to $v = \eta D_k u$ and $h = (\beta^k(x_0) - \beta^k)$, and we set $z_2 = \sum_k z_{k,2}$. For the last function, we write $\tilde{g}$ for the function in $W^{1,p}(\mathbb{R}^n_+)$ with trace equal to $\eta M u$ on $\mathbb{R}^n_0$, and we take $z_3$ to be the function corresponding to $v = \tilde{g}$ and $h \equiv 1$. With $z = z_1 + z_2 + z_3$, it follows that

$$\begin{aligned}\|L_0(v-z)\|_p \le C_1\lambda\left(\zeta\left(\frac{r}{R}\right) + B\left(\frac{r}{R}\right)^\alpha\right)\|D^2v\|_p \\ + C\left(\|f\|_{p,2r} + \frac{\lambda}{r}\|Du\|_{p,2r} + \frac{\lambda}{r^2}\|u\|_{p,2r} + \lambda\|\eta g\|_{1-1/p,p}\right)\end{aligned}$$

and that

$$\begin{aligned}\|D^2z\|_p \le C_1\left(\zeta\left(\frac{r}{R}\right) + B\left(\frac{r}{R}\right)^\alpha\right)\|D^2v\|_p \\ + C\left(\frac{1}{\lambda}\|f\|_{p,2r} + \frac{1}{r}\|Du\|_{p,2r} + \frac{1}{r^2}\|u\|_{p,2r} + \|\eta g\|_{1-1/p,p}\right).\end{aligned}$$

Moreover, $\beta^0 \cdot D(v-z) = 0$ on $\mathbb{R}^n_0$ and $(v-z)(x)$ is zero if $|x - x_1| \ge 2r$, so we can use Proposition 6.21 to conclude that

$$\|D^2(v-z)\|_p \le \frac{C}{\lambda}\|Lv_0\|_p.$$

Combining all these inequalities, we see that

$$\begin{aligned}\|D^2v\|_p \le C_1\left(\zeta\left(\frac{r}{R}\right) + B\left(\frac{r}{R}\right)^\alpha\right)\|D^2v\|_p \\ + C\left(\frac{1}{\lambda}\|f\|_{p,2r} + \frac{1}{r}\|Du\|_{p,2r} + \frac{1}{r^2}\|u\|_{p,2r} + \|\eta g\|_{1-1/p,p}\right).\end{aligned}$$

We now choose $r$ small enough that $C_1\left(\zeta\left(\frac{r}{R}\right) + B\left(\frac{r}{R}\right)^\alpha\right) \le 1/2$ and then argue as in the proof of Theorem 6.23 to obtain (6.39), except that we use Lemma 6.12 in place of Lemma 6.11. □

For domains with curved boundary, the local estimate is easy to prove if we don't try to obtain quite the same simple form for the estimate. Instead, we introduce some notation that is used consistently in [64], so we first give the following definition. We say that a domain $\Omega$ has a $C^{k,\alpha}$ *boundary portion* $T$ if $T \subset \partial\Omega$ and if, for any $x_0 \in T$, there is a open set $S = S(x_0)$ with $x_0 \in S(x_0)$ and a one-to-one mapping $\psi \in H_{k+\alpha}(S)$ such that

$$\psi(S) = B(0,1), \quad \psi(S \cap \Omega) = B^+(0,1), \quad \psi(S) \cap \partial\Omega) = B^0(0,1)$$

and $\psi^{-1} \in H_{k+\alpha}(B(0,1))$). For our purposes, we also note that the function $\psi$ can be taken so that $|D^m\psi| \le Cd^{k+\alpha-m}$ in $S \cap \Omega$ and $|D^m\psi^{-1}| \le C(x^n)^{k+\alpha-m}$ in $B^+(0,1)$ for any given integer $m > k$.

Our local boundary estimate for domains with curved boundary then takes the following form.

**Theorem 6.27.** *Let $\Omega$ be a domain in $\mathbb{R}^n$ with a $C^{1,\alpha}$ boundary portion $T$ for some $\alpha \in (0,1)$. Let $[a^{ij}]$, $b$, and $c$ be defined on $\Omega$ and suppose they satisfy condition* (6.36a) *and*

$$d^{1-\alpha}|b| + |c| \le B\lambda. \tag{6.40}$$

*Suppose further that there is a positive, continuous increasing function $\zeta$ defined on $[0,\infty)$ with $\zeta(0) = 0$ such that*

$$|a^{ij}(x) - a^{ij}(y)| \le \zeta(|x-y|) \tag{6.41}$$

*for all $x$ and $y$ in $\Omega$. Let $\beta$ and $\beta^0$ be defined on $T$ and suppose that*

$$|\beta| \le \mu\beta\cdot\gamma, \quad |\beta^0| \le B\beta\cdot\gamma \tag{6.42a}$$

*on $T$,*

$$|\beta(x) - \beta(y)| + |\beta^0(x) - \beta^0(y)| \le B\beta\cdot\gamma(x) \tag{6.42b}$$

*for all $x$ and $y$ in $T$. Then for any open subset $\Omega'$ of $\Omega$ such that $\overline{\Omega'} \subset \Omega\cap T$ and any $p \in (1, 1/(1-\alpha))$, there is a constant $C$, determined only by $n$, $p$, $\mu$, $B$, $\Omega$, and $\Omega'$ such that*

$$\|D^2u\|_{p,\Omega'} \le C\left(\frac{1}{\lambda}\|Lu\|_{p,\Omega} + \|u\|_{p,\Omega} + \left\|\frac{Mu}{\beta\cdot\gamma}\right\|_{1-1/p,p,T}\right). \tag{6.43}$$

Of course, we can take $T = \partial\Omega$ and $\Omega' = \Omega$ to obtain a global $W^{2,p}$ estimate for $u$, but, rather than state such an estimate directly, we now look at a related global estimate that will be critical to our existence theory.

**Theorem 6.28.** *Let $\Omega$ be a bounded domain in $\mathbb{R}^n$ with $\partial\Omega \in C^{1,\alpha}$ for some $\alpha \in (0,1)$. Let $[a^{ij}]$, $b$, and $c$ be defined on $\Omega$ and suppose they satisfy condition* (6.36a) *and* (6.40). *Suppose further that there is a positive, continuous increasing function $\zeta$ defined on $[0,\infty)$ with $\zeta(0) = 0$ such that* (6.41) *holds for all $x$ and $y$ in $\Omega$. Let $\beta$ and $\beta^0$ be defined on $\partial\Omega$ and suppose that they satisfy* (6.42) *(with $T$ replaced by $\partial\Omega$). Then for any $p \in (1, 1/(1-\alpha))$, there are positive constants $C$ and $\sigma_0$, determined only by $n$, $p$, $\mu$, $B$, and $\Omega$, such that*

$$\|D^2u\|_p \le C\left(\frac{1}{\lambda}\|Lu - \sigma\lambda u\|_p + \left\|\frac{Mu}{\beta\cdot\gamma}\right\|_{1-1/p,p}\right) \tag{6.44}$$

*for any $\sigma \ge \sigma_0$.*

***Proof.*** Define $\Sigma = \Omega \times (-1, 1)$ and define the operator $\mathcal{L}$ on $\Sigma$ by

$$\mathcal{L}h = Lh + \lambda D_{n+1,n+1}h.$$

Then $v$, defined by

$$v(x, x^{n+1}) = u(x)\cos(\sigma^{1/2}x^{n+1}),$$

satisfies $\mathcal{L}v = \cos(\sigma^{1/2}x^{n+1})(Lu - \sigma\lambda u)$ in $\Sigma$ and $Mv = \cos(\sigma^{1/2}x^{n+1})Mu$ on $T = \partial\Omega \times (-1, 1)$. It then follows from Theorem 6.27 with $\Omega' = \Omega \times (-\varepsilon, \varepsilon)$ (for $\varepsilon \in (0, 1/2)$ that there is a constant $C_1$, determined only by $n$, $p$, $\mu$, $B$, and $\Omega$, such that

$$\begin{aligned}\|D_{n+1,n+1}v\|_{p,\Omega'} &\le C_1 \left(\frac{1}{\lambda}\|\mathcal{L}v\|_{p,\Sigma} + \|v\|_{p,\Sigma} + \left\|\frac{Mv}{\beta\cdot\gamma}\right\|_{1-1/p,p}\right)\\ &\le C_1 \left(\frac{1}{\lambda}\|Lu - \sigma\lambda u\|_{p} + \|u\|_{p} + \left\|\frac{Mu}{\beta\cdot\gamma}\right\|_{1-1/p,p}\right).\end{aligned}$$

We now compute

$$\|D_{n+1,n+1}v\|_{p,\Omega'} = \sigma\|v\|_{p,\Omega'} = \sigma\left(\int_{-\varepsilon}^{\varepsilon} |\cos(\sigma^{1/2}t)|^p\,dt\right)^{1/p}\|u\|_p.$$

If we now choose $\varepsilon = \pi/(3\sigma^{1/2})$, then $\cos(\sigma^{1/2}t) \ge 1/2$ for $t \in (-\varepsilon, \varepsilon)$ and hence

$$\|D_{n+1,n+1}v\|_{p,\Omega'} \ge \sigma\left(\frac{2\pi}{3\sigma^{1/2}}\right)^{1/p}\frac{1}{2}\|u\|_p = \frac{1}{2}\left(\frac{2\pi}{3}\right)^{1/p}\sigma^{1-1/(2p)}\|u\|_p.$$

By choosing $\sigma_0$ sufficiently large, we find that

$$\|D_{n+1,n+1}v\|_{p,\Omega'} \ge 2C_1\|u\|_p$$

and hence

$$\|u\|_p \le \frac{1}{\lambda}\|Lu - \sigma\lambda u\|_p + \left\|\frac{Mu}{\beta\cdot\gamma}\right\|_{1-1/p,p}.$$

The proof is completed by combining this inequality with Theorem 6.27 using $T = \partial\Omega$ and $\Omega' = \Omega$. □

## 6.9 Existence of strong solutions for the oblique derivative problem

We are now ready to show that the oblique derivative problem is solvable under suitable hypotheses. There are many ways to use our estimates for this purpose, and we provide here a simple one, namely, the method of continuity. Other methods are suggested in the exercises.

Throughout this section, we assume that the hypotheses of Theorem 6.28 are satisfied and that $\sigma$ is a constant with $\sigma \geq \sigma_0$, the constant of that theorem. We shall show that the problem

$$Lu - \sigma\lambda u = f \text{ in } \Omega, \quad Mu = g \text{ on } \partial\Omega \tag{6.45}$$

has a unique solution $u \in W^{2,p}(\Omega)$ for any $f \in L^p(\Omega)$ and any $g \in W^{1-1/p,p}(\partial\Omega)$. To simplify the proof, we shall assume that $\beta \cdot \gamma \equiv 1$ on $\partial\Omega$. Also, to apply the method of continuity, we write $\mathcal{B}$ for $W^{2,p}(\Omega)$ and $\mathcal{V} = L^p(\Omega) \times W^{1-1/p,p}(\partial\Omega)$.

We begin by setting $\beta_0 = \gamma$, and we define the operators $L_0$ and $M_0$ by

$$L_0 h = \lambda\Delta h - \sigma\lambda h, \quad M_0 h = \beta_0 \cdot Dh.$$

Then, for $t \in [0,1]$, we define the operators $L_t$ and $M_t$ by

$$L_t = tL + (1-t)L_0, \quad M_t = tM + (1-t)M_0.$$

Since $\sigma > 0$, it follows that the problem

$$L_0 u = f \text{ in } \Omega, \quad M_0 u = g \text{ on } \partial\Omega \tag{6.46}$$

has a unique solution in $H_{2+\delta}^{(-1-\alpha)}$ for any $f \in H_1(\Omega)$ and $g \in H_\alpha(\Omega)$, where $\delta \in (0,1)$ is arbitrary. Since $p(1-\alpha) < 1$, it follows that this solution is in $W^{2,p}$. Since $H_1(\Omega)$ is dense in $L^p(\Omega)$ and $H_\alpha(\partial\Omega)$ is dense in $W^{1-1/p,p}(\partial\Omega)$, a simple approximation argument (using Theorem 6.28) then implies that (6.46) has a unique solution $u \in W^{2,p}$ for any $f \in L^p$ and $g \in W^{1-1/p,p}$. Hence $(L_0, M_0)$ maps $\mathcal{B}$ onto $\mathcal{V}$.

In addition, Theorem 6.28 implies that

$$\|u\|_{\mathcal{B}} \leq C\|(L_t u, M_t u)\|_{\mathcal{V}}$$

for all $t \in (0,1)$, so the method of continuity implies that $(L_1, M_1)$ maps $\mathcal{B}$ onto $\mathcal{V}$. In other words, (6.45) is solvable for any $f$ and $g$. The uniqueness of the solution follows from another application of Theorem 6.28.

We therefore have the following basic existence and uniqueness theorem.

**Theorem 6.29.** *Let $\Omega$ be a bounded domain in $\mathbb{R}^n$ with $\partial\Omega \in C^{1,\alpha}$ for some $\alpha \in (0,1)$. Let $[a^{ij}]$, $b$,and $c$ be defined on $\Omega$ and suppose they satisfy condition* (6.36a) *and* (6.40). *Suppose further that there is a positive,* *continuous increasing function* $\zeta$ *defined on* $[0,\infty)$ *with* $\zeta(0)=0$ *such that* (6.41) *holds for all* $x$ *and* $y$ *in* $\Omega$. *Let* $\beta$ *and* $\beta^0$ *be defined on* $\partial\Omega$ *and suppose that they satisfy* (6.42) (*with* $T$ *replaced by* $\partial\Omega$). *Then for any* $p \in (1, 1/(1-\alpha))$, *there is a positive constant* $\sigma_0$, *determined only by* $n$, $p$, $\mu$, $B$, $\zeta$, *and* $\Omega$, *such that* (6.45) *has a unique solution for any* $\sigma \geq \sigma_0$. *In addition there is a constant* $C$, *determined by the same quantities as for* $\sigma_0$, *such that this solution satisfies the estimate*

$$\|u\|_{2,p} \leq C\left(\frac{1}{\lambda}\|f\|_p + \left\|\frac{g}{\beta\cdot\gamma}\right\|_{1-1/p,p}\right). \tag{6.47}$$

Note that the condition $\sigma \geq \sigma_0$ is quite strong and we do not have a simple way of estimating the constant $\sigma_0$. If we use suitable results from Chapter 4, we can prove a uniqueness and existence result for (6.1) assuming that $c$ and $\beta^0$ are nonpositive but not both zero.

**Theorem 6.30.** *Let* $\Omega$ *be a bounded domain in* $\mathbb{R}^n$ *with* $\partial\Omega \in C^{1,\alpha}$ *for some* $\alpha \in (0,1)$. *Let* $[a^{ij}]$, $b$,*and* $c$ *be defined on* $\Omega$ *and suppose they satisfy condition* (6.36a) *and* (6.40). *Suppose further that there is a positive, continuous increasing function* $\zeta$ *defined on* $[0,\infty)$ *with* $\zeta(0)=0$ *such that* (6.41) *holds for all* $x$ *and* $y$ *in* $\Omega$. *Let* $\beta$ *and* $\beta^0$ *be defined on* $\partial\Omega$ *and suppose that they satisfy* (6.42) (*with* $T$ *replaced by* $\partial\Omega$). *Suppose finally that* $c \leq 0$ *in* $\Omega$ *and* $\beta^0 \leq 0$ *on* $\partial\Omega$ *but at least one of these functions is not identically zero. Then for any* $p \in (1, 1/(1-\alpha))$, (6.1) *has a unique solution* $u \in W^{2,p}$ *for any* $f \in L^p(\Omega)$ *and any* $g \in W^{1-1/p,p}(\partial\Omega)$. *In addition there is a constant* $C$, *determined only by* $n$, $c_0$, $p$, $\mu$, $B$, $\zeta$, *and* $\Omega$, *such that this solution satisfies* (6.47).

***Proof.*** Suppose first that $g \in H_1$ and that the functions $a^{ij}$, $b$, $c$, $f$ are all in $H_\alpha(\Omega)$. In this case, Theorem 4.10 gives a unique solution of (6.1) in $H_{2+\alpha}^{(-1-\alpha)}$.

Next, suppose $a^{ij}$, $b$, and $c$ satisfy the hypotheses of the theorem, and let $(a_k^{ij})$, $(b_k)$, and $(c_k)$ be sequences of $H_\alpha$ functions such that $(a_k^{ij})$ converges uniformly to $a^{ij}$, $(b_k)$ converges a.e. to $b$, and $(c_k)$ converges to $c$. We further take the sequences so that each $a_k^{ij}$ satisfies the hypotheses of this theorem, $|b_k| \leq 2B\lambda d^{\alpha-1}$, and $c_k \leq 0$ with $c_k$ nonzero in some open set if $c \not\equiv 0$. We still assume that $f \in H_\alpha$ and that $g \in H_1$, and we write $u_k$ for the solution of

$$L_k u_k = f \text{ in } \Omega, \quad M u_k = g \text{ on } \partial\Omega.$$

We now claim that the sequence $(u_k)$ is uniformly bounded. Suppose to the contrary that this sequence is not uniformly bounded, in which case we may assume that $\|u_k\|_\infty \to \infty$ as $k \to \infty$. Setting $U_k = \|u_k\|_\infty$, $v_k = u_k/U_k$, $f_k = f/U_k$ and $g_k = g/U_k$, we conclude that

$$L_k v_k = f_k \text{ in } \Omega, \quad v_k = g_k \text{ on } \partial\Omega$$

for each $k$ and that $(v_k)$ is bounded in $L^\infty$.

Then Theorem 4.5 gives a uniform $H_{1+\alpha}$ bound for $v_k$ and Lemma 6.31 gives a uniform $W^{2,p}$ bound for $v_k$. It follows that some subsequence converges in $W^{2,q}$ (for any $q < p$) to a limit function $v$, which is therefore a $C^1(\overline{\Omega}) \cap W^{2,n}_{\text{loc}}(\Omega)$ solution of

$$Lv = 0 \text{ in } \Omega, \quad Mv = 0 \text{ on } \partial\Omega.$$

Hence $v \equiv 0$, which contradicts the pair of conditions $\|v_k\|_\infty = 1$ for all $k$ and $(v_k)$ converges uniformly to $v$. Therefore the sequence $(u_k)$ is uniformly bounded, so Lemma 6.31 gives a subsequence which converges in $W^{2,q}$ (for any $q < p$) to a limit function $u$, which is therefore a $W^{2,p}$ solution of (6.1). Since $H_\alpha(\Omega) \times H_1(\partial\Omega)$ is dense in $L^p(\Omega) \times W^{1-1/p,p}(\partial\Omega)$, we conclude that the range of the map $T\colon W^{2,p} \to L^p(\Omega) \times W^{1-1/p,p}(\partial\Omega)$ defined by $Tu = (Lu, Mu)$ is dense in $L^p(\Omega) \times W^{1-1/p,p}(\partial\Omega)$. But Theorem 2.29 and the proof of Theorem 2.30 imply that this range is also closed and hence $T$ is onto, so that (6.1) has a solution for any $f$ and $g$. Another application of Theorem 2.29 implies that the solution is unique. □

We close with a useful local regularity result.

**Lemma 6.31.** *Let $\Omega$ be a domain in $\mathbb{R}^n$ with a $C^{1,\alpha}$ boundary portion $T$ for some $\alpha \in (0,1)$. Let $[a^{ij}]$, $b$, and $c$ be defined on $\Omega$ and suppose they satisfy condition* (6.36a) *and* (6.40). *Suppose further that there is a positive, continuous increasing function $\zeta$ defined on $[0,\infty)$ with $\zeta(0) = 0$ such that* (6.41) *holds for all $x$ and $y$ in $\Omega$. Let $\beta$ and $\beta^0$ be defined on $T$ and suppose that conditions* (6.42) *are satisfied. Let $u \in W^{2,q}_{loc}(\Omega \cup T)$ for some $q \in (1, 1/(1-\alpha))$ and suppose that $Lu \in L^p(\Omega)$ and $Mu \in W^{1-1/p,p}(T)$ for some $p \in (q, 1/(1-\alpha))$. Then $u \in W^{2,p}_{loc}(\Omega \cup T)$.*

***Proof.*** Let us fix a point $x_0 \in T$ and choose $R$ so that $B(x_0, R) \cap \partial\Omega \subset T$. Then there is a set $\Omega^*$ with $\partial\Omega^* \in C^{1,\alpha}$ such that $B(x_0, R/2) \cap \Omega \subset \Omega^* \subset B(x_0, R) \cap \Omega$. Let $\eta \in C^2(\mathbb{R}^n)$ be nonnegative with $\eta \equiv 1$ in $B(x_0, R/2)$ and $\eta \equiv 0$ in $\mathbb{R}^n \setminus B(x_0, 5R/8)$. Then there is a vector field $\bar\beta \in H_\alpha(\partial\Omega^*)$ which is oblique on $\partial\Omega^*$ with $\bar\beta = \beta$ on $\partial\Omega \cap B(x_0, 5R/8)$, so $v = \eta^2 u$ satisfies the conditions

$$a^{ij} D_{ij} v + b^i D_i v - \sigma v = f_1 \text{ in } \Omega^*, \bar\beta \cdot v = g_1 \text{ on } \partial\Omega^* \tag{6.48}$$

for any $\sigma \in \mathbb{R}$, with

$$f_1 = 2\eta a^{ij} D_i\eta D_j u + [a^{ij} D_{ij}(\eta^2) - c\eta^2 - \sigma]u + 2\eta(b^i D_i\eta)u$$

and

$$g_1 = \eta u[2\bar{\beta} \cdot D\eta - \beta^0\eta].$$

It follows from the Sobolev imbedding theorem that $Du \in W^{1,r}(\Omega^*)$ with

$$r = \begin{cases} p & \text{if } q \geq n, \\ \min\{p, nq/(n-q)\} & \text{if } q < n. \end{cases}$$

By using Lemma 6.24 and observing that $D(\eta u|D\eta|) \in L^r$, we see that $f_1 \in L^r(\Omega^*)$ and $g_1 \in W^{1-1/r,r}(\partial\Omega^*)$. If we choose $\sigma$ sufficiently large, then Theorem 6.29 gives a unique solution $v \in W^{2,r}(\Omega^*)$ of (6.48), and the same theorem gives a unique $W^{2,q}(\Omega^*)$ solution of this problem, so these two solutions must be equal. It follows that $v \in W^{2,r}(\Omega^*)$ and hence $u \in W^{2,r}(B(x_0, R/2) \cap \Omega)$. After repeating this argument (with $r$ in place of $q$) finitely many times, we conclude that $u \in W^{2,p}(B(x_0, R/2) \cap \Omega)$.

A similar (but simpler) argument shows that $u \in W^{2,p}(B(x_0, \rho))$ for any $x_0 \in \Omega$ and $\rho \leq d(x_0)/2$, and a covering argument gives the general result.

□

## Notes

Section 6.2 is based on the results in [60], and we refer the interested reader to that paper for a more detailed account of the proof of Theorem 6.2. The main differences between [60] and Section 6.2 are that we have only sketched the proof that $T$ is onto and that we have proved the preliminary results Lemmata 6.3 and 6.4. Lemma 6.3 was first proved by Hardy in [69]; we use a simplified version of the proof in that paper. A more detailed account of the history of this standard result appears in [89]. Lemma 6.4 was proved by Morrey as Lemma 1 in [138] and we have used his proof here. Gagliardo cites [139] as a source for this result, but that source doesn't contain the lemma. Instead, it is stated as Theorem 2.3 in [24], which is part one of [139], and [24] refers to [138] for a proof.

The covering lemmata, labeled as Lemmata 6.7 and 6.10 here, were first proved by Wang [199], and the proof of Lemma 6.7 given here is identical to that in [199]. On the other hand, Lemma 6.8 and Corollary 6.9 are standard results and we refer the reader to page 12 of [175] for further discussion. Lemmata 6.11 and 6.12 are apparently new here, but they are really just minor modifications of results used by Whitney [201] to prove his general extension theorem (Lemma 1.3 is a very special case of this result); however, nowhere in Whitney's paper are the results stated in a form immediately recognizable as related to these lemmata. Instead, we mention Propositions 2 and 3 on page 169 of [175] which have an obvious connection, both in statement and in proof, to our lemmata.

The method of proof for the $L^p$ estimates for constant coefficient problems in Section 6.5 is due to Wang [199], who developed it as an alternative to earlier proofs; the proof in [199] is just presented for interior estimates, but the generalization to boundary value problems is very straightforward. In particular, this generalization consists primarily of proving the estimates in Section 6.4. Applications of Wang's method to the oblique derivative problem were first given in Chapter 7 of [114], and the presentation here streamlines that approach by proving all the local pointwise and $L^2$ estimates at once (in Lemmata 6.13 and 6.14). The corresponding boundary estimates for the Dirichlet problem, under various weak regularity assumptions on the boundary, appear in numerous papers which we shall not list here since additional care must be taken with the boundary condition; however, it would be interesting to know if the Dirichlet problem results can be used to prove estimates for the oblique derivative problem along the lines of the proof of Schauder estimates in Chapter 4. To the best of my knowledge, the only other application of Wang's method to oblique derivative problems is a work by Byun and Wang [19] which gives $W^{1,p}$ estimates for solutions of the conormal problem in Reifenberg flat domains. We shall not give the precise definition of a Reifenberg flat domain, but we point out that they are not directly comparable to Lipschitz domains; all Lipschitz domains with sufficiently small Lipschitz constant are Reifenberg flat, but Lipschitz domains with large Lipschitz constant need not be Reifenberg flat and Reifenberg flat domains need not be Lipschitz.

Although the method of proof for the $L^p$ estimates is relatively recent, the results themselves are not. The basic interior estimate for Poisson's equation is due to Calderón and Zygmund [23], and our Theorem 6.23 was first proved by Greco [67]. Our Theorem 6.26 is a special case of a much more general result, which was first proved by Browder (see Theorem 1 in [16]) and Agmon, Douglis, and Nirenberg (see Theorem 15.1 of [1]). In fact, these results apply to domains with $C^2$ boundaries, and the extension to $H_{1+\alpha}$ boundaries (with $p(1-\alpha) < 1$) seems to have first stated in Theorem 7.35 of [114]. Moreover, the continuity assumption on $[a^{ij}]$ can be relaxed to a smallness condition on the $BMO$ modulus of this matrix. In other words, one needs to know only that

$$\liminf_{R\to 0} \sup_{x\in\Omega} \inf_{L\in\mathbb{R}} R^{-n} \int_{\Omega\cap B(x,R)} |u-L|\, dx$$

is sufficiently small. This fact was apparently first noted explicitly in [21] for nonlinear equations and for $p > n$. A proof for linear equations and all $p > 1$ was first given in [29], and the method developed here was used to prove the same results in [18]. In fact, this smallness condition can be further weakened; see, for example, [79].

Various $L^p$ estimates for derivatives have also been derived in Lipschitz domains by Calderón [22] and by Kenig and Pipher [78]. To state their results, we first fix a positive number $\theta$ and, for each $y \in \partial\Omega$, we write $\Gamma(y)$ for the set of all $x \in \Omega$ with $|x-y| < (1+\theta)d(x)$. We also define the nontangential maximal operator $N$ associated with this number $\theta$ by setting, for any $g$ defined on $\Omega$ and any $y \in \partial\Omega$,

$$N[g](y) = \sup_{\Gamma(y)} |g|.$$

Then, for any continuous oblique unit vector field $\beta$, there are finitely many linear conditions and a positive number $\varepsilon$ (determined only by $\beta$ and $\Omega$) such that if $g \in L^p(\partial\Omega)$ for some $p > 2-\varepsilon$ satisfies the finitely many linear conditions, then there is a solution of the problem

$$\Delta u = 0 \text{ in } \Omega, \quad \beta\cdot Du = g \text{ on } \partial\Omega$$

such that

$$\|N[Du]\|_{p,\partial\Omega} \le C(p,\Omega,\beta)\|g\|_{p,\partial\Omega}.$$

Specifically, Calderón proved this result for $p \in (2-\varepsilon, 2+\varepsilon)$ and Kenig and Pipher proved it for $p > 2$. In fact, uniqueness for this problem is generally not true for $p < 2$; Verchota [194] has shown that there are continuous oblique vector fields $\beta$ such that the homogeneous problem has an arbitrarily large number of linearly independent solutions with $p < 2$ but the homogeneous problem has only the constant solution for $p = 2$. Techniques for these two cases are quite different from each other and from the ones used in this book.

# Chapter 7

# Viscosity Solutions of Oblique Derivative Problems

In this chapter, we provide an extremely brief introduction to viscosity solutions of fully nonlinear elliptic equations with fully nonlinear oblique boundary conditions.

## 7.1 Definitions and notation

Throughout this chapter, we assume that the function $F$, defined on $\Omega \times \mathbb{R} \times \mathbb{R}^n \times \mathbb{S}^n$, satisfies the following monotonicity condition:

$$F(x, z_1, p, r^1) \leq F(x, z_2, p, r^2) \tag{7.1}$$

for all $(x, p) \in \Omega \times \mathbb{R}^n$, all $z_1 \geq z_2$ in $\mathbb{R}$, and all $r^1 \leq r^2$ in $\mathbb{S}^n$. (As before, we use the inequality $r^1 \leq r^2$ to mean that $r^1 - r^2$ is a negative semidefinite matrix.)

In addition, for any subset $S$ of $\mathbb{R}^n$, we write $USC(S)$ for the set of all functions which are upper semicontinuous on $S$ and $LSC(S)$ for the set of all functions which are lower semicontinuous on $S$. If $u$ is defined on $S$, we write $J_S^{2,+}u(x)$ for the set of all $(p, r) \in \mathbb{R}^n \times \mathbb{S}^n$ such that

$$u(y) \leq u(x) + p \cdot (y - x) + \frac{1}{2} r_{ij}(x - y)^i (x - y)^j + o(|x - y|^2)$$

for $y \in S$. In other words, for any $\varepsilon > 0$, there is $\delta > 0$ such that

$$u(y) \leq u(x) + p \cdot (y - x) + \frac{1}{2} r_{ij}(x - y)^i (x - y)^j + \varepsilon |x - y|^2$$

for all $y \in S$ with $|x - y| < \delta$. Similarly, we write $J_S^{2,-}u(x)$ for the set of all $(p, r) \in \mathbb{R}^n \times \mathbb{S}^n$ such that

$$u(y) \geq u(x) + p \cdot (y - x) + \frac{1}{2} r_{ij}(x - y)^i (x - y)^j - o(|x - y|^2)$$

for $y \in S$. A *viscosity subsolution* of $F = 0$ in $\Omega$ is a function $u \in USC(\Omega)$ such that

$$F(x, u(x), p, r) \geq 0$$

for all $x \in \Omega$ and all $(p, r) \in J^{2,+}_{\Omega} u(x)$. A *viscosity supersolution* of $F = 0$ is defined to be a function $u \in LSC(\Omega)$ such that

$$F(x, u(x), p, r) \leq 0$$

for all $x \in \Omega$ and all $(p, r) \in J^{2,-}_{\Omega} u(x)$. (Note that the definition of subsolution puts no restrictions on the function $u$ at a point where $J^{2,+}_{\Omega} u$ is empty.)

Although this formulation of viscosity subsolutions and supersolutions is very convenient for our purposes, we point out that it is related to a more usual one involving comparison functions. Specifically, one can show that $J^{2,+}_{\Omega} u(x)$ is equal to the set of all $(p, r) \in \mathbb{R}^n \times \mathbb{S}^n$ such that $p = D\varphi(x)$ and $r = D^2\varphi(x)$ for some function $\varphi \in C^2(\Omega)$ such that $u - \varphi$ has a local maximum at $x$. Hence $u$ is a viscosity subsolution if and only if

$$F(x, u(x), D\varphi(x), D^2\varphi(x)) \geq 0$$

for any $x$ and $\varphi \in C^2(\Omega)$ such that $u - \varphi$ has a local maximum at $x$. Similar assertions are true for $J^{2,-}_{\Omega}$ and viscosity supersolutions. We don't spell out the details because they won't be important in our brief development of the theory.

We shall discuss the appropriate definition of viscosity subsolutions and viscosity supersolutions for boundary value problems in Section 7.6.

## 7.2 The Theorem of Aleksandrov

In this section, we present a proof of one of the deepest result in the entire book. We show that every convex function is twice differentiable almost everywhere; this result will be used as one step in the analysis of viscosity solutions. This result was first proved in the case of two dimensions by Busemann and Feller [17] and in an arbitrary number of dimensions by Aleksandrov [3].

We proceed via a sequence of results on convex functions and measures. The first result consists of some elementary estimates on convex functions.

**Lemma 7.1.** *Let $f \colon \mathbb{R}^n \to \mathbb{R}$ be a $C^2$ convex function. Then $D^2 f \geq 0$ in*

*$\mathbb{R}^n$ and there is a constant $C$, determined only by $n$, such that*

$$\sup_{B(x,R/2)} |f| \le CR^{-n} \int_{B(x,R)} |f(y)|\,dy, \tag{7.2a}$$

$$\sup_{B(x,R/2)} |Df| \le CR^{-n-1} \int_{B(x,R)} |f(y)|\,dy, \tag{7.2b}$$

*for any $x \in \mathbb{R}^n$ and any $R > 0$.*

***Proof.*** First, we note that, for any $h > 0$ and any $x$ and $y$ in $\mathbb{R}^n$ with $|y| = 1$, we have

$$f(x+hy) + f(x-hy) - 2f(x) \ge 0.$$

Dividing this inequality by $h^2$ and letting $h \to 0$, we see that

$$D_{ij}f(x)y^iy^j \ge 0,$$

and hence $D^2f \ge 0$.

Next, we use Taylor's theorem to conclude that, for any $y$ and $z$ in $\mathbb{R}^n$, there is a point $x^* \in \mathbb{R}^n$ such that

$$f(y) = f(z) + Df(z)\cdot(y-z) + \frac{1}{2}D_{ij}f(x^*)(y-z)^i(y-z)^j,$$

and hence

$$f(y) \ge f(z) + Df(z)\cdot(y-z). \tag{7.3}$$

Now suppose $z \in B(x, R/2)$. Integrating (7.3) with respect to $y$ over $B(z, R/2)$ then gives

$$f(z) \le \frac{1}{|B(z,R/2)|}\int_{B(z,R/2)} |f(y)|\,dy \le 2^n \frac{1}{|B(x,R)|}\int_{B(x,R)} |f(y)|\,dy$$

because

$$\int_{B(z,R/2)} Df(z)\cdot(y-z)\,dy = Df(z)\cdot\int_{B(z,R/2)} (y-z)\,dy = 0.$$

On the other hand, by reversing the roles of $y$ and $z$ in (7.3), we conclude that

$$f(z) \ge f(y) + Df(y)\cdot(z-y). \tag{7.4}$$

There is a function $\varphi \in C^1(\mathbb{R}^n)$ such that $0 \le \varphi \le 1$ and $|D\varphi| \le 4/R$ in $\mathbb{R}^n$ with $\varphi \equiv 1$ on $B(x, R/2)$ and $\varphi \equiv 0$ on $\mathbb{R}^n \setminus B(x, R)$. Multiplying (7.4) by $\varphi(y)$ and integrating with respect to $y$ over $B(x, R)$ yields

$$f(z)\int_{B(x,R)} \varphi(y)\,dy \ge \int_{B(x,R)} f(y)\varphi(y)\,dy + \int_{B(x,R)} \varphi(y)Df(y)\cdot(z-y)\,dy.$$

The divergence theorem implies that

$$\int_{B(x,R)} \varphi(y)Df(y)\cdot(z-y)\,dy = -\int_{B(x,R)} f(y)\,\mathrm{div}[\varphi(y)(z-y)]\,dy,$$

and so

$$\begin{aligned} f(z)\int_{B(x,R)} \varphi(y)\,dy &\geq \int_{B(x,R)} f(y)(\varphi(y)-\mathrm{div}[\varphi(y)(z-y)])\,dy \\ &\geq -(5+n)\int_{B(x,R)} |f(y)|\,dy. \end{aligned}$$

Since

$$\int_{B(x,R)} \varphi(y)\,dy \geq \int_{B(x,R/2)} \varphi(y)\,dy = |B(x,R/2)| = 2^{-n}|B(x,R)|,$$

it follows that

$$f(z) \geq -(5+n)2^n \frac{1}{|B(x,R)|}\int_{B(x,R)} |f(y)|\,dy.$$

In combination with our upper bound for $f(z)$, we infer that

$$|f(z)| \leq C(n)R^{-n}\int_{B(x,R)} |f(y)|\,dy,$$

which implies (7.2a).

Finally, for $z \in B(x,R/2)$, then (7.2b) is immediate if $|Df(z)| = 0$, so suppose $|Df(z)| \neq 0$ and set

$$\xi = \frac{Df(z)}{|Df(z)|}$$

and $y = z - (R/4)\xi$. Then (7.3) implies that

$$f(y) \geq f(z) + \frac{R}{4}Df(x)\cdot\xi = f(z) + \frac{R}{4}|Df(z)|.$$

Hence

$$|Df(z)| \leq \frac{4}{R}|f(y)-f(z)| \leq \frac{8}{R}\sup_{B(z,R/4)}|f|.$$

From (7.2a), we infer that

$$\sup_{B(z,R/4)}|f| \leq C(n)R^{-n}\int_{B(z,R/2)} |f(y)|\,dy$$

and hence

$$\sup_{B(z,R/4)}|f| \leq C(n)R^{-n}\int_{B(x,R)} |f(y)|\,dy$$

Combining this inequality with our estimate for $|Df(z)|$ then yields (7.2b). □

Our next step is to identify the second derivatives of an arbitrary convex function with appropriate signed Radon measures.

**Lemma 7.2.** *Let $f\colon \mathbb{R}^n \to \mathbb{R}$ be convex. Then, for all $i$ and $j$ in $\{1,\dots,n\}$, there is a signed Radon measure $\mu_{ij}$ such that $\mu_{ji} = \mu_{ij}$ and*

$$\int_{\mathbb{R}^n} f(y) D_{ij}\varphi(y)\,dy = \int_{\mathbb{R}^n} \varphi(y)\,d\mu_{ij} \tag{7.5}$$

*for all $\varphi \in C^2(\mathbb{R}^n)$ with compact support. Moreover, $\mu_{ii}$ is a measure for $i = 1,\dots,n$.*

***Proof.*** For each vector $\xi \in \mathbb{R}^n$, define the linear functional $L_\xi$ on $C_c^2$ by

$$L_\xi(\varphi) = \int_{\mathbb{R}^n} f(y) D_{ij}\varphi(y)\xi^i\xi^j\,dy.$$

Now let $\zeta$ be a mollifier, that is, $\zeta \in C^2(\mathbb{R}^n)$ with support in $B(0,1)$, $\zeta \geq 0$ in $\mathbb{R}^n$ and

$$\int_{\mathbb{R}^n} \zeta(y)\,dy = 1.$$

For $\varepsilon > 0$, define $f^\varepsilon$ by

$$f^\varepsilon(x) = \int_{\mathbb{R}^n} f(x - \varepsilon y)\zeta(y)\,dy.$$

Then $f^\varepsilon$ is convex and $C^2$ on $\mathbb{R}^n$, so

$$\int_{\mathbb{R}^n} f^\varepsilon(y) D_{ij}\varphi(y)\xi^i\xi^j\,dy = \int_{\mathbb{R}^n} \varphi(y) D_{ij} f^\varepsilon(y)\xi^i\xi^j\,dy \geq 0$$

by Lemma 7.1 if $\varphi \geq 0$. Since $f^\varepsilon$ converges to $f$ in $L^1(S)$, where $S$ is the support of $\varphi$, it follows that $L_\xi(\varphi) \geq 0$ for any nonnegative $\varphi \in C_c^2$, and then the Riesz representation theorem (in the form of Corollary 1 in Section 1.8 of [43]) provides a Radon measure $\mu_\xi$ such that

$$L_\xi(\varphi) = \int_{\mathbb{R}^n} \varphi(y)\,d\mu_\xi$$

for any $\varphi \in C_c^2$.

To define $\mu_{ij}$, we proceed in two steps. First, for $i = 1,\dots,n$, we define $\mu_{ii}$ to be $\mu_{e_i}$. Then for $i \neq j$, we define

$$\mu_{ij} = \mu_{(e_i+e_j)/\sqrt{2}} - \frac{1}{2}\mu_{e_i} - \frac{1}{2}\mu_{e_j}. \qquad \square$$

Our next step is a regularity result for the gradient of convex functions.

**Corollary 7.3.** *If $f\colon \mathbb{R}^n \to \mathbb{R}$ is convex, then $Df \in BV_{loc}(\mathbb{R}^n)$.*

***Proof.*** By using the function $f^\varepsilon$ from the proof of Lemma 7.2, we see that $|Df^\varepsilon|$ is uniformly bounded on compact subsets of $\mathbb{R}^n$ and hence $Df \in L^\infty_{\text{loc}}(\mathbb{R}^n)$.

Now let $\varphi$ be a vector-valued $C^2_c$ function with $|\varphi| \le 1$ and let $V$ be a bounded open set containing the support of $\varphi$. Then, for $k = 1, \dots, n$, we have

$$\int_{\mathbb{R}^n} D_k f(y) \operatorname{div} \varphi \, dy = -\int_{\mathbb{R}^n} f(y) D_{ik} \varphi^i(y) \, dy.$$

But the definition of $\mu_{ij}$ implies that

$$\int_{\mathbb{R}^n} f(y) D_{ik} \varphi^i(y) \, dy = \int_{\mathbb{R}^n} \varphi(y) \, d\mu_{ik} \le \sum_{i=1}^n |\mu_{ik}|(V) < \infty.$$

Hence $D_k f \in BV_{\text{loc}}(\mathbb{R}^n)$. □

Now the Lebesgue decomposition theorem implies that each $\mu_{ij}$ is the sum of a signed measure $\mu_{ij,ac}$, which is absolutely continuous with respect to Lebesgue measure, and a signed measure $\mu_{ij,s}$, which is singular with respect to Lebesgue measure. We then write $D_{ij} f$ for the Radon-Nikodym derivative of $\mu_{ij,ac}$ with respect to Lebesgue measure. The final step in the proof of Aleksandrov's theorem is to show that this definition of $D_{ij} f$ agrees with the second derivative of $f$ almost everywhere.

**Theorem 7.4.** *If $f$ is convex, then $f$ is twice differentiable almost everywhere. Specifically, if $D_{ij} f$ is the Radon-Nikodym derivative of $\mu_{ij,ac}$ with respect to Lebesgue measure, then there is a measurable set $A \subset \mathbb{R}^n$ with $|\mathbb{R}^n \setminus A| = 0$ such that*

$$|f(x) - f(y) - Df(x) \cdot (y - x) - \frac{1}{2} D_{ij} f(x)(y - x)^i (y - x)^j| = o(|x - y|^2) \quad (7.6)$$

*as $y \to x$ for all $x \in A$.*

***Proof.*** From Lemma 7.2, Corollary 7.3, and the continuity of the $L^1$ norm with respect to translation, it follows that there is a measurable set $A \subset \mathbb{R}^n$ norm with $|\mathbb{R}^n \setminus A| = 0$ such that

$$Df(x) \text{ exists,} \quad (7.7a)$$

$$\lim_{R \to 0} R^{-n} \int_{B(x,R)} |Df(y) - Df(x)| \, dy = 0, \quad (7.7b)$$

$$D_{ij} f(x) \text{ exists for all } i, j, \quad (7.7c)$$

$$\lim_{R \to 0} R^{-n} \int_{B(x,R)} |D_{ij} f(y) - D_{ij} f(x)| \, dy = 0 \quad (7.7d)$$

for all $x \in A$. Moreover, because $\mu_{ij,s}$ is singular with respect to Lebesgue measure, we have

$$\lim_{R\to 0} R^{-n}|\mu_{ij,s}|(B(x,R)) = 0. \tag{7.8}$$

We now fix a point $x \in A$ and assume without loss of generality that $x = 0$. Then we choose $R > 0$ and $y \in B(0,R)$. Finally, we write $f^\varepsilon$ for the mollification of $f$ described in the proof of Lemma 7.2. Taylor's theorem then implies that

$$f^\varepsilon(y) = f^\varepsilon(0) + Df^\varepsilon(0)\cdot y + \int_0^1 (1-s)D_{ij}f^\varepsilon(sy)y^iy^j\,dy.$$

For ease of notation, we now define $F^\varepsilon$ and $F$ by

$$F^\varepsilon(y) = f^\varepsilon(y) - f^\varepsilon(0) - Df^\varepsilon(0)\cdot y - \frac{1}{2}D_{ij}f(0)y^iy^j,$$
$$F(y) = f(y) - f(0) - Df(0)\cdot y - \frac{1}{2}D_{ij}f(0)y^iy^j.$$

Then our Taylor's theorem equation can be rewritten as

$$F^\varepsilon(y) = \int_0^1 (1-s)[D_{ij}f^\varepsilon(sy) - D_{ij}f(0)]y^iy^j\,ds.$$

We now multiply this equation by a $C^2(\mathbb{R}^n)$ function $\varphi$ with support in $B(0,R)$ satisfying $|\varphi| \le 1$ in $\mathbb{R}^n$ and integrate with respect to $y$ over $B(0,R)$ to infer that

$$\int_{B(0,R)} \varphi(y)F^\varepsilon(y)\,dy = \int_{B(0,R)} \varphi(y)\int_0^1 (1-s)[D_{ij}f^\varepsilon(sy) - D_{ij}f(0)]y^iy^j\,ds\,dy.$$

Applying Fubini's theorem and then using the change of variables $z = sy$, we conclude that

$$\int_{B(0,R)} \varphi(y)F^\varepsilon(y)\,dy = \int_0^1 \int_{B(0,sR)} \left(\varphi\left(\frac{z}{s}\right)[D_{ij}f^\varepsilon(z) - D_{ij}f(0)]z^iz^j\,dz\right) ds.$$

Our next step is to take the limit as $\varepsilon \to 0$, and the analysis of this limit procedure is simplified by defining the function $g_\varepsilon$ by

$$g_\varepsilon(s) = \frac{1}{|B(0,Rs)|}\int_{B(0,Rs)} \varphi(z)\left(\frac{z}{s}\right) D_{ij}f^\varepsilon(z)z^iz^j\,dz.$$

Then

$$g_\varepsilon(s) = \frac{1}{|B(0,Rs)|}\int_{B(0,Rs)} D_{ij}\left(\varphi\left(\frac{z}{s}\right)z^iz^j\right) f^\varepsilon(z)\,dz,$$

so

$$\lim_{\varepsilon\to 0} g_\varepsilon(s) = g(s),$$

with

$$g(s) = \frac{1}{|B(0,Rs)|}\int_{B(0,Rs)} f(z)D_{ij}\left(\varphi\left(\frac{z}{s}\right)z^iz^j\right)dz$$
$$= \frac{1}{|B(0,Rs)|}\int_{B(0,Rs)} \varphi\left(\frac{z}{s}\right)z^iz^j\,d\mu_{ij}$$
$$= \frac{1}{|B(0,Rs)|}\int_{B(0,Rs)} \varphi\left(\frac{z}{s}\right)z^iz^jD_{ij}f(z)\,dz$$
$$+ \frac{1}{|B(0,Rs)|}\int_{B(0,Rs)} \varphi\left(\frac{z}{s}\right)z^iz^j\,d\mu_{ij,s}.$$

We wish to show that $g_\varepsilon \to g$ in $L^1(0,1)$. To this end, we prove a uniform upper bound for $g_\varepsilon$, and this upper bound is proved by noting first that

$$g_\varepsilon(s) \le \frac{C(n)}{(Rs)^n}\int_{B(0,Rs)} D_{ij}f^\varepsilon(z)z^iz^j\,dz$$

because $|\varphi| \le 1$ and $D_{ij}f^\varepsilon(z)z^iz^j \ge 0$. Then, we recall that

$$f^\varepsilon(z) = \int_{B(z,\varepsilon)} \zeta_\varepsilon(w-z)f(w)\,dw,$$

where $\zeta_\varepsilon$ is defined by

$$\zeta_\varepsilon(y) = \varepsilon^{-n}\zeta(y/\varepsilon).$$

Hence, if we set

$$I = \int_{B(0,Rs)} D_{ij}f^\varepsilon(z)z^iz^j\,dz,$$

then $g_\varepsilon(s) \le C(n,R)s^{-n}I$ and

$$I = \int_{B(0,Rs)} z^iz^j\int_{B(z,\varepsilon)} D^2\zeta_\varepsilon(w-z)f(w)\,dw\,dz$$
$$= \int_{B(0,Rs)} z^iz^j\int_{B(z,\varepsilon)} \zeta_\varepsilon(w-z)\,d\mu_{ij}\,dz$$

Fubini's theorem then implies that

$$I \le \int_{B(0,Rs+\varepsilon)}\int_{B(0,Rs)\cap B(w,\varepsilon)} \zeta_\varepsilon(w-z)z^iz^j\,dz, d\mu_{ij}$$

because the set

$$\{(w,z) : |z-w| < \varepsilon, z \in B(0,Rs)\}$$

is a subset of

$$\{(w,z) : |z-w| < \varepsilon, |w| < Rs+\varepsilon\}.$$

Recalling also that $0 \le \zeta \le C(n)$, we have

$$I \le |Rs|^2 C(n)\varepsilon^{-n} \min\{|Rs|^n, \varepsilon^n\} \sup_{|\xi|=1} \mu_\xi(B(0, Rs+\varepsilon)).$$

By virtue of (7.7d) and (7.8), it follows that

$$\sup_{|\xi|=1} \mu_\xi(B(0, Rs+\varepsilon)) \le C(Rs+\varepsilon)^n$$

and hence there is a constant $C$, determined only by $n$ and $R$ such that $g_\varepsilon(s) \le C$ for all $\varepsilon$ and $s$ in $(0,1)$.

The dominated convergence theorem then implies that

$$\lim_{\varepsilon\to 0} R^{-n} \int_{B(0,R)} \varphi(y) \int_0^1 (1-s)[D_{ij} f^\varepsilon(sy) - D_{ij} f(0)] y^i y^j \, ds\, dy$$
$$= R^{-n} \int_{B(0,R)} \varphi(y) \int_0^1 (1-s)[D_{ij} f(sy) - D_{ij} f(0)] y^i y^j \, ds\, dy,$$

and therefore

$$\lim_{\varepsilon\to 0} R^{-n} \int_{B(0,R)} \varphi(y) F^\varepsilon(y)\, dy$$

is equal to the same quantity. In other words,

$$R^{-n} \int_{B(0,R)} \varphi(y) F(y)\, dy$$
$$= R^{-n} \int_{B(0,R)} \varphi(y) \int_0^1 (1-s)[D_{ij} f(sy) - D_{ij} f(0)] y^i y^j \, ds\, dy.$$

From (7.7d) and (7.8), we therefore infer that there is a continuous, increasing function $\omega$ with $\omega(0) = 0$ such that

$$\frac{1}{|B(0,R)|} \int_{B(0,R)} \varphi(y) F(y)\, dy \le \omega(R)$$

for any function $\varphi \in C^2$ with compact support in $B(0,R)$ such that $|\varphi| \le 1$ and hence

$$\frac{1}{|B(0,R)|} \int_{B(0,R)} |F(y)|\, dy \le \omega(R). \tag{7.9}$$

Our final step in the proof is to show that this integral estimate can be converted into a pointwise estimate. For this step, we first note that $h$, defined by

$$h(y) = f(y) - f(0) - Df(0) \cdot y,$$

is convex, so Lemma 7.1 gives a constant $C_1(n)$ such that

$$\sup_{B(0,R/2)} |Dh| \le C_1 R^{-n-1} \int_{B(0,R)} |h|\,dy$$

for all $R > 0$. Writing $\Lambda$ for the maximum eigenvalue of the matrix $[D_{ij}f(0)]$ and using (7.9), we conclude that there are constants $C_2(\Lambda, n)$ and $R_1$ such that

$$\sup_{B(0,R/2)} |DF| \le C_2 R$$

if $0 < R < R_1$.

Now, let $\varepsilon \in (0,1)$ and infer from (7.9) that there is a constant $R_0$ such that

$$\int_{B(0,R)} |F|\,dy \le 2^{-n-1}\varepsilon^{n+1}|B(0,R)|$$

if $0 < R < R_0$. Hence, if $0 < R < \min\{R_0, R_1\}$ and $y \in B(0, R/4)$, we have

$$\begin{aligned}\inf_{B(y,\varepsilon R/2)|} |F| &\le \frac{1}{|B(y,\varepsilon R/2)|}\int_{B(y,\varepsilon R/2)} |F|\,dy \\ &\le \frac{2^n\varepsilon^n}{|B(y,\varepsilon R)|}\int_{B(0,R)} |F|\,dy \le \frac{1}{2}\varepsilon R^2.\end{aligned}$$

In particular, there is a point $z \in B(y, \varepsilon R/(1+2C_2))$ such that $|F(z)| \le \frac{1}{2}\varepsilon R^2$, so

$$|F(y)| \le |F(z)| + \left(\sup_{B(0,R)} |DF|\right)|y-z| \le \frac{1}{2}\varepsilon R^2 + C_2 R\frac{\varepsilon}{1+2C_2}R < \varepsilon R^2.$$

Since $\varepsilon$ is arbitrary, we infer (7.6). □

## 7.3 Preliminary results for the comparison theorem for viscosity solutions

We wish to prove a comparison theorem for viscosity subsolutions and viscosity supersolutions. This theorem is a consequence of a number of technical results which we collect in this section, and the comparison theorem itself will be proved in the next section.

Our first preliminary result concerns the behavior of a certain combination of upper semicontinuous functions.

**Lemma 7.5.** *Let $\Omega$ be a subset of $\mathbb{R}^n$ and let $u_1$ and $u_2$ be two upper semicontinuous functions on $\Omega$. For $\alpha > 0$, set*

$$M_\alpha = \sup_{\substack{x\in\Omega\\ y\in\Omega}} \left(u_1(x) + u_2(y) - \frac{\alpha}{2}|x-y|^2\right),$$

*and suppose that there is a nonnegative constant $\alpha_0$ such that $M_{\alpha_0}$ is finite. For $\alpha \geq \alpha_0$, suppose $x_\alpha$ and $y_\alpha$ satisfy*

$$\lim_{\alpha\to\infty} \left(M_\alpha - \left[u_1(x_\alpha) + u_2(y_\alpha) - \frac{\alpha}{2}|x_\alpha - y_\alpha|^2\right]\right) = 0.$$

*Then*

$$\lim_{\alpha\to\infty} \alpha|x_\alpha - y_\alpha|^2 = 0, \tag{7.10}$$

*and, for any increasing sequence $(\alpha_k)$ which diverges to $\infty$ such that $(x_{\alpha_k})$ converges, say to $\hat{x}$, we have*

$$\lim_{\alpha\to\infty} M_\alpha = u_1(\hat{x}) + u_2(\hat{x}) = \sup_{\substack{x\in\Omega\\ y\in\Omega}}(u_1(x) + u_2(y)). \tag{7.11}$$

***Proof.*** We set

$$\delta_\alpha = M_\alpha - \left[u_1(x_\alpha) + u_2(y_\alpha) - \frac{\alpha}{2}|x_\alpha - y_\alpha|^2\right],$$

and note that

$$\lim_{\alpha\to\infty} \delta_\alpha = 0.$$

Since $|x_\alpha - y_\alpha|^2 \geq 0$, it follows that $M_\alpha$ is a decreasing function of $\alpha$ and hence

$$\lim_{\alpha\to\infty} M_\alpha < \infty.$$

For $\alpha \geq 2\alpha_0$, we have

$$\begin{aligned} M_{\alpha/2} &\geq u_1(x_\alpha) + u_2(y_\alpha) - \frac{\alpha}{2}|x_\alpha - y_\alpha|^2 \\ &= u_1(x_\alpha) + u_2(y_\alpha) - \alpha|x_\alpha - y_\alpha|^2 + \frac{\alpha}{2}|x_\alpha - y_\alpha|^2. \end{aligned}$$

It follows that

$$\alpha|x_\alpha - y_\alpha|^2 \leq 2(M_{\alpha/2} - M_\alpha + \delta_\alpha).$$

Since

$$\lim_{\alpha\to\infty} M_{\alpha/2} - M_\alpha = 0$$

and

$$\lim_{\alpha\to\infty} \delta_\alpha = 0,$$

we infer (7.10), and the upper semicontinuity of $u_1(x) + u_2(y)$ in $\Omega \times \Omega$ implies (7.11). □

Our next result is a slight variation of the theorem of Aleksandrov proved in the previous section. To state this variation, we recall that a real-valued function $\varphi$ defined on some subset $S$ of $\mathbb{R}^n$ is *semiconvex* if there is a constant $\lambda \geq 0$ such that the function $\varphi_\lambda$, defined by

$$\varphi_\lambda(x) = \varphi(x) + \frac{\lambda}{2}|x|^2$$

is convex. Any such constant $\lambda$ is called a *semiconvexity constant* for $\varphi$.

**Theorem 7.6.** *Let $\varphi \colon \mathbb{R}^n \to \mathbb{R}$ be semiconvex. Then $\varphi$ is twice differentiable a.e. in $\mathbb{R}^n$.*

***Proof.*** Since $\varphi_\lambda$ is convex for some $\lambda \geq 0$, it follows from Theorem 7.4 that $\varphi_\lambda$ is twice differentiable a.e., and hence so is $\varphi$. □

The Aleksandrov theorem is used to prove an estimate similar to the Aleksandrov-Bakel′man-Pucci estimate.

**Lemma 7.7.** *Let $\varphi$ be a real-valued semiconvex function defined on $\mathbb{R}^n$ and suppose $\varphi$ has a strict local maximum at some $\hat{x}$. For $p \in \mathbb{R}^n$, define $\varphi_p$ by $\varphi_p(x) = \varphi(x) + p \cdot x$. Then, for any positive constants $\rho$ and $\delta$, the set, $K$, of all $x \in B(\hat{x}, \rho)$ for which there is a $p \in B(0, \delta)$ such that $\varphi$ has a local maximum at $x$ has positive measure.*

***Proof.*** There is no loss of generality in assuming that $\rho$ is small enough that $\varphi$ has a unique maximum in $\overline{B(\hat{x}, \rho)}$ at $\hat{x}$.

We suppose now that $\varphi \in C^2$. Then there is a positive number $\delta_0$ such that $|p| < \delta_0$ implies that the maximum value for $\varphi_p$ in $\overline{B(\hat{x}, \rho)}$ can only be attained somewhere in $B(\hat{x}, \rho)$. Since $D\varphi(x) + p = 0$ at such a point, it follows that $B(0, \delta) \subset D\varphi(K)$ if $\delta \leq \delta_0$. We may therefore assume that $\delta \leq \delta_0$.

Now let $\lambda$ be a constant of semiconvexity for $\varphi$. Then $-\lambda I \leq D^2\varphi$ in $B(\hat{x}, \rho)$ and, if $x \in K$, it follows that $D^2\varphi_p(x) \leq 0$ for some $p \in \mathbb{R}^n$ (and, in fact $|p| < \delta_0$). Therefore $-\lambda I \leq D^2\varphi \leq 0$ in $K$, and hence

$$|\det D^2\varphi| \leq \lambda^n$$

in $K$. We now use the change of variables theorem (Theorem 1.7) to infer that

$$|B(0, \delta)| \leq |D\varphi(K)| \leq \int_K |\det D^2\varphi|\, dx \leq |K|\lambda^n.$$

Hence

$$|K| \geq c(n) \left(\frac{\delta}{\lambda}\right)^n. \tag{7.12}$$

If $\varphi$ is an arbitrary semiconvex function, we write $\varphi^{(m)}$ for the mollification of $\varphi$ with parameter $1/m$, that is,

$$\varphi^{(m)}(x) = \int_{\mathbb{R}^n} \varphi(y)\psi\left(x - \frac{y}{m}\right) dy$$

for some (fixed) nonnegative $C^2(\mathbb{R}^n)$ function $\psi$ with support in $B(0,1)$ and $\|\psi\|_{L^1} = 1$. If $\lambda$ is a constant of semiconvexity for $\varphi$, it follows that $\lambda$ is also a constant of semiconvexity for each $\varphi^{(m)}$ and hence

$$|K_m| \geq c(n)\left(\frac{\delta}{\lambda}\right)^n$$

by (7.12), where $K_m$ is the set corresponding to $\varphi^{(m)}$.

Next, suppose that

$$x \in \bigcap_{k=1}^{\infty} \bigcup_{m=k}^{\infty} K_m.$$

Then $x \in B(\hat{x}, \rho)$, so, for each positive integer $m$, there is a point $p^m \in \mathbb{R}^n$ with $|p^m| < \delta$ such that $\varphi^{(m)}_{p^m}$ has a local maximum at $x$. In other words,

$$\varphi^{(m)}(x) + p^m \cdot x \geq \varphi^{(m)}(y) + p^m \cdot y$$

for all $y \in B(\hat{x}, \rho)$. It follows that $(p^m)$ has a convergent subsequence $(p^{m(j)})$ with limit $p$. Since

$$\varphi^{(m(j))}(x) + p^{m(j)} \cdot x \geq \varphi^{(m(j))}(y) + p^{m(j)} \cdot y$$

for all $y \in B(\hat{x}, \rho)$, and $\varphi^{(m)}$ converges pointwise to $\varphi$, we can take the limit as $j \to \infty$ to conclude that

$$\varphi(x) + p \cdot x \geq \varphi(y) + p \cdot y$$

for all $y \in B(\hat{x}, \rho)$; in other words, $x \in K$. Therefore

$$\bigcap_{k=1}^{\infty} \bigcup_{m=k}^{\infty} K_m \subset K,$$

which implies that

$$\left|\bigcap_{k=1}^{\infty} \bigcup_{m=k}^{\infty} K_m\right| \leq |K|.$$

The Borel-Cantelli lemma implies that

$$\left|\bigcap_{k=1}^{\infty} \bigcup_{m=k}^{\infty} K_m\right| = \lim_{k\to\infty} \left|\bigcup_{m=k}^{\infty} K_m\right|.$$

In conjunction with our lower bound for $|K_m|$, we therefore conclude that (7.12) is valid for all semiconvex $\varphi$, and hence $|K|$ has positive measure. $\square$

To continue, we need to introduce a generalization of $J_S^{2,+}$. Specifically, if $u$ is defined on a set $S$ and $x \in S$, we define $\bar{J}_S^{2,+}u(x)$ to be the set of all $(p, r)$ such that there is a sequence $(x_m, p^m, r^m)$ in $S \times \mathbb{R}^n \times \mathbb{S}^n$ such that $(p^m, r^m) \in J_S^{2,+}u(x_m)$ and

$$\lim_{m\to\infty} (x_m, u(x_m), p^m, r^m) = (x, u(x), p, r).$$

We also define $\bar{J}_S^{2,-}u(x)$ by replacing $J_S^{2,+}$ with $J_S^{2,-}$. Finally, if $S = \mathbb{R}^n$, we suppress $S$ from the notation.

**Lemma 7.8.** *Let $f \in C(\mathbb{R}^n)$ and suppose that $f$ is semiconvex with constant of semiconvexity $\lambda$. If there is a matrix $B \in \mathbb{S}^n$ such that*

$$\max_{x\in\mathbb{R}^n} f(x) - \frac{1}{2}B_{ij}x^i x^j = f(0),$$

*then there is a matrix $r \in \mathbb{S}^n$ such that $(0, r) \in \bar{J}^2 f(0)$ and $-\lambda I \le r \le B$.*

***Proof.*** Define $\tilde{f}$ by

$$\tilde{f}(x) = f(x) - \frac{1}{2}B_{ij}x^i x^j - |x|^4.$$

Then $f$ has second derivatives a.e. by Theorem 7.6 and Lemma 7.7 implies that, for any $\delta > 0$, there is $q^\delta \in \mathbb{R}^n$ such that $\tilde{f}_{q^\delta}$ has a maximum at some $\xi_\delta$ with $|\xi_\delta| < \delta$ and $f$ has second derivatives at $\xi_\delta$. Then the differentiability of $f_{q^\delta}$ at $\xi_\delta$ implies that $Df_{q^\delta}(\xi_\delta) = 0$ and hence there is a constant $C_1$, determined only by $n$ and the matrix $B$, such that

$$|Df(\xi_\delta)| \le C\delta. \tag{7.13}$$

Using the explicit description for $\tilde{f}$ and the fact that $D^2 f_{q^\delta}(\xi_\delta) \le 0$, we see that there is a constant $C_2$, determined only by $n$ and the matrix $B$, such that

$$D^2 f(\xi_\delta) \le B + C\delta^2 I. \tag{7.14}$$

Moreover, the semiconvexity of $f$ implies that

$$D^2 f(\xi_\delta) \ge -\lambda I. \tag{7.15}$$

From (7.14) and (7.15), we conclude that there is a sequence $(\delta_m)$ with $\delta_m \to 0$ such that $D^2 f(\xi_{\delta(m)})$ converges to a matrix $r$. Conditions (7.13), (7.14), and (7.15) then imply the desired result. □

Our final result is a property of the sup convolution, defined by (7.16) below.

**Lemma 7.9.** *Let $\lambda > 0$, let $v \in USC(\mathbb{R}^n)$ be bounded above, and define the function $\hat{v}$ by*

$$\hat{v}(\xi) = \sup_{x\in\mathbb{R}^m} \left( v(x) - \frac{\lambda}{2}|x-\xi|^2 \right). \tag{7.16}$$

*If $\eta$ and $q$ are in $\mathbb{R}^n$ and $r \in \mathbb{S}^n$ with $(q,r) \in J^{2,+}\hat{v}(\eta)$, then*

$$(q,r) \in J^{2,+}v\left(\eta + \frac{q}{\lambda}\right), \tag{7.17a}$$

$$\hat{v}(\eta) + \frac{1}{2\lambda}|q|^2 = v\left(\eta + \frac{q}{\lambda}\right). \tag{7.17b}$$

*In particular, if $(0,r) \in \bar{J}^{2,+}\hat{v}(0)$, then $(0,r) \in \bar{J}^{2,+}v(0)$.*

***Proof.*** Fix $\eta$, $q$ and $r$ so that $(q,r) \in J^{2,+}\hat{v}(\eta)$, and let $y$ be a point in $\mathbb{R}^n$ such that

$$\hat{v}(\eta) = v(y) - \frac{\lambda}{2}|y-\eta|^2.$$

Such a point exists because $v$ is upper semicontinuous.

Then for all $x$ and $\xi$ in $\mathbb{R}^n$, we have

$$\begin{aligned} v(x) - \frac{\lambda}{2}|\xi - x|^2 &\le \hat{v}(\xi) \\ &\le \hat{v}(\eta) + q\cdot(\xi-\eta) + \frac{1}{2}r_{ij}(\xi-\eta)^i(\xi-\eta)^j + o(|\xi-\eta|^2) \\ &= v(y) - \frac{\lambda}{2}|y-\eta|^2 + q\cdot(\xi-\eta) + \frac{1}{2}r_{ij}(\xi-\eta)^i(\xi-\eta)^j \\ &\quad + o(|\xi-\eta|^2). \end{aligned}$$

In particular, if $\xi = x - y + \eta$, then $\xi - x = \eta - y$ and $\xi - \eta = x - y$, so

$$v(x) \le q\cdot(x-y) + \frac{1}{2}r_{ij}(x-y)^i(x-y)^j + o(|x-y|^2),$$

and hence $(q,r) \in J^{2,+}v(y)$.

Now, let $\alpha$ be a real parameter at our disposal and take $x = y$ and $\xi = \eta + \alpha(\lambda(\eta - y) + q)$. It follows that

$$\begin{aligned} v(y) &- \frac{\lambda}{2}|(\eta-y) + \alpha(\lambda(\eta-y)+q)|^2 \\ &\le v(y) - \frac{\lambda}{2}|y-\eta|^2 + \alpha q\cdot(\lambda(\eta-y)+q) \\ &\quad + O(|\alpha(\lambda(\eta-y)+q)|^2). \end{aligned}$$

Simple rearrangement of this inequality shows that

$$0 \le \frac{\lambda}{2}(|\xi - y|^2 - |y-\eta|^2) + q \cdot (\xi - \eta) + O(\alpha^2).$$

Then

$$|\xi - y|^2 - |y-\eta|^2 = |\xi - \eta|^2 - 2(y-\eta)\cdot(\xi-\eta),$$

and

$$|\xi - \eta|^2 = \alpha^2 |\lambda(\eta - y) + q|^2 = O(\alpha^2).$$

Hence

$$0 \le \alpha |\lambda(\eta - y) + q|^2 + O(\alpha^2).$$

If $\alpha < 0$, we conclude that

$$|\lambda(\eta - y) + q|^2 = O(|\alpha|)$$

and sending $\alpha \to 0$ yields $\lambda(\eta - y) + q = 0$, which is the same as

$$y = \eta + \frac{q}{\lambda}.$$

Hence

$$\hat{v}(\eta) + \frac{1}{2\lambda}|q|^2 = v\left(\eta + \frac{q}{\lambda}\right).$$

Finally, if $(0, r) \in \bar{J}^{2,+}\hat{v}(0)$, then for any sequence $((\xi_m, q^m, r^m))$ in $\Omega \times \mathbb{R}^n \times S^n$ such that $(q^m, r^m) \in J^{2,+}\hat{v}(\xi_m))$ and $(\xi_m, q^m, r^m) \to (0, 0, r)$, we have that

$$(q^m, r^m) \in J^{2,+}v\left(\xi_m + \frac{q^m}{\lambda}\right)$$

and

$$v\left(\xi_m + \frac{q^m}{\lambda}\right) = \hat{(}\xi_m) + \frac{1}{2\lambda}|q^m|^2.$$

Moreover, we have

$$\begin{aligned} v(0) &\ge \lim_{m\to\infty} v\left(\xi_m + \frac{q^m}{\lambda}\right) \\ &= \lim_{m\to\infty}\left(\hat{v}(\xi_m) + \frac{1}{2\lambda}|q^m|^2\right) \\ &= \hat{v}(0) \ge v(0), \end{aligned}$$

so $\hat{v}(0) = v(0)$ and therefore $(0, r) \in \bar{J}^{2,+}v(0)$. □

Our next result concerns a function of $2n$ variables. In this lemma, we use the matrix norm

$$\|A\| = \sup_{|\xi|=1} \left|A_{ij}\xi^i\xi^j\right|.$$

**Lemma 7.10.** *Let $u_1$ and $u_2$ be upper semicontinuous, real-valued functions defined on $\mathbb{R}^n$ with $u_1(0) = u_2(0) = 0$ and define $w$ on $\mathbb{R}^{2n}$ by $w(x,y) = u_1(x) + u_2(y)$, and abbreviate $(x,y)$ to $X$. If there is a matrix $A \in \mathbb{S}^{2n}$ such that*

$$w(X) \le \frac{1}{2}A_{ij}X^iX^j \tag{7.18}$$

*for all $X \in \mathbb{R}^{2n}$, then, for all $\varepsilon > 0$, there are matrices $r^1$ and $r^2$ in $\mathbb{S}^n$ such that $(0, r^i) \in \bar{J}^{2,+}u_i(0)$ for $i = 1, 2$ and*

$$-\left(\frac{1}{\varepsilon} + \|A\|\right)I \le \begin{pmatrix} r^1 & 0 \\ 0 & r^2 \end{pmatrix} \le A + \varepsilon A^2. \tag{7.19}$$

***Proof.*** Fix $\varepsilon > 0$. Then the Cauchy-Schwarz inequality implies that

$$A_{ij}X^iX^j \le (A + \varepsilon A^2)_{ij}\Xi^i\Xi^j + \left(\frac{1}{\varepsilon} + \|A\|\right)|X - \Xi|^2 \tag{7.20}$$

for all $X$ and $\Xi$ in $\mathbb{R}^{2n}$. If we now set $\lambda = \frac{1}{\varepsilon} + \|A\|$, then it follows form (7.18) and (7.20) that

$$w(X) - \frac{\lambda}{2}|X - \Xi|^2 \le \frac{1}{2}(A + \varepsilon A^2)_{ij}\Xi^i\Xi^j. \tag{7.21}$$

Next, we define $\hat{u}_i$ and $\hat{w}$ by

$$\hat{u}_i(\xi) = \sup_{x\in\mathbb{R}^n}\left(u_i(x) - \frac{\lambda}{2}|x - \xi|^2\right)$$

for $i = 1, 2$,

$$\hat{w}(\Xi) = \sup_{X\in\mathbb{R}^{2n}}\left(w(X) - \frac{\lambda}{2}|X - \Xi|^2\right),$$

and notice that

$$\hat{w}(\xi,\eta) = \hat{u}_1(\xi) + \hat{u}_2(\eta).$$

In addition, we have $\hat{u}_i(0) \ge u_i(0) = 0$ and

$$\hat{u}_1(0) + \hat{u}_2(0) = \hat{w}(0) \le 0$$

by virtue of (7.21), so $\hat{u}_i(0) = 0$.

Moreover, for each $X \in \mathbb{R}^{2n}$, the function $W_X$, defined by

$$W_X(\Xi) = w(X) - \frac{\lambda}{2}|X - \Xi|^2 + \frac{\lambda}{2}|\Xi|^2,$$

satisfies

$$W_X(\Xi) = w(X) - \frac{\lambda}{2}|X|^2 + \lambda X \cdot \Xi,$$

and hence $W_X$ is convex. It follows that $\hat{w}$ and $\hat{u}_i$ are semiconvex with constant $\lambda$.

We can therefore apply Lemma 7.8 to $\hat{w}$ with $B = A + \varepsilon A^2$ to conclude that there is a matrix $R \in \mathbb{S}^{2n}$ such that $(0, R) \in \bar{J}^{2,+}w(0)$ and

$$-\lambda I \le R \le A + \varepsilon A^2.$$

From the form of $w$, it follows that there are matrices $r^1$ and $r^2$ in $\mathbb{S}^n$ such that

$$R = \begin{pmatrix} r^1 & 0 \\ 0 & r^2 \end{pmatrix}$$

and $r^i \in \bar{J}^{2,+}\hat{u}_i(0)$. Since $\hat{u}_i(0) = 0$, it follows from Lemma 7.9 that $r^i \in \bar{J}^{2,+}u_i(0)$. □

Finally, we prove a version of the preceding result for functions defined on more general sets.

**Lemma 7.11.** *Let $S_1$ and $S_2$ be locally compact subsets of $\mathbb{R}^n$, let $u_i$ be upper semicontinuous on $S_i$, let $\alpha \ge 0$, and define $\varphi$ on $\mathbb{R}^{2n}$ by $\varphi(x, y) = \frac{\alpha}{2}|x - y|^2$. Define $w$ on $S = S_1 \times S_2$ by*

$$w(x, y) = u_1(x) + u_2(y)$$

*and suppose that $w - \varphi$ attains a local maximum at some $\hat{X} = (\hat{x}, \hat{y}) \in S$. Then there are matrices $r^1$ and $r^2$ in $\mathbb{S}^n$ such that*

$$(\alpha(\hat{x} - \hat{y}), r^1) \in \bar{J}^{2,+}u_1(\hat{x}), \tag{7.22a}$$

$$(-\alpha(\hat{x} - \hat{y}), r^2) \in \bar{J}^{2,+}u_2(\hat{y}), \tag{7.22b}$$

$$-3\alpha I \le \begin{pmatrix} r^1 & 0 \\ 0 & r^2 \end{pmatrix} \le 3\alpha \begin{pmatrix} I & -I \\ -I & I \end{pmatrix}. \tag{7.22c}$$

***Proof.*** We first observe that a simple translation reduces to the case that $\hat{x} = \hat{y} = 0$. If $S_i$ is not all of $\mathbb{R}^n$, we extend $u_i$ to all of $\mathbb{R}^n$ by setting $u_i(x) = -\infty$ if $x \notin S_i$. For such an extension, we have

$$\bar{J}^{2,+}_{S_i}u_i(x) = \bar{J}^{2,+}u_i(x)$$

for $x \in S_i$. Since $S_1$ and $S_2$ are locally compact, it follows that $u_i \in USC(\mathbb{R}^n)$, so we can apply Lemma 7.10 with

$$A = \alpha \begin{pmatrix} I & -I \\ -I & I \end{pmatrix}$$

and $\varepsilon = 1/\alpha$. We just have to note that $A^2 = 2\alpha I$ and $\|A\| = 2\alpha$. □

## 7.4 The comparison principle for viscosity sub- and supersolutions

A comparison principle follows rather easily from the considerations of the previous section provided $F$ satisfies suitable structure conditions.

**Theorem 7.12.** *Let $\Omega$ be a bounded open subset of $\mathbb{R}^n$ and let $F \in C(\overline{\Omega} \times \mathbb{R} \times \mathbb{R}^n \times \mathbb{S}^n)$. Suppose that there is a positive constant $c_0$ such that*

$$c_0(z_1 - z_2) \le F(x, z_2, p, r^2) - F(x, z_1, p, r^1) \tag{7.23}$$

*for all $(x, z_i, p, r^i) \in \overline{\Omega} \times \mathbb{R} \times \mathbb{R}^n \times \mathbb{S}^n$ with $z_1 \ge z_2$ and $r^1 \le r^2$. Suppose also that there is a continuous increasing function $\omega$ defined on $[0, \infty)$ with $\omega(0) = 0$ such that*

$$F(y, z, \alpha(x-y), r^2) - F(x, z, \alpha(x-y), r^1) \le \omega(\alpha|x-y|^2 + |x-y|) \tag{7.24}$$

*for all $x$ and $y$ in $\overline{\Omega}$, all $z \in \mathbb{R}$, and all $\alpha > 0$ and $r^1$ and $r^2$ in $\mathbb{S}^n$ such that*

$$-3\alpha \begin{pmatrix} I & 0 \\ 0 & I \end{pmatrix} \le \begin{pmatrix} r^1 & 0 \\ 0 & -r^2 \end{pmatrix} \le 3\alpha \begin{pmatrix} I & -I \\ -I & I \end{pmatrix}. \tag{7.25}$$

*If $u \in USC(\overline{\Omega})$ is a viscosity subsolution of $F = 0$ in $\Omega$, and if $v \in LSC(\overline{\Omega})$ is a viscosity supersolution of $F = 0$ in $\Omega$ with $u \le v$ on $\partial\Omega$, then $u \le v$ in $\overline{\Omega}$.*

***Proof.*** If $u \not\le v$ in $\overline{\Omega}$, then there is a point $\tilde{x} \in \Omega$ such that $u - v$ attains its positive maximum value over $\overline{\Omega}$ at $\tilde{x}$. We then use Lemma 7.11 with $u_1 = u$ and $u_2 = -v$, observing that $\bar{J}^{2,-}v(x) = -\bar{J}^{2,+}(-v(x))$ for any $v \in \Omega$ and that $\overline{\Omega}$ is locally compact. □

## 7.5 A test function construction for the oblique derivative problem

In order to apply the arguments of the preceding section to the oblique derivative problem, we first construct a suitable test function. This construction takes several steps.

Our first step is to solve a Hamilton-Jacobi equation. In what follows, we shall look at functions of two multi-dimensional variables, one in $\mathbb{R}^n$ and one in $\mathbb{R}^m$, with $n$ and $m$ positive integers. For ease of notation, we use $x$ as dummy variable in $\mathbb{R}^n$, and we use $q$ as a dummy variable in $\mathbb{R}^m$.

**Lemma 7.13.** *Let $U$ be a bounded open set in $\mathbb{R}^n$, let $m \geq 1$, and let $H$ be a $C^2(U \times \mathbb{R}^m)$ function. Suppose that there is a nonnegative constant $A$ such that*

$$|H(x,q)| \leq A(|q|^2+1)^{1/2}, \quad |H_x(x,q)| \leq A(|q|^2+1)^{1/2}, \tag{7.26a}$$

$$|H_q(x,q)| \leq A, |H_{qq}(x,q)| \leq \frac{A}{(|q|^2+1)^{1/2}}. \tag{7.26b}$$

*Suppose also that there is a nonnegative constant $C_0$ such that*

$$|H_{xx}(x,q)| \leq C_0(|q|+1), \quad |H_{xq}(x,q)| \leq C_0. \tag{7.27}$$

*If $\delta \in (0, 4/A)$, then there is a function $\psi \in C^2(U \times V_\delta)$, where*

$$V_\delta = \{(y,t) \in \mathbb{R}^{m+1} : |t| \leq \delta(|y|^2+1)^{1/2}\} \tag{7.28}$$

*such that*

$$\psi_t(x,y,t) + H(x, \psi_y(x,y,t)) = 0 \tag{7.29}$$

*for all $(x,y,t) \in U \times V_\delta$. Moreover*

$$\psi(x,0,0) \leq 1 \tag{7.30}$$

*for all $x \in U$,*

$$\psi(x,y,t) \geq \frac{1}{4}(|y|^2+1) \tag{7.31}$$

*for all $(x,y,t) \in U \times V_\delta$, and there is a constant $C_1$, determined only by $A$ and $C_0$ such that*

$$|\psi(x,y,t)| + |\psi_x(x,y,t)| + |\psi_{xx}(x,y,t)| \leq C_1(|y|^2+1), \tag{7.32a}$$

$$|\psi_y(x,y,t)| + |\psi_t(x,y,t)| + |\psi_{xy}(x,y,t)| + |\psi_{xt}(x,y,t)| \leq C_1(|y|+1), \tag{7.32b}$$

$$|\psi_{yy}(x,y,t)| + |\psi_{yt}(x,y,t)| + |\psi_{tt}(x,y,t)| \leq C_1. \tag{7.32c}$$

***Proof.*** First, we fix $(x,y,t) \in U \times V_\delta$ and define $J$ on $\mathbb{R}^m$ by

$$J(q) = \frac{1}{2}|q-y|^2 + tH(x,q).$$

Since $J$ is continuous, and $H$ grows linearly as $|q| \to \infty$, it follows that there is a $p \in \mathbb{R}^m$ at which $J$ attains its minimum. In addition $J_q(p) = 0$, so

$$y = p + tH_q(x,p). \tag{7.33}$$

It follows from (7.26) and the definition of $V_\delta$ that

$$|y-p| = |t||H_q(x,p)| \leq A|t| \leq A\delta(|y|^2+1)^{1/2}. \tag{7.34}$$

In addition,

$$\left|(|p|^2+1)^{1/2}-(|y|^2+1)^{1/2}\right|\le|y-p|,$$

so

$$(1-A\delta)(|y|^2+1)^{1/2}\le(|p|^2+1)^{1/2}\le(1+A\delta)(|y|^2+1)^{1/2}. \tag{7.35}$$

Our restriction on $\delta$ then implies that

$$\frac{3}{4}(|y|^2+1)^{1/2}\le(|p|^2+1)^{1/2}\le\frac{5}{4}(|y|^2+1)^{1/2}. \tag{7.36}$$

From (7.26), we infer that

$$|tH_{qq}(x,p)|\le\frac{\delta(|y|^2+1)^{1/2}A}{(|p|^2+1)^{1/2}},$$

and then from (7.36) and our restriction on $\delta$, we have

$$|tH_{qq}(x,p)|\le\frac{1}{2}, \tag{7.37a}$$

$$\left|(I+H_{qq}(x,p))^{-1}\right|\le 2. \tag{7.37b}$$

Before using these calculations to construct $\psi$, we observe that, if $r$ is another element of $\mathbb{R}^m$ that minimizes $J$, then

$$r+tH_q(x,r)=p+tH_q(x,p).$$

It follows that

$$|p-r|=|t||H_q(x,p)-H_q(x,r)|\le\frac{1}{2}|p-r|$$

by virtue of (7.37a), and hence $p$ is unique. In other words, there is a function $p(x,y,t)$ which solves (7.33), and the implicit function theorem implies that $p\in C^1(U\times V_\delta)$.

We now define $\psi$ by

$$\psi(x,y,t)=\frac{1}{2}(|y|^2+1)-\frac{1}{2}|p(x,y,t)-y|^2-tH(x,p(x,y,t)).$$

Direct calculation of the partial derivatives of $\psi$ along with (7.33) gives us (7.29) and

$$\psi_x(x,y,t)=-tH_x(x,y,t), \tag{7.38a}$$

$$\psi_y(x,y,t)=p(x,y,t). \tag{7.38b}$$

Moreover, we have

$$\psi(x,0,0)=\frac{1}{2}-\frac{1}{2}|p(x,0,0)|^2$$

which easily implies (7.30). In addition, $J(p) \le J(y) = tH(x,y)$, so (7.26) implies that

$$\psi(x,y,t) \ge \frac{1}{2}(|y|^2+1) - tH(x,y) \ge \left(\frac{1}{2} - A\delta\right)(|y|^2+1),$$

which implies (7.31) because of our restriction on $\delta$.

Since

$$\psi(x,y,t) \le \frac{1}{2}(|y|^2+1) - tH(x,p(x,y,t)),$$

the estimates on $\psi$, $\psi_x$, $\psi_y$, and $\psi_t$ in (7.32) now follow from (7.29), (7.38), and (7.35).

Finally, we differentiate (7.33) with respect to $x$, $y$, and $t$ to deduce that

$$\begin{aligned} p_x(x,y,t) &= -t(I + tH_{qq}(x,p(x,y,t))^{-1}H_{qx}(x,p(x,y,t)), \\ p_y(x,y,t) &= (I + tH_{qq}(x,y,t))^{-1}, \\ p_t(x,y,t) &= -(I + tH_{qq}(x,y,t))^{-1}H_q. \end{aligned}$$

It follows that

$$|p_x| \le 2C_0(|y|+1), \quad |p_y| \le 2, \quad |p_x| \le 2A,$$

and the second derivative estimates for $\psi$ are obtained by differentiating the equations for the first derivatives of $\psi$ and combining the first derivative estimates for $p$ with (7.27), (7.26), and our restriction on $\delta$. □

Our next step is to construct a function like $\psi$ but defined for all $(y,t) \in \mathbb{R}^{m+1}$. To simplify writing, we abbreviate $(y,t)$ to $\xi$.

**Lemma 7.14.** *With $U$, $H$, $\delta$, and $m$ as in Lemma 7.13, there is a function $\varphi \in C^2(U \times \mathbb{R}^{m+1})$ such that*

$$\varphi(x,0) \le 1 \tag{7.39}$$

*for all $x \in U$,*

$$\varphi(x,\xi) \ge \frac{1}{8}|\xi|^2 \tag{7.40}$$

*for all $(x,\xi) \in U \times \mathbb{R}^{m+1}$, and there are constants $\tilde\delta$ and $C_2$, determined only by $A$ and $C_0$, such that*

$$\varphi_t(x,\xi) + H(x,\varphi_\xi(x,\xi)) \ge 0 \text{ if } t \ge -\tilde\delta|y|, \tag{7.41a}$$

$$\varphi_t(x,\xi) + H(x,\varphi_\xi(x,\xi)) \le 0 \text{ if } t \le \tilde\delta|y|, \tag{7.41b}$$

*and*

$$|\varphi(x,\xi)| + |\varphi_x(x,y,t)| + |\varphi_{xx}(x,y,t)| \le C_2(|\xi|^2+1), \tag{7.42a}$$

$$|\varphi_\xi(x,\xi)| + |\varphi_{x\xi}(x,\xi)| \le C_2(|y|+1), \tag{7.42b}$$

$$|\varphi_{\xi\xi}(x,y,t)| \le C_2. \tag{7.42c}$$

***Proof.*** Let $f \in C^\infty(\mathbb{R})$ satisfy $0 \le f(r) \le 1$ and $rf'(r) \le 0$ for all $r \in \mathbb{R}$, $f(r) = 1$ if $|r| \le 1$, and $f(r) = 0$ if $|r| \ge 2$. With $g$ defined by $g(r) = 1 - f(2r)$ and $\alpha = \min\{1, \delta/3\}$, we define $f_\alpha$ and $g_\alpha$ by

$$f_\alpha(y,t) = f\left(\frac{t}{\alpha(|y|^2+1)^{1/2}}\right), \qquad g_\alpha(y,t) = g\left(\frac{t}{\alpha(|y|^2+1)^{1/2}}\right).$$

Note in particular that $f_\alpha$ has support in $V_\delta$. With $\beta > 0$ to be chosen, we then set

$$\varphi(x,y,t) = f_\alpha(y,t)\psi(x,y,t) + g_\alpha(y,t)\beta t^2.$$

Since $g_\alpha(0,0) = 0$ and $f_\alpha(0,0) = 1$, (7.39) follows from (7.30).

Next, if $t \le \alpha(|y|^2+1)^{1/2}$, then $f_\alpha(y,t) = 1$, so

$$\varphi(x,y,t) \ge \psi(x,y,t) \ge \frac{1}{4}(|y|^2+1) \ge \frac{1}{8}|y|^2 + \frac{1}{8\alpha^2}t^2 \ge \frac{1}{8}|\xi|^2$$

because $\alpha \le 1$. On the other hand, if $t > \alpha(|y|^2+1)^{1/2}$, then $g_\alpha(y,t) = 1$, so

$$\varphi(x,y,t) \ge \beta t^2 \ge \frac{\beta}{2}t^2 + \frac{\beta\alpha^2}{2}|y|^2 \ge \frac{1}{8}|\xi|^2$$

provided $\beta \ge 1/(4\alpha^2)$. With this restriction on $\beta$, (7.40) holds.

To verify the remaining inequalities, we first compute

$$\frac{\partial f_\alpha}{\partial t}(y,t) = \frac{1}{\alpha(|y|^2+1)^{1/2}}f', \quad \frac{\partial g_\alpha}{\partial t}(y,t) = \frac{1}{\alpha(|y|^2+1)^{1/2}}g',$$
$$\frac{\partial f_\alpha}{\partial y}(y,t) = -\frac{ty}{\alpha(|y|^2+1)^{3/2}}f', \quad \frac{\partial g_\alpha}{\partial y}(y,t) = -\frac{ty}{\alpha(|y|^2+1)^{3/2}}g',$$

where, here and below, $f'$ and $g'$ are always evaluated at $t/(\alpha(|y|^2+1)^{1/2})$.

When $|t| \le \alpha(|y|^2+1)^{1/2}/2$, then $\varphi(x,y,t) = \psi(x,y,t)$ so $\psi_t(x,y,t) + H(x,\psi_y(x,y,t)) = 0$, so (7.41a) and (7.41b) hold in this case.

When $\alpha(|y|^2+1)^{1/2}/2 \le t \le \alpha(|y|^2+1)^{1/2}$, we have

$$\begin{aligned}
\varphi_t(x,y,t) + H(x,\varphi_y(x,y,t)) &= \psi_t(x,y,t) + 2\beta t g_\alpha(y,t) + \frac{\beta t^2}{\alpha(|y|^2+1)^{1/2}} \\
&\quad + H(x, \psi_y(x,y,t) - \beta t^2 \frac{ty}{\alpha(|y|^2+1)^{3/2}}g') \\
&\ge \psi_t(x,y,t) + H(x,\psi_y(x,y,t)) \\
&\quad + \frac{\beta t^2}{\alpha(|y|^2+1)^{1/2}}g'[1 - A\frac{t|y|}{|y|^2+1}]
\end{aligned}$$

because $|H_q| \le A$ and $g_\alpha$ and $g'$ are nonnegative. By construction, $\psi_t(x,y,t) + H(x,\psi_y(x,y,t)) = 0$, and we have

$$A\frac{t|y|}{|y|^2+1} \le A\alpha \le A\delta \le 1,$$

so (7.41a) holds in this case as well. A similar argument shows that (7.41b) holds for $\alpha(|y|^2+1)^{1/2}/2 \le -t \le \alpha(|y|^2+1)^{1/2}$.

If $t \ge \alpha(|y|^2+1)^{1/2}$, then $g_\alpha(y,t)=1$ and $g'=0$, so

$$\begin{aligned}\varphi_t(x,y,t)+H(x,\varphi_y(x,y,t)) &= \psi_t(x,y,t)+\frac{1}{\alpha(|y|^2+1)^{1/2}}\psi f'+2\beta t\\ &\quad + H(x,f_\alpha(y,t)\psi_y(x,y,t)+f'\psi(x,y,t))\\ &\ge (|y|^2+1)^{1/2}[2\beta\alpha - C_1 - \frac{F}{\alpha} - 8C_1F]\end{aligned}$$

by virtue of (7.32). (Here $F=\sup|f'|$; note that we can take $F\le 2$.) If $\beta$ is sufficiently large (determined only by $\alpha$, $C_1$ and $\sup|f'|$), we obtain (7.41a) in this case, and a similar argument gives (7.41b) when $t \le -\alpha(|y|^2+1)^{1/2}$. We therefore have (7.41) with $\tilde\delta = \alpha/2$.

Finally, since the derivatives of $f_\alpha$ and $g_\alpha$ vanish if $|t| \le \frac{1}{2}\alpha(|y|^2+1)^{1/2}$ or $|t| \ge 2\alpha(|y|^2+1)^{1/2}$, it follows that

$$\left|\frac{\partial f_\alpha}{\partial \xi}(x,y,t)\right| + \left|\frac{\partial g_\alpha}{\partial \xi}(x,y,t)\right| \le \frac{C}{|y|+|t|+1}$$

with a corresponding estimate for the second derivatives of $f_\alpha$ and $g_\alpha$. Combining these estimates with (7.32) gives (7.42). □

From this construction, we now produce our test function. At this stage, it will be convenient to work with a function more like our function $b$ in the boundary condition.

**Theorem 7.15.** *Let $\Omega$ be a bounded open subset of $\mathbb{R}^n$ with $\partial\Omega \in C^1$, let $x_0 \in \partial\Omega$, let $U$ be a neighborhood of $x_0$, and suppose that $\hat b$ is a real-valued function defined on $U \times \mathbb{R}^n$ such that*

$$\gamma(x_0)\cdot \hat b_p(x,p) \ge 1, \tag{7.43a}$$

$$|\hat b_p(x,p)| \le B_0, \tag{7.43b}$$

$$|\hat b_x(x,p)| + |\hat b_{xx}(x,p)| \le B_0(|p|+1), \tag{7.43c}$$

$$|\hat b_{xp}(x,p)| \le B_0,\ |\hat b_{pp}(x,p)| \le \frac{B_0}{|p|+1}, \tag{7.43d}$$

*for some positive constant $B_0$ and all $(x,p) \in U \times \mathbb{R}^n$. Then there are positive constants $C$ and $\tilde\delta$, determined only by $\Omega$ and $B_0$ such that, if $|\gamma(x)-\gamma(x_0)| \le \tilde\delta$ for all $x \in U \cap \partial\Omega$, then, for any $\varepsilon > 0$, there is a function $w_\varepsilon \in C^2(U\times U)$ such that*

$$w_\varepsilon(x,x) \le \varepsilon \tag{7.44}$$

*for all* $x \in U$,

$$w_\varepsilon(x,y) \geq \frac{|x-y|^2}{8\varepsilon} \tag{7.45}$$

*for all* $(x,y) \in U \times U$,

$$\hat{b}(x, w_{\varepsilon,x}(x,y)) \geq -C\left(\frac{|x-y|^2}{\varepsilon} + \varepsilon\right) \tag{7.46}$$

*for all* $x \in \partial\Omega \cap U$ *and* $y \in \overline{\Omega} \cap U$,

$$\hat{b}(x, -w_{\varepsilon,y}(x,y)) \leq C\left(\frac{|x-y|^2}{\varepsilon} + \varepsilon\right) \tag{7.47}$$

*for all* $x \in \overline{\Omega} \cap U$ *and* $y \in \partial\Omega \cap U$,

$$|w_{\varepsilon,y}(x,y)| \leq C\left(\frac{|x-y|}{\varepsilon} + 1\right), \tag{7.48a}$$

$$|w_{\varepsilon,x}(x,y) + w_{\varepsilon,x}(x,y)| \leq C\left(\frac{|x-y|^2}{\varepsilon} + \varepsilon\right), \tag{7.48b}$$

$$\left(Dw_\varepsilon(x,y), \frac{C}{\varepsilon}\begin{pmatrix} I & -I \\ -I & I \end{pmatrix} + C\left(\frac{|x-y|^2}{\varepsilon} + \varepsilon\right)\begin{pmatrix} I & 0 \\ 0 & I \end{pmatrix}\right) \in J^{2,+}w_\varepsilon(x,y) \tag{7.48c}$$

*for all* $(x,y) \in U \times U$.

***Proof.*** First, by rotation, we may assume that $\gamma(x_0) = e_n$. Then, conditions (7.43a) and (7.43b) imply that we can define a function $H$ implicitly by the equation

$$\hat{b}(x, p', -H(x,p')) = 0,$$

where $p' = (p_1, \ldots, p_{n-1})$, noting that conditions (7.43) imply that $H$ satisfies the hypotheses of Lemma 7.13 with $A$ and $C_0$ determined only by $B_0$ and $m = n-1$. We now take $\tilde{\delta}$ to be the constant given by Lemma 7.14, $\psi$ to be the function from Lemma 7.14, and define $w_\varepsilon$ by

$$w_\varepsilon(x,y) = \varepsilon\varphi\left(x, \frac{x-y}{\varepsilon}\right).$$

(Here, $y$ denotes a point in $\mathbb{R}^n$. In this proof, $\xi$ will be the same $n$-dimensional quantity as in Lemma 7.14, but $y$ is not the first $n-1$-dimensional part of $\xi$.)

With this choice, (7.44) follows from (7.39) and (7.45) follows form (7.40).

Next, we observe that $\hat{b}(x,p) \ge 0$ if and only if $p_n + H(x,p') \ge 0$. It then follows from (7.43b) that

$$\begin{aligned}\hat{b}(x, w_{\varepsilon,x}(x,y)) &= \hat{b}(x, \varepsilon\varphi_x\left(x, \frac{x-y}{\varepsilon}\right) + \varphi_\xi\left(x, \frac{x-y}{\varepsilon}\right)) \\ &\ge \hat{b}(x, \varphi_\xi\left(x, \frac{x-y}{\varepsilon}\right)) - B_0\varphi_x\left(x, \frac{x-y}{\varepsilon}\right).\end{aligned}$$

Since $|\gamma(x) - \gamma(x_0)| \le \tilde{\delta}$, it follows that

$$\frac{x^n - y^n}{\varepsilon} \ge -\tilde{\delta}\left|\frac{x' - y'}{\varepsilon}\right|$$

if $x \in \partial\Omega \cap U$ and $y \in \overline{\Omega} \cap U$, so (7.41a) implies that

$$\hat{b}(x, w_{\varepsilon,x}(x,y)) \ge -B_0\varphi_x\left(x, \frac{x-y}{\varepsilon}\right),$$

and (7.46) is immediate. A similar argument gives (7.47).

We also infer (7.48a) and (7.48b) from (7.42) along with the observation that

$$w_{\varepsilon,x}(x,y) + w_{\varepsilon,y}(x,y) = \varepsilon\varphi_x\left(x, \frac{x-y}{\varepsilon}\right).$$

For our final estimate, we first infer from Taylor's theorem and (7.42) that

$$\begin{aligned}\varphi(x+h, \xi+k) \le p\cdot h + q\cdot k &\\ &+ C[(|\xi|^2+1)|h|^2 + (|\xi|+1)|h||k| + |k|^2] \\ &+ o(|h|^2 + |k|^2)\end{aligned}$$

for

$$p = \varphi_x(x,\xi), \quad q = \varphi_\xi(x,\xi).$$

Taking $\xi = (x-y)/\varepsilon$ and $k = (h-K)/\varepsilon$ in this inequality yields

$$\begin{aligned}w_\varepsilon(x+h, y+K) - w_\varepsilon(x,y) &\le p\cdot h + q\cdot(h-K) \\ &\quad + C[s|h|^2 + \frac{|h-K|^2}{\varepsilon}] + o(|h|^2 + |h-K|^2),\end{aligned}$$

for

$$s = \frac{|x-y|^2}{\varepsilon} + \varepsilon.$$

We then observe that

$$s|h|^2 + \frac{|h-K|^2}{\varepsilon} = s(|h|^2 + |K|^2) + \frac{1}{\varepsilon}\begin{pmatrix} I & -I \\ -I & I\end{pmatrix}\begin{pmatrix} h \\ K\end{pmatrix}\begin{pmatrix} h \\ K\end{pmatrix}$$

to infer (7.48c). □

## 7.6 The comparison principle for oblique derivative problems

We are now almost ready to prove the comparison principle for viscosity sub- and supersolutions of oblique derivative problems. We just need two more steps. The first step is a modified version of Lemma 7.11.

**Lemma 7.16.** *Let $S_1$ and $S_2$ be locally compact subsets of $\mathbb{R}^n$, let $u_i \in USC(S_i)$ $(i = 1, 2)$, and define $w$ on $S = S_1 \times S_2$ by*

$$w(x, y) = u_1(x) + u_2(y).$$

*Suppose that there are vectors $p^1$ and $p^2$ in $\mathbb{R}^n$, points $\hat{x}_1$ and $\hat{x}_2$ in $\mathbb{R}^n$ and positive constants $\theta_1$ and $\theta_2$ such that*

$$\left(p^1, p^2, \theta_1 \begin{pmatrix} I & -I \\ -I & I \end{pmatrix} + \theta_2 \begin{pmatrix} I & 0 \\ 0 & I \end{pmatrix}\right) \in J^{2,+}w(\hat{x}_1, \hat{x}_2).$$

*Then there are matrices $r^1$ and $r^2$ in $\mathbb{S}^n$ such that*

$$-3\theta_1 \begin{pmatrix} I & 0 \\ 0 & I \end{pmatrix} \le \begin{pmatrix} r^1 - \theta_1 I & 0 \\ 0 & r^2 - \theta_2 I \end{pmatrix} \le 3\theta_1 \begin{pmatrix} I & -I \\ -I & I \end{pmatrix} \tag{7.49}$$

*and*

$$(p^i, r^i) \in \overline{J}^{2,+} u_i(\hat{x}_i)$$

*for $i = 1, 2$.*

***Proof.*** As in the proof of Lemma 7.11, we may assume that $\hat{x}_1 = \hat{x}_2 = 0$. We then define $\hat{u}_i$ by $\hat{u}_i(x_i) = u_i(x_i) - p^i \cdot x_i$ and, for each $\eta > 0$, we define $v_{i,\eta}$ for $\eta > 0$ by

$$v_{i,\eta}(x) = \hat{u}_i(x) - \frac{\eta}{2}|x|^2.$$

It's easy to see that $v_{i,\eta}(0) = 0$ for all $i$ and $\eta$ and that $w$, defined by

$$w(x_1, x_2) = v_1(x_1) + v(x_2) - \frac{1}{2} A_{ij} x^i x^j$$

has a local minimum at 0. It follows from Lemma 7.10 that, for each $\eta > 0$, there are matrices $r^{i,\eta}$ such that (7.49) holds with $r^{i,\eta}$ in place of $r^i$ and $(0, r^{i,\eta}) \in \overline{J}^{2,+} v_i(0)$. Using (7.49) again, we infer that there is a sequence $(\eta(j))$ with $\eta(j) \to 0$ as $j \to \infty$ such that $(r^{i,\eta(j)})$ converges to some matrix $r^i$, which must also satisfy (7.49). It is easily verified that $(p^i, r^i) \in \overline{J}^{2,+} u_i(\hat{x}_i)$. □

Our next step is define viscosity subsolutions and supersolutions. We say that $u \in USC(\overline{\Omega})$ is a *viscosity subsolution* of $F = 0$ in $\Omega$, $b = 0$ on $\partial\Omega$ if

$$F(x, u(x), p, r) \geq 0$$

for all $x \in \Omega$ and all $(p, r) \in J^{2,+}_{\overline{\Omega}} u(x)$, and

$$\max\{F(x, u(x), p, r), b(x, u(x), p)\} \geq 0$$

for all $x \in \partial\Omega$ and all $(p, r) \in J^{2,+}_{\overline{\Omega}} u(x)$. (Note that for $x \in \Omega$, we have $J^{2,+}_{\overline{\Omega}} u(x) = J^{2,+}_{\Omega} u(x)$.) Similarly, we say that $u \in LSC(\overline{\Omega})$ is a *viscosity supersolution* of $F = 0$ in $\Omega$, $b = 0$ on $\partial\Omega$ if

$$F(x, u(x), p, r) \leq 0$$

for all $x \in \Omega$ and all $(p, r) \in J^{2,-}_{\overline{\Omega}} u(x)$, and

$$\min\{F(x, u(x), p, r), b(x, u(x), p)\} \leq 0$$

for all $x \in \partial\Omega$ and all $(p, r) \in J^{2,-}_{\overline{\Omega}} u(x)$. The important feature of this definition is that the boundary condition need not be satisfied on the boundary.

Our comparison principle now follows via a suitable modification of the argument used to prove Theorem 7.12.

**Theorem 7.17.** *Let $\Omega$ be a bounded open subset of $\mathbb{R}^n$ with $\partial\Omega \in C^1$, let $b \in C(\partial\Omega \times \mathbb{R} \times \mathbb{R}^n)$ and let $F \in C(\overline{\Omega} \times \mathbb{R} \times \mathbb{R}^n \times \mathbb{S}^n)$. Suppose that $b$ is twice continuously differentiable with respect to $x$ and $p$. Suppose also that, for each $(x_0, z) \in \partial\Omega \times \mathbb{R}$, there are a neighborhood $U$ of $x_0$ and a positive constant $\chi$ such that*

$$\gamma(x_0) \cdot b_p(x, z, p) \geq \chi, \tag{7.50}$$

*and that there is a nonnegative constant $B_0$ such that*

$$|b_x(x, z, p)| + |b_{xx}(x, z, p)| \leq B_0\chi[|p| + 1], \tag{7.51a}$$

$$|b_p(x, z, p)| + |b_{px}(x, z, p)| \leq B_0, \tag{7.51b}$$

$$|b_{pp}(x, z, p)| \leq \frac{B_0\chi}{|p| + 1} \tag{7.51c}$$

*for all $(x, p) \in U \times \mathbb{R}^n$ such that $b(x, z, p) = 0$. Suppose further that there is a positive constant such that condition (7.23) holds for all $(x, z_i, p, r^i) \in \overline{\Omega} \times \mathbb{R} \times \mathbb{R}^n \times \mathbb{S}^n$ with $z_1 \geq z_2$ and $r^1 \leq r^2$. Suppose finally that there are a continuous increasing function $\omega$, defined on $[0, \infty)$ with $\omega(0) = 0$ and a positive constant $\delta_0$ such that*

$$F(y, z, p, r^2) - F(x, z, p, r^1) \leq \omega(|x - y|[|p| + 1] + \alpha|x - y|^2) \tag{7.52}$$

*for all $x$ and $y$ in $\overline{\Omega}$, all $z \in \mathbb{R}$, and all $\alpha > 0$ and $r^1$ and $r^2$ satisfying (7.25), and*

$$|F(x, z, p^1, r^1) - F(x, z, p^2, r^2)| \le \omega(|p^1 - p^2| + |r^1 - r^2|) \tag{7.53}$$

*for all $x \in \overline{\Omega}$ with $d(x) < \delta_0$, all $z \in \mathbb{R}$, all $p^1$ and $p^2$ in $\mathbb{R}^n$, and all $r^1$ and $r^2$ in $\mathbb{S}^n$. If $u \in USC(\overline{\Omega})$ is a viscosity subsolution of $F = 0$ in $\Omega$, $b = 0$ on $\partial\Omega$ and if $v \in LSC(\overline{\Omega})$ is a viscosity supersolution of $F = 0$ in $\Omega$, $b = 0$ on $\partial\Omega$, then $u \le v$ in $\overline{\Omega}$.*

***Proof.*** By virtue of Theorem 7.12, if $u - v$ has a positive maximum, it must occur at some $x_0 \in \partial\Omega$. We use $\sigma$ to denote the positive maximum value.

To proceed, we fix $z \in (v(x_0), u(x_0))$, rotate axes so that $\gamma(x_0) = e_n$, and define $H$ implicitly by

$$b(x, z, p', -H(x, p')) = 0.$$

We then decrease $U$ so that it satisfies the conditions of Theorem 7.15. With $\theta$ and $\varepsilon$ positive constants at our disposal, we define functions $\bar{u}$, $\bar{v}$, and $\Phi$ by

$$\begin{aligned}
\bar{u}(x) &= u(x) + \theta^2\gamma(x_0) \cdot (x - x_0) - \theta|x - x_0|^2, \\
\bar{v}(x) &= v(x) - \theta^2\gamma(x_0) \cdot (x - x_0), \\
\Phi(x, y) &= \bar{u}(x) - \bar{v}(y) - w_\varepsilon(x, y).
\end{aligned}$$

If $(x_1, y_1)$ is a point in $(U \cap \overline{\Omega}) \times (U \cap \overline{\Omega})$ at which $\Phi$ attains its maximum, then

$$\sigma - \varepsilon \le \Phi(x_0, x_0) \le \Phi(x_1, y_1).$$

Since

$$-\theta|x_1 - x_0|^2 = -\theta|x_1 - x_0 + \theta\gamma(x_0)|^2 - \theta|\theta\gamma(x_0)|^2 - 2\theta^2(x_1 - x_0) \cdot \gamma(x_0)$$

and $\theta|\theta\gamma(x_0)|^2 = \theta^3$, it follows that

$$\Phi(x_1, y_1) = u(x_1) - v(y_1) - \theta|x_1 - x_0 + \theta\gamma(x_0)|^2 - \theta^3 + \theta(y_1 - x_1) \cdot \gamma(x_0) - w_\varepsilon(x_1, y_1).$$

Moreover, because $\hat{u} - \hat{v}$ is bounded from above on $(U \cap \overline{\Omega}) \times (U \cap \overline{\Omega})$, it follows that, for fixed $\theta$, we have

$$\lim_{\varepsilon \to 0^+} |x_1 - y_1| = 0$$

and

$$\limsup_{\varepsilon \to 0^+} \frac{|x_1 - y_1|^2}{\varepsilon} \le \theta^3, \tag{7.54a}$$

$$\lim_{\varepsilon \to 0^+} x_1 = \theta\gamma(x_0) + x_0. \tag{7.54b}$$

Simple algebra now shows that, if $x \in \Omega$ satisfies $|x - x_0 - \theta\gamma(x_0)| \le \theta$, then $|x - x_0| = o(\theta)$ as $\theta \to 0^+$. Therefore, there is a positive $\rho$ such that, for $x$ and $y$ in $\overline{\Omega} \cap B(x_0, \rho)$, the inequality $u(x) - v(y) > \sigma - \rho$ implies that $u(x) \ge z \ge v(y)$.

We now take $\varepsilon$ sufficiently small that $|x_1 - x_0| = o(\theta)$ if $x_1 \in \partial\Omega$ and $|y_1 - x_0| = o(\theta)$ if $y_1 \in \partial\Omega$. We also take $\theta$ sufficiently small that $x_1$ and $y_1$ are in the interior of $U$ and $u(x_1) \le z \le v(y_1)$. Since $\Phi(x, y) \le \Phi(x_1, y_1)$ for $x$ and $y$ in $U \cap \overline{\Omega}$, we conclude that

$$\bar{u}(x) - \bar{v}(y) \le \bar{u}(x_1) - \bar{v}(y_1) + w_\varepsilon(x, y) - w_\varepsilon(x_1, y_1)$$

for $x$ and $y$ in $U \cap \overline{\Omega}$. Setting

$$p = w_{\varepsilon,x}(x_1, y_1), \quad q = -w_{\varepsilon,y}(x_1, y_1),$$
$$s = \frac{|x_1 - y_1|^2}{\varepsilon} + \varepsilon,$$

we then infer from (7.48c) that

$$\begin{aligned}\bar{u}(x) - \bar{v}(y) \le \bar{u}(x_1) &- \bar{v}(y_1) + p \cdot (x - x_1) + q \cdot (y - y_1)\\ &+ \frac{C}{2}\left(\frac{|(x - x_1) - (y - y_1)|^2}{\varepsilon} + s[|x - x_1|^2 + |y - y_1|^2]\right)\\ &+ o(|x - x_1|^2 + |y - y_1|^2).\end{aligned}$$

If we define $\bar{w}$ by $\bar{w}(x, y) = \bar{u}(x) - \bar{v}(y)$, it follows that

$$\left(p, q, \frac{C}{\varepsilon}\begin{pmatrix} I & -I \\ -I & I\end{pmatrix} + Cs\begin{pmatrix} I & 0 \\ 0 & I\end{pmatrix}\right) \in J^{2,+}\bar{w}(x_1, y_1),$$

and hence Lemma 7.16 gives matrices $r^1$ and $r^2$ such that

$$-\frac{C}{\varepsilon}\begin{pmatrix} I & 0 \\ 0 & I\end{pmatrix} \le \begin{pmatrix} r^1 - CsI & 0 \\ 0 & r^2 - CsI\end{pmatrix} \le \frac{C}{\varepsilon}\begin{pmatrix} I & -I \\ -I & I\end{pmatrix} \tag{7.55}$$

and

$$(p, r^1) \in \overline{J}^{2,+}\bar{u}(x_1), \quad (-q, -r^2) \in \overline{J}^{2,-}\bar{v}(y_1).$$

It follows that

$$(p + \theta^2\gamma(x_0) + 2\theta(x_1 - x_0), r^1 + 2\theta I) \in \overline{J}^{2,+}u(x_1) \tag{7.56}$$

and

$$(-q - \theta^2\gamma(x_0), -r^2) \in \overline{J}^{2,-}v(y_1). \tag{7.57}$$

By using (7.51), (7.50), (7.46), and (7.54), we infer that

$$b(x_1, u(x_1), p + \theta^2\gamma(x_0) + 2\theta(x_1 - x_0)) \ge \theta^2 - C\theta o(\theta) - C\varepsilon$$

if $x_1 \in \partial\Omega$, and hence, if we choose $\theta$ sufficiently small and then $\varepsilon$ sufficiently small, we conclude that

$$b(x_1, u(x_1), p + \theta^2\gamma(x_0) + 2\theta(x_1 - x_0)) > 0.$$

A similar argument shows that, if $y_1 \in \partial\Omega$ and if we chose first $\theta$ and then $\varepsilon$ sufficiently small, then

$$b(y_1, u(y_1), -q - \theta^2\gamma(x_0)) < 0.$$

The definitions of viscosity subsolution and viscosity supersolution then imply that

$$F(x_1, u(x_1), p + \theta^2\gamma(x_0) + 2\theta(x_1 - x_0), r^1) \leq 0 \tag{7.58}$$

and

$$F(y_1, u(y_1), -q - \theta^2\gamma(x_0), -r^2) \geq 0 \tag{7.59}$$

whether or not $x_1$ or $y_1$ is in $\partial\Omega$.

We now use (7.53) and (7.58), along with the observation that (7.48b) implies that $|p + q| \leq Cs$, to infer that

$$\omega(Cs + C\theta) \geq F(x_1, u(x_1), -q - \theta^2\gamma(x_0), r^1 - CsI).$$

In addition, we infer from (7.53) and (7.59) that

$$F(y_1, u(y_1), -q - \theta^2\gamma(x_0), -r^2 + CsI) \leq \omega(Cs + C\theta).$$

Combining these two inequalities with (7.23), (7.52), and the inequality $r^1 - CsI \leq r^2 + CsI$ (which follow from (7.55)), we conclude that

$$c_0[u(x_1) - v(y_1)] \leq 2\omega(Cs + C\theta).$$

Once $\varepsilon$ and $\theta$ are sufficiently small, the left side of this inequality is greater than $c_0\sigma/2$ (a fixed positive number) while the right side can be made arbitrarily small. This contradiction shows that $u - v$ cannot have a positive maximum and hence $u \leq v$. □

## 7.7 Existence and uniqueness of viscosity solutions

In this section, we prove that a viscosity solution of

$$F = 0 \text{ in } \Omega, \quad b = 0 \text{ on } \partial\Omega \tag{7.60}$$

exists and is unique. The method of proof is a simple variant of the Perron process described in previous sections for linear problems.

To begin, if $u$ is a function defined on $\overline{\Omega}$ with values in $[-\infty, \infty]$, we define functions $u^*$ and $u_*$ by

$$u^*(x) = \lim_{\delta \to 0+} \sup\{u(y) : y \in \overline{\Omega} \cap B(x,\delta)\},$$
$$u_*(x) = \lim_{\delta \to 0+} \inf\{u(y) : y \in \overline{\Omega} \cap B(x,\delta)\}.$$

Since the idea of the Perron process is to generate a solution as a supremum of subsolutions, we now show that the supremum of a family of viscosity subsolutions generates a viscosity subsolution.

**Lemma 7.18.** *Let $\Omega$ be a bounded, Lipschitz domain, let $F \in USC(\overline{\Omega} \times \mathbb{R} \times \mathbb{R}^n \times \mathbb{S}^n)$, let $b \in USC(\partial\Omega \times \mathbb{R} \times \mathbb{R}^n)$, and let $\mathcal{F}$ be a family of subsolutions of* (7.60). *Define $w$ on $\overline{\Omega}$ by*

$$w(x) = \sup\{u(x) : u \in \mathcal{F}\},$$

*and suppose that $w$ is everywhere finite. Then $w^*$ is a subsolution of* (7.60).

***Proof.*** Fix $x \in \overline{\Omega}$ and let $(p, r) \in J^{2,+}_{\overline{\Omega}} w^*(x)$. It follows that there is a sequence $(x_m, u_m)$ in $\overline{\Omega} \times \mathcal{F}$ such that

$$\lim_{m\to\infty} (x_m, u_m(x_m)) = (x, w^*(x)),$$

and hence, for any $\varepsilon > 0$, there is a $\delta > 0$ such that

$$w^*(y) \le w^*(x) + p \cdot (y - x) + \frac{1}{2} r_{ij}(x-y)^i(x-y)^j + \varepsilon|x-y|^2 \qquad (7.61)$$

for all $y \in \Omega[x, \delta]$.

We now define $v_m$ by

$$v_m(y) = u_m(y) - p \cdot (y - x) + \frac{1}{2} r_{ij}(x-y)^i(x-y)^j + 2\varepsilon|x-y|^2$$

and, for each $m$, there is a point $y_m \in \Omega[x, \delta]$ at which $v_m$ attains its maximum. Since $v_m(y) \le v_m(y_m)$ for all $y \in \Omega[x, \delta]$, it follows that

$$\begin{aligned} u_m(y) \le u_m(y_m) &+ p \cdot (y - y_m) \\ &+ \frac{1}{2}[r_{ij}(y-x)^i(y-x)^j - r_{ij}(y_m-x)^i(y_m-x)^j] \quad (7.62) \\ &+ 2\varepsilon(|y-x|^2 - |x-y_m|^2). \end{aligned}$$

Since $\Omega[x, \delta]$ is compact, it follows that, after possibly extracting a subsequence, the sequence $(y_m)$ has a limit point $y_0 \in \Omega[x, \delta]$. If we now take $y = x_m$ in (7.62) and then send $m \to \infty$, we infer that

$$w^*(x) \le \liminf_{m\to\infty}(u_m(x_m)) - p \cdot (y_0 - x) - \frac{1}{2} r_{ij}(y_0-x)^i(y_0-x)^j - 2\varepsilon|x-y_0|^2.$$

Since

$$\liminf_{m\to\infty}(u_m(x_m)) \le w^*(y_0),$$

we infer, after using (7.61), that

$$w^*(x) \le w^*(x) - \varepsilon|x - y_0|^2,$$

and hence $y_0 = x$. Furthermore, $u_m(y_m) \to w^*(x)$.

Since (7.62) implies that

$$(p^m, r + 4\varepsilon I) \in J^{2,+}_{\overline{\Omega}} u_m(y_m),$$

where $p^m$ is the vector with components

$$p^m_i = p_i + 4\varepsilon(y_m - x)^i + r_{ij}(y_m - x)^j,$$

we conclude that $S$, the set of all $(q, s) \in \mathbb{R}^n \times \mathbb{S}^n$ such that there are sequences $(x_m)$ in $\overline{\Omega}$ and $(p^m, r^m) \in J^{2,+}_{\overline{\Omega}} u_m(x_m)$ with

$$\lim_{m\to\infty}(x_m, u_m(x_m), p^m, r^m) = (x, w^*(x), q, s),$$

contains $(p, r + 4\varepsilon I)$ for any $\varepsilon > 0$. Since $S$ is also closed, it follows that $S$ also contains $(p, r)$.

We now consider two possibilities determined by the sequences $(x_m)$ and $(p^m, r^m)$ such that

$$\lim_{m\to\infty}(x_m, u_m(x_m), p^m, r^m) = (x, w^*(x), p, r).$$

If there are infinitely many values of $m$ for which $F(x_m, u_m(x_m), p^m, r^m) \ge 0$, it follows that $F(x, w^*(x), p, r) \ge 0$. On the other hand, if there are only finitely many values of $m$ for which $F(x_m, u_m(x_m), p^m, r^m) \ge 0$, then only finitely many $x_m$s are in $\Omega$, and hence $b(x_m, u_m(x_m), p^m) \ge 0$ for infinitely many values of $m$, and hence $b(x, w^*(x), p) \ge 0$. It follows that $w^*$ is a subsolution of (7.60). □

We are now ready to prove our existence and uniqueness theorem.

**Theorem 7.19.** *Let $\Omega$ be a bounded open subset of $\mathbb{R}^n$ with $\partial\Omega \in C^1$, let $b \in C(\partial\Omega \times \mathbb{R} \times \mathbb{R}^n)$ and let $F \in C(\overline{\Omega} \times \mathbb{R} \times \mathbb{R}^n \times \mathbb{S}^n)$. Suppose that $b$ is twice continuously differentiable with respect to $x$ and $p$. Suppose also that, for each $(x_0, z) \in \partial\Omega \times \mathbb{R}$, there are a neighborhood $U$ of $x_0$ and a positive constant $\chi$ such that (7.50) holds and that there is a nonnegative constant $B_0$ such that (7.51) holds for all $(x, p) \in U \times \mathbb{R}^n$ such that $b(x, z, p) = 0$. Suppose further that there is a positive constant such that condition (7.23) holds for all $(x, z_i, p, r^i) \in \overline{\Omega} \times \mathbb{R} \times \mathbb{R}^n \times \mathbb{S}^n$ with $z_1 \ge z_2$ and $r^1 \le r^2$.*

*Suppose finally that there are a continuous increasing function $\omega$, defined on $[0,\infty)$ with $\omega(0)=0$ and a positive constant $\delta_0$ such that (7.52) holds for all $x$ and $y$ in $\overline{\Omega}$, all $z\in\mathbb{R}$, and all $\alpha>0$ and $r^1$ and $r^2$ satisfying (7.25), and (7.53) holds for all $x\in\overline{\Omega}$ with $d(x)<\delta_0$, all $z\in\mathbb{R}$, all $p^1$ and $p^2$ in $\mathbb{R}^n$, and all $r^1$ and $r^2$ in $\mathbb{S}^n$. Then there is a unique viscosity solution of (7.60).*

***Proof.*** From (7.23) and the continuity of $F$, we infer that there is a positive constant $M$ such that $F(x,M,0,0)<0<F(x,-M,0,0)$ for all $x\in\overline{\Omega}$. Hence $M$ is a supersolution and $-M$ is a subsolution, so $\mathcal{F}$, the set of all viscosity subsolutions of (7.60) $w$ with $w\ge -M$ in $\overline{\Omega}$ is nonempty, and $W$, given by

$$W(x)=\sup\{w(x):w\in\mathcal{F}\}$$

satisfies $-M\le W\le M$ on $\overline{\Omega}$. It then follows from Lemma 7.18 that $W^*$ is a subsolution of (7.60). Since $W^*\in\mathcal{F}$, it also follows that $W=W^*$.

Our next step is to show that $W_*$ is a viscosity supersolution of (7.60). Suppose to the contrary that $W_*$ is not a viscosity supersolution of (7.60). Then there is $(x,p,r)\in\overline{\Omega}\times\mathbb{R}^n\times\mathbb{S}^n$ with $(p,r)\in J^{2,-}_{\overline{\Omega}}W_*(x)$ such that

$$F(x,W_*(x),p,r)>0$$

(and

$$b(x,W_*(x),p)>0$$

if $x\in\partial\Omega$). By continuity, there is a positive constant $\delta_0$ such that, if $\kappa$, $\delta$, and $\varepsilon$ are in $(0,\delta_0)$, then $u_{\delta,\varepsilon}$, defined by

$$u_{\delta,\varepsilon}(y)=W_*(x)+\delta+p\cdot(y-x)+\frac{1}{2}r_{ij}(y-x)^i(y-x)^j-\varepsilon|y-x|^2,$$

is a viscosity subsolution of $F=0$ in $\Omega[x,\kappa]$, $b=0$ on $\partial\Omega[x,\kappa]$. But

$$\begin{aligned}W(y)&\ge W_*(y)\\&\ge W_*(x)+p\cdot(y-x)+\frac{1}{2}r_{ij}(y-x)^i(y-x)^j+o(|y-x|^2),\end{aligned}$$

so if $\varepsilon$ is sufficiently small, if $\kappa$ taken sufficiently small (determined by $\varepsilon$) and if $\delta=\varepsilon\kappa^2/8$, then

$$W(y)>u_{\delta,\varepsilon}(y)$$

for $|y-x|\in[\kappa/2,\kappa]$. With $\varepsilon$, $\kappa$, and $\delta$ now fixed this way, we infer from Lemma 7.18 that $U$, defined by

$$U(y)=\begin{cases}\max\{W(y),u_{\delta,\varepsilon}(y)\} & \text{if } y\in\overline{\Omega}[x,\kappa],\\ W(y) & \text{if } y\in\overline{\Omega}[x,\kappa],\end{cases}$$

is in $\mathcal{F}$. Moreover, if $(x_m)$ is a sequence in $\overline{\Omega}$ such that

$$x_m \to x, \qquad W(x_m) \to W_*(x),$$

then

$$\lim_{m\to\infty} U(x_m) - W(x_m) = \delta > 0,$$

so there is a point in $\overline{\Omega}$ at which $U > W$. This inequality contradicts the inference that $U \in \mathcal{F}$, and hence $W_*$ is a viscosity supersolution of (7.60).

Theorem 7.17 implies that $W_* \geq W$ and the definition of $W_*$ implies that $W_* \leq W$, so $W = W_*$, and hence $W$ is a viscosity solution of (7.60). Another application of Theorem 7.17 shows that this viscosity solution is unique. □

## Notes

The main part of this chapter was taken from Section 3 of [34], and we refer the reader to that work for a more complete description of the theory and for a good bibliography (up to 1992). Two places where we stray from [34] are in our proof of Aleksandrov's Theorem and the comparison principle for oblique derivative problems. Our proof of Aleksandrov's Theorem is essentially taken from Sections 6.3 and 6.4 of [43] (which, in turn, is based on Rešetnjak's work [160]); we have expanded the proof slightly and made some inconsequential changes in the arguments.

Our proof of the comparison principle is a slight reworking of the argument in [77], although we have taken the proof of Lemma 7.16 from [33].

We also mention several other works related to viscosity solutions of oblique derivative problems. First are two works of Dupuis and Ishii: In [40], the authors study oblique derivative problems in general bounded Lipschitz domains, while, in [41], they study oblique derivative problems in domains which can be represented as the intersection of finitely many smooth domains. Both papers assume that $b$ has the special form

$$b(x, z, p) = \beta(x) \cdot p + f(x, z).$$

with $\beta$ a suitable oblique vector field and $f$ a scalar function satisfying appropriate structure conditions. In [6] and [7], Barles studies oblique derivative problems with $b$ in the same form as in this chapter. He only makes assumptions about the first derivatives of $b$ with respect to $(x, z, p)$ but, as compensation, his domains must have $H_3$ boundary. The primary difference between these two papers is that [6] studies general (fully nonlinear) elliptic equations and oblique boundary conditions while [7] mostly studies problems with quasilinear equations and, hence, condition (7.53) is relaxed.

In addition, the comparison principles in this chapter (and related ones proved in other sources) can be used to prove *a priori* estimates. Hölder estimates for viscosity solutions of uniformly elliptic equations with general oblique boundary conditions have been proved in [8]. Higher regularity (but only for solutions of the Neumann problem) has been proved in [132].

# Chapter 8

# Pointwise Bounds for Solutions of Problems with Quasilinear Equations

## Introduction

In this chapter, we study various pointwise estimates for solutions of the boundary value problem

$$Qu = 0 \text{ in } \Omega, \quad Nu = 0 \text{ on } \partial\Omega \tag{8.1}$$

under two different basic structures. In Sections 8.1 and 8.2, we assume that $Q$ and $N$ are given by

$$Qu = a^{ij}(x, u, Du)D_{ij}u + a(x, u, Du), \tag{8.2a}$$

$$Nu = b(x, u, Du), \tag{8.2b}$$

assuming that $[a^{ij}]$ is a positive definite matrix with minimum eigenvalue $\lambda$ and $b$ is oblique. For our present purposes, we say that $b$ is oblique if $\partial\Omega$ is Lipschitz and if, for all $(x, z) \in \partial\Omega \times \mathbb{R})$, there is an inward pointing vector $p_0(x, z)$ such that

$$b(x, z, p + tp_0(x, z)) \leq b(x, z, p + sp_0(x, z))$$

for all $p \in \mathbb{R}^n$ and all $t \leq s$ in $\mathbb{R}$.

In Sections 8.3 and 8.4, we assume that

$$Qu = \operatorname{div} A(x, u, Du) + B(x, u, Du), \tag{8.3a}$$

$$Nu = A(x, u, Du) \cdot \gamma + \psi(x, u) \tag{8.3b}$$

for some vector function $A$ and scalar functions $B$ and $\psi$.

## 8.1 Maximum estimates for nondivergence equations

A crucial role will be played in the theory by the Bernstein $\mathcal{E}$ function, defined by

$$\mathcal{E}(x, z, p) = a^{ij}(x, z, p)p_i p_j. \tag{8.4}$$

Our first estimate is a quasilinear version of Lemma 1.2.

**Lemma 8.1.** *Let $u \in C^1(\overline{\Omega}) \cap C^2(\Omega)$ be a solution of (8.1) with $Q$ and $N$ given by (8.2). Suppose there are nonnegative constants $\mu_1$, $\mu_2$, $M_1$, and $R$ such that* $\operatorname{diam}\Omega \le R$ *and*

$$\frac{(\operatorname{sgn} z)a(x,z,p)}{\mathcal{E}(x,z,p)} \le \frac{\mu_1|p| + \mu_2}{|p|^2} \tag{8.5a}$$

*for all $(x,z,p) \in \Omega \times \mathbb{R} \times \mathbb{R}^n$ with $|z| \ge M_1$ and $0 < |p| \le \mu_2 \exp((1+\mu_1)R)$,*

$$(\operatorname{sgn} z)b(x,z,p) < 0 \tag{8.5b}$$

*for all $(x,z,p) \in \partial\Omega \times \mathbb{R} \times \mathbb{R}^n$ with $|z| \ge M_1$ and $|p| \le \mu_2 \exp((1+\mu_1)R)$. Then*

$$\sup_\Omega |u| \le M_1 + \frac{\mu_2}{1+\mu_1} \exp((1+\mu_1)R).$$

***Proof.*** We use the function $\varphi$ from Lemma 1.2 with a slight change. Setting $K = \mu_1 + 1$, we define

$$\varphi(x) = -\frac{\mu_2}{K} \exp(K(R - x^1))$$

and we assume, without loss of generality, that $0 < x^1 < R$ in $\Omega$. We then let $x_0$ be a point in $\overline{\Omega}$ at which $u - \varphi$ attains its maximum. Our first step is to show that $u(x_0) \le M_1$.

Suppose first that $x_0 \in \Omega$, and define the operator $\bar{Q}$ by

$$\bar{Q}v = a^{ij}(x,u,Dv)D_{ij}v + a(x,u,Dv).$$

Since $D\varphi(x_0) = Du(x_0)$, it follows that

$$\begin{aligned}\bar{Q}\varphi(x_0) &= a^{ij}(x_0,u,Du)D_{ij}\varphi(x_0) + a(x_0,u,Du)\\ &\le a^{ij}(x_0,u,Du))D_{ij}u(x_0) + a(x_0,u,Du)\\ &= 0\end{aligned}$$

because $D^2(u-\varphi)(x_0)$ is negative semidefinite. But $\mathcal{E}(x_0,u,D\varphi) = a^{11}(x_0,u,D\varphi))|D\varphi(x_0)|^2$, so, if $u(x_0) > M_1$, then we have

$$\begin{aligned}\bar{Q}\varphi(x_0) &= -a^{11}(x_0,u,D\varphi)\mu_2 K \exp(K(R-x_0^1)) + a(x_0,u,D\varphi)\\ &< a^{11}(x_0,u,D\varphi)\exp(K(R-x_0^1))[-\mu_2 K + \mu_2\mu_1 + \mu_2]\end{aligned}$$

because $\exp(K(R-x^1)) > 1$ in $\Omega$. Since $-\mu_2 K + \mu_2\mu_1 + \mu_2 = 0$, it follows that $u(x_0) \le M_1$ if $x_0 \in \Omega$.

If $x_0 \in \partial\Omega$, we use the obliqueness condition to conclude that

$$b(x_0,u,D\varphi) \ge b(x,u,Du) = 0.$$

Since $|D\varphi(x_0)| \le \mu_2 \exp(KR)$, it follows that $u(x_0) \le M_1$ in this case as well.

Hence,

$$\begin{aligned} u(x) &\le u(x) - \varphi(x) \le u(x_0) - \varphi(x_0) \le M_1 - \varphi(x_0) \\ &\le M_1 + \frac{\mu_2}{K} \exp(KR) \end{aligned}$$

for any $x \in \overline{\Omega}$. This gives the desired upper bound for $u$ and a lower bound is proved similarly. □

Note that Lemma 8.1 applies, for example, to the problem

$$a^{ij}(x, Du)D_{ij}u = f(x) \text{ in } \Omega, \quad \frac{\partial u}{\partial \gamma} = u + g(x) \text{ on } \partial\Omega, \tag{8.6}$$

provided the minimum eigenvalue $\lambda(x,p)$ satisfies the lower bound $\lambda \ge 1/(1+|p|)$ and if $f$ and $g$ are bounded. We can modify the proof to obtain a slightly better result along the lines of Theorem 1.9. To proceed, we recall the definition $\mathcal{D} = \det[a^{ij}]$.

**Lemma 8.2.** *Suppose that there are positive, measurable functions $g \in L^n_{loc}(\mathbb{R}^n)$ and $h \in L^n(\Omega)$ such that*

$$\int_\Omega h(x)^n \, dx < \int_{\mathbb{R}^n} g(p)^n \, dx \tag{8.7}$$

*Suppose also that*

$$\frac{(\operatorname{sgn} z)a(x,z,p)}{(n\mathcal{D})^{1/n}} \le \frac{h(x)}{g(p)} \tag{8.8a}$$

*for all $(x,z,p) \in \Omega \times \mathbb{R} \times \mathbb{R}^n$ and that there is a continuous, increasing function $M_1$ such that*

$$(\operatorname{sgn} z)b(x,z,p) < 0 \tag{8.8b}$$

*for all $(x,z,p) \in \partial\Omega \times \mathbb{R} \times \mathbb{R}^n$ with $|z| \ge M_1(|p|)$. Then there is a number $t > 0$, determined only by $h$ and $g$ such that*

$$\int_\Omega h(x)^n \, dx = \int_{\{|p| \le t\}} g(p)^n \, dx. \tag{8.9}$$

*Moreover,*

$$\sup |u| \le M_1(t) + t \operatorname{diam} \Omega. \tag{8.10}$$

***Proof.*** We recall from the proof of Theorem 1.9 that the contact set $\Gamma(u)$ is defined to be the set of all $x \in \Omega$ such that there is a $p \in \mathbb{R}^n$ with

$$u(y) \le u(x) + p \cdot (y - x) \tag{8.11}$$

for all $y \in \overline{\Omega}$. We also write $\chi$ for the set of all $p \in \mathbb{R}^n$ such that, for some $x \in \Omega$ and all $y \in \overline{\Omega}$, (8.11) is satisfied. It then follows that

$$\det(-D^2u) \le \frac{(-a^{ij}D_{ij}u)^n}{n\mathcal{D}} \le \frac{h(x)^n}{g(Du)^n}$$

on $\Gamma(u)$. Hence

$$\int_\chi g(p)^n\,dp = \int_{\Gamma(u)} g(Du)^n \det(-D^2u)\,dx \le \int_{\Gamma(u)} h(x)^n\,dx.$$

Hence, for any $T > t$, there is a $p \in \mathbb{R}^n$ with $p \notin \chi$, which means that $u - p \cdot x$ must take its maximum at a point $x_0 \in \overline{\Omega}$. If $x_0 \in \Omega$, then $u(x_0) \le M_1(T)$ by the definition of $\chi$. If $x_0 \in \partial\Omega$, then

$$b(x_0, u, p) \ge b(x_0, u, Du) = 0$$

so condition (8.8b) implies that $u(x_0) \le M_1$ in this case as well. Then we argue as in Theorem 1.9 to infer (8.10). □

Note that Lemma 8.2 applies to the problem (8.6) if $\lambda(x,p) \ge \Psi(p)$ for some positive function $\Psi$, if $f$ and $g$ are bounded provided $\sup|f|$ is sufficiently small, determined only by $\Psi$ and $\Omega$. If we wish to consider the slightly different problem

$$a^{ij}(x, Du)D_{ij}u = u + f(x) \text{ in } \Omega, \quad \frac{\partial u}{\partial \gamma} = g(x) \text{ on } \partial\Omega,$$

we need to make slightly different hypotheses. To state them more easily, we suppose that $\partial\Omega \in C^2$, and we take advantage of Lemma 5.18. Specifically, we recall that, if $\partial\Omega \in C^2$, then there is a $C^2(\overline{\Omega})$ function $\rho$ such that $\rho = 0$ and $D\rho = \gamma$ on $\partial\Omega$. Moreover, $|D\rho| \le 1$ in $\Omega$, and there are constants $R_0$ and $\rho_2$ such that $|D^2\rho| \le \rho_2$ and $0 < \rho \le R_0$ in $\Omega$.

**Lemma 8.3.** *Suppose $\partial\Omega \in C^2$. Suppose also that there are positive constant $\mu_1$ and $M_1$ such that*

$$\mu_1\rho_2\Lambda(x,z,p) + (\operatorname{sgn} z)a(x,z,p) < 0 \tag{8.12a}$$

*for all $(x,z,p) \in \Omega \times \mathbb{R} \times \mathbb{R}^n$ with $|z| \ge M_1$ and $|p| \le \mu_1$,*

$$(\operatorname{sgn} z)b(x,z,-(\operatorname{sgn} z)\mu_1\gamma) < 0 \tag{8.12b}$$

*for all $(x,z,p) \in \partial\Omega \times \mathbb{R} \times \mathbb{R}^n$ with $|z| \ge M_1$ and $|p| \le \mu_1$. Then*

$$\sup_\Omega |u| \le M_1 + R_0\mu_1.$$

***Proof.*** Set $w = -\mu_1\rho$, $v_0 = \sup(u - w)$, and $v = w + v_0$. Then $\sup(u - v) = 0$, so there is a point $x_0 \in \overline{\Omega}$ with $u(x_0) = v(x_0)$.

If $x_0 \in \Omega$, then $Du(x_0) = Dv(x_0)$ and the matrix $D^2(u - v)(x_0)$ is negative semidefinite. It follows that

$$\begin{aligned}
Qv(x_0) &= a^{ij}(x_0, v, Dv)D_{ij}v(x_0) + a(x_0, v, Dv) \\
&= a^{ij}(x_0, u, Du)D_{ij}v(x_0) + a(x_0, u, Du) \\
&\geq a^{ij}(x_0, u, Du)D_{ij}u(x_0) + a(x_0, u, Du) \\
&= 0.
\end{aligned}$$

Since $|D\rho| \leq 1$, it follows that $|Dv| \leq \mu_1$, so (8.12a) implies that $v(x_0) \leq M_1$.

If $x_0 \in \partial\Omega$, then $Du(x_0) - Dv(x_0)$ is some nonpositive multiple of $\gamma$, so

$$b(x_0, v, Dv) = b(x_0, u, Dv) \geq b(x_0, u, Du) = 0.$$

Since $Dv = -\mu_1\gamma$ on $\partial\Omega$, it follows from (8.12b) that $v(x_0) \leq M_1$ in this case as well.

We therefore have $-\mu_1\rho(x_0) + v_0 \leq M_1$ and hence

$$v_0 \leq M_1 + \sup \mu_1\rho = M_1 + \mu_1 \sup \rho = M_1 + \mu_1 R_0/2.$$

In addition,

$$u(x) \leq -\mu_1\rho(x) + v_0 \leq v_0$$

for any $x \in \overline{\Omega}$, so $\sup u \leq M_1 + \mu_1 R_0/2$. The lower bound is proved similarly. □

## 8.2 Hölder estimates for nondivergence equations

In later chapters, it will be useful to have a Hölder estimate for solutions of nonlinear oblique derivative problems. We begin by recalling the notation from Chapter 1. Let $\omega$ be a Lipschitz function defined on the $(n-1)$–dimensional ball $B_{n-1}(0, R)$ (for some $R > 0$), and set

$$\begin{aligned}
\Omega[R] &= \{x \in \mathbb{R}^n : |x| < R,\ x^n > \omega(x')\}, \\
\Sigma[R] &= \{x \in \mathbb{R}^n : |x| < R,\ x^n = \omega(x')\}.
\end{aligned}$$

Let $\omega_0$ be a nonnegative constant such that

$$|\omega(x') - \omega(y')| \leq \omega_0|x' - y'|$$

for all $x'$ and $y'$ in $B_{n-1}(0,R)$, and let $\varepsilon \in (0,1)$. We say that a vector field $\beta$ defined on $\Sigma[R]$ has *modulus of obliqueness at least* $\varepsilon$ if

$$\omega_0|\beta'(x)| \le (1-\varepsilon)\beta^n(x) \tag{8.13}$$

for all $x \in \Sigma[R]$.

Our Hölder continuity result is then stated in the following form.

**Lemma 8.4.** *Let $[a^{ij}]$ be a positive-definite matrix-valued function defined on $\Omega[R]$ with minimum eigenvalue $\lambda$, and suppose that there is a positive constant $\mu$ such that*

$$|a^{ij}(x)| \le \mu\lambda(x)$$

*for all $x \in \Omega[R]$. Define $Lu = a^{ij}D_{ij}u$. Let $\beta$ have modulus of obliqueness at least $\varepsilon$, and let $u \in C^2(\Omega[R]) \cap C(\overline{\Omega[R]})$ satisfy*

$$|Lu| \le \lambda(b_0|Du|^2 + b_1) \text{ in } \Omega[R], \tag{8.14a}$$

$$|\beta \cdot Du| \le G\beta^n \text{ on } \Sigma[R] \tag{8.14b}$$

*for some nonnegative constants $b_0$, $b_1$ and $G$. Then there are constants $C$, $\alpha$, and $\delta$ determined only by $\varepsilon$, $\mu$, $\omega_0$, and $b_0 \sup|u|$ such that*

$$\operatorname*{osc}_{\Omega[r]} u \le C\left(\frac{r^\alpha}{R^\alpha}\operatorname*{osc}_{\Omega[R]} u + (b_1R+G)r^\delta R^{1-\delta}\right) \tag{8.15}$$

*for all $r \in (0,R)$.*

***Proof.*** We start with a suitable weak Harnack inequality. Suppose that $w$ is nonnegative and satisfies

$$Lw \le \lambda(b_0|Dw|^2 + b_1) \text{ in } \Omega[R], \quad \beta \cdot Dw \le G\beta^n \text{ on } \Sigma[R],$$

and set

$$v = \frac{1-e^{-b_0 w}}{b_0}.$$

It's easy to check that $Lv \le b_1\lambda$ in $\Omega[R]$ and $\beta \cdot Dv \le G\beta^n$ on $\Sigma[R]$ (recalling that $e^{-b_0 w} \in (0,1]$), so Theorem 1.20 implies that there are positive constants $C$ and $p$, determined by $n$, $\varepsilon$, $\mu$, and $\omega_0$ such that

$$\left(R^{-n}\int_{\Omega[R/2]} v^p\,dx\right)^{1/p} \le C(\inf_{\Omega[R/2]} v + b_1R^2 + GR). \tag{8.16}$$

Since $v \le w \le \exp(b_0 \sup w)v$, this estimate is also valid with $w$ in place of $v$ (and $C$ depending also on $b_0 \sup w$). The proof of Theorem 1.26 immediately gives (8.15). □

We observe here that the inequality $|Lu| \le \lambda(b_0|Du|^2 + b_1)$ is satisfied if $Qu = 0$ with the structure conditions

$$|a^{ij}(x,z,p)| \le \mu(|z|)\lambda(x,z,p),$$
$$|a(x,z,p)| \le (b_0(|z|)|p|^2 + b_1(|z|))\lambda(x,z,p)$$

for increasing, nonnegative functions $\mu$, $b_0$, and $b_1$. The connection of the linear boundary condition in this theorem with the nonlinear structure of current interest will be clarified in later chapters.

## 8.3 Maximum estimates for conormal problems

In this section, we prove a maximum estimate for solutions of the conormal problem under quite general hypotheses. All of our structure conditions are based on a positive, increasing function $g$ defined on $[0,\infty)$ and its integral $G$, which is defined by

$$G(s) = \int_0^s g(t)\,dt. \tag{8.17}$$

There are two inequalities we shall use repeatedly. First, since $g$ is increasing, we have

$$ag(b) \le ag(a) + bg(b) \tag{8.18}$$

for any nonnegative $a$ and $b$. Also

$$G(s) \le \int_0^s g(s)\,dt = sg(s). \tag{8.19}$$

We also take $G_1$ to be an increasing $C^1([t_0,\infty))$ function such that

$$G_1(t)tg(t) \le G(t)^2, \quad tG_1'(t) \le G(t) \tag{8.20}$$

for all $t \ge t_0$, where $t_0$ is some positive constant. (If there is a positive constant $g_0$ such that $G$ satisfies the inequality $G(s) \ge g_0 sg(s)$ for all $s > 0$, then we can take $G_1$ to be a suitable multiple of $G$. We shall return to this inequality in the next section.)

**Lemma 8.5.** *Suppose that there are nonnegative constants $a_1$ and $b_1$ and positive constants $c_0 < 1$, $M$, and $R$ such that*

$$p \cdot A(x,z,p) \ge |p|g(|p|) - a_1\frac{RG(|z|/R)^2}{|z|g(|z|/R)}, \tag{8.21a}$$

$$zB(x,z,p) \le b_0 p \cdot A(x,z,p) + b_1 G(|z|/R) \tag{8.21b}$$

*for all* $(x, z, p) \in \Omega \times \mathbb{R} \times \mathbb{R}^n$ *with* $|z| \geq M$,

$$z\psi(x, z) \leq Rc_0 G_1(|z|/R) \tag{8.21c}$$

*for all* $(x, z, p) \in \partial\Omega \times \mathbb{R} \times \mathbb{R}^n$ *with* $|z| \geq M$. *Suppose also that*

$$M/R \geq t_0, \quad G\left(\frac{M}{R}\right) \geq 1. \tag{8.22}$$

*Then there is a constant* $c_1$, *determined only by* $n$, $\Omega$, $a_1$, $b_0$, $b_1$, *and* $c_0$, *such that*

$$\sup_\Omega \left(1 - \frac{M}{|u|}\right)_+^{2n} G\left(\frac{|u|}{R}\right) \leq c_1 \int_\Omega G\left(\frac{|u|}{R}\right) dx. \tag{8.23}$$

***Proof.*** With $q \geq 1 + b_0$ a parameter, we define the function $\varphi$ by

$$\varphi(u) = \left(1 - \frac{M}{|u|}\right)_+^{2nq-2n+1} G\left(\frac{|u|}{R}\right)^{q-1} u.$$

Since

$$\varphi'(u) = (2nq - 2n + 1)\left(1 - \frac{M}{|u|}\right)_+^{2nq-2n} G\left(\frac{|u|}{R}\right)^{q-1} \frac{M}{|u|}$$
$$+ \frac{\varphi(u)}{u}$$
$$+ (q-1)\left(1 - \frac{M}{|u|}\right)_+^{2nq-2n+1} G\left(\frac{|u|}{R}\right)^{q-2} \frac{|u|}{R} g\left(\frac{|u|}{R}\right),$$

it follows that $\varphi'(u) \geq q\varphi(u)/u$, and hence

$$\varphi'(u) \geq b_0 \frac{\varphi(u)}{u}. \tag{8.24}$$

Now, we use $\varphi(u)$ as test function in the weak form of the boundary value problem to infer that

$$\int_\Omega \varphi'(u) Du \cdot A\, dx = \int_\Omega \varphi(u) B\, dx + \int_{\partial\Omega} \varphi(u)\psi\, ds.$$

From (8.21) and (8.24), we infer (after some rearrangement) that

$$\int_\Omega \left(\varphi'(u) - b_0 \frac{\varphi(u)}{u}\right) g(|Du|)|Du|\, dx \leq I_1 + I_2 + I_3,$$

where

$$I_1 = a_1 \int_\Omega \left(\varphi'(u) - b_0 \frac{\varphi(u)}{u}\right) \frac{RG(|u|/R)^2}{|u| g(|u|/R)}\, dx$$

$$I_2 = b_1 \int_\Omega \left(1 - \frac{M}{|u|}\right)_+^{2nq-2n+1} G\left(\frac{M}{|u|}\right)^q dx,$$

$$I_3 = Rc_0 \int_{\partial\Omega} \left(1 - \frac{M}{|u|}\right)_+^{2nq-2n+1} G\left(\frac{|u|}{R}\right)^{q-1} G_1\left(\frac{|u|}{R}\right) ds.$$

By direct computation, we also obtain

$$\varphi'(u) \le 2(n+1)q\left(1-\frac{M}{|u|}\right)_+^{2nq-2n} G\left(\frac{|u|}{R}\right)^{q-2}\frac{|u|}{R}g\left(\frac{|u|}{R}\right),$$

so, after noting that $(1-M/|u|)_+ \le 1$, we have

$$I_1+I_2 \le [2a_1(n+1)q+b_1]\int_\Omega \left(1-\frac{M}{|u|}\right)_+^{2nq-2n} G\left(\frac{|u|}{R}\right)^q\,dx.$$

Next, we define the function $\varphi_1$ by

$$\varphi_1(t) = \left(1-\frac{M}{t}\right)_+^{2nq-2n+1} G\left(\frac{t}{R}\right)^{q-1} G_1\left(\frac{t}{R}\right)$$

and observe that

$$\varphi_1'(t) = \Phi_1(t)+\Phi_2(t)+\Phi_3(t),$$

where

$$\Phi_1(t) = (2nq-2n+1)\left(1-\frac{M}{t}\right)_+^{2nq-2n} G\left(\frac{t}{R}\right)^{q-1} G_1\left(\frac{t}{R}\right)\frac{M}{t^2},$$
$$\Phi_2(t) = (q-1)\left(1-\frac{M}{t}\right)_+^{2nq-2n+1} G\left(\frac{t}{R}\right)^{q-2} G_1\left(\frac{t}{R}\right)\frac{g(t/R)}{R},$$
$$\Phi_3(t) = \left(1-\frac{M}{t}\right)_+^{2nq-2n+1} G\left(\frac{t}{R}\right)^{q-1} G_1'\left(\frac{t}{R}\right)\frac{1}{R}.$$

Then Lemma 5.17 implies that there is a constant $K_0$, determined only by $\Omega$, such that

$$I_3 \le Rc_0\int_\Omega |Du|\varphi_1'(|u|)\,dx + K_0c_0R\int_\Omega \varphi_1(|u|)\,dx.$$

We now use (8.18) to conclude that

$$\int_\Omega |Du|\varphi_1'(|u|)\,dx \le I_{31}+I_{32},$$

with

$$I_{31} = \int_\Omega |Du|g(|Du|)\frac{\varphi_1'(|u|)}{g(|u|/R)}\,dx,$$
$$I_{32} = \int_\Omega \frac{|u|}{R}\varphi_1'(|u|),dx.$$

Next, we infer from the first inequality of (8.20) and the first inequality of (8.22) that $G_1(t) \le G(t)$ for $t \ge M/R$. In addition, the second inequality of (8.20) and (8.19) imply that $G_1'(t) \le g(t)$ for $t \ge M/R$. Comparing the explicit expressions for the derivatives of $\varphi$ and $\varphi_1$ and using (8.19) and the second inequality of (8.22), we see that

$$\varphi_1'(|u|) \le \left(1 + \frac{1}{q-1}\right) \varphi'(|u|) \frac{g(|u|/R)}{R}.$$

If $q$ is large enough (determined only by $c_0$) that

$$c_0 \left(1 + \frac{1}{q-1}\right) \le \frac{1+c_0}{2},$$

we infer that

$$I_{31} \le \frac{1+c_0}{2} \int_\Omega |Du| g(|Du|) \varphi'(|u|)\, dx.$$

Using (8.20) and the inequality $G_1 \le G$, we conclude that

$$\varphi_1'(|u|) \le 2(n+1)q \left(1 - \frac{M}{|u|}\right)_+^{2nq-2n} G\left(\frac{|u|}{R}\right)^{q-2} \frac{1}{R}.$$

It therefore follows that

$$I_3 \le \frac{1+c_0}{2} \int_\Omega g(|Du|)|Du|\varphi'(u)\, dx$$
$$+ C(n)c_0 q \int_\Omega \left(1 - \frac{M}{|u|}\right)_+^{2nq-2n} G\left(\frac{|u|}{R}\right)^q dx.$$

Combining this inequality with our estimates for $I_1$ and $I_2$, we find that there is a constant $C$, determined only by $a_1$, $b_0$, $b_1$, and $c_0$, such that

$$\int_\Omega \left(\frac{1-c_0}{2}\varphi'(u) - b_0 \frac{\varphi(u)}{u}\right) |Du| g\left(\frac{|u|}{R}\right) dx$$
$$\le Cq \int_\Omega \left(1 - \frac{M}{|u|}\right)^{2q-2n} G\left(\frac{|u|}{R}\right)^q dx.$$

To proceed, we note that

$$\frac{1-c_0}{2}\varphi(u) - b_0 \frac{\varphi(u)}{u} \ge \left(1 - \frac{M}{|u|}\right)_+^{2nq-2n+1} G\left(\frac{|u|}{R}\right)^{q-1}$$

if $q$ is sufficiently large (now determined by $b_0$ and $c_0$), and we define the function $H$ by

$$H(x) = \left(1 - \frac{M}{|u|}\right)_+^{2q-2n+2} G\left(\frac{|u|}{R}\right)^q.$$

Then

$$|DH| \le \frac{3q}{R} \left(1 - \frac{M}{|u|}\right)_+^{2nq-2n+1} G\left(\frac{|u|}{R}\right)^{q-1} g\left(\frac{|u|}{R}\right),$$

so

$$\int_\Omega |DH| + |H|\,dx \le Cq^2 \int_\Omega \left(1 - \frac{M}{|u|}\right)^{2nq-2n} G\left(\frac{|u|}{R}\right)^q dx.$$

Applying the Sobolev imbedding theorem, we find that

$$\left(\int_\Omega \left(1 - \frac{M}{|u|}\right)_+^{2n\kappa q-2n} G\left(\frac{|u|}{R}\right)^{\kappa q} dx\right)^\kappa \le Cq^2 \int_\Omega \left(1 - \frac{M}{|u|}\right)_+^{2q-2n} G\left(\frac{|u|}{R}\right)^q dx,$$

where $\kappa = n/(n-1)$. If we set

$$w = \left(1 - \frac{M}{|u|}\right)_+^{2n} G\left(\frac{|u|}{R}\right), \quad d\mu = \left(1 - \frac{M}{|u|}\right)_+^{-2n} dx,$$

then this inequality takes the standard form

$$\left(\int_\Omega w^{\kappa q}\,d\mu\right)^{1/\kappa} \le Cq^2 \int_\Omega w^q\,d\mu,$$

and then the Moser iteration scheme, Lemma 5.30, implies that

$$\sup_\Omega w \le C \int_\Omega w\,d\mu.$$

Rewriting this inequality in terms of $u$ and $dx$ yields (8.23). □

Estimates of the right hand side of (8.23) require somewhat stronger conditions on the coefficients. The most important condition is that $zB - b_0 p \cdot A$ is negative for large $z$. More precisely, we have the following result.

**Lemma 8.6.** *Suppose there are nonnegative constants $a_1$, $b_0$, and $c_0$, along with positive constants $b_2$, $M$, and $R$ (with $M/R \ge t_0$) such that condition* (8.21a) *holds for all $(x, z, p) \in \Omega \times \mathbb{R} \times \mathbb{R}^n$ with $|z| \ge M$, condition* (8.21c) *holds for all $(x, z) \in \partial\Omega \times \mathbb{R}$ with $|z| \ge M$, and*

$$zB \le b_0 p \cdot A - b_2 G\left(\frac{|z|}{R}\right) \tag{8.25}$$

*for all $(x, z, p) \in \Omega \times \mathbb{R} \times \mathbb{R}^n$ with $|z| \ge M$. Suppose also that there is a $q \ge \max\{1, b_0 - 1\}$ and $\varepsilon \in (0, 1)$ such that*

$$c_0(q+1) \le \min\{q, q+1-b_0\}, \tag{8.26a}$$

$$a_1 \max\{q, q+1-b_0\} + c_0(q+1+K_0R) \le (1-\varepsilon)b_2. \tag{8.26b}$$

*Then*

$$\int_\Omega G\left(\frac{|u|}{R}\right) dx \le \varepsilon^{-1/q} G\left(\frac{M}{R}\right) |\Omega|. \tag{8.27}$$

***Proof.*** To simplify the notation, we set $\tau = G(M/R)$, we write $\Omega'$ for the subset of $\Omega$ on which $|u| \ge M$, and we suppress the argument $|u|/R$ from $G$, $G_1$, and $G_1'$ . If we use the test function

$$\varphi = (G^q - \tau^q)_+ u$$

and apply (8.25), we can write the resulting inequality as

$$I_0 + I_1 \le I_2$$

where

$$\begin{aligned} I_0 &= \int_{\Omega'} [(1-b_0)(G^q - \tau^q) + qG^{q-1}\frac{|u|g(|u|/R)}{R}]Du \cdot A\,dx, \\ I_1 &= b_2 \int_{\Omega'} (G^q - \tau^q)G\,dx, \\ I_2 &= c_0 R \int_{\partial\Omega} G(G^q - \tau^q)_+\,ds. \end{aligned}$$

From Lemma 5.17, we conclude that

$$\begin{aligned} I_2 \le K_0 c_0 R &\int_{\Omega'} (G^q - \tau^q)G_1\,dx \\ &+ c_0 \int_{\Omega'} |Du|((G^q - \tau^q)G_1' + qG^{q-1}G_1 g(|u|/R))\,dx. \end{aligned}$$

We infer from (8.20) that $G_1(t) \le G(t)$ and $G_1'(t) \le g(t)$ for all $t > 0$, and hence, because $\tau \ge 0$, we have

$$\begin{aligned} I_2 \le K_0 c_0 R &\int_{\Omega'} G^{q+1}\,dx \\ &+ c_0 q \int_{\Omega'} |Du| g(|u|/R) G^{q-1} G_1\,dx \\ &+ c_0 \int_{\Omega'} |Du| G^q G_1'\,dx. \end{aligned}$$

Applying (8.20), (8.18), and (8.19), we find that

$$\begin{aligned} I_2 \le c_0(q+1) &\int_{\Omega'} |Du| g(|Du|) G^q\,dx \\ &+ c_0(q+1+K_0R) \int_{\Omega'} G^{q+1}\,dx. \end{aligned}$$

Our next step is to obtain a lower bound for $I_0$. We set

$$q_1 = \min\{q, q+1-b_0\}, \quad q_2 = \max\{q, q+1-b_0\},$$

and use (8.19) along with the assumption $q \ge b_0 - 1$ to conclude that

$$q_1 G^q \le (1-b_0)(G^q - \tau^q)_+ + qG^{q-1}\frac{|u|g(|u|/R)}{R} \le q_2 G^{q-1}\frac{|u|g(|u|/R)}{R}.$$

It follows that

$$I_0 \geq q_1 \int_{\Omega'} G^q |Du| g(|Du|)\, dx - a_1 q_2 \int_{\Omega'} G^{q+1}\, dx.$$

Taking (8.26a) into account, we conclude that

$$\begin{aligned} b_2 \int_{\Omega'} (G^q - \tau^q) G\, dx &\leq \int_{\Omega'} G^{q+1} \left(a_1 q_2 + c_0(q+1+K_0 R)\right) dx \\ &\leq (1-\varepsilon) b_2 \int_{\Omega'} G^{q+1}\, dx. \end{aligned}$$

Simple rearrangement yields

$$\varepsilon \int_{\Omega'} G^{q+1} \leq \int_{\Omega'} \tau^q G\, dx,$$

and hence

$$\int_{\Omega'} G\, dx \leq \left(\varepsilon^{-1} \tau^q \int_{\Omega'} G\, dx\right)^{1/(q+1)} |\Omega'|^{q/(q+1)}$$

by Hölder's inequality. One more rearrangement then yields (8.27). □

It is possible to prove an estimate like that in Lemma 8.6 with the negativity assumption on $zB - b_0 p \cdot A$ replaced by one on $z\psi$, but we shall not do so here.

## 8.4 Hölder estimates for conormal problems

In this section, we make a slightly stronger assumption on $g$, namely, we assume that there are positive constants $\delta$ and $m$, with $1+\delta \leq m$ such that

$$(1+\delta) G(t) \leq t g(t) \leq m G(t) \tag{8.28}$$

for all $t > 0$. As a simple consequence, we observe that

$$\frac{G(t)}{t^{1+\delta}} \geq \frac{G(s)}{s^{1+\delta}}, \tag{8.29a}$$

$$\frac{G(t)}{t^m} \leq \frac{G(s)}{s^m} \tag{8.29b}$$

for all $s \in (0, t]$.

To see the significance of condition (8.28), we note first that it is satisfied for $g(t) = t^m$ with $\delta = 1 - m$ for any $m > 1$. Moreover, if we take

$$g(t) = r t^{r-1} (\ln(t+e))^s + \frac{s t^r}{t+e} (\ln(t+e))^{s-1}$$

for constants $r > 1$ and $s \geq 0$, then

$$G(t) = t^r(\ln(t+e))^s,$$

so (8.28) is satisfied with $\delta = r - 1$ and $m = r + s$. On the other hand,

$$\lim_{t\to\infty} \frac{tg(t)}{G(t)} = r,$$

so this example can be considered only a small perturbation of a power function. To see the full strength of our structure condition, we let $\delta > 0$ and $m > 1 + \delta$ to be arbitrary and set

$$a = \frac{1+\delta+m}{2}, \quad b = \frac{m-1-\delta}{2\sqrt{2}}.$$

Then, for

$$g(t) = \frac{1}{t}[a + b(\cos(\ln(\ln t)) + \sin(\ln(\ln t))] \exp(\ln t(a + b\sin(\ln(\ln t))),$$

we have

$$G(t) = \exp(\ln t(a + b\sin(\ln(\ln t))) = t^{(a+b\sin(\ln(\ln t)))}$$

and (8.28) is satisfied. Moreover, we have

$$\liminf_{t\to\infty} \frac{tg(t)}{G(t)} = 1 + \delta, \quad \limsup_{t\to\infty} \frac{tg(t)}{G(t)} = m.$$

Hence condition (8.28) is a genuine improvement on the assumption that $g$ is a power function.

We next prove a weak Harnack inequality under the structure conditions:

$$p \cdot A \geq |p|g(|p|) - a_1 g(a_1), \tag{8.30a}$$
$$|A| \leq a_2 g(|p|) + g(a_3), \tag{8.30b}$$
$$B \geq -b_0|p|g(|p|) - g(b_1)/R, \tag{8.30c}$$
$$\psi \geq -g(c_0). \tag{8.30d}$$

which are assumed to hold for all $z \geq 0$, all $x \in \Omega[R]$, and all $p \in \mathbb{R}^n$.

**Lemma 8.7.** *Set $\chi = a_1 + a_3 + b_1 + c_0$ and $M = \sup u$. Then there are positive constants $p$ and $C$, determined only by $n$, $a_2$, $b_0M$, and $K_1$, such that*

$$\left(R^{-n}\int_{\Omega[R/2]} u^p\,dx\right)^{1/p} \leq C(\inf_{\Omega[R]} u + \chi R) \tag{8.31}$$

*for any nonnegative supersolution of*

$$\operatorname{div} A(x,u,Du) + B(x,u,Du) = 0 \text{ in } \Omega[R], \tag{8.32a}$$
$$A(x,u,Du)\cdot\gamma + \psi(x,u) = 0 \text{ on } \Sigma[R]. \tag{8.32b}$$

***Proof.*** Our proof consists of three steps. The main part of the proof is to show that $u$ lies in a space which is almost a De Giorgi class, but based on $G$ rather than a power function. Then, we show that $u$ is in a De Giorgi class, and then the proof is easily finished.

For the first step, we let $\eta$ be a cut-off function (so $0 \le \eta \le 1$ in $B(R)$ and $\eta = 0$ on $\partial B(R)$). With $k$ at our disposal, we use the test function

$$\varphi = \eta^m \exp(-b_0 u)(u-k)_-.$$

From our structure conditions, we have that

$$I_1 \le I_2 + I_3 + I_4 + I_5 + I_6,$$

where

$$I_1 = \int_{\Omega[R]} e^{-b_0 u}\eta^m g(|Dw|)|Dw|\,dx,$$
$$I_2 = \int_{\Omega[R]} e^{-b_0 u}\eta^m g(b_1)\frac{w}{R}\,dx,$$
$$I_3 = \int_{A_k} e^{-b_0 u}\eta^m a_1 g(a_1)(1+b_0 w)\,dx,$$
$$I_4 = \int_{\Omega[R]} e^{-b_0 u} m\eta^{m-1} w a_2|D\eta| g(|Dw|)\,dx,$$
$$I_5 = \int_{\Omega[R]} e^{-b_0 u} m\eta^{m-1} g(a_3) w|D\eta|\,dx,$$
$$I_6 = \int_{\Sigma[R]} e^{-b_0 u}\eta^m w g(c_0)\,ds.$$

Here, we have set $w = (u-k)_-$ and we have written $A_k$ for the subset of $\Omega[R]$ on which $w$ is nonzero. Let us suppose first that $k \le M$, and we write $C$ for any constant determined only by $n$, $a_2$, $\omega_0$, $b_0 M$.

Then (8.18) implies that

$$I_2 \le m\int_{\Omega[R]} e^{-b_0 u}\eta^m G\left(\frac{k}{R}\right)dx + m\int_{\Omega[R]} e^{-b_0 u}\eta^m G(b_1)\,dx.$$

Also

$$I_3 \le \int_{\Omega[R]} e^{-b_0 u}\eta^m C G(a_1)\,dx$$

because $w \le k \le M$. Two more applications of (8.18) (along with the monotonicity of $g^{-1}$ and conditions (8.28) and (8.29)) give

$$\begin{aligned} I_4 &\le \int_{\Omega[R]} e^{-b_0 u}\eta^m \frac{1}{4}|Dw| g(\frac{1}{4}|Dw|)\,dx + \int_{\Omega[R]} e^{-b_0 u}\eta^m G(4a_2 m w|D\eta|\eta^{-1}))\,dx \\ &\le \frac{1}{4}\int_{\Omega[R]} e^{-b_0 u}\eta^m |Dw| g(|Dw|)\,dx + \int_{\Omega[R]} e^{-b_0 u} G(Ca_2 w|D\eta|)\,dx \end{aligned}$$

and

$$I_5 \le \int_{\Omega[R]} e^{-b_0 u} mG(w|D\eta|)\,dx + \int_{\Omega[R]} e^{-b_0 u} MG(\eta^{m-1} a_3 m)\,dx.$$

By repeating the proof of Lemma 5.17 with $\bar\gamma$ a unit vector in the $x^n$ direction, we find that $I_6 \le I_7 + I_8$, where

$$I_7 = Cc_0 \int_{\Omega[R]} e^{-b_0 u} \eta^m [1 + b_0 w]|Du|\,dx,$$

$$I_8 = Cc_0 \int_{\Omega[R]} e^{-b_0 u} m\eta^{m-1} w|D\eta|\,dx.$$

As we have already noted, $b_0 w \le C$, so (8.18) implies that

$$I_7 \le \int_{\Omega[R]} e^{-b_0 u} \eta^m \frac{1}{2}|Dw| g(|Dw|)\,dx + \int_{\Omega[R]} e^{-b_0 u} \eta^m G(Cc_0)\,dx,$$

and

$$I_8 \le \int_{\Omega[R]} e^{-b_0 u} \eta^{m-1} G(mw|D\eta|)\,dx + \int_{\Omega[R]} e^{-b_0 u} \eta^{m-1} G(mCc_0),\,dx.$$

Combining all these estimates (and using (8.29)) leads us to the inequality

$$\int_{\Omega[R]} \eta^m G(|Dw|)\,dx \le \int_{A(k)} G\left(C\left[w|D\eta| + \frac{k}{R} + \chi\right]\right)\,dx.$$

In fact, this inequality is valid for $k > M$ as well. To see why this is so, we note that $(u-k)_- = k - u$ if $k \ge M$ so $A(k) = \Omega[R]$ and $Dw = -Du$. Therefore the left hand side of this inequality is independent of $k$ for $k \ge M$ while the right hand side is an increasing function of $k$, and hence we may assume that $k \ge \chi R$ to obtain

$$\int_{\Omega[R]} \eta^m G(|Dw|)\,dx \le \int_{A(k)} G\left(C\left[w|D\eta| + \frac{k}{R}\right]\right)\,dx$$

for all $k \ge \chi R$.

If we choose $\eta$ so that $\eta \equiv 1$ in $B(\sigma R)$ and $|D\eta| \le 2/(1-\sigma)R$, we obtain

$$\int_{A(k,\Omega[\sigma R])} G(|Du|)\,dx \le G\left(\sup_{A(k)} \frac{(u-k)_-}{(1-\sigma)R} + \frac{k}{R}\right)|A(k)|.$$

Note that this inequality is the analog of (5.62) with $G(t)$ in place of $t^p$.

To proceed, we define

$$h(t) = \frac{G(t)}{t}, \quad H(t) = \int_0^t h(s)\,ds.$$

Then (8.28) implies that $g(t)/m \le h(t) \le g(t)/(1+\delta)$ for all $t > 0$ and hence $G(t)/m \le H(t) \le G(t)/(1+\delta)$ as well. Moreover, if we define $\Psi$ by $\Psi(t) = H(t^{1/(1+\delta)})$, then

$$\int_{A(k,\Omega[\sigma R])} \Psi(|Du|^{1+\delta})\,dx \le \Psi\left(\left(\sup_{A(k)} \frac{(u-k)_-}{(1-\sigma)R} + \frac{k}{R}\right)^{1+\delta}\right)|A(k)|. \tag{8.33}$$

We now set $A'(k) = A(k,\Omega[\sigma R])$, $v = |Du|^{1+\delta}$, and

$$J = \left(\sup_{A(k)} \frac{(u-k)_-}{(1-\sigma)R} + \frac{k}{R}\right)^{1+\delta}.$$

With these abbreviations, (8.33) becomes

$$\int_{A'(k)} \Psi(v)\,dx \le \Psi(J)|A(k)|.$$

We now observe that $\Psi$ is increasing and

$$\Psi''(t) = \frac{t^{-(2+\delta)/(1+\delta)}}{(1+\delta)^2}(-\delta H'(t^{1/(1+\delta)}) + t^{1/(1+\delta)}H''(t^{1/(1+\delta)})).$$

But

$$sH''(s) - \delta H'(s) = sh'(s) - \delta h(s) = g(s) - (1+\delta)h(s) \ge 0,$$

so $\Psi$ is convex. It therefore follows from Jensen's inequality that

$$\Psi\left(\frac{1}{|A'(k)|}\int v\,dx\right) \le \frac{|A(k)|}{|A'(k)|}\Psi(J) \le \Psi\left(\frac{|A(k)|}{|A'(k)|}J\right).$$

In the second inequality, we have used the estimate $K\Psi(J) \le \Psi(KJ)$ for $K \ge 1$, which follows from the observation that $\Psi(t)/t$ is an increasing function of $t$. Since $\Psi$ is strictly increasing, we conclude that

$$\int_{A'(k)} v\,dx \le J|A(k)|,$$

and this inequality tells us that $u \in DG^-(1+\delta, \Omega[R], \chi, 1/n, K)$ for some $K$ determined only by $m$, $\delta$, $b_0M$, $a_2$, and $\omega_0$. The weak Harnack inequality is then an immediate consequence of Lemma 5.40 via the argument in Theorem 1.20. □

Finally, we obtain a Hölder estimate from this weak Harnack inequality by following the argument of Theorem 1.26.

**Theorem 8.8.** *Let $g$ satisfy* (8.28) *for some positive constants $\delta$ and $m$ with $1+\delta \le m$. Let $\Omega$ be a bounded Lipschitz domain and let $u$ be a* *bounded weak solution of* (8.1) *with* $Q$ *and* $N$ *defined by* (8.3). *Suppose also that there are nonnegative constants* $a_1$, $a_2$, $a_3$, $b_0$, $b_1$, *and* $c_0$ *such that conditions* (8.30a), (8.30b),

$$|B(x,z,p)| \leq b_0|p|g(|p|) + \frac{g(b_1)}{\operatorname{diam}\Omega} \tag{8.34a}$$

*are satisfied for all* $x \in \Omega$, $z = u(x)$, *and all* $p \in \mathbb{R}^n$, *and*

$$|\psi(x,z)| \leq g(c_0) \tag{8.34b}$$

*for all* $x \in \partial\Omega$, $z = u(x)$, *and all* $p \in \mathbb{R}^n$. *Then there are positive constants* $\alpha$ *and* $C$, *determined only by* $a_2$, $b_0 M$, $\delta$, $m$, *and* $\Omega$, *such that*

$$\operatorname*{osc}_{\Omega[r]} u \leq C\left(\frac{r^\alpha}{R^\alpha} + (a_1 + a_3 + b_1 + c_0)r\right) \tag{8.35}$$

*for all* $R \in (0, \operatorname{diam}\Omega)$ *and all* $r \in (0, R)$.

## Notes

The results in this chapter are based on well-known results for the Dirichlet problem although certain elements of the oblique derivative problem require a more careful analysis of the maximum estimates. For example, Lemma 8.1 is essentially the same as Theorem 10.3 of [64], and Lemma 8.2 is essentially the same as Theorem 10.5 of [64]. Both results (for the Dirichlet problem) are described in [64] as "basically variants of the Hopf maximum principle". Lemma 8.3 is a slight rewording of Lemma 3.2 from [98], which, in turn, is based on Theorem 1 in [31]; in this case, all results are for oblique derivative problems.

The Hölder estimate Lemma 8.4 first appears as an interior Hölder estimate in [162] (under the additional assumption that $b_0 = 0$) and in Theorem 1 from [93] (with arbitrary $b_0$). For $C^2$ domains, the estimate first appears as Theorem 2.3 in [120], and, for Lipschitz domains, it was first proved as Corollary 3.5 in [116].

The maximum estimates for conormal problems, Lemmata 8.5 and 8.6, first appear (in slightly different form) as Lemmata 2.4 and 2.5 in [112]. For the Dirichlet problem and with $g$ being a power function, these results appear in Theorem 7.1 from Chapter 4 of [91]; see also Lemma 2.2 from Chapter 10 of [91] for a version of our Lemma 8.5 with $g$ a power function.

Finally, the Hölder estimate Lemma 8.7 is stated here for the first time, in this exact form. As demonstrated in Section 1.9, it is a straightforward consequence of the weak Harnack inequality, and various other versions have been stated. For example, when $g$ is a power function, it is proved on page 467 of [91] as a simple modification of the corresponding interior estimate for solutions of such equations, and the interior Hölder estimate for equations of the general type considered here is just Corollary 1.5 of [112]. The proof given here is based on the results of Klimov [81], specifically, Lemma 2 from [81], which asserts that any function satisfying (8.33) is in the De Giorgi class $DG^-(1+\delta,\Omega[R],\chi,1/n,K)$. A further alternative approach (under slightly different assumptions) is given in [66].

# Chapter 9

# Gradient Estimates for General Form Oblique Derivative Problems

In this chapter, we study the gradient estimate for the boundary value problem

$$Qu = 0 \text{ in } \Omega, \quad Mu = 0 \text{ on } \partial\Omega, \tag{9.1}$$

with $Q$ and $M$ defined by

$$Qu = a^{ij}(x, u, Du)D_{ij}u + a(x, u, Du), \tag{9.2a}$$

$$Mu = b(x, u, Du). \tag{9.2b}$$

We study a somewhat general differential equation, based on the structure in Chapter 15 of [64]. It is important for the purposes of this chapter that boundary condition (9.2) NOT be restricted to the conormal condition for the differential equation. The main assumption (except in Section 9.7) is that the principal coefficients of $Q$ can be decomposed as follows:

$$a^{ij}(x, z, p) = a_*^{ij}(x, z, p) + \tau(x, z, p)p_i p_j \tag{9.3}$$

for some positive-definite matrix $[a_*^{ij}] \in C^1(\overline{\Omega} \times \mathbb{R} \times \mathbb{R}^n)$ and some nonnegative function $\tau \in C^1(\overline{\Omega} \times \mathbb{R} \times \mathbb{R}^n)$. The model problem here is the false mean curvature equation:

$$\Delta u + D_i u D_j u D_{ij} u = 0,$$

which corresponds to $a_*^{ij} = \delta^{ij}$ and $\tau \equiv 1$. We begin by providing an interior gradient estimate of a very special form, which will then be used to prove gradient estimates near the boundary.

## 9.1 Interior gradient bounds

Our first estimate is an interior gradient bound. Some notation will prove very useful in this chapter. First, we use numerical superscripts to denote differentiation with respect to the corresponding component of $p$, numerical subscripts to denote differentiation with respect to the corresponding component of $x$, and the subscript $z$ to denote differentiation with respect to $z$. (In other words,

$$a^{ij,k} = \frac{\partial a^{ij}}{\partial p_k}, \quad a_z = \frac{\partial a}{\partial z},$$

and so on.) We also define the operators $\delta$ and $\bar{\delta}$ by

$$\delta f(x,z,p) = f_z(x,z,p) + \frac{p}{|p|^2} \cdot f_x(x,z,p), \quad \bar{\delta} f(x,z,p) = p \cdot f_p(x,z,p).$$

To prove our gradient estimate, we differentiate the equation $Qu = 0$ with respect to $x^k$ for $k = 1, \dots, n$, multiply the resultant equation by $D_k u$, and then sum over $k$. After performing this calculation, we find that $w_0 = |Du|^2$ satisfies the differential equation

$$a^{ij} D_{ij} w_0 - 2a_*^{ij} D_{ik} u D_{jk} u + B_0^i D_i w_0 + C_0 w_0 = 0, \tag{9.4}$$

where

$$B_0 = a_p^{jk} D_{jk} u + a_p \tag{9.5}$$

and

$$C_0 = 2[\delta a_*^{ij} D_{ij} u + \delta a] + \delta \tau Du \cdot Dw_0.$$

For future reference, we note that

$$B_0 = a_{*,p}^{jk} D_{jk} u + a_p + \frac{1}{2} Du \cdot Dw_0 \tau_p + \tau Dw_0.$$

It will be convenient, however, to perform one more algebraic step here. We write our equation as

$$a^{ij} D_{ij} w - 2a_*^{ij} D_{ik} u D_{jk} u + B_1^i D_i w_0 + C_1 w_0 = 0 \tag{9.6}$$

with

$$B_1 = B_0 + \delta \tau w_0 Du$$

and

$$C_1 = 2[\delta a_*^{ij} D_{ij} u + \delta a].$$

Although we can prove a global gradient estimate (under suitable hypotheses on the coefficients of $Q$) by manipulating this equation appropriately, we refer the reader to Section 15.2 of [64] for a complete description of the procedure, and, instead, we turn to interior estimates.

Let $B$ be a subset of the ball $B(y,R)$ for some $y \in \mathbb{R}^n$ and some $R > 0$, and suppose that $u \in C^3(B)$ satisfies $Qu = 0$ in $B$. If we define $\eta$ by

$$\eta(x) = 1 - \frac{|x-y|^2}{R^2},$$

then $w_1 = \eta^2 w_0$ satisfies the differential equation

$$a^{ij}D_{ij}w_1 - 2\eta^2 a_*^{ij}D_{ik}uD_{jk}u + B_2^iD_iw_1 + C_2w_1 - 2\eta w_0 B_1^iD_i\eta = 0,$$

where

$$B_2^i = B_1^i - \frac{2}{\eta}a^{ij}D_j\eta,$$

and

$$C_2 = C_1 - \frac{2}{\eta^2}a^{ij}D_{ij}\eta + \frac{2}{\eta^2}a^{ij}D_i\eta D_j\eta.$$

To continue, we rewrite this equation to separate terms involving second derivatives of $u$. Specifically, we have

$$B_1^iD_i\eta = D_k\eta a_*^{ij,k}D_{ij}u + a^iD_i\eta + \frac{1}{2\eta^2}(\tau_p \cdot D\eta Du + 2\tau D\eta)\cdot Dw_1$$
$$- \frac{1}{\eta}(\tau_p \cdot D\eta Du \cdot D\eta + 2\tau w_0|D\eta|^2),$$

and

$$C_2w_1 = 2\delta a_*^{ij}D_{ij}u\, w_1 + 2\delta a w_1 + \frac{2}{\eta^2}[a^{ij}D_i\eta D_j\eta - a^{ij}D_i\eta D_j\eta].$$

and hence

$$\begin{aligned} &a^{ij}D_{ij}w_1 - 2\eta^2 a_*^{ij}D_{ik}uD_{jk}u + B_3^iD_iw_1 \\ &\qquad = w_1\left(2\delta a_*^{ij} - \frac{2}{\eta}D_r\eta a_*^{ij,r}\right)D_{ij}u + 2\delta a w_1 + \tilde{I}w_1, \end{aligned} \tag{9.7}$$

where

$$B_3^i = B_2^i - \frac{1}{\eta}|Du|^2D\eta \cdot \tau_p Du - \frac{2}{\eta}|Du|^2\tau D\eta - \frac{2}{\eta}a^{ij}D_j\eta$$

and

$$\tilde{I} = \frac{2}{\eta^2}[\eta a^{ij}D_{ij}\eta - a^{ij}D_i\eta D_j\eta] + \frac{2}{\eta}(\delta\tau|Du|^2 Du \cdot D\eta)$$
$$+ \frac{2}{\eta}D\eta \cdot a_p - \frac{2}{\eta^2}[\tau_p \cdot DuDu \cdot D\eta + \tau|D\eta|^2].$$

For further computation, we set $M_1 = \sup_B w_1$ and we introduce a $C^2$ function $h$ (defined on some interval containing the range of $u$) such that $0 \le h \le 1$, and we set

$$w = w_1 + \frac{1}{2}M_1h(u).$$

We observe that $w_1 \le w \le w_1 + M_1/2$, so $M_1/2 \le \sup_B w \le 3M_1/2$. These comparisons on the maxima of $w$ and $w_1$ will be very useful.

To take advantage of the function $w$, we first recall

$$\mathcal{E}(x,z,p) = a^{ij}(x,z,p)p_i p_j$$

from (8.4), and we write

$$a^{ij}D_{ij}h = h'a^{ij}D_{ij}u + h''a^{ij}D_iuD_ju = -h'a + h''\mathcal{E}$$

and

$$\begin{aligned}B_2^iD_iu = \bar{\delta}a_*^{ij}D_{ij}u &+ \frac{1}{\eta^2}(\frac{1}{2}\bar{\delta}\tau + \tau)D_kuD_kw + \bar{\delta}a - \frac{2}{\eta}a^{ij}D_iuD_j\eta \\ &\frac{1}{2\eta}(\bar{\delta}\tau + 2\tau)w_0M_1h' - \frac{1}{\eta}(\bar{\delta}\tau + 2\tau)Du\cdot D\eta \\ &+ \delta\tau w_0^2 - \frac{2}{\eta}a^{ij}D_iuD_j\eta - \frac{w_0}{\eta}(\tau_p\cdot D\eta|Du|^2 + \tau Du\cdot D\eta).\end{aligned}$$

It then follows that

$$\begin{aligned}-2\eta^2a_*^{ij}D_{ik}uD_{jk}u + a^{ij}D_{ij}w + B_4^iD_iw = S \\ + \frac{1}{2}M_1[\mathcal{E}h'' + H_0(h')^2 + (J_0 + \tilde{J})h'] + (K_0 + \tilde{K})w_1,\end{aligned}$$

where

$$B_4^i = B_3^i - \frac{M_1h'}{2\eta^2}[\bar{\delta}\tau - \frac{1}{2}\tau]D_iu, \tag{9.8a}$$

$$S = -2w_1\delta a_*^{ij}D_{ij}u - \frac{2w_1}{\eta}D_k\eta a_*^{ij,k}D_{ij}u + \frac{1}{2}M_1h'\bar{\delta}a_*^{ij}D_{ij}u, \tag{9.8b}$$

$$H_0 = -\frac{M_1}{4\eta^2}|Du|^2(\bar{\delta}\tau + 2\tau), \tag{9.8c}$$

$$J_0 = (\bar{\delta} - 1)a + \frac{1}{2}\delta\tau|Du|^4, \tag{9.8d}$$

$$\tilde{J} = -\frac{4}{\eta}a^{ij}D_iuD_j\eta - \frac{1}{\eta}(\bar{\delta}\tau + 2\tau)Du\cdot D\eta, \tag{9.8e}$$

$$K_0 = -2\delta a, \tag{9.8f}$$

$$\tilde{K}_0 = \tilde{I} - \frac{w_0}{\eta}(D\eta\cdot\tau_p + \tau Du\cdot D\eta). \tag{9.8g}$$

(Note that $S$ contains all the terms in which second derivatives of $u$ appear linearly. The terms $H_0$, $J_0$, and $K_0$ will be important in determining the nature of the gradient estimate and they do not contain derivatives of $\eta$. The terms $\tilde{J}$ and $\tilde{K}_0$ contain derivatives of $\eta$; they will be treated as perturbation terms, so our eventual structure conditions will guarantee that they are small, in an appropriate sense, wherever $\eta^2|Du|^2$ is large.)

Our next step is to use the Cauchy-Schwarz inequality to estimate the terms in $S$:

$$2w_1\delta a_*^{ij}D_{ij}u \le \frac{1}{4}\eta^2 a_*^{ij}D_{ik}uD_{jk}u + 4|Du|^2 w_1\sum_{i,j}\frac{(\delta a_*^{ij})^2}{\lambda_*},$$

$$\frac{2w_1}{\eta}D_r\eta a_*^{ij,r}D_{ij}u \le \frac{1}{4}\eta^2 a_*^{ij}D_{ik}uD_{jk}u + 4|Du|^2 w_1\frac{|D\eta|^2}{\eta^2}\sum_{i,j,r}\frac{(a^{ij,r})^2}{\lambda_*},$$

$$-\frac{1}{2}M_0h'\bar\delta a_*^{ij}D_{ij}u \le \frac{1}{4}\eta^2 a_*^{ij}D_{ik}uD_{jk}u + \frac{1}{4}\frac{M_0^2}{\eta^2}(h')^2\sum_{i,j}\frac{(\bar\delta a_*^{ij})^2}{\lambda_*}$$

and hence

$$a^{ij}D_{ij}w + B_4^iD_iw \ge \frac{1}{2}M_1[\mathcal{E}h'' + H_1(h')^2 + J_0h'] + K_1w_1 + \frac{1}{2}M_1\tilde{J}h' + \tilde{K}w_1 \tag{9.9}$$

where

$$H_1 = H_0 - \frac{1}{2}\frac{M_1}{\eta^2}\sum_{i,j}\frac{(\bar\delta a_*^{ij})^2}{\lambda_*}, \tag{9.10a}$$

$$K_1 = K_0 - 4|Du|^2\sum_{i,j}\frac{(\delta a_*^{ij})^2}{\lambda_*}, \tag{9.10b}$$

$$\tilde{K} = \tilde{K}_0 - 4|Du|^2\frac{|D\eta|^2}{\eta^2}\sum_{i,j,r}\frac{(a_*^{ij,r})^2}{\lambda_*}. \tag{9.10c}$$

Next, we suppose that

$$\delta a \le O(\mathcal{E}), \tag{9.11a}$$

$$(\bar\delta - 1)a \le O(\mathcal{E}), \tag{9.11b}$$

$$|\delta a_*^{ij}| + |\bar\delta a_*^{ij}| = O(\sqrt{\lambda_*\mathcal{E}}), \tag{9.11c}$$

$$\bar\delta\tau|p|^4 \le O(\mathcal{E}) \tag{9.11d}$$

as $|p| \to \infty$, uniformly with respect to $x$ and $z$. It then follows that there are nonnegative constants $\bar{M}$, $H$, $J$, and $K$ (all determined only by the limit behavior in (9.11)) such that

$$H_1 \ge -H\mathcal{E}, \tag{9.12a}$$

$$J_0 \le J\mathcal{E}, \tag{9.12b}$$

$$K_1 \ge -K\mathcal{E} \tag{9.12c}$$

wherever $|Du| \ge \bar{M}$ and $w_1 \ge M_1/2$. The constants $H$, $J$, and $K$ will be used to determine the nature of the gradient estimate.

We are now ready to state our interior gradient estimate. In order to obtain a precise form of the estimate, we consider two possibilities. First, if the oscillation of $u$ is sufficiently small, then we have the following result.

**Theorem 9.1.** *Let $u \in C^3(B)$, where $B = B(y,R)$ for some $y \in \mathbb{R}^n$ and $R > 0$, and suppose that $Qu = 0$ in $B$. Suppose $a^{ij}$ can be decomposed as in (9.3) and that conditions (9.11) are satisfied. Let $\bar{M}$, $H$, $J$, $K$ be the constants from (9.12) and suppose that there is a nonnegative constant $\mu_1$ such that*

$$\Lambda|p|^2 \le \mu_1 \mathcal{E}, \tag{9.13a}$$

$$|p|^2|a^{ij}_{*,p}| \le \mu_1\sqrt{\lambda_* \mathcal{E}}, \tag{9.13b}$$

$$|p||a_p| \le \mu_1 \mathcal{E}, \tag{9.13c}$$

$$|p||\tau_p| \le \mu_1 \tau \tag{9.13d}$$

*whenever $|p| \ge \bar{M}$. If*

$$\operatorname{osc} u \le \min\{\frac{1}{1+2J}, \frac{1}{\sqrt{e(1+2K)(1+H)}}\}. \tag{9.14}$$

*then*

$$\sup_{B(y,R/2)} |Du| \le 2\bar{M} + C(n,\mu_1,H,K)\frac{\operatorname{osc} u}{R}. \tag{9.15}$$

***Proof.*** If $M_1 \le \bar{M}^2$, then the conclusion is immediate, so we assume that $M_1 > \bar{M}^2$. We set $\sigma = \operatorname{osc} u$ and choose

$$h(z) = \frac{1}{1+H}\exp(-(z - \inf_B u)^2/\sigma^2).$$

Note that

$$\frac{1}{(1+H)e} \le h \le \frac{1}{1+H}, \tag{9.16}$$

and hence

$$h''\mathcal{E} + H_1(h')^2 + J_0 h' \ge \mathcal{E}(h'' - H(h')^2 + Jh') = \frac{2}{\sigma^2}h\mathcal{E}H^*$$

for

$$H^* = 1 + 4\frac{(\sup u - u)^2}{\sigma^2}(1 - Hh) - J(\sup u - u)$$

wherever $w_1 \ge \frac{1}{2}M_1$. From the first inequality of (9.14), we have $\operatorname{osc} u \le 1/(2J)$, and the the second inequality of (9.16) gives $Hh \le 1$. Hence $H^* \ge 1/2$, and therefore

$$h''\mathcal{E} + H_1(h')^2 + J_0 h' \ge M_1 \frac{h}{\sigma^2}\mathcal{E}.$$

then the second inequality of (9.14) and the first inequality of (9.16) imply that

$$h''\mathcal{E} + H_1(h')^2 + J_0 h' \geq M_1\mathcal{E}\left(\frac{1}{2} + K\right).$$

Moreover, from (9.12c), we have

$$K_1 w_1 \geq -KM_1\mathcal{E},$$

and the second inequality of (9.14) along with the first inequality of (9.16) imply that

$$K_1 w_1 \geq \frac{-K}{1+2K}\frac{h}{\sigma^2}M_1\mathcal{E}.$$

Since $h \geq 1/(e(1+H))$, we infer that

$$\frac{1}{2}M_1[\mathcal{E}h'' + H_1(h')^2 + J_0 h'] + K_1 w_1 \geq \frac{1}{4e(1+H)(1+2K)\sigma^2}M_1\mathcal{E}$$

wherever $w_1 \geq M_1/2$.

Our next step is to estimate

$$\tilde{I} = \frac{1}{2}M_1\tilde{J}h' + \tilde{K}w_1.$$

We first observe that there is a positive constant $C_1$, determined only by $n$ and $\mu_1$, such that

$$\tilde{J} \geq -C_1\frac{1}{\eta R|Du|}\mathcal{E}$$

and

$$\tilde{K} \geq -C_1\left(\frac{1}{\eta R|Du|} + \frac{1}{\eta^2 R^2|Du|^2}\right)\mathcal{E}$$

whenever $w_1 \geq \bar{M}$. With $C_2$ a constant at our disposal, we assume that $M_1 \geq C_2\sigma^2/R^2$. Then

$$\frac{1}{2}M_1\tilde{J}h' \geq -C_1\frac{\sigma}{Rw_1^{1/2}}\frac{M_1 h}{\sigma^2}\mathcal{E} \geq -\frac{\sqrt{2}C_1 h}{\sqrt{C_2}2(1+H)\sigma^2}M_1\mathcal{E}$$

and

$$\tilde{K}w_1 \geq -C_1\left(\frac{\sqrt{2}}{C_2\sigma} + \frac{2}{C_2\sigma^2}\right)\frac{M_0 h}{\sigma^2}\mathcal{E}.$$

By choosing $C_2$ sufficiently small (determined only by $H$, $K$, and $C_1$) and noting that $\sigma \leq 1$, which follows from (9.14), we obtain

$$a^{ij}D_{ij}w + B_3^i D_i w > 0$$

on $S_0$, the set on which $w_1 \geq M_1/2$. It follows from the maximum principle that $w$ cannot attain its maximum in $S$ and hence it attains its maximum on $\partial S_0$. On $\partial B$, we have $w_1 = 0$ and hence $\sup w \leq M_1/2$ which implies that $\sup w_1 \leq M_1/2$, which can't happen. On $\partial S_0 \cap B$, we have $w_1 = M_1/2$, so $w \leq M_1(2+H)/(2(1+H)) < M_1$, which contradicts our inequality $\sup w \geq M_1$. Therefore we must have

$$M_1 \leq C_2 \frac{\sigma^2}{R^2}$$

and then (9.15) follows since $\eta \geq \frac{3}{4}$ in $B(y, R/2)$. □

If the oscillation of $u$ is not small, then a gradient estimate is true provided we make an additional hypothesis on the coefficients $H$, $J$, and $K$. In our next theorem, we assume that $K$ is sufficiently small and we refer the reader to the Notes for a discussion of other possibilities.

**Theorem 9.2.** *Let $u \in C^3(B)$, where $B = B(y, R)$ for some $y \in \mathbb{R}^n$ and $R > 0$, and suppose that $Qu = 0$ in $B$. Suppose $a^{ij}$ can be decomposed as in* (9.3) *and that conditions* (9.11) *are satisfied. Let $\bar{M}$, $H$, $J$, $K$ be the constants from* (9.12) *and suppose that there is a nonnegative constant $\mu_1$ such that condition* (9.13) *holds whenever $|p| \geq \bar{M}$. If*

$$\operatorname*{osc}_B u \geq \min\{\frac{1}{1+2J}, \frac{1}{\sqrt{e(1+2K)(1+H)}}\} \tag{9.17}$$

*and*

$$K \leq \frac{1}{2}(J+1)\exp(-(J+1)(H+1)\operatorname*{osc}_B u), \tag{9.18}$$

*then*

$$\sup_{B(y,R/2)} |Du| \leq 2\bar{M} + C(n, \mu_1, H, J, K)\frac{\operatorname{osc} u}{R}. \tag{9.19}$$

***Proof.*** Again, we assume that $M_1 \geq \bar{M}^2$, and we set $\sigma = \operatorname{osc}_B u$. With $A = (1+J)(1+H)$, we choose

$$h(z) = \frac{1}{1+H}\exp(A(\inf_B u - z)).$$

It then follows that

$$\frac{1}{2}M_1(h''\mathcal{E} + H_1(h')^2 + J_0 h') + K_1 w_1 \geq \frac{1}{4}Ah\mathcal{E}$$

wherever $w_1 \geq M_1/2$.

The estimate of $\tilde{I}$ in the proof of Theorem 9.1 is easily modified to show that there is a constant $C_2(n, \mu_1, H, J, K)$ such that

$$\tilde{I} \geq -\frac{1}{8}AhM_0\mathcal{E}$$

whenever $w_1 \geq M_1/2$ provided $M_1 \geq C_2\sigma^2/R^2$.

Again, we conclude that

$$a^{ij}D_{ij}w + B_3^iD_iw > 0$$

whenever $w_1 \geq M_1/2$ provided $M_1 \geq C_2\sigma^2/R^2$, and hence (9.19) follows. □

For future reference, we note that a gradient bound can also be proved if we have a bound for the gradient only on the boundary of $\Omega$.

**Corollary 9.3.** *Let $\Omega$ be an open set and let $B = B(y,R) \cap \Omega$ for some $y \in \mathbb{R}^n$ and some $R > 0$. Let $u \in C^3(B) \cap C^1(B(y,R) \cap \overline{\Omega})$ and suppose that $Qu = 0$ in $B$. Under the hypotheses of Theorem* 9.15, *we have*

$$\sup_{B(y,R/2)\cap\Omega} |Du| \leq 2(\bar{M} + \sup_{\partial\Omega\cap B(y,R)} |Du|) + C(n, \mu_1, H, K)\frac{\operatorname{osc}_B u}{R}. \quad (9.20)$$

*Under the hypotheses of Theorem* 9.19, *we have*

$$\sup_{B(y,R/2)\cap\Omega} |Du| \leq 2(\bar{M} + \sup_{\partial\Omega\cap B(y,R)} |Du|) + C(n, \mu_1, H, J, K)\frac{\operatorname{osc}_B u}{R}. \quad (9.21)$$

***Proof.*** Just note that $w$ can't have a maximum on $\partial\Omega \cap \partial B$ if $M_1 \geq \sup_{\partial\Omega\cap B(y,R)} |Du|^2$. □

## 9.2 A simple boundary value problem

We now observe that the arguments of the preceding section yield a gradient bound when the boundary condition is

$$Du \cdot \gamma = 0 \text{ on } \partial\Omega \quad (9.22)$$

and $\Omega$ is convex. This boundary condition implies that $Du$ is a tangential vector field on $\partial\Omega$ and hence we can differentiate the boundary condition in the direction of $Du$ to obtain $Du \cdot D(Du \cdot \gamma) = 0$ or

$$D_iuD_{ij}u\gamma^j = -D_juD_iuD_i\gamma^j. \quad (9.23)$$

If we set

$$w_0 = |Du|^2$$

(as before), then $D_j w = 2D_i u D_{ij} u$, and hence

$$Dw \cdot \gamma = 2D_i u D_{ij} u \gamma^j = -2D_j u D_i u D_i \gamma^j$$

by virtue of (9.23). Since $D\gamma$ is negative semidefinite, we infer that

$$Dw_0 \cdot \gamma \geq 0 \text{ on } \partial\Omega. \tag{9.24}$$

If we repeat the proof of Theorem 9.2 with $\eta \equiv 1$, then we infer under the hypotheses (9.11),

$$\operatorname*{osc}_{\Omega} u \geq \min\{\frac{1}{1+2J}, \frac{1}{\sqrt{e(1+2K)(1+H)}}\}, \tag{9.25}$$

and (9.18) that

$$a^{ij} D_{ij} w + B_4^i D_i w > 0$$

wherever $w_0 \geq M_0/2$, where

$$w = w_0 + \frac{1}{2} M_0 h(u)$$

and $M_0 = \sup w_0$. It's also easy to check that

$$\gamma \cdot Dw \geq 0$$

on $\partial\Omega$. If $w$ has a maximum at a point where $|Du| \leq \bar{M}$, then we have a bound for $Du$ everywhere. Otherwise, the strong maximum principle implies that $w_0$ must be a constant. Now let $x_0$ be a point on $\partial\Omega$ at which $u$ attains its maximum over $\partial\Omega$. It follows from (9.24) that $Du(x_0) = 0$ because the tangential derivatives of $u$ must be zero at $x_0$, which implies that $|Du| \equiv 0$, which also implies a gradient bound.

## 9.3 Gradient estimates for general boundary conditions. General considerations

If we want to study oblique boundary value problems under more general hypotheses, then the considerations of the previous section are no longer adequate for obtaining gradient bounds. It is not hard to see that the choice of estimating $w = |Du|^2$ in Section 9.2 is closely tied to the specific boundary condition $Du \cdot \gamma = 0$. Hence, we want to estimate a function $w$ with two basic properties:

(1) $w$ satisfies a suitable differential inequality in the domain and a suitable boundary inequality which allow the application of the maximum principle.

(2) Estimates on $w$ imply estimates on $Du$.

There are several considerations that are implied by the general description of these two properties. First, the function $w$ must take the boundary condition into account, and, second, $w$ should be a convex function of $Du$ in order to control the differential inequality for $w$. (The convexity of $w = |Du|^2$ was used in the previous two sections to obtain a lower bound for $a_*^{ij} D_{ik}uD_{jk}u$.) Ideally, $w$ should even be strictly convex in order to estimate $|Du|$ in terms of $w$. In fact, we shall now use a strictly convex $w$, and we shall not estimate $|Du|$ directly. For the simple boundary condition $Du \cdot \gamma = 0$, any estimate for the tangential gradient of $u$ on $\partial\Omega$ implies a corresponding estimate for $|Du|$ on $\partial\Omega$; then we can the argument in Corollary 9.3 to obtain an estimate for $|Du|$ in all of $\Omega$. This is the basis for the argument in the next section. We make some preliminary observations in this section.

As a first step, we recall from Lemma 5.18 that, if $\partial\Omega \in C^3$, then there is a positive constant $R_0$, determined only by $\Omega$ such that $d \in C^3(\overline{\Omega_*})$, where $\Omega_*$ is the subset of $\Omega$ on which $d < R_0$. It follows that $\gamma = Dd$ is a $C^2(\overline{\Omega_*})$ unit vector field. In fact, many of the results in this chapter can be proved (even for $\partial\Omega \in H_{2+\alpha}$) by using other unit vector fields. We also define

$$c^{ij} = \delta^{ij} - \gamma^i\gamma^j,$$

and we set

$$w_T = c^{ij} D_iuD_ju.$$

Under some reasonable assumptions, we can estimate $|Du \cdot \gamma|$ in terms of $v$ on $\partial\Omega$ just by using the boundary condition.

**Lemma 9.4.** *Let $b \in C^0(\partial\Omega \times I \times \mathbb{R}^n)$ for some set $I$ of real numbers, and suppose that $b$ is differentiable with respect to the third variable with $b_p \cdot \gamma > 0$ and*

$$\lim_{t\to\infty} \pm b(x, z, p \pm t\gamma) > 0. \tag{9.26}$$

*Suppose also that there are positive constants $\mu_2$, $M_3$ and $M_4$ such that if $|p - (p \cdot \gamma(x))\gamma(x)| \le M_3$ and $b(x, z, p) = 0$, then $|p \cdot \gamma| \le M_4$ and*

$$|b_p(x, z, p)| \le \mu_2 b_p(x, z, p) \cdot \gamma(x) \tag{9.27}$$

*for all $(x, z, p) \in \partial\Omega \times \mathbb{R} \times \mathbb{R}^n$ with $b(x, z, p) = 0$. If $u \in C^1(\overline{\Omega})$ satisfies $b(x, u, Du) = 0$ on $\partial\Omega$ and $u(x) \in I$ for all $x \in \partial\Omega$, then*

$$|Du \cdot \gamma| \le M_4 + \mu_2 w_T^{1/2}$$

*on $\partial\Omega$.*

***Proof.*** We set $p' = p - (p\cdot\gamma)\gamma$ and $D'u = Du - (Du\cdot\gamma)\gamma$ (so $D'_i u = c^{ij}D_j u$), we fix $x \in \partial\Omega$, and we consider two cases. If $|D'u| \le M_3$, then we have $|Du\cdot\gamma| \le M_4$.

If $|D'u| > M_3$, then we note that the equation $b(x,u,p'+p_\gamma\gamma) = 0$ gives a function $F$ such that $p_\gamma = F(p')$. In other words, if $p$ is any vector such that $|p'| \ge M_2$, then there is a unique $p_\gamma \in \mathbb{R}$ such that $b(x,u,p'+p_\gamma\gamma) = 0$. Noting that $p'$ lies in an $(n-1)$-dimensional hyperplane, we see that

$$\left|\frac{\partial F}{\partial p'}\right| = \frac{|b'_p|}{b_p\cdot\gamma} \le \mu_2.$$

Since $|p_\gamma| \le M_4$ if $|p'| = M_3$, it follows from the mean value theorem that $|p_\gamma| \le M_4 + \mu_2|p'|$ and hence $|Du\cdot\gamma| \le M_4 + \mu_2|D'u|$ in this case. Combining these inequalities and observing that $|D'u| = w_T^{1/2}$ completes the proof. □

A number of other functions will also be useful in our gradient bounds. We write $\tilde{c}$ for the vector with components $\tilde{c}_i$ defined by

$$\tilde{c}_i(x,p) = D_i(c^{km})p_k p_m.$$

Since there will be no danger of confusion, we abbreviate $\tilde{c}_i(x,Du)$ to $\tilde{c}_i$. Similarly, we define the vector-valued functions $Tu$ and $T'u$ with components given by

$$T_i u = c^{ij}D_j u, \quad T'_i u = D_i\gamma^j D_j u,$$

and we set

$$\Gamma u = Du\cdot\gamma.$$

Because $\gamma$ is a unit vector, it follows that $\gamma^j D_i\gamma^j = 0$ and that $\tilde{c}_i = -2\Gamma u T'_i u$. We also define the matrix $[\bar{c}_{ij}]$ with entries given by

$$\bar{c}_{ij} = |Tu|^{-2}D_{ij}(c^{km})D_k u D_m u.$$

We also define two operators $\delta_T$ and $\delta_N$, acting on functions of the variables $(x,z,p)$, by

$$\delta_T g = g_z + |p'|^{-2}p'\cdot g_x - \frac{1}{2}|p'|^{-2}\tilde{c}\cdot g_p, \quad \delta_N g = \gamma\cdot g_p.$$

Another structure functions that will be very useful is

$$\mathfrak{C}_T^2 = a^{ij}c^{km}D_{ik}uD_{jm}u.$$

For the estimates, we also use the abbreviations

$$v = \sqrt{1+|Du|^2}, \quad \nu = \frac{Du}{v}, \quad g^{ij} = \delta^{ij} - \nu^i\nu^j.$$

In addition, if $[f^{ij}]$ is a matrix, we write $F$ in place of $f^{ij}\gamma_i\gamma_j$. The only exception to this abbreviation is that we reserve the letter $C$ to denote a constant. Since $c^{ij}\gamma_i\gamma_j = 0$, such an exception is natural.

Although it is possible to prove a number of global gradient estimates directly, we prove our global gradient estimates via local gradient estimates because the additional complications introduced by this step are relatively minor compared to those generated by the general boundary condition we consider and any smooth extension of $\gamma$ cannot be a unit vector in all of $\Omega$. In fact, the second difficulty is not significant in the proof since the extension can be taken to be a vector with length no greater than one.

## 9.4 Global gradient estimates for general boundary conditions and false mean curvature equations I

We again assume that the decomposition (9.3) is valid for some positive definite matrix $[a_*^{ij}]$ with minimum eigenvalue $\lambda_*$ and some nonnegative scalar function $\tau$. We also assume that $a_*^{ij}$ and $\tau$ are $C^1$ with respect to $(x, z, p)$.

To illustrate the use of our decomposition, we begin by deriving a gradient estimate under fairly strong assumptions on the coefficients of $Q$ and $M$. Because the calculations, even in this case, are delicate, we first derive a global estimate for $w_T$ under strong hypotheses. To derive a differential inequality for $w_T$, we first differentiate the equation $Qu = 0$ in the direction $x^m$ and then multiply the resulting equation by $c^{km}D_k u$. In this way, we obtain

$$a^{ij}D_{ij}w_T - 2\mathfrak{C}_T^2 + B_0^r D_r w_T = S + I_1, \tag{9.28}$$

with

$$\begin{aligned} S &= -2\delta_T a^{ij}D_{ij}u w_T + 2a^{ij}D_{im}uD_j(c^{km})D_k u, \\ I_1 &= -2\delta_T a w_T + a^{ij}\bar{c}_{ij}w_T. \end{aligned}$$

Our goal is now easy to state. As before, the second derivatives of $u$ appear only linearly in $S$, so we want to use Cauchy's inequality to estimate $S$ in terms of $\mathfrak{C}_T^2$ (which is quadratic in the second derivatives) and then use appropriate structure conditions to show that we obtain a differential inequality for $w$ which yields a maximum principle. Unfortunately, $\mathfrak{C}_T^2$ is only quadratic with respect to some of the second derivatives. For example, if $B^0$ lies in the hyperplane $x^n = 0$ and if $a^{ij} = \delta^{ij}$, then

$$\mathcal{C}_T^2 = \sum_{i=1}^{n}\sum_{j=1}^{n-1}(D_{ij}u)^2.$$

In general, we expect the terms in $S$ to include $D_{nn}u$ in this special situation, so we must estimate this second derivative, and this estimate is accomplished because the differential equation for $u$ can be solved for $D_{nn}u$. We use the decomposition (9.3) of $a^{ij}$ to achieve this subgoal by rewriting some of the terms of $S$ using $Dw_T$.

We begin by showing how to estimate various linear combinations of second derivatives of $u$. Our first estimate will be used to estimate terms involving derivatives of $a_*^{ij}$, so we obtain bounds in terms of $\lambda_*$. To simplify notation, if $\mathcal{F} = [f^{ij}]$ is a matrix, we define

$$\|\mathcal{F}\| = \left(\sum_{i,j=1}^{n}(f^{ij})^2\right)^{1/2}. \tag{9.29}$$

**Lemma 9.5.** *If $Qu = 0$, then for any matrix $[f^{ij}]$ and any $\varepsilon > 0$, we have*

$$\begin{aligned} f^{ij}D_{ij}u \le \varepsilon\mathcal{C}_T^2 &+ \frac{2F^2}{\varepsilon A\lambda_*}\tau w_T \\ &+ \frac{4\|\mathcal{F}\|^2}{\lambda_*\varepsilon} + (n+1)\frac{F^2\Lambda_*}{\varepsilon A\lambda_*} - \frac{Fa}{A}. \end{aligned} \tag{9.30}$$

***Proof.*** We set

$$\begin{aligned} f_1 &= f^{ij}c^{jk}D_{ik}u, \\ f_2 &= f^{im}\gamma_m c^{ij}D_{jk}u\gamma^k, \\ f_3 &= -\frac{F}{A}a^{ij}c^{jk}D_{ik}u, \\ f_4 &= -\frac{F}{A}a^{im}\gamma_m c^{ij}D_{jk}u\gamma^k. \end{aligned}$$

It follows that

$$f^{ij}D_{ij}u = f_1 + f_2 + f_3 + f_4 + \frac{F}{A}a^{ij}D_{ij}u,$$

so the equation $Qu = 0$ implies that

$$f^{ij}D_{ij}u = f_1 + f_2 + f_3 + f_4 - \frac{F}{A}a. \tag{9.31}$$

Since

$$a^{ij}\xi_i\xi_j \ge a_*^{ij}\xi_i\xi_j \ge \lambda_*|\xi|^2$$

for any vector $\xi$, we conclude directly from the Cauchy-Schwarz inequality that

$$f_1 \le \frac{1}{8}\varepsilon\mathcal{C}_T^2 + \frac{2\|\mathcal{F}\|^2}{\lambda_*\varepsilon},$$

$$f_2 \le \frac{1}{8}\varepsilon\mathcal{C}_T^2 + \frac{2\|\mathcal{F}\|^2}{\lambda_*\varepsilon}.$$

Next, we use the Cauchy-Schwarz inequality another time to infer that

$$f_3 \le \frac{1}{4}\varepsilon\mathcal{C}_T^2 + \frac{1}{\varepsilon}\frac{F^2}{A^2}a^{ij}c^{ij},$$

and we also have

$$a^{ij}c^{ij} = a_*^{ij}c^{ij} + \tau D_i u c^{ij} D_j u = a_*^{ij}c^{ij} + \tau w_T \le n\Lambda_* + \tau w_T. \tag{9.32}$$

Hence,

$$f_3 \le \frac{1}{4}\varepsilon\mathcal{C}_T^2 + \frac{1}{\varepsilon}\frac{F^2}{A^2}[n\Lambda_* + \tau w_T].$$

Using the decomposition (9.3), we see that

$$f_4 = -\frac{F}{A}a_*^{ij}\gamma_m c^{ij} D_{jk}u\gamma_k - \frac{F}{A}\tau\Gamma u D_i u c^{ij} D_{jk}u\gamma^k.$$

Schwarz's inequality gives

$$\begin{aligned} -\frac{F}{A}a_*^{im}\gamma_m c^{ij} D_{jk}u\gamma^k &\le \frac{|F|}{A}(a_*^{ij}\gamma_i\gamma_j)^{1/2}(\Lambda_*\sum_{i,k}(c^{ij}D_{jk}u)^2)^{1/2} \\ &\le \frac{|F|}{A^{1/2}}\left(\frac{\Lambda_*}{\lambda_*}a_*^{ij}c^{km}D_{ik}uD_{jm}u\right)^{1/2} \end{aligned}$$

because $a_*^{ij}\gamma_j\gamma_j = A_* \le A$, and hence Cauchy's inequality implies that

$$-\frac{F}{A}a_*^{im}\gamma_m c^{ij} D_{jk}u\gamma^k \le \frac{1}{4}\varepsilon\mathcal{C}_T^2 + \frac{F^2\Lambda_*}{\varepsilon\lambda_* A}.$$

A final application of the Cauchy-Schwarz inequality yields

$$-\frac{F}{A}\tau D_i u D_m u\gamma_m c^{ij} D_{jk}u\gamma^k \le \frac{1}{4}\varepsilon\mathcal{C}_T^2 + \frac{F^2\tau^2(\Gamma u)^2}{\varepsilon A^2\lambda_*}w_T$$

and hence, since $\tau(\Gamma u)^2 \le A$,

$$f_4 \le \frac{1}{2}\varepsilon\mathcal{C}_T^2 + \frac{\Lambda_* F^2}{\varepsilon A\lambda_*} + \frac{F^2\tau}{\varepsilon A\lambda_*}w_T.$$

The proof is completed by combining the estimates on $f_1$, $f_2$, $f_3$, and $f_4$ and noting that $A \ge \lambda_*$. □

The estimate for terms involving $\tau$ and its derivatives is similar.

**Lemma 9.6.** *Let $\xi$ be a vector and set $\zeta = \xi \cdot \gamma$. If $Qu = 0$, then for any $\varepsilon > 0$, we have*

$$\begin{aligned}\xi^r D_{ir}uD_iu \le \varepsilon \mathcal{C}_T^2 + \frac{1}{\varepsilon\tau}c^{ij}\xi_i\xi_j + \frac{(n+1)\Lambda_*}{\varepsilon\tau A}\zeta^2 \\ + \frac{\zeta^2 A_*^2 w_T}{\varepsilon A^2\lambda_*} - \frac{\zeta\Gamma u}{A}a.\end{aligned} \tag{9.33}$$

***Proof.*** We begin by noting that

$$\xi^r D_{ir}uD_iu = \xi^k c^{kr} D_{ir}uD_iu + \zeta\gamma^r D_{ir}uD_iu,$$

and the Cauchy-Schwarz inequality implies that

$$\xi^k c^{kr} D_{ir}uD_iu \le \frac{1}{4}\varepsilon\mathcal{C}_T^2 + \frac{1}{\varepsilon\tau}c^{ij}\xi_i\xi_j.$$

To continue, we set

$$\begin{aligned}Z_1 &= -\frac{\zeta\Gamma u}{A}[a^{ij}c^{ir}D_{rj}u], \\ Z_2 &= -\frac{\zeta\Gamma u}{A}[a_*^{ij}\gamma_i\gamma_r c^{jm}D_{rm}u], \\ Z_3 &= -\frac{\zeta\Gamma u}{A}[\tau\Gamma u D_j u\gamma_r c^{jm}D_{rm}u], \\ Z_4 &= \zeta\gamma^i c^{jr}D_{ij}uD_ru,\end{aligned}$$

and we observe that

$$\zeta\gamma^r D_{ir}uD_iu = Z_1 + Z_2 + Z_3 + Z_4 - \frac{\zeta\Gamma u}{A}a.$$

We then use the Cauchy-Schwarz inequality and (9.32) to infer that

$$\begin{aligned}Z_1 &\le \frac{\varepsilon}{4}\mathcal{C}_T^2 + \frac{\zeta^2(\Gamma u)^2}{\varepsilon A^2}(n\Lambda_* + \tau w_T) \\ &\le \frac{\varepsilon}{4}\mathcal{C}_T^2 + \frac{\zeta^2}{\varepsilon\tau A}(n\Lambda_* + \tau w_T)\end{aligned}$$

because $\tau(\Gamma u)^2 \le A$.

By applying the Cauchy-Schwarz inequality and the inequality $\tau(\Gamma u)^2 \le A$ again, we see that

$$Z_2 \le \frac{\varepsilon}{4}\mathcal{C}_T^2 + \frac{\zeta^2}{\varepsilon\tau A}\Lambda_*.$$

Next, we observe that

$$\begin{aligned}Z_3 + Z_4 &= \zeta\left(1 - \frac{\tau(\Gamma u)^2}{A}\right)\gamma^i c^{jr}D_{ij}uD_ru \\ &= \zeta\frac{A_*}{A}\gamma^i c^{jr}D_{ij}uD_ru.\end{aligned}$$

Another application of the Cauchy-Schwarz inequality leads to

$$Z_3 + Z_4 \le \frac{\varepsilon}{4}\mathcal{C}_T^2 + \frac{\zeta^2 A_*^2 w_T}{\varepsilon A^2 \lambda_*}.$$

The proof is completed by combining all these estimates. □

We are now ready to estimate $S$. First, we write

$$w_T(\delta_T a^{ij})D_{ij}u = w_T(\delta_T a_*^{ij})D_{ij}u + (w_T\delta_T\tau D_i u - \tau\tilde{c}^i)D_j u D_{ij}u.$$

We estimate $(\delta_T a_*^{ij})D_{ij}uw_T$ by using $f^{ij} = w_T\delta_T a_*^{ij}$ in Lemma 9.5 with $\varepsilon = 1/8$, and we estimate $(w_T\delta_T\tau D_i u - \tau\tilde{c}^i)D_j u D_{rj}u$ by using $\xi = w_T\delta_T\tau Du - \tau\tilde{c}$ and $\varepsilon = 1/8$ in Lemma 9.6. Combining these estimates with the inequalities $A \ge A_* \ge \lambda_*$ yields

$$-(\delta_T a^{ij})D_{ij}uw_T \ge -\frac{1}{4}\mathcal{C}_T^2 + A_1 w_T, \tag{9.34}$$

with

$$\begin{aligned} A_1 = &- \frac{16(\delta_T A_*)^2\tau w_T^2}{\lambda_*^2} - \frac{32\|\delta_T \mathcal{A}_*\|^2}{\lambda_*} - \frac{16(\delta_T\tau)^2 w_T^2}{\tau} \\ &- 8(n+1)\Lambda_* w_T\left(\frac{(\delta_T A_*)^2}{\lambda_*^2} + \frac{(\delta_T\tau)^2}{\tau^2}\right) - \frac{16\tau|\tilde{c}|^2}{w_T} \\ &- |a|\left(\frac{|\delta_T A_*|}{\lambda_*} + \frac{|\delta_T\tau|}{\tau}\right). \end{aligned} \tag{9.35}$$

To obtain a useful lower bound for the second term,

$$2a^{ij}D_{im}uD_j(c^{km})D_k u,$$

in $S$, we note that $D_j(c^{km}) = -D_j\gamma^m\gamma^k - D_j\gamma^k\gamma^m$ and therefore (using also (9.3))

$$-a^{ij}D_{im}uD_j(c^{km})D_k u = I_2 + I_3 + I_4,$$

where

$$\begin{aligned} I_2 &= a^{ij}D_{im}uD_j\gamma^m\Gamma u, \\ I_3 &= a_*^{ij}D_{im}u\gamma^m D_j\gamma^k D_k u, \\ I_4 &= \tau D_i u D_{im}u\gamma^m D_j\gamma^k D_k u D_j u. \end{aligned}$$

We begin by using the Cauchy-Schwarz inequality to infer that

$$I_2 \le \frac{1}{8}\mathcal{C}_T^2 + 2(\Gamma u)^2 a^{ij}D_i\gamma^m D_j\gamma^m.$$

Next, we use Lemma 9.5 with $f^{ij} = a_*^{im} D_m\gamma^k D_k u\gamma^j$ and $\varepsilon = 1/8$ to infer that

$$I_3 \le \frac{1}{16}\mathcal{C}_T^2 + \frac{32(a_*^{im}D_m\gamma^k D_k u\gamma^i)^2}{A\lambda_*}\tau w_T + \frac{64a_*^{im}D_m\gamma^k D_k u a_*^{ir} D_r\gamma^j D_j u}{\lambda_*}$$
$$+ 16(n+1)\frac{(a_*^{im}D_m\gamma^k D_k u\gamma^i)^2\Lambda_*}{\lambda_*^2} - \frac{a_*^{im}D_m\gamma^k D_k u\gamma^i}{A}a.$$

To estimate $I_4$, we first note that $D_j\gamma^k D_k u D_j u = T'u \cdot Du$. It then follows from Lemma 9.6 with $\xi = \tau T'u \cdot Du\gamma$ and $\varepsilon = 1/16$ that

$$I_4 \le \frac{1}{16}\mathcal{C}_T^2 + \frac{16(n+1)\tau\Lambda_*(T'u \cdot Du)^2}{\lambda_*} - \frac{\tau(T'u \cdot Du)\Gamma u}{A}a$$
$$+ \frac{16\tau^2(T'u \cdot Du)^2}{\lambda_*}.$$

Combining these estimates with the inequality $\tau(\Gamma u)^2 \le A$ yields

$$a^{ij}D_{im}uD_j(c^{km})D_k u \ge -\frac{1}{4}\mathcal{C}_T^2 + A_2 w_T$$

with

$$A_2 = -\frac{2}{w_T}(\Gamma u)^2 a^{ij}D_i\gamma^m D_j\gamma^m - \frac{32(a_*^{im}D_m\gamma^k D_k u\gamma^i)^2\tau}{\lambda_*^2}$$
$$- \frac{64a_*^{im}D_m\gamma^k D_k u a_*^{ir}D_r\gamma^j D_j u}{\lambda_* w_T}$$
$$- \frac{16(n+1)\Lambda_*}{\lambda_* w_T}((a_*^{im}D_m\gamma^k D_k u\gamma^i)^2 + \tau(T'u \cdot Du)^2)$$
$$- \frac{16\tau^2(T'u \cdot Du)^2}{\lambda_* w_T}$$
$$+ a\left(\frac{a_*^{im}D_m\gamma^k D_k u\gamma^i + \tau(\Gamma u)T'u \cdot Du}{Aw_T}\right).$$

We now collect these estimates to obtain

$$-\mathcal{C}_T^2 + a^{ij}D_{ij}w_T + B_0^i D_i w_T \ge A_3 w_T, \tag{9.36}$$

where

$$A_3 = 2A_1 + 2A_2 - 2\delta_T a + a^{ij}\bar{c}_{ij}. \tag{9.37}$$

For later use, the $\mathcal{C}_T^2$ term will be useful, but, for now, we note that

$$a^{ij}D_{ij}w_T + B_0^i Dw_T \ge A_3 w_T. \tag{9.38}$$

To apply the maximum principle, we need to introduce hypotheses that imply $A_3 > 0$ and the complicated form of this quantity make its analysis somewhat unwieldy. We start by taking advantage of a few simple inequalities. Specifically, if we assume that there is a positive constant $\mu_3$ such that

$$|\delta_T a_*^{ij}| \le \mu_3\lambda_*, \quad |\delta_T\tau| \le \mu_3\tau \tag{9.39a}$$

$$\Lambda_* \le \mu_3\lambda_* \tag{9.39b}$$

wherever $|Du| \ge 1$, then there is a constant $C_4(n,\Omega,\mu_3)$ such that

$$A_1 \ge -C_4(\lambda_*|Du|^2 + \tau|Du|^4 + |a|) \tag{9.40}$$

wherever $w_T \ge 1$.

Similarly, if there is a positive constant $\mu_3$ such that condition (9.39b) holds whenever $|Du| \ge 1$, then there is a constant $C_5(n,\Omega,\mu_3)$ such that

$$A_2 \ge -C_5\left[\lambda_*|Du|^2 + \tau|Du|^4 + |a|\left(1 + \frac{\tau^{1/2}}{A^{1/2}}\right)\right] \tag{9.41}$$

whenever $w_T \ge 1$.

At this point, we can state a gradient bound under fairly strong hypotheses. The most stringent one is that $\delta_T a$ must be large and negative.

**Theorem 9.7.** *Let $u \in C^3(\Omega) \cap C^2(\overline{\Omega})$ and suppose that there are a positive definite matrix $[a_*^{ij}]$ and a nonnegative scalar $\tau$ such that $[a^{ij}]$ can be decomposed as in* (9.3)*. Suppose conditions* (9.11) *are satisfied and let $M_1$, $H$, $J$, and $K$ be the constants from* (9.12)*. Suppose that there is a nonnegative constant $\mu_1$ such that* (9.13) *holds. Suppose also that $b_p \cdot \gamma > 0$,* (9.26) *is satisfied, $\delta_T b < 0$, and that there are positive constants $\mu_2$, $M_3$ and $M_4$ such that* (9.27) *holds for all $(x,z,p)$ with $b(x,z,p) = 0$ and that if $b(x,z,p) = 0$ and $|p-(p\cdot\gamma)\gamma| \le M_3$, then $|p\cdot\gamma| \le M_4$. Set $R = \min\{1, R_0\}$. Suppose further that, for any $y \in \Omega$ with $d(y) \ge R$, either condition* (9.14) *or conditions* (9.17) *and* (9.18) *are satisfied. Suppose finally that there are positive constants $\mu_3$, $\mu_4$, and $\bar{\mu}_1$ such that conditions* (9.39) *and*

$$\tau \le \mu_4\lambda_*, \tag{9.42a}$$

$$|a| \le \bar{\mu}_1\delta_T a, \quad \mathcal{E} \le \bar{\mu}_1\delta_T a \tag{9.42b}$$

*are satisfied for $|Du| \ge \bar{M}$. If*

$$\bar{\mu}_1(C_4 + C_5(1 + \mu_4^{1/2}) + 1) \le 2$$

*(for the constants $C_4$ and $C_5$ from* (9.40) *and* (9.41)*), then*

$$\sup_\Omega |Du| \le C(n,\Omega,\mu_1,\ldots,\mu_4,\bar{M},M_3,M_4,H,J,K,\operatorname{osc}_\Omega u).$$

***Proof.*** We first fix $y \in \Omega$ with $d(y) \geq R$. If condition (9.14) holds, then Theorem 9.1 gives a constant $C_6$ (determined only by $n$, $\mu_1$, $\operatorname{osc} u$, $H$, $M_1$, $M_2$, and $\Omega$) such that $|Du(y)| \leq C_6$. On the other hand, if conditions (9.17) and (9.18) hold, then Theorem 9.2 gives a constant $C_7$ (determined by $J$, $K$, and the same quantities as for $C_6$) such that $|Du(y)| \leq C_7$. It then follows from Corollary 9.3 and Lemma 9.4 that

$$\sup |Du| \leq C(C_6, C_7, \mu_2, M_3, \sup_{\Gamma_R} w_T),$$

and hence we only need to estimate $w_T$ in $\Gamma_R$. Moreover, if we set $M_6 = \max\{C_6, C_7\}$, then

$$w_T \leq M_6 \text{ on } \partial\Gamma_R \cap \Omega.$$

Next, because $\delta_T b < 0$ on $\partial\Omega$ and

$$b^r D_r w_T + (\delta_T b) w_T = 0$$

on $\partial\Omega$, it follows that $w_T$ can't have a maximum on $\partial\Omega$.

If $w_T$ has a maximum at some $x_0 \in \Gamma_R$, then we use (9.38) to infer that $A_3(x_0) \leq 0$. If $w_T(x_0) \geq \max\{1, M_1, \ldots, M_5\}$, then we have

$$A_3(x_0) \geq -\delta_T a[2 - (2C_4 + 2C_5(1 + \mu_4^{1/2}) + C(\Omega) w_T^{-1/2})\bar{\mu}_1].$$

But this inequality implies that $A_3(x_0) > 0$ if $w_T(x_0) > C(\Omega)^2$. It follows that

$$\sup_{\Gamma_R} w_T \leq \max\{2, 2\bar{M}, 2M_4, C(\Omega)^2\},$$

and the desired estimate follows. □

## 9.5 Global gradient estimates for general boundary conditions and false mean curvature equations II

If we want to relax the hypotheses $\delta_T a < 0$ and $\delta_T b < 0$, we need to apply the maximum principle to a function similar to $w$ from Section 9.1. We set $M_T = \sup_{\Gamma_R} w_T$ and

$$w = w_T + \frac{1}{4k_0} M_T v d + \frac{1}{4} M_T h(u)$$

with $k_0$ is a constant at our disposal such that

$$dv \leq k_0, \tag{9.43}$$

and $h$ a function to be further specified but satisfying $0 \leq h \leq 1$. (Hypotheses on the operator $Q$ which guarantee such an inequality will be made later.) Using the notation from Section 9.4, we infer from (9.36) that

$$-\mathcal{C}_T^2 + a^{ij}D_{ij}w + B_0^iD_iw \geq A_3w_T + (A_4 + A_5)M_T,$$

where

$$A_4 = \frac{1}{4k_0}[v(a^{ij}D_{ij}d + B_0^iD_id) + d(a^{ij}D_{ij}v + B_0^iD_iv) + 2a^{ij}D_ivD_jd],$$

$$A_5 = \frac{1}{4}[h''\mathcal{E} - h'a + h'B_0^iD_iu].$$

We rewrite the terms in $A_4$ in two steps. The first step is to differentiate the differential equation $Qu = 0$ to infer that

$$a^{ij}D_{ij}v + B_0^iD_iv = a^{ij}g^{km}D_{ik}uD_{jm}u - \frac{|Du|^2}{v}(\delta a^{ij}D_{ij}u + \delta a).$$

For the second step, we recall that $Dd = \gamma$ and then define the operator $\delta_N$ by

$$\delta_N g = \gamma \cdot \frac{\partial g}{\partial p},$$

so

$$B_0^iD_id = \delta_N a^{ij}D_{ij}u + \delta_N a.$$

For $A_5$, we note that

$$B_0^iD_iu = \bar{\delta}a^{ij}D_{ij}u + \bar{\delta}a,$$

and hence (since $a^{ij}g^{km}D_{ik}uD_{jm}u \geq 0$)

$$a^{ij}D_{ij}w + B_0^iD_iw \geq \mathcal{C}_T^2 + A_3w_T + \frac{1}{4}M_T[h''\mathcal{E} + J_1h' + K_2] + S_2,$$

where

$$J_1 = (\bar{\delta} - 1)a,$$

$$K_2 = \frac{1}{k_0}(va^{ij}D_{ij}d + v\delta_N a - \frac{d|Du|^2}{v}\delta a),$$

$$S_2 = \frac{M_T}{4k_0}[-\frac{d|Du|^2}{v}\delta a^{ij}D_{ij}u - 2a^{ij}D_ivD_jd + v\delta_N a^{ij}D_{ij}u] + \frac{M_T}{4}h'\bar{\delta}a^{ij}D_{ij}u.$$

The terms in $S_2$ are now estimated using Lemmata 9.5 and 9.6. First, we note that

$$\frac{M_Td|Du|^2}{4k_0v}\delta a^{ij}D_{ij}u = \frac{M_Td|Du|^2}{4k_0v}\delta a_*^{ij}D_{ij}u + \frac{M_Td|Du|^2}{4k_0v}\delta_T D_iuD_{ij}uD_ju.$$

We estimate the first term in this expression via Lemma 9.5 with $\varepsilon = 1/16$ and $f^{ij} = M_T d\delta a_*^{ij}/(4k_0 v)$, and we estimate the second term via Lemma 9.6 with $\varepsilon = 1/16$ and $\xi = M_T d\delta\tau Du/(4k_0 v\tau)$. From these inequalities, (9.43), and

$$\tau(\Gamma u)^2 \le A, \quad \lambda_* \le A, \quad |Du| \le v \tag{9.44}$$

we infer that

$$-\frac{M_T d|Du|^2}{4k_0 v}\delta a^{ij} D_{ij}u \ge -\frac{1}{8}\mathcal{C}_T^2 + \frac{1}{4}M_T K_3,$$

where

$$\begin{aligned} K_3 = &-\frac{8M_T(\delta A_*)^2}{\lambda_*^2}\tau w_T - 16M_T\frac{\|\delta\mathcal{A}_*\|^2}{\lambda_*} \\ &- 4(n+1)M_T\Lambda_*\left(\frac{(\delta A_*)^2}{\lambda_*^2} + \frac{(\delta\tau)^2}{\tau^2}\right) \\ &- |a|\left(\frac{|\delta A_*|}{\lambda_*} + \frac{|\delta\tau|}{\tau}\right). \end{aligned}$$

Next, we write $D_i v = \frac{1}{v}D_{ik}uD_k u$ and hence

$$\frac{M_T}{2k_0}2a^{ij}D_i vD_j d = \frac{M_T}{2k_0 v}a_*^{ik}\gamma_k D_j uD_{ij}u + \frac{M_T}{2k_0 v}\tau(\Gamma u)D_i uD_{ij}uD_j u.$$

Using $\varepsilon = 1/16$, $f^{ij} = M_T a_*^{ik}\gamma_k D_j u/(2k_0 v)$, and $\xi = M_T(\Gamma u)Du/(2k_0 v)$ (along with (9.43) and (9.44) and the definition of $\Lambda_*$), we find that

$$-\frac{M_T}{4k_0}2a^{ij}D_i vD_j d \ge -\frac{1}{8}\mathcal{C}_T^2 + \frac{1}{4}M_T K_4,$$

with

$$K_4 = -\frac{24}{k_0^2}\left(1 + \frac{\tau}{\lambda_*}\right)\frac{\Lambda_*^2}{\lambda_*^2}\tau M_T w_T - \frac{24n+32}{k_0^2}\frac{\Lambda_*^3}{\lambda_*^2} - \frac{2|a|}{k_0}.$$

For the next term, we have

$$\frac{M_T v}{4k_0}\delta_N a^{ij}D_{ij}u = \frac{M_T v}{4k_0}\delta_N a_*^{ij}D_{ij}u + \frac{M_T v}{4k_0}[\delta_N\tau D_i u + 2\tau\gamma^i]D_{ij}uD_j u$$

Using $\varepsilon = 1/16$, $f^{ij} = -M_T v\delta_N a_*^{ij}/(4k_0)$, and $\xi = -M_T v[\delta_N\tau Du + 2\tau\gamma]/(4k_0)$, we then have

$$\frac{M_T v}{4k_0}\delta_N a^{ij}D_{ij}u \ge -\frac{1}{8}\mathcal{C}_T^2 + \frac{1}{4}M_T K_5$$

with

$$K_5 = -\frac{8M_T v^2(\delta_N A_*)^2 \tau w_T}{k_0^2\lambda_*^2} - \frac{16M_T v^2\|\delta_N \mathcal{A}_*\|^2}{k_0^2\lambda_*}$$
$$- \frac{4(n+1)\Lambda_* M_T v^2}{k_0^2}\left(\frac{\|\delta_N A_*\|^2}{\lambda_*^2} + \frac{2(\delta_N\tau)^2}{\tau^2} + \frac{8\tau}{\lambda_*}\right)$$
$$- \frac{2\Lambda_* M_T v^2(\delta_N\tau)^2}{k_0^2\tau\lambda_*} - \frac{8\tau^2 v^2 M_T}{k_0^2\lambda_*}$$
$$- \frac{|a|}{k_0}\left(\frac{v|\delta_N A_*|}{\lambda_*} + \frac{v|\delta_N\tau|}{\tau} + \frac{2\tau^{1/2}v}{A^{1/2}}\right).$$

Next, we write

$$\frac{M_T}{4}h'\bar\delta a^{ij}D_{ij}u = \frac{M_T}{4}h'\bar\delta a_*^{ij}D_{ij}u + \frac{M_T}{4}h'(\bar\delta+2)\tau D_iuD_{ij}uD_ju.$$

Using $\varepsilon = 1/16$, $f^{ij} = M_T h'\bar\delta a_*^{ij}/4$ and $\xi = M_T h'(\bar\delta+2)\tau Du/4$, we find that

$$\frac{M_T}{4}h'\bar\delta a^{ij}D_{ij}u \ge -\frac{1}{8}\mathcal{C}_T^2 + \frac{1}{4}M_T[H_2(h')^2 + J_2h']$$

with

$$H_2 = -\frac{2(\bar\delta A_*)^2\tau M_T w_T}{\lambda_*^2} - \frac{16M_T\|\bar\delta\mathcal{A}_*\|^2}{\lambda_*}$$
$$- 4(n+1)\Lambda_* M_T\left(\frac{(\bar\delta A_*)^2}{\lambda_*^2} + \frac{[(\bar\delta+2)\tau]^2}{\tau^2}\right)$$

and

$$J_2 = -\frac{a}{A}(\bar\delta A_* + (\bar\delta+2)\tau(\Gamma u)^2).$$

Combining all these estimates, we see that

$$a^{ij}D_{ij}w + B_0^iD_iw \ge A_3w_T + \frac{1}{4}M_T[h''\mathcal{E} + H_2(h')^2 + J_3h' + K_6], \quad (9.45)$$

where

$$J_3 = J_1 + J_2, \quad K_6 = K_2 + K_3 + K_4 + K_5.$$

We are now ready to state our next gradient bound.

**Theorem 9.8.** *Suppose that*

$$\Lambda_* = O(\lambda_*), \quad (9.46a)$$
$$|\delta a_*^{ij}| = o(\lambda_*), \quad |\delta\tau| = o(\tau), \quad (9.46b)$$
$$|p|\left|\frac{\partial a_*^{ij}}{\partial p}\right| = O(\lambda_*), \quad |p|\left|\frac{\partial\tau}{\partial p}\right| = O(\tau), \quad (9.46c)$$
$$|p|\left|\frac{\partial a}{\partial p}\right| = O(\mathcal{E}), \quad (9.46d)$$
$$\delta a \le O(\mathcal{E}), \quad (9.46e)$$
$$(\bar\delta - 1)a \le O(\mathcal{E}). \quad (9.46f)$$

*Suppose further that*

$$|\delta_T a_*^{ij}| = o(\lambda_*), \quad |\delta_T \tau| = o(\tau), \tag{9.47a}$$
$$|\delta_T a| = o(\mathcal{E}), \tag{9.47b}$$
$$\tau = O(\lambda_*), \tag{9.47c}$$
$$\left(1 + \frac{\tau^{1/2}|p|}{\lambda_*^{1/2}}\right)|a| = O(\mathcal{E}). \tag{9.47d}$$

*Suppose also that* $b_p \cdot \gamma > 0$, (9.26) *is satisfied and that there are positive constants* $\mu_2$, $M_3$, *and* $M_4$ *such that* (9.27) *holds for all* $(x, z, p)$ *with* $b(x, z, p) = 0$ *and that if* $b(x, z, p) = 0$ *and* $|p - (p \cdot \gamma)\gamma| \le M_3$, *then* $|p \cdot \gamma| \le M_4$. *Suppose finally that*

$$\delta_T b, \; \bar{\delta} b \le o(|p| b_p \cdot \gamma) \tag{9.48}$$

*wherever* $b(x, z, p) = 0$. *Then there is a constant* $C$ *determined only by* $\operatorname{osc} u$, $\Omega$, $\mu_2$, $M_3$, $M_4$, *and the limit behavior in* (9.46), (9.47), *and* (9.48) *such that* $|Du| \le C$ *in* $\Omega$.

***Proof.*** We first observe from Theorems 9.1 and 9.2 that there is a constant $C_1$, determined only by $\operatorname{osc} u$, $\Omega$, and the limit behavior in (9.46) such that $|Du(x)| \le C_1(1 + 1/d(x))$. We now fix $R = \min\{R_0, 1\}$. It follows that $|Du| \le C_1/R$ if $d \ge R$ and that $|Du| \le 2C_1/d$ if $d < R$.

Our next step is to estimate $w_T$ where $d < R$. To this end, we observe that there are positive constants $\bar{H}$, $\bar{J}$, and $M_6$ (determined by the limit behavior in (9.46)) such that

$$H_2 \ge -\bar{H}\mathcal{E}, \quad J_3 \le \bar{J}\mathcal{E} \tag{9.49}$$

whenever $w_T \ge M_6$ and $w_T \ge \frac{1}{2} M_T$. Our next step is to proceed as in the proof of Theorem 9.2 and set

$$\bar{A} = (1 + \bar{J})(1 + \bar{H}), \tag{9.50a}$$
$$h(z) = \frac{1}{1 + \bar{H}} \exp(\bar{A}(\inf_\Omega u - z)), \tag{9.50b}$$
$$\bar{K} = \exp(-\bar{A} \operatorname{osc} u). \tag{9.50c}$$

We then conclude that

$$h''\mathcal{E} + H_2(h')^2 + J_3(h') \ge \bar{K}\mathcal{E}$$

as long as $w_T \ge \frac{1}{2} M_T$ and $M_T \ge M_6^2$. Using Theorems 9.1 and 9.2 again, we also see that there are constants $C_2$ and $M_7$ determined only by $\operatorname{osc} u$,

$\Omega$, and the limit behavior in (9.46) and (9.47) such that $k_0 \geq C_2$ implies that

$$A_3 w_T + \frac{1}{4} M_T K_6 > -\bar{K}\mathcal{E}$$

provided (9.43) holds, $w_T \geq \frac{1}{2} M_T$, and $w_T \geq M_7$. In particular, this inequality holds if $k_0 = 1 + C_1 + C_2$. Hence $w_T$ can't have an interior maximum greater than $M_6 + M_7 + C_1/R$. Moreover, on $\partial\Omega$, we have

$$b_p \cdot Dw + \delta_T b w_T - \frac{1}{4} M_T \left[ \frac{1}{k_0} v b_p \cdot \gamma + h' \bar{\delta} b \right] = 0,$$

and there is a constant $M_8$ (determined only by $\operatorname{osc} u$, $\Omega$, and the limit behavior in (9.46), (9.47), and (9.48)) such that

$$\delta_T b w_T - \frac{1}{4} M_T \left[ \frac{1}{k_0} v b_p \cdot \gamma + h' \bar{\delta} b \right] < 0$$

wherever $w_T \geq \frac{1}{2} M_T$ and $w_T \geq M_8$. As before, we infer a bound on $w_T$ where $d \leq R$, and the proof is completed by using Lemma 9.4 and Corollary 9.3. □

The hypothesis (9.47d) is stronger than the usual assumption $|a| = O(\mathcal{E})$ unless $\tau |p|^2 = O(\lambda_*)$, but it is not clear how to obtain a gradient bound in this generality (specifically with $O(\mathcal{E})$ growth for $a$); see Theorem 5.1 in [101] for an attempt at weakening (9.47d) by strengthening the assumptions on $b$.

We now look at our hypotheses on $b$ in a representative case. Suppose that

$$b(x, z, p) = F(v) Du \cdot \gamma + \psi(x, z)$$

for some $C^1$ functions $F$ and $\psi$; we assume that $sF'(s) \geq -F(s)$ for all $s \geq 1$. Then

$$b_p \cdot \gamma = F(v) + F'(v) \frac{(Du \cdot \gamma)^2}{v} \geq [1 - (\nu \cdot \gamma)^2] F(v) \geq c_0 F(v)$$

for some positive constant $c_0$ by virtue of Lemma 9.4, and

$$\bar{\delta} b = Du \cdot \gamma \left[ F(v) + F'(v) \frac{|Du|^2}{v} \right] = -\psi(x, z) \left[ 1 + \frac{F'(v)}{F(v)} \frac{|Du|^2}{v} \right].$$

When $F(v) = v^{q-1}$ for some nonnegative constant $q$, we conclude that

$$\bar{\delta} b = \psi \left[ 1 + (q-1) \frac{|Du|^2}{v^2} \right].$$

For $q = 0$, we have

$$1 + (q-1)\frac{|Du|^2}{v^2} = \frac{1}{v^2}$$

and hence $\bar{\delta} b = O(b_p \cdot \gamma/|p|)$. For $q > 0$, we see that $\bar{\delta} b = O(|p|^{1-q} b_p \cdot \gamma)$. It follows that $\bar{\delta} b \le o(|p| b_p \cdot \gamma)$ for all $q \ge 0$. In addition,

$$\delta_T b = \psi_z + \frac{p' \cdot \psi_x}{|p'|^2} + \frac{(Du \cdot \gamma)^2 F'(v)}{|p'|^2 v} \le \psi_z + O\left(\frac{1}{|p|}\right) + O(F)$$

(using $\gamma \cdot T'u = 0$ and Lemma 9.4). It follows that $\delta_T b = o(|p| b_p \cdot \gamma)$ if $q > 0$ or if $\psi_z \le 0$.

Hence, we obtain a gradient bound via Theorem 9.8 if $q > 1$ or if $q = 0$ and $\psi_z \le 0$. If $q = 0$ and $\psi_z$ is bounded from above by a sufficiently small positive constant, then the argument of Theorem 9.8 is easily modified to obtain a gradient bound. (See Exercise 9.2 for details.)

## 9.6 Local gradient estimates

To proceed, we fix a point $y \in \partial\Omega$ and a number $R \in (0, R_0)$ (to be further specified), and we write

$$B = B(y,R), \quad B^+ = B \cap \Omega, \quad B^0 = B \cap \partial\Omega.$$

(Note that this is somewhat different notation from what we have used, but it should not cause any confusion.) We wish to combine the argument of the previous section with the one used to derive an interior gradient bound. The interior gradient bound was proved by using a simple cut-off function $\eta$, but, here, we need a more carefully chosen function. In particular, we introduce an important restriction on $R$. First, we assume that condition (9.27) holds, and we assume that $\partial\Omega$ is locally $C^1$ near $y$. (Of course, for our gradient estimate, we shall assume more regularity of $\partial\Omega$.) Then there is a positive constant $R_1$ such that

$$|\gamma(x) - \gamma(y)| \le \frac{1}{(4\mu_2 + 1)(\mu_2 + 1)}$$

if $x \in B^0$ and $R \le R_1$. For any such $x$, we therefore have

$$\begin{aligned} b_p(x,z,p) \cdot \gamma(y) &= b_p(x,z,p) \cdot \gamma(x) + b_p(x,z,p) \cdot [\gamma(y) - \gamma(x)] \\ &\ge b_p(z,x,p) \cdot \gamma(x) - |b_p(x,z,p)| \frac{1}{(4\mu_2+1)(\mu_2+1)} \\ &\ge b_p(x,z,p) \cdot \gamma(x)[1 - \frac{\mu_2}{(4\mu_2+1)(\mu_2+1)}] \\ &\ge b_p(x,z,p) \cdot \gamma(x) \frac{4\mu_2}{4\mu_2 + 1}. \end{aligned}$$

It therefore follows that

$$|b_p(x,z,p)| \le (\mu_2 + \frac{1}{4})b_p(x,z,p)\cdot\gamma(y).$$

To simplify notation, we now rotate and translate variables so that $\gamma(y)$ points along the positive $x^n$-axis and $y = 0$, and we define

$$\eta(x) = 1 - \frac{2|x'|^2}{R^2} - \frac{(R - 4(2\mu_2+1)x^n)^2}{4R^2}. \tag{9.51}$$

It's easy to check that $\eta(x) \le 0$ if $|x'| \ge R/\sqrt{2}$ or if $|x^n| \ge R/\sqrt{2}$. Also, simple algebra shows that $|x| \ge R$ implies $|x'| \ge R/\sqrt{2}$ or $|x^n| \ge R/\sqrt{2}$, and hence $\eta$ is nonpositive outside $B(y,R)$ with $\eta(y) = 3/4$. We now write $E$ and $E^0$ for the subsets of $\Omega$ and $\partial\Omega$, respectively on which $\eta > 0$, and we set $E^+ = \partial E \setminus E^0$. It follows that $\eta \in C^2(\overline{E})$ with $|D\eta| \le C(\mu_2)/R$ and $|D^2\eta| \le C(\mu_2)/R^2$. Moreover, on $E^0$, we have

$$\begin{aligned} b_p\cdot D\eta &= \frac{2}{R}[-2b_p\cdot\frac{x'}{R} + (2\mu_2+1)b_p\cdot\gamma(y) - 2(2\mu_2+1)^2\frac{x^n}{R}b_p\cdot\gamma(y)] \\ &\ge \frac{2b_p\cdot\gamma(y)}{R}\left(-(2\mu_2+\frac{1}{2})\frac{|x'|}{R} + 2\mu_2 + 1 - 8(2\mu_2+1)^2\frac{x^n}{R}\right). \end{aligned}$$

Next, we note that there is a positive constant $R_2$ such that $x^n \le |x'|/16(2\mu_2+1)^2$ if $x \in E^0$ and $R \le R_2$. With this additional restriction on $R$, we have

$$b_p\cdot D\eta \ge \frac{b_p\cdot\gamma(y)}{r}\left(-(2\mu_2+1)\frac{|x'|}{R} + 2\mu_2 + 1\right).$$

Since $|x'| \le R$ wherever $\eta > 0$, it follows that $b_p\cdot D\eta \ge 0$ on $E^0$.

Next, we set

$$M_T = \sup_{B^+} \eta^2 w_T,$$

and, with a nonnegative function $h$ and a constant $k_0$ also to be further determined, we set

$$w = \eta^2 w_T + \frac{1}{4k_0}M_T v d + \frac{1}{4}M_T h(u).$$

As before, we assume that condition (9.43) is satisfied and $h \le 1$. Although most of our effort will be directed toward estimating terms in the differential equation for $w$, we note here that our choice of $\eta$ gives the boundary condition

$$b_p\cdot Dw + \delta_T b\eta^2 w_T - \frac{1}{4}M_T[v b_p\cdot\gamma + h'\bar{\delta}b] \ge 0. \tag{9.52}$$

From our previous calculations and the identity

$$B_0^i D_i\eta = D_k\eta a^{ij,k} D_{ij}u + D_i\eta a^i,$$

we see that

$$a^{ij}D_{ij}w + B_0^i D_i w \ge 2\mathcal{C}_T^2 + A_6\eta^2 w_T + \frac{1}{4}M_T[h''\mathcal{E} + J_1h' + K_2] + S_4,$$

where

$$A_6 = A_3 + \frac{2a^{ij}D_{ij}\eta}{\eta} + \frac{2a^iD_i\eta}{\eta},$$
$$S_4 = S_2 + 2\eta w_T D_k\eta a^{ij,k}D_{ij}u + 4\eta a^{ij}D_iw_TD_j\eta.$$

By arguing as in the previous section (except that $\varepsilon = \eta^2/16$), we can estimate the terms in $S_2$ just as before. In this way, we find that

$$-\frac{M_Td|Du|^2}{4k_0v} \ge -\frac{1}{8}\eta^2\mathcal{C}_T^2 + \frac{1}{4}M_TK_{10},$$
$$-\frac{M_T}{4k_0v}2a^{ij}D_ivD_jd \ge -\frac{1}{8}\eta^2\mathcal{C}_T^2 + \frac{1}{4}M_TK_{11},$$
$$\frac{M_Tv}{4k_0}\delta_Na^{ij}D_{ij}u \ge -\frac{1}{8}\eta^2\mathcal{C}_T^2 + \frac{1}{4}M_TK_{12},$$
$$\frac{M_T}{4}h'\bar{\delta}a^{ij}D_{ij}u \ge -\frac{1}{8}\eta^2\mathcal{C}_T^2 + \frac{1}{4}M_T[H_3(h')^2 + J_2h']$$

wherever $\eta w_T \ge \frac{1}{2}M_T$, with

$$\begin{aligned} K_{10} = &-\frac{16(\delta A_*)^2}{\lambda_*^2}\tau w_T^2 - 32w_T\frac{\|\delta\mathcal{A}_*\|^2}{\lambda_*} \\ &- 8(n+1)w_T\Lambda_*\left(\frac{(\delta A_*)^2}{\lambda_*^2} + \frac{(\delta\tau)^2}{\tau^2}\right) \\ &- |a|\left(\frac{|\delta A_*|}{\lambda_*} + \frac{|\delta\tau|}{\tau}\right), \\ K_{11} = &-\frac{48}{k_0^2}\left(1+\frac{\tau}{\lambda_*}\right)\frac{\Lambda_*^2}{\lambda_*^2}\tau w_T^2 - \frac{48n+64}{k_0^2}\frac{\Lambda_*^3}{\lambda_*^2} - \frac{2|a|}{k_0}, \\ K_{12} = &-\frac{8M_Tv^2(\delta_NA_*)^2\tau w_T}{k_0^2\lambda_*^2} - \frac{16M_Tv^2\|\delta_N\mathcal{A}_*\|^2}{k_0^2\lambda_*} \\ &- \frac{4(n+1)\Lambda_*M_Tv^2}{k_0^2}\left(\frac{\|\delta_NA_*\|^2}{\lambda_*^2} + \frac{2(\delta_N\tau)^2}{\tau^2} + \frac{8\tau}{\lambda_*}\right) \\ &- \frac{2\Lambda_*M_Tv^2(\delta_N\tau)^2}{k_0^2\tau\lambda_*} - \frac{8\tau^2v^2M_T}{k_0^2\lambda_*} \\ &- \frac{|a|}{k_0}\left(\frac{v|\delta_NA_*|}{\lambda_*} + \frac{v|\delta_N\tau|}{\tau} + \frac{2\tau^{1/2}v}{A^{1/2}}\right), \end{aligned}$$

$$H_3 = -\frac{4(\bar{\delta}A_*)^2\tau w_T^2}{\lambda_*^2} - \frac{32w_T\|\bar{\delta}\mathcal{A}_*\|^2}{\lambda_*}$$
$$- 8(n+1)\Lambda_* w_T\left(\frac{(\bar{\delta}A_*)^2}{\lambda_*^2} + \frac{[(\bar{\delta}+2)\tau]^2}{\tau^2}\right).$$

Similarly, we can estimate the remaining terms in $S_4$. Using $\varepsilon = \eta^2/16$, $f^{ij} = -2w_T\eta D_k\eta a_*^{ij,k}$, and $\xi = -2w_T\eta[D_k\eta\tau^k p + 2\tau D\eta]$, we find that

$$2w_T\eta D_k\eta a^{ij,k}D_{ij}u \geq -\frac{1}{8}\eta^2\mathcal{C}_T^2 + \tilde{A}_1\eta^2 w_T,$$

where

$$\tilde{A}_1 = -128\frac{(D_i\eta A_*^i)^2}{\eta^2\lambda_*^2}\tau w_T^2 - 256w_T\frac{\sum_{i,j}(a_*^{ij,k}D_k\eta)^2}{\eta^2\lambda_*}$$
$$- 64(n+1)\Lambda_* w_T\left(\frac{(D_i\eta A_*^i)^2}{\eta^2\lambda_*^2} + \frac{2(D_k\eta\tau^k)^2}{\eta^2\tau^2} + \frac{8\tau|D\eta|^2}{\eta^2\lambda_*}\right)$$
$$- 128\left(\frac{(D_k\eta\tau^k)^2 w_T}{\eta^2\tau} + 4\tau\frac{|D\eta|^2}{\eta^2}\right) - 128\frac{\Lambda_*(D_k\eta\tau^k)^2 w_T}{\eta^2\lambda_*\tau} - 512\frac{\tau^2|D\eta|^2}{\eta^2\lambda_*}$$
$$- 2|a|\left(\frac{|D_i\eta A_*^i|}{\eta\lambda_*} + \frac{|D_k\eta\tau^k|}{\eta\tau} + \frac{|D\eta|\tau^{1/2}}{\eta\lambda_*^{1/2}}\right).$$

Finally, we use the Cauchy-Schwarz inequality to see that

$$4\eta a^{ij}D_i w_T D_j\eta \geq -\frac{\eta^2}{8} + \tilde{A}_2\eta^2 w_T$$

with

$$\tilde{A}_2 = -512\frac{a^{ij}D_i\eta D_j\eta}{\eta^2}.$$

Combining all these estimates, we find that

$$a^{ij}D_{ij}w + B_0^i D_i w \geq [A_3 + \tilde{A}_3]\eta^2 w_T + \frac{1}{4}M_T[h''\mathcal{E} + H_3(h')^2 + J_3 h' + K_{13}]$$

where

$$\tilde{A}_3 = \frac{2a^{ij}D_{ij}\eta}{\eta} + \frac{2a^i D_i\eta}{\eta} + \tilde{A}_1 + \tilde{A}_2,$$
$$K_{13} = K_{10} + K_{11} + K_{12},$$

From these estimates, we can now prove our basic local gradient estimate.

**Theorem 9.9.** *Let $y \in \mathbb{R}$, let $r > 0$, and let $u$ be a solution of*

$$Qu = 0 \text{ in } \Omega \cap B(y,r), Nu = 0 \text{ on } \partial\Omega \cap B(y,r),$$

*with $\partial\Omega \cap B(y,r) \in C^3$. Suppose that conditions (9.46) and (9.47) are satisfied. Suppose also that $b_p \cdot \gamma > 0$, (9.26) is satisfied and that there are positive constants $\mu_2$, $M_3$, and $M_4$ such that (9.27) holds for all $(x,z,p)$ with $b(x,z,p) = 0$ and that if $b(x,z,p) = 0$ and $|p - (p\cdot\gamma)\gamma| \le M_3$, then $|p\cdot\gamma| \le M_4$. Suppose finally that (9.48) is satisfied wherever $b(x,z,p) = 0$. Then there are positive constants $R$ (determined only by $r$, $\mu_2$ and $\Omega$) and $C$ (determined only by $\Omega$, $\mu_2$, $M_3$, $M_4$, and the limit behavior in (9.46), (9.47), and (9.48)) such that*

$$|Du(y)| \le C\left(1 + \frac{\operatorname{osc}_{\Omega\cap B(y,R)} u}{R}\right).$$

***Proof.*** We choose $R = \min\{r, R_1, R_2, 1, R_0\}$. It follows (as in Theorem 9.8) that there is a constant $C_1$ such that $d|Du| \le C_1$ in $B(y,R)$. In addition, there are positive constants $\hat{H}$, $\hat{J}$, and $M_8$ so that

$$H_3 \ge -\hat{H}\mathcal{E}, \quad J_3 \le \hat{J}\mathcal{E}, \quad K_{13} \ge -\mathcal{E}$$

if $\eta^2 w_T \ge M_8$ and $\eta^2 w_T \ge \frac{1}{2}M_T$.

We now distinguish two cases, using the same definition $\sigma = \operatorname{osc}_E u$, as before. First, if

$$\sigma \le \min\{\frac{1}{1+2\hat{J}}, \frac{1}{\sqrt{16}e(1+\hat{H})}\},$$

we take

$$h(z) = \frac{1}{1+\hat{H}}\exp(-(z - \inf_E u)^2/\sigma^2).$$

In this case, we argue as in Theorem 9.1 to infer that there is a further constant $M_9$ so that $w$ can't have an interior maximum if $M_T \ge M_9(1+\frac{\sigma}{R})$. On the other hand, if

$$\sigma > \min\{\frac{1}{1+2\hat{J}}, \frac{1}{\sqrt{16}e(1+\hat{H})}\},$$

we take

$$h(z) = \frac{1}{1+\hat{H}}\exp((1+\hat{J})(1+\hat{H})(\inf_E u - z))$$

and argue as in Theorem 9.2 to infer that there is a constant $M_9$ so that $w$ can't have an interior maximum if $M_T \ge M_9(1+\frac{\sigma}{R})$.

Since $\eta^2 w_T = 0$ on $E^+$, it follows that $w$ can't attain its maximum there.

Finally, by virtue of (9.52), there is a constant $M_{10}$ so that $w$ can't have a maximum on $E^0$ if $M_T \ge M_{10}$. The theorem is now proved with $C = \max\{M_9, M_{10}\}$. □

When the oscillation of $u$ is sufficiently small, we can combine the proof of the preceding theorem with that of Theorem 9.1 to obtain a local gradient bound. In view of Lemma 8.4, such a result will be useful in studying uniformly elliptic problems.

**Theorem 9.10.** *Let $y \in \mathbb{R}$, let $r > 0$, and let $u$ be a solution of*

$$Qu = 0 \text{ in } \Omega \cap B(y,r), Nu = 0 \text{ on } \partial\Omega \cap B(y,r),$$

*with $\partial\Omega \cap B(y,r) \in C^3$. Suppose that*

$$\Lambda_* = O(\lambda_*), \tag{9.53a}$$

$$|\delta a_*^{ij}| = O(\lambda_*), \quad |\delta\tau| = O(\tau), \tag{9.53b}$$

$$|p|\left|\frac{\partial a_*^{ij}}{\partial p}\right| = O(\lambda_*), \quad |p|\left|\frac{\partial \tau}{\partial p}\right| = O(\tau), \tag{9.53c}$$

$$|p|\left|\frac{\partial a}{\partial p}\right| = O(\mathcal{E}), \tag{9.53d}$$

$$\delta a \le O(\mathcal{E}), \tag{9.53e}$$

$$(\bar{\delta} - 1)a \le O(\mathcal{E}), \tag{9.53f}$$

$$|\delta_T a_*^{ij}| = O(\lambda_*), \quad |\delta_T \tau| = O(\tau), \tag{9.53g}$$

$$|\delta_T a| = O(\mathcal{E}), \tag{9.53h}$$

$$\tau = O(\lambda_*), \tag{9.53i}$$

$$\left(1 + \frac{\tau^{1/2}|p|}{\lambda_*^{1/2}}\right)|a| = O(\mathcal{E}). \tag{9.53j}$$

*Suppose also that $b_p \cdot \gamma > 0$, (9.26) is satisfied and that there are positive constants $\mu_2$, $M_3$, and $M_4$ such that (9.27) holds for all $(x,z,p)$ with $b(x,z,p) = 0$ and that if $b(x,z,p) = 0$ and $|p - (p\cdot\gamma)\gamma| \le M_3$, then $|p\cdot\gamma| \le M_4$. Suppose finally that*

$$\delta_T b, \ \bar{\delta} b \le O(|p| b_p \cdot \gamma) \tag{9.54}$$

*wherever $b(x,z,p) = 0$. Then there are positive constants $\sigma$ (determined only by the limit behavior in (9.53) and (9.54)), $R$ (determined only by $r$, $\mu_2$ and $\Omega$) and $C$ (determined only by $\Omega$, $\mu_2$, $M_3$, $M_4$, and the limit behavior in (9.53), and (9.54)) such that, if*

$$\operatorname*{osc}_{\Omega\cap B(y,R)} u \le \sigma,$$

*then*

$$|Du(y)| \le C\left(1 + \frac{\operatorname{osc}_{\Omega\cap B(y,R)} u}{R}\right). \tag{9.55}$$

***Proof.*** We now take

$$w = \eta^2 w_T + \frac{1}{4k_0} vd + \frac{M_T}{4} h(u)$$

with $k_0 \in (0,1)$ to be chosen and $h$ defined by

$$h(s) = \frac{1}{\alpha_1} \exp(\alpha_1 (\inf_{\Omega \cap B(y,R)} u - s))$$

for $R$ and $\alpha_1$ also to be chosen. By following the calculations in the proof of Theorem 9.9, we find that there are positive constants $M_1$, $\bar{H}$, $\bar{J}$, $\bar{K}$, $\mu_1$ (determined by the limit behavior in (9.53)) such that

$$a^{ij} D_{ij} w + B_0^i D_i w \ge \frac{1}{4} M_T \mathcal{E}[h'' - \bar{H}(h')^2 - \bar{J}h' - (k_0)^{-2}\bar{K}]$$

wherever $w \ge M_1$ and $\eta w_T \ge \frac{1}{2} M_T$. Using the specific form of $h$ and setting $m = \inf_{\Omega \cap B(y,R)} u$, we infer that

$$h'' - \bar{H}(h')^2 - \bar{J}h' - (k_0)^{-2}\bar{K} \ge \exp(\alpha_1(m-u))[\alpha_1 - \bar{H} - \bar{J}] - (k_0)^{-2}\bar{K}.$$

Assuming that $\alpha_1 \ge \bar{H} + \bar{J}$ and

$$\alpha_1 \operatorname*{osc}_{\Omega \cap B(y,R)} u \le \ln 2, \tag{9.56}$$

we conclude that

$$h'' - \bar{H}(h')^2 - \bar{J}h' - (k_0)^{-2}\bar{K} \ge \frac{1}{2}[\alpha_1 - \bar{H} - \bar{J}] - (k_0)^{-2}\bar{K}.$$

Moreover, by virtue of (9.52), we also have

$$b_p \cdot Dw \ge v b_p \cdot \gamma \left[ \frac{1}{4k_0} - \mu_1 \right] M_T$$

for some constant $\mu_1$ determined only by the limit behavior in (9.54). We now choose $k_0 = 4/(1+\mu_1)$ and then choose $\alpha_1 \ge \bar{H} + \bar{J} + (k_0)^{-2}\bar{K}$. Finally, we choose $R$ (according to Theorem 9.1 and our assumed modulus of continuity) so small that (9.56) holds and $vd \le k_0$ in $\Omega \cap B(y,R)$. With these choices for $R$, $k_0$, and $\alpha_1$, we infer (9.55). □

## 9.7 Gradient estimates for capillary-type problems

In this section, we assume that there are a positive function $\Lambda$ and a positive constant $\mu_1$ such that

$$\Lambda g^{ij}\xi_i\xi_j \le a^{ij}\xi_i\xi_j, \quad |a^{ij}| \le \mu_1 \Lambda. \tag{9.57}$$

(Of course, the two inequalities can't both be true if $\mu_1 \le 0$.)

As before, it is crucial to estimate terms which are linear in the second derivatives as a first step in obtaining gradient bounds. This time, we must use the identities

$$\delta^{ij} = c^{ij} + \gamma^i\gamma^j, \quad \delta^{ij} = g^{ij} + \nu^i\nu^j.$$

Furthermore, we take advantage of a change of independent variable in place of the function $h$ used in the preceding sections. Specifically, we write $I = [\inf_\Omega u, \sup_\Omega u]$ and let $\Psi \in C^3(I)$ be a strictly increasing function with $\Psi'$ never zero. It will be convenient to set $\psi = \Psi^{-1}$, $\omega = \psi''/(\psi')^2$, and $\tilde{u} = \Psi(u)$. We also suppress the argument $\tilde{u}$ from $\psi$ and $\omega$, and we suppress the argument $(x, u, Du)$ from $a^{ij}$, $a$, $\mathcal{E}$, and their derivatives. For notational simplicity, we then set

$$\begin{aligned}\tilde{C}_T^2 &= a^{ij}c^{km}D_{ik}\tilde{u}D_{jm}\tilde{u},\\ \tilde{c}_i &= D_i(c^{km})D_k\tilde{u}D_m u,\\ \tilde{w}_T &= c^{ij}D_i\tilde{u}D_j\tilde{u}.\end{aligned}$$

We also observe that

$$a^{ij}D_{ij}\tilde{u} + \frac{1}{\psi'}(a + \omega\mathcal{E}) = 0. \tag{9.58}$$

Our bound on terms that are linear in second derivatives will actually be a bound on terms of the form $f^{ij}D_{ij}\tilde{u}$, which we now state.

**Lemma 9.11.** *Suppose $\mathcal{F} = [f^{ij}]$ is a matrix. Then, for any $\varepsilon > 0$, we have*

$$\begin{aligned} f^{ij}D_{ij}\tilde{u} \le \varepsilon\tilde{\mathcal{C}}_T^2 &+ \frac{1}{\varepsilon}\left(\frac{5|\mathcal{F}|^2}{\Lambda G} + \frac{9n\mu_1\Lambda F^2}{2A^2}\right)\\ &+ \psi'\left(\frac{1}{2v}f^{ij}\nu_j + \varepsilon\Lambda\frac{\nu\cdot\gamma}{v}\gamma_r D_{mr}\tilde{u}c^{im}\right)D_i\tilde{w}_T\\ &+ \frac{2\varepsilon\Lambda}{Gv^2}|\tilde{c}|^2 + \frac{|F||a|}{A\psi'} + \frac{\omega|F|\mathcal{E}}{A\psi'} - \frac{1}{v}f^{ij}\nu_j\tilde{c}_i. \end{aligned} \tag{9.59}$$

***Proof.*** We start by setting

$$\begin{aligned}F_1 &= f^{ij}c^{jk}D_{ik}\tilde{u},\\ F_2 &= f^{ij}\gamma_j\gamma^k D_{ik}\tilde{u},\end{aligned}$$

so $f^{ij}D_{ij}\tilde{u} = F_1 + F_2$.

To estimate $F_1$, we write $F_1 = F_{11} + F_{12} + F_{13}$ with

$$\begin{aligned} F_{11} &= f^{ij} g^{jk} c^{kr} D_{rm}\tilde{u} c^{im}, \\ F_{12} &= f^{ij} \gamma_i g^{jk} c^{kr} D_{rm}\tilde{u} \gamma^m, \\ F_{13} &= f^{ij} \nu^j \nu^k c^{kr} D_{ir}\tilde{u}. \end{aligned}$$

Also, to simplify some later calculations, we set

$$\begin{aligned} G_0 &= 2g^{jk}\gamma_k\gamma_r D_{mr}\tilde{u} c^{im} c^{js} D_{is}\tilde{u}, \\ G_1 &= g^{jk} c^{kr} D_{mr}\tilde{u} c^{im} c^{js} D_{is}\tilde{u}, \\ G_2 &= G\gamma^m D_{im}\tilde{u} c^{ij} D_{jk}\tilde{u}\gamma^k, \\ \tilde{\mathcal{G}}_T^2 &= g^{ij} c^{km} D_{ik}\tilde{u} D_{jm}\tilde{u}. \end{aligned}$$

Note that

$$\tilde{\mathcal{G}}_T^2 = G_0 + G_1 + G_2. \tag{9.60}$$

We then have that

$$F_{11} \le (g^{ij} f^{ik} f^{jk})^{1/2} (G_1)^{1/2}$$

and

$$F_{12} \le (c^{kr} f^{ij} \gamma_i g^{jk} f^{ms} \gamma_m g^{is})^{1/2} \left(\frac{G_2}{G}\right)^{1/2}$$

by Cauchy's inequality, while

$$F_{13} = \frac{\psi'}{2v} f^{ij} \nu_j D_i \tilde{w}_T - \frac{1}{v} f^{ij} \nu_j \tilde{c}_i.$$

Since $0 < G \le 1$, it follows that, for any $\varepsilon_1 > 0$, we have

$$F_1 \le \varepsilon_1 G_1 + \frac{\varepsilon_1}{2} G_2 + \frac{3}{4\varepsilon_1 G} |\mathcal{F}|^2 + \frac{\psi'}{2v} f^{ij} \nu_j D_i \tilde{w}_T - \frac{1}{v} f^{ij} \nu_j \tilde{c}_i. \tag{9.61}$$

The estimate of $F_2$ is slightly more elaborate in that we must use the differential equation. First, we write $F_2 = F_{21} + F_{22}$ with

$$\begin{aligned} F_{21} &= f^{ij} \gamma_j c^{ir} D_{rk}\tilde{u}\gamma^k, \\ F_{22} &= \frac{F}{A}(AD_{ij}\tilde{u}\gamma^i\gamma^j). \end{aligned}$$

As before, Cauchy's inequality gives

$$F_{21} \le (f^{ij}\gamma_j c^{ir} f^{rk}\gamma_k)^{1/2} \left(\frac{G_2}{G}\right)^{1/2}.$$

We now define

$$A_1 = a^{ij} c^{jk} D_{ik}\tilde{u}, \quad A_2 = a^{ij}\gamma_i\gamma_k D_{km}\tilde{u} c^{im}, \quad A_3 = AD_{ij}\tilde{u}\gamma_i\gamma_j.$$

It's easy to check that $a^{ij}D_{ij}\tilde{u} = A_1 + A_2 + A_3$. Moreover, if we set

$$A_4 = a^{ij}c^{jk}c^{im}D_{km}\tilde{u},$$

then we have $A_1 = A_4 + A_2$, and hence

$$a^{ij}D_{ij}\tilde{u} = 2A_1 - A_4 + A_3.$$

From the equation (9.58) and this identity, we infer that

$$F_{22} = \frac{F}{A}\left[-\frac{a+\omega\mathcal{E}}{\psi'} - 2a^{ij}c^{jk}D_{ik}\tilde{u} + a^{ij}c^{jk}c^{im}D_{im}\tilde{u}\right].$$

Then Cauchy's inequality yields

$$-2\frac{F}{A}a^{ij}c^{jk}D_{ik}\tilde{u} \le 2\frac{|F|}{A}(n\mu_1\Lambda)^{1/2}(\tilde{\mathcal{C}}_T^2)^{1/2}$$

and

$$\frac{F}{A}a^{ij}c^{jk}c^{im}D_{im}\tilde{u} \le \frac{|F|}{A}(n\mu_1\Lambda)^{1/2}(\tilde{\mathcal{C}}_T^2)^{1/2}.$$

It follows that, for any $\varepsilon_1 > 0$ (which will be equal to the $\varepsilon_1$ above), we have

$$F_2 \le \frac{1}{2}\varepsilon_1 G_2 + \frac{1}{2}\varepsilon\tilde{\mathcal{C}}_T^2 + \frac{1}{4\varepsilon_1 G}|\mathcal{F}|^2 + \frac{9}{2\varepsilon}n\mu_1\Lambda\frac{F^2}{A^2} + \frac{|F||a|}{A\psi'} + \frac{|F|\omega\mathcal{E}}{A\psi'} \tag{9.62}$$

Combining (9.61) and (9.62) (and noting that $0 < G \le 1$) gives

$$\begin{aligned} f^{ij}D_{ij}\tilde{u} \le{}& \varepsilon_1(G_1+G_2) + \frac{5}{4\varepsilon_1 G}|\mathcal{F}|^2 + \frac{1}{2}\varepsilon\tilde{\mathcal{C}}_T^2 + \frac{9}{2\varepsilon}n\mu_1\Lambda\frac{F^2}{A^2} \\ &+ \frac{|F||a|}{A\psi'} + \frac{|F|\omega\mathcal{E}}{A\psi'} + \frac{\psi'}{2v}f^{ij}\nu_j D_i\tilde{w}_T - \frac{1}{v}f^{ij}\nu_j\tilde{c}_i. \end{aligned} \tag{9.63}$$

Our next step is to estimate $G_1 + G_2$ in terms of $\tilde{\mathcal{G}}_T^2$. (We refer the reader to Exercise 9.3 for an explanation of why this estimate is needed.) To this end, we note that $g^{jk}\gamma_k = \gamma_j - \nu^j\nu\cdot\gamma$ and hence

$$\begin{aligned} G_0 &= -\frac{2}{v}\nu\cdot\gamma\gamma_r D_{rm}\tilde{u}c^{im}D_{is}\tilde{u}c^{ks}D_k u \\ &= -\frac{\psi'\nu\cdot\gamma}{v}\gamma_r D_{mr}\tilde{u}c^{im}D_i\tilde{w}_T + \frac{\nu\cdot\gamma}{v}\gamma_r D_{mr}uc^{im}\tilde{c}_i. \end{aligned}$$

From Cauchy's inequality, we infer that

$$\frac{\nu\cdot\gamma}{v}\gamma_r D_{mr}\tilde{u}c^{im}\tilde{c}_i \ge -\frac{|\nu\cdot\gamma|}{v}\left(\frac{|\tilde{c}|^2}{G}\right)^{1/2}(G_2)^{1/2},$$

and hence, since $G_1 \ge 0$, we have

$$G_0 \ge -\frac{\psi'\nu\cdot\gamma}{v}\gamma_r D_{mr}\tilde{u}c^{im}D_i\tilde{w}_T - \frac{(\nu\cdot\gamma)^2}{2v^2G}|\tilde{c}|^2 - \frac{1}{2}(G_1+G_2).$$

If we add $G_1 + G_2$ to both sides of this inequality and then multiply the resulting inequality by 2, we conclude that

$$G_1 + G_2 \le 2\tilde{\mathcal{G}}_T^2 + \frac{(\nu\cdot\gamma)^2}{Gv^2}|\tilde{c}|^2 + 2\frac{\psi'\nu\cdot\gamma}{v}\gamma_r D_{mr}\tilde{u}c^{im}D_i\tilde{w}_T.$$

the proof is completed by using this inequality in (9.63) with choosing $\varepsilon_1 = \varepsilon\Lambda/4$ and noting that $\Lambda\tilde{\mathcal{G}}_T^2 \le \tilde{\mathcal{C}}_T^2$. □

From this second derivative estimate, we can derive a gradient bound for solutions of (9.1) if $a^{ij}$ satisfies (9.57) along with other appropriate structure conditions (also on $a$ and $b$), which we describe in detail in the next theorem. Although the basic method of proof is identical to that for Theorem 9.8, there are several important differences in the details. We have already mentioned the change of dependent variable, but another important difference is a much stronger use of an estimate like that in Lemma 9.4 (but valid over $\Omega$ instead of just $\partial\Omega$); we state this estimate as structure condition (9.66a) below.

**Theorem 9.12.** *Suppose, in addition to* (9.57), *that*

$$(\bar{\delta}-1)a + \delta_T\mathcal{E} \le O(\mathcal{E}), \tag{9.64a}$$

$$\delta_T a \le O(\mathcal{E}), \tag{9.64b}$$

$$\Lambda = O(\mathcal{E}), \tag{9.64c}$$

$$|\bar{\delta}a^{ij}| = o((\Lambda\mathcal{E})/|p|), \tag{9.64d}$$

$$|\delta_T a^{ij}| = O((\Lambda\mathcal{E})/|p|), \tag{9.64e}$$

$$|\delta_N a^{ij}| = O(\Lambda\mathcal{E})/|p|), \tag{9.64f}$$

$$\delta_T b \le O(b_p\cdot\gamma), \tag{9.64g}$$

$$\bar{\delta}b \le o(b_p\cdot\gamma). \tag{9.64h}$$

*Suppose also that*

$$\limsup_{|p|\to\infty}\sup_{x,z}\frac{\bar{\delta}\mathcal{E}}{\mathcal{E}} < 1, \tag{9.65}$$

*Suppose finally that there are positive constants $\mu_0$ and $\mu_1$, and a decreasing function $M_1$ such that*

$$\sup_{d\le s}|Du| \le \mu_0(\sup_{d\le s}|D'u| + 1), \tag{9.66a}$$

$$\sup_{d\ge s}|Du| \le M_1(s) \tag{9.66b}$$

*for all $s \in (0,\mu_1)$. Then there is a constant $C$, determined only by $\mu_0$, $\mu_1$, $M_1$, and the limit behavior in* (9.64) *and* (9.65) *such that, if $u$ solves* (9.1), *then $|Du| \le C$ in $\Omega$.*

***Proof.*** By differentiating (9.58) with respect to $x^k$, multiplying by $c^{km}D_m\tilde{u}$, and summing on $k$, we find that

$$0 = \frac{1}{2}a^{ij}D_{ij}\tilde{w}_T + B_4^i D_i\tilde{w}_T - \tilde{\mathcal{C}}_T^2 + C^*\tilde{w}_T + S, \tag{9.67}$$

where

$$\begin{aligned}
B_4 &= \omega\mathcal{E}_p + \psi' a_p^{ij}D_{ij}\tilde{u} + a_p,\\
C^* &= \frac{\omega'}{\psi'}\mathcal{E} + \omega^2 H_{10} + \omega J_{10} + K_{20},\\
S &= S_{11} + S_{12} + S_{13},\\
H_{10} &= (\bar{\delta} - 1)\mathcal{E},\\
J_{10} &= (\bar{\delta} - 1)a + \delta_T\mathcal{E},\\
K_{20} &= \delta_T a - \frac{1}{2}\frac{a^{ij}D_{ij}(c^{km})D_k u D_m u}{w_T},\\
S_{11} &= -2a^{ij}D_i(c^{km})D_{jk}\tilde{u}D_m\tilde{u},\\
S_{12} &= \psi'\omega\bar{\delta}a^{ij}D_{ij}\tilde{u}\tilde{w}_T,\\
S_{13} &= \delta_T a^{ij}D_{ij}\tilde{u}\tilde{w}_T.
\end{aligned}$$

By using Lemma 9.11 with

$$\begin{aligned}
f^{ij} &= 2a^{jk}D_k(c^{im})D_m\tilde{u},\\
f^{ij} &= -\psi'\omega\tilde{w}_T\bar{\delta}a^{ij},\\
f^{ij} &= -\psi'\tilde{w}_T\delta_T a^{ij}
\end{aligned}$$

to estimate $S_{11}$, $S_{12}$, and $S_{13}$, respectively, we obtain

$$0 \le \frac{1}{2}a^{ij}D_{ij}\tilde{w}_T + B_5^i D_i\tilde{w}_T + C_1^*\tilde{w}_T - \frac{1}{4}\tilde{\mathcal{C}}_T^2,$$

where

$$\begin{aligned}
B_5^i = B_4^i + \frac{1}{2v}[-a^{jk}D_k(c^{rm})D_m\tilde{u}\nu_j + \psi'\omega\tilde{w}_T\bar{\delta}a^{ij}\nu_j + \psi'\tilde{w}_T\delta_T a^{ij}\nu_j]&\\
+\frac{3}{4v}\Lambda\nu\cdot\gamma_k D_{km}\tilde{u}c^{mi}&,
\end{aligned}$$

$$H_{11} = H_{10} + \frac{20w_T|\bar{\delta}\mathcal{A}|^2}{\Lambda G} + \frac{18n\mu_1\Lambda w_T|\bar{\delta}A|^2}{A^2} + \frac{|\bar{\delta}A|\mathcal{E}}{|A|},$$

$$J_{11} = J_{10} + \frac{|a^{jm}D_k\gamma^m D_m u\gamma^j|\mathcal{E}}{Aw_T} + \frac{|a||\bar{\delta}A|}{A} + \frac{\psi'}{v}\bar{\delta}a^{ij}\nu_j\tilde{c}_i + \frac{|\delta_T A|\mathcal{E}}{A}$$

and

$$K_{21} = K_{20} + \frac{20\sum_{i,j=1}^n |a^{jk}D_k(c^{im})D_m u|^2}{\Lambda G w_T} + \frac{18nu_1\Lambda(a^{jk}D_k\gamma^m D_m u\gamma_j)^2}{A^2 w_T}$$
$$+ \frac{24\Lambda|\tilde{c}|^2}{Gv^2\tilde{w}_T} + \frac{|a||a^{jk}D_k\gamma^m D_m u\gamma_j|}{Aw_T} - \frac{a^{jk}D_m(c^{im})D_m uc_i}{vw_T}$$
$$+ \frac{20w_T|\delta_T\mathcal{A}|^2}{20\Lambda G} + \frac{18n\mu_1\Lambda|\delta_T A|^2}{A^2}$$
$$+ \frac{|a||\delta_T A|}{A} + \frac{\delta_T a^{ij}\nu_j c_i}{v}.$$

To proceed, with $s \in (0, \mu_1)$ and $k_2$ to be determined and $\tilde{M}_T = \sup_{d\leq s}\tilde{w}_T$, we set

$$w = \tilde{w}_T + k_2\tilde{M}_T d.$$

It then follows (after noting that $c^{im}\gamma_m = 0$) that

$$0 \leq \frac{1}{2}a^{ij}D_{ij}w + B_5^i D_i w + C_1^*\tilde{w}_T - \frac{1}{4}\tilde{\mathcal{C}}_T^2 + k_2 C_2^*\tilde{M}_T + S_{14},$$

where

$$C_2^* = J_{12}\omega + K_{22},$$

with

$$J_{12} = -\delta_N\mathcal{E} - \frac{w_T}{2v}\bar{\delta}a^{ij}\nu_j\gamma_i,$$
$$K_{22} = -\delta_N a - \frac{1}{2v}a^{jk}D_k\gamma^m D_m u\nu_j = w_T\delta_T a^{ij}\nu_j\gamma_i,$$

and

$$S_{14} = -k_2\tilde{M}_T\psi'\delta_N a^{ij}D_{ij}\tilde{u}.$$

Applying Lemma (9.11) one last time, we find that

$$\frac{1}{2}a^{ij}D_{ij}w + B_6^i D_i w + C_1^*\tilde{w}_T + C_3^* M_T$$

with

$$B_6^i = B_5^i + k_2\psi'\left(\frac{\psi'}{2v}\tilde{M}_T\delta_N a^{ij}\nu_j + \frac{\Lambda\nu\cdot\gamma}{4v}\gamma^r D_{rm}\tilde{u}c^{im}\right),$$
$$C_3^* = k_2 C_2^* + \omega J_{14} + K_{23},$$
$$J_{14} = k_2\frac{|\delta_N\mathcal{A}|\mathcal{E}}{A},$$

and

$$K_{23} = \frac{k_2^2(\psi')^2\tilde{M}_T}{2v}\delta_N a^{ij}\nu_j\gamma_i + \frac{\Lambda|\tilde{c}|^2}{2Gv^2\tilde{w}_T} + \frac{k_2|\delta_N\mathcal{A}||a|}{A} - k_2\delta_N a^{ij}\nu_j c_i + \frac{20(\psi')^2\tilde{M}_T k_2^2}{\Lambda G}|\delta_N\mathcal{A}|^2 + \frac{18n\mu_1\Lambda k_2^2\tilde{M}_T}{A^2}|\delta_N A|^2.$$

Our next step is to determine $k_2$. Since $\delta_T b = O(b_p\cdot\gamma)$, there are constants $M_2$ and $\beta_1$ such that $\delta_T b \le \beta_1 b_p\cdot\gamma$ wherever $|p'| \ge M_2$. From the boundary condition, we infer that

$$b_p\cdot Dw = k_2\tilde{M}_T b_p\cdot\gamma - 2(\delta_T b + \omega\bar{\delta}b)\tilde{w}_T,$$

so choosing $k_2 = 1+\beta_1$ gives us

$$b_p\cdot Dw \ge \tilde{M}_T b_p\cdot\gamma - 2\omega\bar{\delta}b\tilde{w}_T$$

wherever $|D'u| \ge M_2$. We then choose $s$ so that $k_2 s \le \frac{1}{2}$.

We now write $\Omega_s(M_2)$ for the subset of $\Omega$ where $d \le s$ and $|D'u| \ge M_2$. Taking (9.66) into account, we infer that there are constants $\hat{J}$ and $\hat{K}$ such that

$$\frac{1}{2}a^{ij}D_{ij}w + B_5^i D_i w + C_4^*\mathcal{E}\tilde{w}_T + C_5^*\mathcal{E}M_T \ge 0$$

on $\Omega_s(M_2)$, where

$$C_4^* = \frac{\omega'}{\psi'} + \omega^2\frac{H_{11}}{\mathcal{E}} + \omega\hat{J} + \hat{K},$$
$$C_5^* = \omega\hat{J} + \hat{K}.$$

Then conditions (9.65) and (9.64d) imply that there are positive constants $M_3 \ge M_2$ and $e_1$ such that

$$\frac{H_{11}}{\mathcal{E}} \le -e_1$$

on $\Omega_s(M_3)$.

Next, we take $\Omega'$ to be the subset of $\Omega$ on which $d \le s$, $|D'u| \ge M_3$ and $\tilde{w}_T \ge \frac{1}{2}\tilde{M}_T$. then

$$\frac{1}{2}a^{ij}D_{ij}w + B_5^i D_i w + C_6^*\mathcal{E}\tilde{w}_T \ge 0$$

on $\Omega'$, where

$$C_6^* = \frac{\omega'}{\psi'} - e_1\omega^2 + 3\omega\hat{J} + 3\omega\hat{K}.$$

It follows that we can take $\omega$ to be a constant in such a way that $C_6^* > 0$ on $\Omega'$.

With this choice for $\omega$ (and the corresponding function $\psi$), we finally choose $M_4 \geq M_3$ so that $2\omega\bar{\delta}b < b_p \cdot \gamma$ wherever $|D'u| \geq M_4$. Then we write $\Omega''$ for the subset of $\Omega$ on which $d \leq s$, $|D'u| \geq M_4$ and $\tilde{w}_T \geq \frac{1}{2}\tilde{M}_T$. All these choices imply that

$$\frac{1}{2}a^{ij}D_{ij}w + B_5^i D_i w > 0 \text{ in } \Omega'', \quad b_p \cdot Dw > 0 \text{ on } \partial\Omega'' \cap \partial\Omega.$$

Since we have an upper bound for $w$ on $\partial\Omega'' \setminus \partial\Omega$ (in terms of $M_4$ and $M_1(s)$), it follows from the maximum principle that we have an upper bound for $w$ in $\Omega''$ and hence a bound for $|Du|$ in $\Omega$. □

Note that the strongest condition in our list of hypotheses is probably $\bar{\delta}b \leq o(b_p \cdot \gamma)$ since, in our example

$$b(x, z, p) = v^{q-1} Du \cdot \gamma + \psi(x, z),$$

this hypothesis is only satisfied if $q = 0$ or $q > 1$. Presumably this condition can be relaxed to $\delta b \leq O(b_p \cdot \gamma)$ by using a function of the form

$$w = \tilde{w}_T + \tilde{M}_T f(v)d$$

with $f$ a suitable function which goes to infinity sufficiently slowly as its argument goes to infinity but the verification of this assertion has not yet been attempted. Moreover, unlike the situation considered in previous sections, there is no reason to expect $M_1(s)$ to behave nicely. For example, Finn has shown in Section V of [45] that the gradient of a solution of the minimal surface equation grows exponentially with respect to distance to the boundary.

Our next step is to obtain hypotheses under which conditions (9.66) are satisfied. In the next chapter, we shall derive a number of gradient bounds for equations similar to the prescribed mean curvature equation under the additional assumption that the coefficients $a^{ij}$ are given by $a^{ij}(x, z, p) = \partial A^i(x, z, p)/\partial p_j$ for some vector-valued function $A$. Here, we show, via the maximum principle, that such an estimate is valid if $a^{ij} = g^{ij}$.

**Lemma 9.13.** *Suppose $a$ is differentiable with respect to $(x, z, p)$ with*

$$\delta a \leq O(1), \tag{9.68a}$$

$$(\bar{\delta} - 1)a \leq O(1/|p|), \tag{9.68b}$$

$$|a_p| = O(1/|p|). \tag{9.68c}$$

*If* $u \in C^3(B(0,R)) \cap L^\infty(B(0,R))$ *for some* $R > 0$ *with*

$$g^{ij}D_{ij}u + a(x,u,Du) = 0 \text{ in } B(0,R), \tag{9.69}$$

*then there is a constant* $C_0$, *determined only by the maximum of* $|u|$, $R$, *and the limit behavior in* (9.68) *such that*

$$|Du(0)| \le C_0. \tag{9.70}$$

***Proof.*** First, to simplify notation, we assume without loss of generality that $u < 0$ in $B(0,R)$. With this assumption and positive constants $u_0$ and $C_1$ to be chosen, we define functions $\varphi$, $f$, and $\eta$ by

$$\varphi(x,z) = \left(\frac{z}{2u_0} + 1 - \frac{|x|^2}{R^2}\right)_+,$$
$$f(s) = \exp(C_1 s) - 1,$$
$$\eta(x) = f(\varphi(x,u(x))).$$

Our first step is to derive a suitable differential equation for $w = \eta v$. It follows from our previous arguments that

$$g^{ij}D_{ij}v + (g^{ij,r}D_{ij}u + a^r)D_r v + Cv - \mathfrak{C}^2 = 0,$$

with

$$C = v^{-2}(a_z|Du|^2 + a_x \cdot Du)$$

and

$$\mathfrak{C}^2 = \frac{1}{v}g^{ij}g^{km}D_{ik}uD_{jm}u.$$

But a direct calculation shows that

$$g^{ij,r}D_{ij}u = -\frac{2}{v^2}D_{rj}uD_ju + \frac{2}{v^4}D_iuD_juD_{ij}uD_ru = -2g^{rj}D_jv,$$

and hence

$$g^{ij}D_{ij}v + a^rD_rv - 2g^{ij}D_ivD_jv + Cv - \mathfrak{C}^2 = 0.$$

If we multiply this equation by $\eta$, we find that

$$g^{ij}D_{ij}w + (a^r + \frac{\eta}{v}a^{rj}D_jv)D_rw + (C - g^{ij}D_{ij}\eta - a^jD_j\eta)v - \mathfrak{C}^2 = 0.$$

From the definition of $\eta$, we see that

$$g^{ij}D_{ij}\eta + a^jD_j\eta = f'(-2(n - \frac{|p|^2}{v^2}) - \frac{1}{2u_0}(\bar{\delta} - 1)a - 2a_p \cdot \frac{x}{R})$$
$$+ f''[g^{ij}\frac{x^ix^j}{R^2} + \frac{Du \cdot x}{u_0v^2R} + \frac{|Du|^2}{4u_0^2v^2}].$$

From the limit behavior in (9.68), we conclude that there are constants $M_0$ and $\mu_0$ such that

$$-2(n - \frac{|p|^2}{v^2}) - \frac{1}{2u_0}(\bar{\delta} - 1)a - 2a_p \cdot \frac{x}{R} \ge -\mu_0$$

wherever $|Du| \ge M_0$. In addition, because $g^{ij}x^i x^j \ge 0$, we have

$$g^{ij}\frac{x^i x^j}{R^2} + \frac{Du \cdot x}{u_0 v^2 R} + \frac{|Du|^2}{4u_0^2 v^2} \ge \frac{|Du|^2}{8u_0^2 v^2}$$

wherever $|Du| \ge 8u_0/R$. Therefore

$$g^{ij}D_{ij}\eta + a^j D_j\eta \ge C_1 \exp(C_1\varphi(x,u))[-\mu_0 + C_1\frac{|Du|^2}{8u_0^2 v^2}].$$

We now choose $u_0 = -u(0)$ and observe that $|Du|^2/v^2 \ge 1/10$ if $|Du| \ge 3$. Hence,

$$g^{ij}D_{ij}\eta + a^j D_j\eta \ge C_1$$

wherever $|Du| \ge \max\{M_0, 3, 8u_0/R\}$ if $C_1$ is sufficiently large. Moreover, there are constants $M_1$ and $\mu_1$ such that $C \le \mu_1$ wherever $|Du| \ge M_1$. We now set

$$M_2 = \max\{M_0, M_1, 3, 8u_0/R\}.$$

By further increasing $C_1$ so that $C_1 > \mu_1$, we conclude that

$$g^{ij}D_{ij}w + (a^r + \frac{\eta}{v}g^{rj}D_j v)D_r w \ge 0$$

wherever $|Du| \ge M_2$ and $w > 0$. It follows that $w$ cannot have a maximum at such a point, and therefore, because $0 \le \eta \le C_1 \exp(C_1)$,

$$w \le \exp(C_1)(1 + M_2^2)^{1/2}.$$

Since $\varphi(0, u(0)) = 1/2$, it follows that $\eta(0) = \exp(C_1/2) - 1 > 0$ and hence

$$|Du(0)| \le \frac{\exp(C_1)}{\exp(C_1/2) - 1}(1 + M_2^2)^{1/2}.$$

□

## Notes

The underlying idea in the gradient estimates presented here goes back to Bernstein [9], and we refer the reader to [64] (specifically the Notes to Chapter 15) for further discussion of this approach to gradient estimates which are purely interior (like those in Section 9.1) or which estimate the gradient inside the domain from a gradient estimate on the boundary (like our Corollary 9.3). However, we do point out that the results quoted here are based on corresponding, but more general, results in [92] and [169]. One notable exception to this general rule is our Lemma 9.13, which is a slight reworking of the gradient estimate in Section 1 of [82].

A simple version of the gradient for problem (9.1) appears in Section 4 of [98], and the results in Sections 9.3 through 9.6 come from [101] with one exception: Theorem 9.10 is a combination of Theorem 9.9 with the gradient bound Theorem 3.3 from [120]. The approach for capillary-type problems in Section 9.7 comes from [107] (except for Lemma 9.13 as already pointed out). An alternative maximum principle approach, but restricted to the capillary boundary condition, was presented in [84].

In many ways, the current state of the gradient bound for quasilinear elliptic equations and nonlinear boundary conditions is still in an unsatisfactory form. The bounds in this chapter are the most general ones known to the author and the assumptions in these theorems, especially the decomposition (9.3), seem very restrictive compared to the corresponding ones in, for example, Chapter 15 of [64]. In addition, the computations are rather intensive. An alternative approach for uniformly parabolic equations was presented by Dong in [38] and we discuss this approach more fully in the notes to Chapter 11. It seems that Dong's approach could be used to simplify some of the proofs in this chapter, but no one has attempted to do so yet.

On the other hand, the gradient bounds require very little geometric structure, unlike the corresponding gradient bounds for the Dirichlet problem. In particular, our gradient bound Theorem 9.12 provides a gradient bound for the differential equation $g^{ij}D_{ij}u = 0$ and the boundary condition

$$\frac{Du \cdot \gamma}{v} + \psi(x) = 0$$

for any smooth domain $\Omega$ as long as $|\psi| < 1$. For the same differential equation with Dirichlet data $u = \varphi$ on $\partial\Omega$, a gradient bound is only possible (for arbitrary $\varphi$) if the mean curvature of $\partial\Omega$ is everywhere nonnegative. (See Theorem 14.14 of [64] for more information on this situation.)

## Exercises

**9.1** Suppose that $a^{ij}$ can be decomposed as in (9.3). Suppose also that there are functions $r$ and $s$ and positive constants $k_2$ and $k_2$ such that

$$|(\bar{\delta}+r+1)a_*^{ij}| + |(\delta+s)a_*^{ij}| \le k_1\sqrt{\lambda_*\mathcal{E}}/|p|,$$
$$\max\{\bar{\delta}\mathcal{E}, \delta\mathcal{E}, (\bar{\delta}+r)a, (\delta+s)a\} \le k_1\mathcal{E},$$

and set

$$A_1 = (\bar{\delta}+r)\mathcal{E} + \frac{|p|^2}{4\lambda_*}\sum_{i,j}|(\bar{\delta}+r+1)a_*^{ij}|^2,$$

$$B_1 = (\delta+s)\mathcal{E} + (\bar{\delta}+r)a + \frac{|p|^2}{2\lambda_*}\sum_{i,j}[(\bar{\delta}+r+1)a_*^{ij}][(\delta+s)a_*^{ij}],$$

$$C_1 = (\delta+s)a + \frac{|p|^2}{4\lambda_*}\sum_{i,j}|(\bar{\delta}+r+1)a_*^{ij}|^2,$$

and

$$a_\infty = \limsup_{|p|\to\infty}\ \sup_{(x,z)\in\Omega\times[-M_0,M_0]}\frac{A_1}{\mathcal{E}},$$
$$b_\infty = \limsup_{|p|\to\infty}\ \sup_{(x,z)\in\Omega\times[-M_0,M_0]}\frac{B_1}{\mathcal{E}},$$
$$c_\infty = \limsup_{|p|\to\infty}\ \sup_{(x,z)\in\Omega\times[-M_0,M_0]}\frac{C_1}{\mathcal{E}}.$$

Suppose also that $a_\infty \le 0$, $c_\infty \le 0$, or $b_\infty \le -2\sqrt{a_\infty c_\infty}$. Prove a gradient bound for solutions of $Qu = 0$ in $\Omega$, $Du \cdot \gamma = 0$ on $\partial\Omega$ provided $\Omega$ is convex and $\partial\Omega \in C^2$. (Hint: Use the calculations from the proof of Theorem 15.2 in [64] along with the argument in Section 9.2.)

**9.2** Verify the comment at the end of Section 9.5 by showing that Theorem 9.8 remains valid if we assume that there is a sufficiently small positive constant $\bar{\mu}$ such that

$$\delta_T b, \bar{\delta} b \le \bar{\mu}|p|b_p \cdot \gamma$$

for large $|p|$.

**9.3** Define the matrices $\mathcal{F}$ and $\mathcal{U}$ by $f^{ij} = 1$ for $i,j = 1,2$ and $u^{ij} = (-1)^{i+j}$ for $i,j = 1,2$. Observe that these matrices are symmetric. Show that they are positive semi-definite with $f^{11}u_{11} = f^{22}u_{22} = 1$ but $f^{ij}u_{ij} = 0$. Use this set of matrices to show that we can't have $G_1 + G_2 \le c\tilde{\mathcal{G}}_T^2$ for any positive constant $c$.

## Chapter 10

# Gradient Estimates for the Conormal Derivative Problems

In this chapter, we prove gradient bounds for the conormal derivative problem

$$\operatorname{div} A(x,u,Du) + B(x,u,Du) = 0 \text{ in } \Omega, \tag{10.1a}$$

$$A(x,u,Du)\cdot\gamma + \psi(x,u) = 0 \text{ on } \partial\Omega \tag{10.1b}$$

under varying hypotheses on $A$, $B$, and $\psi$ provided $\partial\Omega \in C^2$. The model problems here are the capillary problem, in which $A(x,z,p) = p/\sqrt{1+|p|^2}$ and $B(x,z,p) = -\kappa z$ for some positive constant $\kappa$, and the Neumann problem in which $A(x,z,p) = p$ and $B(x,z,p) = f(x)$ for some function $f$. In both cases, we have obtained gradient bounds in Chapter 9, provided $\partial\Omega \in C^3$ (and $f \in C^1(\overline{\Omega})$ for the Neumann problem). Here, we weaken the regularity hypotheses on the data of the problem when the problem has this special conormal form. The basic techniques here were presented in Chapter 5, and we shall look at various ideas to extend the applicability of those techniques.

### 10.1 The Sobolev inequality of Michael and Simon

For the reader's convenience, we begin by stating the Sobolev inequality of Michael and Simon in its general form. A proof can be found in Theorem 11.8 of [114].

**Lemma 10.1.** *Let $m$ and $n$ be positive integers with $2 \le n \le m$, let $U$ be an open subset of $\mathbb{R}^m$ and let $M$ be a subset of $U$. Let $\mu$ be a measure defined on all sets of the form $M \cap B$, where $B$ is a Borel subset of $\mathbb{R}^m$, which is finite on any set of the form $M \cap K$ with $K$ compact. Let $\gamma$ be a* *continuous, symmetric-matrix-valued function with*

$$\sum_{i=1}^{m} \gamma^{ii}(x) = n, \tag{10.2a}$$

$$0 \le \gamma^{ij}(x)\xi_i\xi_j \le |\xi|^2 \tag{10.2b}$$

*for all $x \in M$ and all $\xi \in \mathbb{R}^m$, and define the operator $\delta$ on $C^1(U)$ by $\delta h = (\delta^1 h, \dots, \delta^m h)$ and $\delta^i h = \gamma^{ij} D_j h$. Suppose also that*

$$\limsup_{\rho \to 0+} \frac{\mu(M \cap B(x,\rho))}{\omega_n \rho^n} \ge 1 \tag{10.3}$$

*for all $x \in M$. If there is a vector-valued function $\mathcal{H}$ such that*

$$\int_M (\delta h + h\mathcal{H})\, d\mu = 0 \tag{10.4}$$

*for all $h \in C^1(U)$ with compact support, then*

$$\left( \int_M |h|^{n/(n-1)}\, d\mu \right)^{(n-1)/n} \le C(n) \int_M (|\delta h| + |h||\mathcal{H}|)\, d\mu \tag{10.5}$$

*for all $h \in C^{0,1}(U)$ with compact support.*

For our applications, we let $n \ge 2$ be a positive integer, and let $\Omega$ be an open subset of $\mathbb{R}^n$. We then set

$$v = (1 + |Du|^2)^{1/2}, \quad \nu = Du/v, \tag{10.6a}$$

$$g^{ij} = \delta^{ij} - \nu^i \nu^j, H = D_i(\nu^i). \tag{10.6b}$$

The Sobolev inequality we then use is as follows.

**Lemma 10.2.** *Let $h \in C^1(\overline{\Omega})$ have compact support. Then*

$$\left( \int_\Omega |h|^{n/(n-1)} v\, dx \right)^{(n-1)/n} \le C(n) \int_\Omega (|\delta h| + |h||H|) v\, dx. \tag{10.7}$$

*If $N > 2$ is a constant with $N \ge n$ and if $\kappa = N/(N-2)$, then*

$$\left( \int_\Omega h^{2\kappa} v\, dx \right)^{1/\kappa} \le C(n,N) \left( \int_\Omega h^2 v\, dx \right)^{(N-n)/N} \times \left( \int_\Omega (g^{ij} D_i h D_j h + h^2 H^2) v\, dx \right)^{n/N}. \tag{10.8}$$

***Proof.*** The main step is to show that the hypotheses of Lemma 10.1 are satisfied with $m = n + 1$, $U = \Omega \times \mathbb{R}$, $M$ equal to the graph of $u$, and a suitable choice of $\gamma$, $\mathcal{H}$, and $\mu$.

We begin by defining $\mu$. For any Borel set $B$, there is a Lebesgue measurable subset $S_B$ of $\Omega$ such that

$$M \cap B = \{(x, u(x)) : x \in S_B\},$$

and we define

$$\mu(M \cap B) = \int_{S_B} v\, dx.$$

Next, we set $\nu^{n+1}(x) = -1/v$, and, for $i, j = 1, \ldots, n + 1$, we define $\gamma^{ij}(x, u(x)) = \delta^{ij} - \nu^i(x)\mu^j(x)$. Then

$$\gamma^{ij}(x, u(x)) = \begin{cases} g^{ij}(x) & \text{if } i, j \le n, \\ \frac{\nu^i}{v} & \text{if } i \le n, j = n + 1, \\ \frac{\nu^j}{v} & \text{if } j \le n, i = n + 1, \\ \frac{|Du|^2}{v^2} & \text{if } i = j = n + 1. \end{cases}$$

Finally, we set $\mathcal{H}(x, u(x)) = \nu(x)H(x)$.

It's easy to check (10.2) and (10.3). To verify (10.4), we define $\varphi$ by $\varphi(x) = h(x, u(x))$, and we use $k$ to denote an integer parameter that only takes values in $\{1, \ldots, n\}$. It follows that

$$\gamma^{ik} D_k\varphi = \gamma^{ik} D_k h + \gamma^{ik} D_k u D_{n+1} h.$$

From the expression for $\gamma^{ij}$, we also have that $\gamma^{ik} D_k u = \gamma^{i,n+1}$, so

$$\gamma^{ik} D_k\varphi = \delta_i h.$$

Therefore

$$\int_M \delta_i h\, d\mu = \int_\Omega \gamma^{ik} D_k\varphi v\, dx = -\int_\Omega D_k(v\gamma^{ik})\varphi\, dx.$$

We now evaluate $D_k(v\gamma^{ik}) = -vH\nu^i$, so (10.4) is satisfied.

Next, we observe that, if $\xi \in \mathbb{R}^n$ and $\Xi \in \mathbb{R}^{n+1}$ is defined by $\Xi^i = \gamma^{ik}\xi_k$, then

$$|\Xi|^2 = g^{ij}\xi_i\xi_j. \tag{10.9}$$

Hence $|\delta h|^2 = g^{ij} D_i h D_j h$ if $h \in C^{0,1}(\Omega)$, and (10.7) follows from Lemma 10.1.

The proof of (10.8) from (10.7) is the same as the proof of Theorem 5.8. □

For our gradient estimate, one further inequality will be useful, which estimates $H^2$ by a specific combination of second derivatives of $u$. To prove this estimate, we define the functions

$$\tilde{g}^{ij} = \delta^{ij} - \frac{v-1}{v}\frac{p_i p_j}{|p|^2}.$$

It is easy to check that $[\tilde{g}^{ij}]$ is the square root of the matrix $[g^{ij}]$, that is, $g^{ij} = \tilde{g}^{ik}\tilde{g}^{kj}$. Hence, if we define

$$b^{ij} = \tilde{g}^{ik}\tilde{g}^{jm}D_{km}u$$

then

$$vH = \delta_{ij}b^{ij},$$

so

$$|vH| \le (\delta_{ij}\delta_{ik}\delta_{jk})^{1/2}\left(\delta_{ij}b^{ik}b^{jk}\right)^{1/2} = n^{1/2}\left(g^{ij}g^{km}D_{ik}uD_{jm}u\right)^{1/2}.$$

it follows that

$$H^2 \le \frac{n}{v^2}g^{ij}g^{km}D_{ik}uD_{jm}u. \tag{10.10}$$

Alternatively, this inequality can be proved by noting that $H$ is the sum of the principal curvatures of the graph of $u$ while $v^{-2}g^{ij}g^{km}D_{ik}uD_{jm}u$ is the sum of the squares of these principal curvatures.

## 10.2 The interior gradient bound

In this section, we prove an interior gradient bound for a large class of divergence structure equations. A key assumption is related to the derivatives of $A$ and $B$. We assume that $A$ is differentiable with respect to $x$, $z$, and $p$ and that, for all $i$ and $k$ in $\{1,\dots,n\}$, there are functions $C^i_k$ and $D^i_k$ such that $D^i_k$ is differentiable with respect to $x$, $z$, and $p$ and

$$\frac{\partial A^i}{\partial z}p_k + \frac{\partial A^i}{\partial x^k} + B\delta^i_k = C^i_k + D^i_k. \tag{10.11}$$

For notational convenience, we define

$$\mathcal{D}^{ij} = \nu^k\frac{\partial D^i_k}{\partial p_j}, \quad \mathcal{F} = \nu^k\left(p_i\frac{\partial D^i_k}{\partial z} + \frac{\partial D^i_k}{\partial x^i}\right). \tag{10.12}$$

We also assume that there are functions $\Lambda_0$ and $\bar{\mu}$ and nonnegative constants $\beta_1$ and $\tau_0$ such that

$$C_k^i \nu^k \xi_i \le \beta_1 \Lambda_0^{1/2} (a^{ij}\xi_i\xi_j)^{1/2}, \tag{10.13a}$$

$$C_k^i g^{jk} \eta_{ij} \le \beta_1 \Lambda_0^{1/2} (a^{ij} g^{km} \eta_{ik}\eta_{jm})^{1/2}, \tag{10.13b}$$

$$v\mathcal{D}^{ij}\eta_{ij} \le \beta_1 \Lambda_0^{1/2} (a^{ij}\delta^{km}\eta_{ik}\eta_{jm})^{1/2}, \tag{10.13c}$$

$$v\mathcal{F} \le \beta_1^2 \Lambda_0, \tag{10.13d}$$

$$va^{ij}\hat{\xi}_i\xi_j \le (a^{ij}\xi_i\xi_j)^{1/2}(\bar{\mu}|\hat{\xi}|^2)^{1/2} \tag{10.13e}$$

for all vectors $\xi$ and $\hat{\xi}$ and all matrices $\eta$ wherever $v \ge \tau_0$. (Note that there is no loss of generality in assuming that $\tau_0 \ge 1$.)

For our estimates, we shall also use the following two quantities:

$$\mathcal{C}_0^2 = a^{ij} g^{km} D_{ik}u D_{jm}u, \tag{10.14a}$$

$$\mathcal{E}_0 = a^{ij} D_i v D_j v. \tag{10.14b}$$

With these definitions and assumptions, we have our first integral inequality, which is the basis for our estimates.

**Lemma 10.3.** *Suppose conditions* (10.13) *are satisfied, and let $\chi$ be an increasing function on $[\tau_0, \infty)$. Suppose that there are constants $\tau \ge \tau_0$ and $c_\chi$ such that*

$$0 \le (s-\tau)\frac{\chi'(s)}{\chi(s)} \le c_\chi \tag{10.15}$$

*for all $s \ge \tau$. If $\zeta \in C^1(\overline{\Omega})$ vanishes on $\partial\Omega$, and if $Qu = 0$ in $\Omega$, then*

$$\int_{\Omega_\tau} [(1-\frac{\tau}{v})\mathcal{C}_0^2 + \mathcal{E}_0]\chi\zeta^2\,dx \le 18(1+c_\chi)\int_{\Omega_\tau} (\beta_1^2\Lambda_0\zeta^2 + \bar{\mu}|D\zeta|^2)\chi\,dx, \tag{10.16}$$

*where $\Omega_\tau$ is the subset of $\Omega$ on which $v \ge \tau$.*

***Proof.*** Let $\theta$ be a $C^2(\overline{\Omega})$ vector-valued function with compact support in $\Omega$ and use $D_k\theta^k$ as test function in the weak form of the differential equation $Qu = 0$. It follows that

$$\int_\Omega A^i D_{ik}\theta^k\,dx = \int_\Omega BD_k\theta^k,$$

and an integration by parts yields

$$\int_\Omega (D_k A^i + B\delta_k^i) D_i\theta_k\,dx = 0.$$

We now observe that

$$D_k A^i + B\delta_k^i = a^{ij}D_{jk}u + C_k^i + D_k^i,$$

so another integration by parts yields

$$\int_\Omega (a^{ij} D_{jk} u C_k^i) D_i \theta_k - D_i(D_k^i)\theta^k \, dx = 0.$$

By approximation, this equality is true for any $\theta \in C^1(\overline{\Omega})$ that vanishes on $\partial\Omega$. In particular, we can use

$$\theta = \zeta^2 (v - \tau)_+ \chi \nu.$$

Since

$$D_i \theta^k = [\chi + (v-\tau)\chi']\zeta^2 \nu^k D_i v + \zeta^2 (v-\tau)\chi g^{jk} D_{ij} u + 2\zeta(v-\tau)\chi\nu^k D_i\zeta$$

on $\Omega_\tau$, it follows that

$$\begin{aligned} 0 = &\int ([\chi + (v-\tau)\chi']\mathcal{E}_0 + (1 - \frac{\tau}{v})\chi \mathcal{C}_0^2)\zeta^2 \, dx \\ &+ \int 2(v-\tau)\chi a^{ij} D_j v D_i \zeta\zeta \, dx \\ &+ \int [\chi + (v-\tau)\chi'] C_k^i \nu^k D_i v \zeta^2 + (1 - \frac{\tau}{v}) C_k^i g^{jk} D_{ij} u \zeta^2 \, dx \\ &+ \int 2(v-\tau)\chi C_k^i \nu^k D_i \zeta\zeta \, dx \\ &- \int \mathcal{D}^{ij} D_{ij} u (v-\tau)\chi\zeta^2 + \mathcal{F}(v-\tau)\chi\zeta^2 \, dx, \end{aligned}$$

where here and in the remainder of this proof, all integrals are integrated over $\Omega_\tau$.

We now estimate the integrals in this identity, using the abbreviations

$$\begin{aligned} I_0 &= \int \chi \mathcal{E}_0 \zeta^2 \, dx, \\ I_1 &= \int [\chi + (v-\tau)\chi']\mathcal{E}_0 \zeta^2 \, dx, \\ I_2 &= \int (1 - \frac{\tau}{v})\chi \mathcal{C}_0^2 \zeta^2 \, dx, \\ I_3 &= \int \beta_1^2 \Lambda_0 \chi \zeta^2 \, dx, \\ I_4 &= \int \chi \bar{\mu} |D\zeta|^2 \, dx. \end{aligned}$$

Using (10.13e), we have that

$$\int 2(v-\tau)\chi a^{ij} D_j v D_i \zeta\zeta \, dx \geq -\frac{1}{8} I_0 - 8 I_4.$$

Then (10.13a) and (10.15) imply that

$$\int [\chi + (v-\tau)\chi'] C^i_k \nu^k D_i v \zeta^2\, dx \ge -\frac{1}{8} I_1 - 2(1+c_\chi) I_3$$

and (10.13b) implies that

$$\int (1 - \frac{\tau}{v}) \chi \zeta^2 C^i_k g^{jk} D_{ij} u\, dx \ge -\frac{1}{4} I_2 - 4 I_3.$$

Next, (10.13a) gives

$$\int 2(v-\tau)\chi C^i_k \nu^k D_i \zeta \zeta\, dx \ge -I_3 - I_4.$$

Finally, (10.13c) gives

$$-\int \mathcal{D}^{ij} D_{ij} u (v-\tau)\chi\zeta^2\, dx \ge -\frac{1}{4} I_1 - \frac{1}{4} I_2 - I_3$$

and (10.13d) implies that

$$-\int \mathcal{F}(v-\tau)\chi\zeta^2\, dx \ge -I_3.$$

The proof is completed by combining these inequalities and noting that $I_0 \le I_1$. □

Note that conditions (10.13) are exactly the hypotheses which give (10.16). For this form of the estimate, the use of $\beta_1$ is not necessary, but its utility will be seen when we present our gradient estimates.

Our proof of the gradient bound is effected through suitable choices of the function $\chi$. It will be useful to introduce an auxiliary positive, real-valued function $w \in C^1([\tau_0, \infty))$. We assume that there is a positive number $\beta$ such that

$$0 \le s w'(s) \le \beta w(s) \tag{10.17}$$

for all $s \ge \tau_0$. (Equivalently, $s^{-\beta} w(s)$ is a decreasing function of $s$.) Instead of estimating $v$ directly, we shall estimate $w(v)$.

It will also be useful to introduce three more positive, real-valued structure functions $\Lambda \in C(\Omega)$ and $\lambda$ and $\tilde{\lambda}$ in $C^1([\tau_0, \infty))$. We suppose that they are related to the operator $Q$ (and to each other) by the following relations

$$\Lambda_0 \le \Lambda v, \tag{10.18a}$$

$$\lambda(v(x)) \le \Lambda(x), \tag{10.18b}$$

$$\bar{\mu} \le \Lambda v, \tag{10.18c}$$

$$\lambda(v) \left( 1 + \left( \frac{v\lambda'(v)}{\lambda(v)} \right)^2 \right) g^{ij} \xi_i \xi_j \le v a^{ij} \xi_i \xi_j \tag{10.18d}$$

for all $\xi \in \mathbb{R}^n$,

$$(\Lambda(x)/\lambda(v(x)))^{N/2}/\Lambda(x) \le \tilde{\lambda}(v), \tag{10.18e}$$

where $N > 2$, $N \ge n$, and $n \ge 2$. We further suppose that $\tilde{\lambda}$ satisfies the monotonicity conditions

$$\beta w'(s)\tilde{\lambda}(s) + \lambda'(s)w(s) \ge 0, \tag{10.19a}$$

$$s\tilde{\lambda}'(s) \le \beta\tilde{\lambda}(s) \tag{10.19b}$$

for all $s \ge \tau_0$.

Under these hypotheses, we can estimate $\sup w$ in terms of an integral.

**Lemma 10.4.** *Suppose conditions* (10.13), (10.17), (10.18), *and* (10.19) *are satisfied. If $Qu = 0$ in some ball $B(y,R)$ and $\tau \ge \tau_0$, then*

$$\sup_{B(y,R/2)} \left(1 - \frac{\tau}{v}\right)_+^N w(v) \le c_1 R^{-n} \int_{B_\tau(y,R)} w(v)\Lambda(x)\tilde{\lambda}(v)v\,dx, \tag{10.20}$$

*where $B_\tau(y,R)$ is the subset of $B(y,R)$ on which $v \ge \tau$, and $c_1$ is determined only by $\beta$, $\beta_1 R$, $n$, and $N$.*

***Proof.*** With $q > 1 + \beta$ a parameter, we set

$$\chi(s) = \tilde{\lambda}(s)w^q(s)\left(1 - \frac{\tau}{s}\right)_+^{Nq-N}.$$

Then

$$\begin{aligned}\chi'(s) &= w(s)^{q-1}\left(1 - \frac{\tau}{s}\right)_+^{Nq-N}[\tilde{\lambda}'(s)w(s) + q\tilde{\lambda}(s)w'(s)] \\ &\quad + N(q-1)\tilde{\lambda}(s)w^q(s)\left(1 - \frac{\tau}{s}\right)_+^{Nq-N-1}\frac{\tau}{s^2}.\end{aligned}$$

Since $q \ge \beta$ and $w' \ge 0$, it follows from (10.19a) that $\tilde{\lambda}'w + q\tilde{\lambda}w' \ge 0$, and hence $\chi' \ge 0$. Moreover, (10.19b) and (10.17) imply that

$$\begin{aligned}(s-\tau)[\tilde{\lambda}'(s)w(s) + q\tilde{\lambda}(s)w'(s)] &\le s\tilde{\lambda}'(s)w(s) + q\tilde{\lambda}(s)sw'(s) \\ &\le \beta(1+q)\tilde{\lambda}(s)w(s)\end{aligned}$$

and

$$(s-\tau)\left(1 - \frac{\tau}{s}\right)_+^{Nq-N-1}\frac{\tau}{s^2} = \left(1 - \frac{\tau}{s}\right)_+^{Nq-N}\frac{\tau}{s}.$$

Our choice of $q$ now implies that $\chi$ satisfies (10.15) with $c_\chi = C(n,N,\beta)q$, and hence we can use $\chi$ in Lemma 10.3. If we also replace $\zeta^2$ by $\zeta^{Nq-N+2}$ in Lemma 10.3 and set

$$I_1 = \int_{B(y,R)} [\mathcal{E}_0 + (1 - \frac{\tau}{v})\mathcal{C}_0^2]\chi(v)\zeta^{Nq-N+2}\,dx,$$

we obtain

$$I_1 \le C(n,\beta)q \int_{B(y,R)} [\beta_1^2 \Lambda_0 \zeta^2 + \bar{\mu}|D\zeta|^2]\chi(v)\zeta^{Nq-N}\,dx.$$

To proceed, we choose

$$\zeta(x) = \left(1 - \frac{|x-y|^2}{R^2}\right)_+$$

and observe that $0 \le \zeta \le 1$ and $|D\zeta| \le 2/R$. Setting also

$$I_2 = \int_{B_\tau(y,R)} w^q \tilde{\lambda}\Lambda[(1-\frac{\tau}{v})\zeta]^{Nq-N} v\,dx,$$

we then obtain from (10.18a) and (10.18c) that

$$I_1 \le C(n,\beta,\beta_1 R)qR^{-2}I_2. \tag{10.21}$$

We next define the function $\Psi$ by

$$\Psi(v) = \chi(v)\lambda(v)(1-\frac{\tau}{v})^2$$

and the function $h$ by

$$h(x) = \Psi(v)^{1/2}\zeta(x)^{[Nq-N]/2},$$

and note that

$$h^2 = \tilde{\lambda}\lambda w^q[\zeta(1-\frac{\tau}{v})]^{Nq-N+2}$$

and

$$\tilde{\lambda}^\kappa \lambda^\kappa \ge \tilde{\lambda}\Lambda$$

by virtue of (10.18e). Then the Sobolev inequality (10.8) yields

$$\left(\int \tilde{\lambda}\Lambda w^{\kappa q}[\zeta(1-\frac{\tau}{v})]^{\kappa Nq-N} v\,dx\right)^{1/\kappa} \le C(n,N)\left(\int h^2 v\,dx\right)^{(N-n)/N} I_3^{n/N},$$

where

$$I_3 = \int [g^{ij}D_ihD_jh + h^2H^2]v\,dx.$$

Our next step is to estimate the integrals on the right of this inequality. First, we have

$$\int h^2 v\,dx \le \int \tilde{\lambda}\Lambda w^q[\zeta(1-\frac{\tau}{v})]^{Nq-N} v\,dx$$

because $\lambda \le \Lambda$ and $\zeta(1-\frac{\tau}{v}) \le 1$.

Next,

$$2h\delta h = [(\chi'\lambda + \chi\lambda')(1-\frac{\tau}{v})^2 + 2\chi\lambda(1-\frac{\tau}{v})\frac{\tau}{v^2}]\zeta^{Nq-N+2}\delta v + \chi\lambda(1-\frac{\tau}{v})^2\zeta^{Nq-N+1}\delta\zeta,$$

and hence

$$2h|\delta h| \le C(n,N,\beta)q\frac{1}{v}\left(1+\frac{|v\lambda'|}{\lambda}\right)[\chi\lambda(1-\frac{\tau}{v})]|\delta v|\zeta^{Nq-N+2} + C(n,N)\frac{q}{R}\chi\lambda(1-\frac{\tau}{v})\zeta^{Nq-N+1}.$$

We now square this inequality and divide by $h$ to conclude that

$$\begin{aligned}|\delta h|^2 &\le C(n,N,\beta)q^2\frac{\lambda}{v^2}(1+\frac{(v\lambda')^2}{\lambda^2})\chi|\delta v|^2\zeta^{Nq-N+2} + C(n,N)\frac{q^2}{R^2}\chi\lambda\zeta^{Nq-N} \\ &\le C(n,N,\beta)q^2\mathcal{E}_0\chi\zeta^{(n+2)q-n}\frac{1}{v} + C(n,N)\frac{q^2}{R^2}\tilde\lambda\Lambda w^q[\zeta(1-\frac{\tau}{v})]^{Nq-N}\end{aligned}$$

by virtue of (10.18d). Using (10.21), we conclude that

$$\int g^{ij}D_ihD_jhv\,dx \le C(n,N,\beta,\beta_1R)q^2R^{-2}\int_{B_\tau(y,R)} w^q\tilde\lambda\Lambda[(1-\frac{\tau}{v})\zeta]^{Nq-N}v\,dx.$$

It follows from (10.10) that

$$\int h^2H^2v\,dx \le C(n)I_1.$$

Using our estimate for $I_1$ once more, we find that

$$\left(\int w^{\kappa q}[\zeta(1-\frac{\tau}{v})]^{\kappa Nq-N}\tilde\lambda\Lambda v\,dx\right)^{1/\kappa} \le Cq^3R^{-2n/N}\int w^q[\zeta(1-\frac{\tau}{v})]^{Nq-N}\tilde\lambda\Lambda v\,dx,$$

where here, and in the remainder of the proof, we use $C$ to denote a constant determined only by $n$, $N$, $\beta$, and $\beta_1R$. This inequality is now easily rewritten in a form suitable for Moser iteration. We set

$$\bar w = w[\zeta(1-\frac{\tau}{v})]^N$$

and we define the measure $\mu$ by

$$d\mu = R^{-n}\tilde\lambda\Lambda[\zeta(1-\frac{\tau}{v})_+]^{-N}v\,dx.$$

It follows that

$$\left(\int \bar w^{\kappa q}\,d\mu\right)^{1/\kappa} \le C(n,\beta,\beta_1R)q^3\int \bar w^q\,d\mu.$$

Our usual iteration scheme then yields

$$\sup\bar w \le C(n,\beta,\beta_1R)\int \bar w\,d\mu,$$

and rewriting this inequality in terms of the original structure functions gives (10.20) because $\zeta \ge \frac{1}{2}$ in $B(y,R/2)$. □

To continue, we relate the functions $\Lambda$ and $\tilde{\lambda}$ to the operator $Q$ via the assumption that there are positive constants $\beta_2$ and $\beta_3$ such that

$$\Lambda(x)\tilde{\lambda}(v)v \le \beta_2 w^{\beta_3} Du \cdot A. \tag{10.22}$$

With this assumption, our gradient estimate is reduced to an estimate of

$$\int w^q Du \cdot A\,dx$$

for $q = 1+\beta_3$. This integral can be estimated if there are positive constants $\beta_4$, $\beta_5$ and $\beta_6$ and a decreasing function $\varepsilon$ such that

$$w' A \cdot \xi \le \beta_4 (a^{ij}\xi_i\xi_j)^{1/2}(Du \cdot A)^{1/2} \tag{10.23a}$$

for all $\xi \in \mathbb{R}^n$,

$$w|A| \le \beta_4 Du \cdot A, \tag{10.23b}$$

$$\Lambda_0 \le \varepsilon(v) w^2 Du \cdot A, \tag{10.23c}$$

$$\bar{\mu} \le \beta_5 Du \cdot A, \tag{10.23d}$$

$$B \le \beta_6 Du \cdot A \tag{10.23e}$$

wherever $v \ge \tau_0$. In fact, we also need a condition connecting the function $\varepsilon$ to the oscillation of $u$.

**Lemma 10.5.** *Suppose conditions* (10.13) *and* (10.23) *are satisfied, and set* $\sigma = \operatorname{osc}_{B(y,R)} u$ *and* $E = \exp(\beta_6\sigma)$, *and let* $q \ge 2$ *and* $\tau \ge \tau_0$. *Finally, suppose* $w$ *satisfies* (10.17). *If there is a constant* $\tau_1 \ge \tau$ *such that*

$$72(1+\beta q)\beta_1^2\beta_4^2\sigma^2 q^2 E\varepsilon(\tau_1) \le 1, \tag{10.24}$$

*then we have*

$$\begin{aligned}\int_{\Omega_\tau(y,R/2)} & w^q Du \cdot A\,dx \\ & \le C(q,\beta,\beta_5)E^q\left(w(\tau_1) + \frac{\beta_4\sigma}{R}\right)^q \int_{\Omega_\tau(y,R)} Du \cdot A\,dx.\end{aligned} \tag{10.25}$$

***Proof.*** We begin by setting $m = \min_{\Omega(y,R)} u$ and defining functions $f$ and $f_1$ by

$$f(s) = (s-m)\exp(\beta_6(s-m)), \quad f_1(s) = \exp(\beta_6(s-m)).$$

We now set

$$I_1 = \int_\Omega (w^q - w(\tau)^q)_+ \zeta^q f'(u) Du \cdot A\,dx.$$

Since $f'(u)Du = D(f(u))$, an integration by parts shows that

$$I_1 = I_2 + I_3 + I_4,$$

where

$$I_2 = -\int_\Omega qf(u)w^{q-1}\zeta^q w' Dv \cdot A\,dx,$$
$$I_3 = -\int_\Omega qf(u)(w^q - w(\tau)^q)_+\zeta^{q-1}D\zeta \cdot A\,dx,$$
$$I_4 = -\int_\Omega (w^q - w(\tau)^q)_+\zeta^q \operatorname{div} A\,dx.$$

On the other hand, if we set

$$I_5 = \int_\Omega (w^q - w(\tau)^q)_+\zeta^q f_1(u)Du \cdot A\,dx,$$
$$I_6 = \int_\Omega (w^q - w(\tau)^q)_+\zeta^q \beta_6(u-m) f_1(u)Du \cdot A\,dx$$

then $I_1 = I_5 + I_6$, so

$$I_5 = I_2 + I_3 + (I_4 - I_6).$$

Moreover, we have $-\operatorname{div} A = B$ and hence

$$I_4 = \int_\Omega (w^q - w(\tau)^q)_+\zeta^q f(u)B\,dx \le I_6$$

by virtue of (10.23e). We conclude that

$$I_5 \le I_2 + I_3. \tag{10.26}$$

For notational convenience, we set

$$I_6 = \int_{\Omega_\tau} w^q\zeta^q f_1(u)Du \cdot A\,dx,$$
$$I_0 = \int_{B_\tau(y,R)} f_1(u)Du \cdot A\,dx,$$

and note that $I_5 = I_6 - w(\tau)^q I_0$.

It follows from (10.23a) that

$$I_2 \le q\beta_4 \int_\Omega w^{q-1}\zeta^q(\sigma^2 E\mathcal{E}_0)^{1/2}\left(\frac{f(u)}{\sigma}Du \cdot A\right)^{1/2} dx$$
$$\le \frac{1}{4}I_6 + q^2\beta_4^2\sigma^2 E\int_{\Omega_\tau} w^{q-2}\zeta^q\mathcal{E}_0\,dx.$$

Since $\chi = w^{q-2}$ satisfies (10.15) with $c_\chi = \beta q$, it follows from Lemma 10.3 that

$$\int_{\Omega_\tau} w^{q-2}\zeta^q \mathcal{E}_0\,dx \le 18(1+\beta q)\int_{\Omega\tau} \beta_1^2\Lambda_0 w^{q-2}\zeta^q\,dx + \frac{72}{R^2}(1+\beta q)\int_{\Omega_\tau} \bar{\mu} w^{q-2}\zeta^{q-2}\,dx$$

and then (10.23c) and (10.23d) imply that

$$I_2 \le (\frac{1}{4} + 18(1+\beta q)\beta_1^2\beta_4^2\sigma^2 E\varepsilon(\tau))I_6 + C(\beta,\beta_5,q)\frac{\beta_4^2\sigma^2}{R^2}\int_{\Omega_\tau} w^{q-2}\zeta^{q-2} Du\cdot A\,dx.$$

Next (10.23b) implies that

$$\begin{aligned} I_3 &\le \beta_4 q\sigma\int_{\Omega_\tau} w^{q-1}\zeta^{q-1}|D\zeta|\exp(\beta_6(u-m))Du\cdot A\,dx \\ &\le \frac{1}{4}I_6 + C(q)\frac{\beta_4^2\sigma^2}{R^2}E\int_{B_\tau(y,R)} w^{q-2}\zeta^{q-2}Du\cdot A\,dx. \end{aligned}$$

If $\tau \ge \tau_1$, then (after using (10.24)) we have

$$I_5 \le \frac{3}{4}I_6 C(q)\frac{\beta_4^2\sigma^2}{R^2}\int_{\Omega_\tau} w^{q-2}\zeta^{q-2}Du\cdot A\,dx.$$

A slight rearrangement then yields

$$I_6 \le C\frac{\beta_4^2\sigma^2}{R^2}\int_{\Omega_\tau} w^{q-2}\zeta^{q-2}Du\cdot A\,dx + 2w(\tau)^2 I_0,$$

where here and, in the remainder of the proof, we use $C$ to denote a constant determined only by $n$, $q$, $\beta$, and $\beta_5$. From Young's inequality along with the inequality $f_1(u) \ge 1$, we then infer that

$$\int_{B_\tau(y,R)} w^q\zeta^q Du\cdot A\,dx \le CE^q(w(\tau) + \frac{\beta_4\sigma}{R})^q\int_{B_\tau(y,R)} Du\cdot A\,dx. \quad (10.27)$$

Note that this inequality is true with $\tau_1$ in place of $\tau$ as well.

If $\tau < \tau_1$, we use the preceding inequality with $\tau_1$ in place of $\tau$ and note that

$$\int_{B_\tau(y,R)\setminus B_{\tau_1}(y,R)} w^q\zeta^q Du\cdot A\,dx \le w(\tau_1)^q\int_{B_\tau(y,R)} Du\cdot A\,dx$$

to see that (10.27) is valid in the form

$$\int_{B_\tau(y,R)} w^q\zeta^q Du\cdot A\,dx \le CE^q(w(\tau_1) + \frac{\beta_4\sigma}{R})^q\int_{B_{\tau_1}(y,R)} Du\cdot A\,dx.$$

The proof is completed by observing that

$$\int_{B_\tau(y,R/2)} w^q Du\cdot A\,dx \le 2^q \int_{B_\tau(y,R)} w^q\zeta^q Du\cdot A\,dx$$

and

$$\int_{B_{\tau_1}(y,R)} Du\cdot A\,dx \le \int_{B_\tau(y,R)} Du\cdot A\,dx.$$

□

Our final step in the proving an interior gradient bound is to estimate $\int Du\cdot A\,dx$. Such an estimate is possible under very weak conditions on $A$ and $B$.

**Lemma 10.6.** *Let $u$ be a bounded weak solution of $Qu = 0$ in $B(y,R)$ and suppose that there are a decreasing function $\varepsilon_1$ and a nonnegative constant $\beta_6$ such that*

$$|A| \le \varepsilon_1(v) Du\cdot A, \tag{10.28a}$$

$$B \le \beta_6 Du\cdot A \tag{10.28b}$$

*for $v \ge \tau_0$. Set $\sigma = \operatorname{osc}_{B(y,R)} u$ and $E = \exp(\beta_6\sigma)$ and let $\tau \ge \tau_0$. If there is a constant $\tau_2 \ge \tau$ such that*

$$4\varepsilon_1(\tau_2)\frac{\sigma}{R} \le 1, \tag{10.29}$$

*then*

$$\int_{B_\tau(y,R/2)} Du\cdot A\,dx \le C(n)R^n E\Delta_1,$$

*where*

$$\Delta_1 = \sup_{S_0}\left(\sigma(B-\beta_6 Du\cdot A)_+ + (Du\cdot A)_- + \frac{\sigma|A|}{R}\right)$$

*and $S_0$ is the subset of $B(y,R)$ on which $v < \tau_2$.*

***Proof.*** For this proof, we need a function $\zeta$ which is identically one on $B(y,R/2)$. This can be achieved with, for example,

$$\zeta(x) = \begin{cases} 1 & \text{if } |x-y| \le R/2, \\ 1-\frac{2|x-y|}{R} & \text{if } R/2 < |x-y| \le R, \\ 0 & \text{if } |x-y| \ge R, \end{cases}$$

and this is the function we shall use in the proof.

We now set $M = \sup_{B(y,R)} u$ and define functions $\eta_1$ and $\eta$ by

$$\eta_1(x) = (M-u(x)-\sigma(1-\zeta(x)))_+, \quad \eta(x) = \exp(\beta_6(M-u(x)))\eta_1(x).$$

Using $\eta$ as test function in the weak form of the differential equation for $u$, we find that

$$\int_{\operatorname{supp}\eta} \exp(\beta_6(M-u))Du\cdot A\,dx = \int_\Omega (B-\beta_6 Du\cdot A)\eta\,dx + \int_{\operatorname{supp}\eta} \exp(\beta_6(M-u))\sigma D\zeta\cdot A\,dx.$$

To simplify notation, we introduce two sets: $\Sigma$, which is the subset of $\operatorname{supp}\eta$ on which $v \geq \tau_2$, and $S$, which is the subset of $\operatorname{supp}\eta$ on which $v < \tau_2$. Then

$$\int_\Omega (B-\beta_6 Du\cdot A)\eta\,dx \leq \int_S (B-\beta_6 Du\cdot A)\eta\,dx \leq \omega_n R^n E\sigma \sup_{S_0} \sigma(B-\beta_6 Du\cdot A)_+.$$

Since $|D\zeta| \leq 2/R$ and $Du\cdot A \geq 0$ if $v \geq \tau$, we conclude that

$$\int_\Sigma \exp(\beta_6(M-u))\sigma D\zeta\cdot A\,dx \leq \frac{1}{2}\int_\Sigma \exp(\beta_6(M-u))Du\cdot A, dx \leq \frac{1}{2}\int_{\Sigma_0} \exp(\beta_6(M-u))Du\cdot A\,dx,$$

where $\Sigma_0$ is the subset of $\operatorname{supp}\eta$ on which $v \geq \tau$, while

$$\int_S \exp(\beta_6(M-u))\sigma D\zeta\cdot A\,dx \leq 2\omega_n R^n E\frac{\sigma}{R}\sup_{S_0}|A|.$$

Finally,

$$\int_{\operatorname{supp}\eta} \exp(\beta_6(M-u))Du\cdot A\,dx \geq \int_{\Sigma_0} \exp(\beta_6(M-u))Du\cdot A\,dx + \int_S \exp(\beta_6(M-u))Du\cdot A\,dx$$

and

$$-\int_S \exp(\beta_6(M-u))Du\cdot A\,dx \leq \omega_n R^n E \sup_{S_0}(Du\cdot A)_-.$$

Combining these estimates with the observations that $B_\tau(y,R/2) \subset \Sigma_0$ and that $\exp(\beta_6(M-u)) \geq 1$ gives the desired result. □

To see the applicability of these results, we consider three examples.

First, we look at prescribed mean curvature equations. We suppose that $A=\nu$ and that $B(x,z,p) = B^*(x,z,p) + B_0(x,z,p)$, where $B^*$ is differentiable with respect to $(x,z,p)$. We also suppose that there are nonnegative constants $\theta_1$ and $\theta_2$ such that

$$v|B_0| \leq \theta_1, \quad v(|B^*_p| + |p\cdot B^*_p|) \leq \theta_1, \quad vB^*_z + |B^*_x| \leq \theta_1^2, \quad B \leq \theta_2.$$

If we choose $C^i_k = B_0\delta^i_k$ and $D^i_k = B^*\delta^i_k$, then conditions (10.13) are satisfied with $\beta_1 = \theta_1$, $\Lambda_0 = \bar{\mu} = v$, and conditions (10.17) are satisfied with $w = (\ln v)^2$ and $\beta = 1$ provided we take $\tau_0 = 3$. In addition (10.18) and (10.19) hold with $\Lambda = \lambda = \tilde{\lambda} = 1$. It follows from Lemma 10.4 (with $R/4$ in place of $R$) that, for any $\tau \geq 3$, we have

$$\sup_{B(y,R/8)} \left(1 - \frac{\tau}{v}\right)^N (\ln v)^2 \leq C(n, N, \theta_1 R) R^{-n} \int_{B_\tau(y,R/4)} (\ln v)^2 v\, dx.$$

To continue, we observe that (10.22) holds with $\beta_2 = 2$ and $\beta_3 = 0$. We now set $c_1 = \sqrt{1728e}\theta_1 R$ and $c_2 = 2\theta_2 R$, and we chose $\tau_1$ so that $(c_1 + c_2)\theta/R \leq \ln \tau_1$. Then conditions (10.23) and (10.24) are satisfied with $\beta_4 = 1$, $\beta_5 = 2$, $\beta_6 = 2\theta_2/\ln \tau_1$, $w = \ln v$, and $\varepsilon(t) = 2/(\ln t)^2$. Hence we can apply Lemma 10.5 (with $R/2$ in place of $R$) to conclude that

$$\int_{B_\tau(y,R/4)} (\ln v)^2 v\, dx \leq C(n, \theta_1 R, \theta_2 R) \left(\ln \tau_1 + \frac{\sigma}{R}\right)^2 \int_{B_\tau(y,R/2)} Du \cdot A\, dx.$$

This final integral is estimated via Lemma 10.6; conditions (10.28) and (10.29) are satisfied with $\varepsilon_1(v) = 1/(2v)$, $\tau_2 = 3 + 8\sigma/R$, $\beta_6 = 2\theta_2/\tau_2$. It follows that $E \leq C(\theta_2 R)$ and $\Delta_2 \leq C(\theta_2 R)\sigma/R$ and hence

$$\int_{B_\tau(y,R/2)} Du \cdot A\, dx \leq C(n, \theta_2 R) R^n \left(\tau + \frac{\sigma}{R}\right).$$

Combining our three estimates and taking $\tau = 3$, we conclude that

$$v(y) \leq \exp\left( C(n, \theta_1 R, \theta_2 R) \left(1 + \frac{\sigma}{R}\right)^{3/2}\right).$$

Next, we look at uniformly elliptic equations, which we define in terms of a positive $C^1(1,\infty)$ function $\Psi$ such that

$$\frac{\Psi(s)}{\theta_0} \leq s\Psi'(s) \leq \theta_0 \Psi(s) \tag{10.30}$$

for some positive constant $\theta_0 \geq 1$. (In particular, we can take $\Psi$ to be any power function.) We assume that there are positive constants $\theta_1, \dots \theta_4$ such that

$$Du \cdot A \geq v\Psi(v) - \theta_1, \quad |A| \leq \theta_2 \Psi, \tag{10.31a}$$

$$\frac{\Psi(v)}{v}|\xi|^2 \leq a^{ij}\xi_i\xi_j \leq \theta_3 \frac{\Psi(v)}{v}|\xi|^2 \tag{10.31b}$$

for all $\xi \in \mathbb{R}^n$,

$$v|A_z| + |A_x| + |B| \leq \theta_4 v\Psi(v). \tag{10.31c}$$

For future reference, we observe that $\Psi(s) \to \infty$ as $s \to \infty$, so we may assume that $\tau_0$ is large enough that $\tau_0\Psi(\tau_0) \geq 1 + 2\theta_1$. Then conditions (10.13) are satisfied with $D^i_k = 0$, $\Lambda_0 = v^3\Psi(v)$, $\bar{\mu} = \theta_3 v\Psi(v)$, and $\beta_1 = \theta_4$. Furthermore, if we take $w = v$, $\Lambda = (1+\theta_3)v^2\Psi(v)$, $\lambda = \Psi(v)/(1+\theta_0^2)$, and $\tilde{\lambda} = (1+\theta_3)^{(N-2)/2}(1+\theta_0^2)^{N/2}v^{N-1}$, then conditions (10.17), (10.18), and (10.19) hold with $\beta = N + \theta_0$. Hence Lemma 10.4 implies that

$$\begin{aligned}\sup_{B_\tau(y,R/8)} \left(1 - \frac{\tau}{v}\right)^{n+2} v &\leq C(n, \theta_0, \theta_1, \theta_2\theta_3, \theta_4 R)R^{-n} \int_{B_\tau(y,R/4)} v^{N+3}\Psi(v)\,dx \\ &\leq CR^{-n} \int_{B_\tau(y,R/4)} v^{N+2} Du \cdot A\,dx.\end{aligned}$$

Next, conditions (10.23) are satisfied with $\beta_4 = 2\theta_2$, $\beta_5 = 2\theta_3$, $\beta_6 = 2\theta_4$, and $\varepsilon(v) = 2$. Since $u$ is Hölder continuous (by virtue of Theorem 8.8), it follows that there is a constant $R_0$ (determined only by $\theta_0$, $\theta_1$, $\theta_2$, $\theta_4 R$, and $n$) such that (10.24) holds and $\beta_6\sigma \leq 1$ as long as $R \leq R_0$. Hence

$$\int_{B_\tau(y/R/4)} v^{N+2} Du \cdot A\,dx \leq C\left(\tau_0 + \frac{\sigma}{R}\right)^{N+2} \int_{B_\tau(y,R/2)} Du \cdot A\,dx.$$

Finally, conditions (10.28) hold with $\varepsilon_1(v) = 2\theta_2/v$ and $\beta_6 = 2\theta_4$, so condition (10.29) holds with $\tau_2 = \tau_0 + 8\theta_2\frac{\sigma}{R}$. It follows that

$$\Delta_1 \leq C(\theta_0, \theta_1, \theta_2, \tau_0)\left(1 + \frac{\sigma}{R}\right)\Psi\left(1 + \frac{\sigma}{R}\right)$$

and hence

$$\int_{B_\tau(y,R/2)} Du \cdot A\,dx \leq CR^n\left(1 + \frac{\sigma}{R}\right)\Psi\left(1 + \frac{\sigma}{R}\right).$$

Combining our inequalities, we obtain

$$v(y) \leq C(n, \theta_0, \ldots, \theta_3, \theta_4 R)\left(1 + \frac{\sigma}{R}\right)^{N+3}\Psi\left(1 + \frac{\sigma}{R}\right).$$

Finally, we look at a problem with exponential growth, the false mean curvature equation. Specifically, we assume that $A = \exp(\theta v^2)p$ for some positive constant $\theta$. We also assume that there is a positive constant $\theta_1$ such that

$$|B(x, z, p)| \leq \theta_1 \exp(\theta v^2)v^2.$$

As in the case of uniformly elliptic equations, we choose $D^i_k = 0$ and note that $C^i_k = B\delta^i_k$. Then conditions (10.13) are satisfied with $\Lambda_0 = \exp(\theta v^2)v^3$, $\bar{\mu} = \exp(\theta v^2)v^2$, and $\beta_1 = \theta_1$. Furthermore, if we take $w = v$, $\Lambda = \exp(\theta v^2)v^3$, $\lambda = \exp(\theta v^2)v^{-3}$, and $\tilde{\lambda} = v^{-2}$, then conditions (10.17),

(10.18), and (10.19) hold with $\beta = 1$ provided $\theta_0$ is sufficiently large. We therefore obtain

$$\sup_{B_\tau(y,R/4)} \left(1 - \frac{\tau}{v}\right)^{n+2} \leq C(n, \theta_1 R) R^{-n} \int_{B_\tau(y/R/2)} \exp(\theta v^2) v^2 \, dx.$$

Since $\exp(\theta v^2)v^2 \leq 2Du \cdot A$ for $v \geq 2$, we can use Lemma 10.6 to estimate this integral. Conditions (10.28) are satisfied with $\varepsilon_1(v) = 1/(2v)$ and $\beta_6 = 2\theta_1$, so (10.29) holds with $\tau_2 = \tau_0 + 2\frac{\sigma}{R}$. We therefore infer that

$$v(y) \leq C(n, \theta, \theta_1 R) \left(1 + \frac{\sigma}{R}\right)^2 \exp(\theta \left(1 + \frac{\sigma}{R}\right)^2).$$

## 10.3 Preliminaries for estimates

To prove a gradient bound for conormal problems, we must modify the preceding argument in a suitable fashion. In particular, we can no longer estimate $w(v)$ because this function does not satisfy a suitable boundary condition. Instead, we use a construction originally due to Dong to create an analog of the function $v$.

As a preliminary step, we recall from Lemma 5.18 that the distance function in a $C^2$ domain is $C^2$ near the boundary. From this lemma, we can generate a function $v_1(x, z, p)$ which behaves like $v$ and which behaves appropriately on $\partial\Omega$. To simplify some writing, we introduce some notation. First, we extend $\gamma$ into $\Gamma$ by taking $\gamma(x) = Dd(x)$. Then, we define $c^{ij}(x) = \delta^{ij} - \gamma_i(x)\gamma_j(x)$, and for a vector $p$, we define $p'$ to be the vector with components $p'_i = c^{ij}(x)p_j$. (Although $p'$ is really a function of $x$ and $p$, we suppress this fact from the notation.) We also set $v' = \left(1 + |p'|^2\right)^{1/2}$. Finally, for $\tau \geq 1$, we write $\Sigma(\tau)$ and $\Sigma^*(\tau)$ for the subsets of $\partial\Omega \times \mathbb{R} \times \mathbb{R}^n$ and $\overline{\Gamma} \times \mathbb{R} \times \mathbb{R}^n$, respectively, on which $v \geq \tau$.

As a first step, we show how to extend a function defined on $\Sigma_\tau$ to one defined on $\overline{\Gamma} \times \mathbb{R} \times \mathbb{R}^n$ with important properties preserved. The extended function (denoted by $g$ in this lemma) will be used to define a suitable $v_1$ in the next lemma.

**Lemma 10.7.** *Let $\partial\Omega \in C^2$, and let $R_0$ and $\Gamma$ be as in Lemma 5.18. Let $\tau_0 \geq 1$ be given and let $G \in C^1(\Sigma(\tau_0)$ with*

$$\lim_{t\to\infty} G(x, z, p - t\gamma(x)) < 0 < \lim_{t\to\infty} G(x, z, p + t\gamma(x)) \tag{10.32}$$

*for all $(x, z, p) \in \Sigma(\tau_0)$. Write $\Sigma_0(\tau_0)$ for the subset of $\Sigma(\tau_0)$ on which*

*$G = 0$, and suppose that there are positive constants $\beta_0$ and $c_0$ such that*

$$|G_p| \le \beta_0 G_p \cdot \gamma, \tag{10.33a}$$

$$|p \cdot \gamma| \le c_0 v' \tag{10.33b}$$

*on $\Sigma_0(\tau_0)$. Suppose further that there are decreasing functions $\varepsilon_z$ and $\varepsilon_x$ (defined on $[\tau_0, \infty)$) such that*

$$s\varepsilon_z(s) \le t\varepsilon_z(t), \quad s^2\varepsilon_x(s) \le t^2\varepsilon_x(t) \tag{10.34}$$

*for $\tau_0 \le s \le t$ and*

$$|G_z| \le \varepsilon_z(v)vG_p \cdot \gamma, \tag{10.35a}$$

$$|G_x| \le \varepsilon_x(v)v^2 G_p \cdot \gamma, \tag{10.35b}$$

*on $\Sigma_0(\tau_0)$. Then there is a function $g \in C^1(\overline{\Gamma} \times \mathbb{R} \times \mathbb{R}^n)$ such that*

$$g(x, z, p_1) = g(x, z, p_2) \tag{10.36a}$$

*if $(x, z) \in \overline{\Gamma} \times \mathbb{R}$ and $p_1 \cdot \gamma(x) = p_2 \cdot \gamma(x)$,*

$$p \cdot \gamma(x) = g(x, z, p) \text{ if } G(x, z, p) = 0 \tag{10.36b}$$

*for all $(x, z, p) \in \partial\Omega \times \mathbb{R} \times \mathbb{R}^n$ with $v > (1 + c_0^2)^{1/2} v$,*

$$|g_z| \le \varepsilon_z(v)v, \quad |g_x| \le \varepsilon_x(v)v^2 + \frac{v}{R_0}, \quad |g_p| \le \beta_0 \tag{10.36c}$$

*on $\overline{\Gamma} \times \mathbb{R} \times \mathbb{R}^n$, where $\varepsilon_z$ and $\varepsilon_x$ are defined for $s \in [1, \tau_0)$ by*

$$\varepsilon_z(s) = \frac{s}{\tau_0}\varepsilon_z(\tau_0), \quad \varepsilon_x(s) = \frac{s^2}{\tau_0^2}\varepsilon_x(\tau_0).$$

***Proof.*** We begin by observing that the implicit function theorem gives a function $g_0$, defined for all $(x, z, p) \in \partial\Omega \times \mathbb{R} \times \mathbb{R}^n$ with $v' \ge \tau_0$ such that $G(x, z, p) = 0$ if and only if $p \cdot \gamma = g_0(x, z, p')$. Moreover $g_0$ is continuously differentiable with respect to its arguments, provided we only differentiate $x$ in tangential directions. We then define $g$, for $v' > \tau_0$ by $g(x, z, p) = g_0(x, z, p')$. Implicitly differentiating the equation

$$G(x, z, p + \gamma[g(x, z, p) - p \cdot \gamma]) = 0$$

shows that the conditions in (10.36) are satisfied for $v' > \tau_0$ and $x \in \partial\Omega$. It then follows that we can extend $g$ so that these conditions are satisfied in all of $\partial\Omega \times \mathbb{R} \times \mathbb{R}^n$. The important observation is that $G(x, z, p)$ is never zero if $|v'| \le \tau_0$ and $v \ge (1 + c_0^2)^{1/2}\tau_0$.

Finally, we extend $g$ to $\overline{\Gamma} \times \mathbb{R} \times \mathbb{R}^n$ by setting $g(x, z, p) = g(y(x), z, p)$, where $y(x)$ is the closest point in $\partial\Omega$ to $x$. Since $\gamma(y(x)) = \gamma(x)$, all the conditions are easily verified. □

**Lemma 10.8.** *Under the hypotheses of Lemma* 10.7, *there is a* $C^1(\overline{\Gamma}\times\mathbb{R}\times\mathbb{R}^n)$ *function* $N$ *such that*

$$\frac{1}{2}\le N_p\cdot\gamma\le\frac{3}{2}, \tag{10.37a}$$

$$|N_p|\le 4\beta_0, \tag{10.37b}$$

$$|N_z|\le 8\varepsilon_z(v)v, \tag{10.37c}$$

$$|N_x|\le\left(24\varepsilon_x(v)v+\frac{2}{R_0}\right)v \tag{10.37d}$$

*on* $\Sigma^*$ *and*

$$N=0 \text{ on } \Sigma_0((1+c_0^2)^{1/2}\tau_0). \tag{10.37e}$$

*Moreover, if we set*

$$v_1=(1+|p|^2-(p\cdot\gamma)^2+\varepsilon N^2)^{1/2} \tag{10.38}$$

*for some* $\varepsilon\in(0,1/(c_0+1)^2)$, *then*

$$\frac{\sqrt{\varepsilon}}{2}v\le v_1\le 2v. \tag{10.39}$$

*If we also define*

$$\nu_1=\frac{\partial v_1}{\partial p}, \tag{10.40}$$

*then*

$$|\nu_1|\le 2 \tag{10.41a}$$

*in* $\Omega$ *and*

$$\nu_1\cdot Du\ge v_1/8 \tag{10.41b}$$

*wherever* $v_1\ge 3$. *Moreover, if we define*

$$b^{km}=v_1\frac{\partial^2 v_1}{\partial p_k\partial p_m}+\nu_1^k\nu_1^m, \tag{10.42}$$

*then there are positive constants* $\varepsilon_0(\beta_0,c_0)$ *and* $c_1(\varepsilon,\beta_0,n)$ *such that* $\varepsilon<\varepsilon_0$ *implies that*

$$c_1|\xi|^2\le b^{km}\xi_k\xi_m \tag{10.43}$$

*for all* $\xi\in\mathbb{R}^n$.

***Proof.*** The basic idea is to obtain $N$ as a regularized distance to the surface $G(x, z, p) = 0$ once we have extended $G$ to a function defined on $\overline{\Gamma} \times \mathbb{R} \times \mathbb{R}^n$. For the extension, we first recall that $\gamma = Dd$ in $\overline{\Gamma}$ and take $g$ from Lemma 10.7. Then we define $\bar{g}$ as follows: Let $\varphi$ be a nonnegative $C^\infty(\mathbb{R}^n)$ function having support in $B(0, 1)$ and

$$\int_{\mathbb{R}^n} \varphi(x)\, dx = 1,$$

and set

$$\bar{g}(x, z, p, s) = \int_{\mathbb{R}^n} g(x, z, p + \frac{sy}{2\beta_0 + 2c_0})\varphi(y)\, dy.$$

Note that $|\bar{g}_p| \le \beta_0$, $\bar{g}_p \cdot \gamma = 0$, and $|\bar{g}_s| \le \frac{1}{2}$. Since $|\bar{g}_s| \le \frac{1}{2}$, there is a function $\rho(x, z, p)$ such that

$$p \cdot \gamma - \bar{g}(x, z, p, \rho) = \rho.$$

Since $\bar{g}(x, z, p, 0) = g(x, z, p)$ and $|\bar{g}(x, z, p, \rho) - \bar{g}(x, z, p, 0) \le \frac{1}{2}|\rho|$, we conclude that

$$\frac{1}{2}|\rho| \le |p \cdot \gamma - g(x, z, p)| \le \frac{3}{2}|\rho|.$$

In particular, we have

$$|\rho| \le (1 + c_0)v.$$

Hence, if $|y| \le 1$, we have

$$\left|\frac{\rho(x, z, p)y}{2\beta_0 + 2c_0}\right| \le \frac{1}{2}v$$

because $\beta_0 \ge 1$. Moreover, it's easy to check that

$$|\rho_p| \le 4\beta_0$$

because $\beta_0 \ge 1$,

$$|\rho_z| \le 8\varepsilon_z(v)v,$$
$$|\rho_x| \le 24\varepsilon_x(v)v^2 + \frac{v}{R_0}.$$

For $\varepsilon'$ a positive constant to be further specified, we then define $N$ as

$$N(x, z, p) = p \cdot \gamma - \bar{g}(x, z, p, \varepsilon'\rho(x, z, p)).$$

To simplify notation, we write $X = (x, z, p, \varepsilon'\rho(x, z, p))$. From our previous estimates, we then infer that

$$|\bar{g}_x(X)| \le 24\varepsilon_x(v)v^2 + \frac{v}{R_0},$$
$$|\bar{g}_z(X)| \le 4\varepsilon_z(v)v.$$

Since $\bar{g}_p \cdot \gamma = 0$, conditions (10.37) are then easily verified if $\varepsilon' < 1/(2\beta_0)$.

To prove (10.39), we write

$$|N(x,z,p)| = |p \cdot \gamma - \bar{g}(X)| \le |p \cdot \gamma - g(x,z,p)| + |g(x,z,p) - \bar{g}(X)|$$
$$\le \frac{3}{2}|\rho| + \frac{\varepsilon'}{2}|\rho| \le 2(1+c_0)v$$

if $\varepsilon < 1$. It follows that

$$v_1 \le (1 + 4\varepsilon(1+c_0)^2)^{1/2} v \le 2v$$

if $\varepsilon \le 3(1+c_0)^2/4$. Also,

$$|N(x,z,p) - p \cdot \gamma| = |\bar{g}(X)| \le |g(x,z,p)| + |g(x,z,p) - \bar{g}(X)|$$
$$\le c_0 v' + \frac{\varepsilon'}{2}(1+c_0)v \le (c_0+1)v' + \frac{1}{2}|p \cdot \gamma|$$

if $\varepsilon < 1/(c_0+1)$, so

$$|N(x,z,p)| \ge \frac{1}{2}|p \cdot \gamma| - (c_0+1)v',$$

and hence

$$v_1^2 = (v')^2 + \varepsilon N^2 \ge (v')^2(1 - 2\varepsilon(1+c_0)^2) + \frac{1}{8}\varepsilon(p \cdot \gamma)^2.$$

If $\varepsilon \le 1/(4(c_0+1)^2)$, it follows that

$$v_1^2 \ge \frac{\varepsilon}{8}v^2,$$

and hence $v_1 \ge \sqrt{\varepsilon}v/3$.

Now, we write $p' = p - (p \cdot \gamma)\gamma$ and then

$$\nu_1 = \frac{p' + \varepsilon N N_p}{v_1}.$$

It follows that

$$|\nu_1|^2 \le \frac{1}{v_1^2}[2|p'|^2 + 2\varepsilon^2 N^2 |N_p|^2] \le \frac{1}{v_1^2}[2|p'|^2 + 32\beta_0^2\varepsilon^2 N^2].$$

If $16\beta_0^2\varepsilon \le 1$, we conclude that $|\nu_1|^2 \le 2$ and hence (10.41a) is verified.

Next, we have

$$\nu_1 \cdot p = \frac{1}{v_1}(p' \cdot p + \varepsilon N N_p \cdot p) = \frac{1}{v_1}(|p'|^2 + \varepsilon N^2) + \varepsilon\frac{N}{v_1}(N_p \cdot p - N).$$

From the definition of $N$ and our observation that $\bar{g}_p \cdot \gamma = 0$, we conclude that

$$N_p \cdot p - N = -\bar{g}(X) + \bar{g}_p(X) \cdot p' + \varepsilon' \bar{g}_s(X) p \cdot \rho_p,$$

and hence

$$|N_p \cdot p - N| \le (c_0 + \beta_0)v' + \frac{\varepsilon'}{2}(c_0 + \beta_0 + 1)v.$$

From these inequalities, we conclude that

$$\begin{aligned}\nu_1 \cdot p &\ge \frac{1}{v_1}(|p'|^2 + \varepsilon N^2 - \varepsilon k_1|N|(v' + \varepsilon' v)) \\ &= \frac{1}{v_1}((v')^2 + \varepsilon N^2 - 1 - \varepsilon k_1|N|(v' + \varepsilon' v))\end{aligned}$$

with $k_1 = 2(1 + c_0 + \beta_0)$. From Cauchy's inequality, we have

$$\varepsilon k_1|N|(v' + \varepsilon' v) \le \frac{1}{2}\varepsilon N^2 + k_1^2\varepsilon(v')^2 + k_1^2\varepsilon(\varepsilon')^2v^2,$$

and (10.39) implies that $\varepsilon v^2 \le 3v_1^2$. If $\varepsilon \le 1/(2k_1^2)$ and $\varepsilon' \le 1/(4k_1)$, we conclude that

$$\varepsilon k_1|N|(v' + \varepsilon' v) \le \frac{3}{4}v_1^2,$$

and hence we infer (10.41b).

To prove (10.43), we write $N^i = \partial N/\partial p_i$, $N^{ij} = \partial^2 N/\partial p_i\partial p_j$ and note that

$$N^{ij}\xi_i\xi_j = -\frac{\partial^2\bar{g}}{\partial p_i\partial p_j}\xi_i\xi_j - 2\varepsilon'\frac{\partial^2\bar{g}}{\partial p_i\partial s}\xi\rho_p\cdot\xi - (\varepsilon')^2\frac{\partial^2\bar{g}}{\partial s^2}(\rho_p\cdot\xi)^2 - \varepsilon'\bar{g}_s\frac{\partial^2\rho}{\partial p_i\partial p_j}\xi_i\xi_j,$$

where $\bar{g}$ and its derivatives are evaluated at $X$ and $\rho$ and its derivatives are evaluated at $(x, z, p)$. Since

$$|\bar{g}_{pp}(x, z, p, s)| + |\bar{g}_{ps}(x, z, p, s)| + |\bar{g}_{ss}(x, z, p, s)| \le C(n, \beta_0, c_0)/s$$

and

$$\frac{\partial^2\bar{g}}{\partial p_i\partial p_j}\gamma^i = \frac{\partial^2\bar{g}}{\partial p_i\partial s}\gamma^i = 0,$$

it follows that there is a constant $c_3$ determined only by $n$, $\beta_0$, and $c_0$ such that

$$|N^{ij}\xi_i\xi_j| \le c_3\left(\frac{1}{\varepsilon' N}|\xi'|^2 + \frac{\varepsilon'}{N}|\xi\cdot\gamma|^2\right).$$

In addition,

$$b^{km}\xi_k\xi_m = |\xi'|^2 + \varepsilon(N_p\cdot\xi)^2 + \varepsilon N N^{km}\xi_k\xi_m,$$

and hence

$$b^{km}\xi_k\xi_m \ge |\xi'|^2\left(1 - \varepsilon\left(\beta_0^2 + \frac{c_3}{\varepsilon'}\right)\right) + \varepsilon|\xi\cdot\gamma|^2\left(\frac{1}{8} - c_3\varepsilon'\right).$$

If $\varepsilon' \le 1/8(c_3 + 1)$ and then $\varepsilon \le 1/2(\beta_0^2 + c_3/\varepsilon')$, we thus infer (10.43). □

So far, we have presented results that apply to any oblique boundary condition. We now turn to results that are specific to the cornormal boundary condition. Our first step extends our previous Sobolev inequalities.

**Lemma 10.9.** *Let $\partial\Omega \in C^2$, let $R_0$ be the constant from Lemma* 5.18 *and suppose $u \in C^2(\overline{\Omega})$. Suppose also that there are constants $c_0 \in (0,1)$ and $\tau > 1$ such that*

$$|\nu \cdot \gamma| \le c_0 \text{ on } \Sigma(\tau). \tag{10.44}$$

*Let $N > 2$ be a constant such that $N \ge n$ and set $\kappa = N/(N-2)$. Then there is a constant $C$, determined only by $c_0$, $n$, and $N$, such that*

$$\int_{\partial\Omega} hv\,ds \le C\int_\Omega (|\delta h| + h|H| + h/R_0)v\,dx, \tag{10.45a}$$

$$\left(\int_\Omega h^{2\kappa} v\,dx\right)^{1/\kappa} \le C\left(\int_\Omega h^2 v\,dx\right)^{(N-n)/N}\left(\int_\Omega (|\delta h|^2 + h^2H^2 + h^2/R_0^2)v\,dx\right)^{n/N} \tag{10.45b}$$

*for all nonnegative Lipschitz functions $h$ which vanish whenever $v \le \tau$ and $d > R_0$.*

***Proof.*** From (10.44), we see that

$$\int_{\partial\Omega} hv\,ds \le \frac{1}{1-c_0^2}\int_{\partial\Omega} h(1-(\nu\cdot\gamma)^2)v\,ds,$$

and then

$$\int_{\partial\Omega} h(1-(\nu\cdot\gamma)^2)v\,ds = \int_{\partial\Omega} h(\gamma - (\nu\cdot\gamma)\nu)\cdot\gamma v\,ds.$$

If we define $\gamma = Dd$, then $\gamma$ is $C^1$ on the support of $h$, and the divergence theorem gives

$$\int_{\partial\Omega} h(1-(\nu\cdot\gamma)^2)v\,ds = \int_\Omega \operatorname{div}(hv(\gamma-(\nu\cdot\gamma)\nu))\,dx.$$

Moreover, we have

$$\operatorname{div}(hv(\gamma-(\nu\cdot\gamma)\nu)) = (\gamma^i - (\nu\cdot\gamma)\nu^i)D_ihv + h\operatorname{div}(v(\gamma-(\nu\cdot\gamma)\nu)).$$

Since $\gamma^i - (\nu\cdot\gamma)\nu^i = g^{ij}\gamma_j$, it follows that $(\gamma^i - (\nu\cdot\gamma)\nu^i)D_ih \le |\delta h|$ and

$$\operatorname{div}(v(\gamma-(\nu\cdot\gamma)\nu)) = D_i(vg^{ij}\gamma_j) = \gamma_jD_i(vg^{ij}) + vg^{ij}D_i\gamma_j = -\nu^ivH + vg^{ij}D_i\gamma_j.$$

Combining all these estimates with the observation that $g^{ij}D_i\gamma_j \le n/R_0$ yields

$$\operatorname{div}(h(\gamma - (\nu \cdot \gamma)\nu)) \le (|\delta h| + h|H| + nh/R_0)v,$$

and then (10.45a) follows.

To prove (10.45b), we first define $\eta_k = \min\{1, kd\}$ for $k$ a positive integer and then replace $h$ by $h\eta_k$ in (10.7) to obtain

$$\left(\int_\Omega (h\eta_k)^{n/(n-1)} v\,dx\right)^{1/\kappa} \le C(n)\int_\Omega (|\delta(h\eta_k)| + h\eta_k|H|)\,dx.$$

From the monotone convergence theorem we conclude that

$$\int_\Omega (h\eta_k)^{n/(n-1)} v\,dx \to \int_\Omega h^{n/(n-1)} v\,dx$$

and

$$\int_\Omega h\eta_k|H|v\,dx \to \int_\Omega h|H|v\,dx$$

as $k \to \infty$. It is easy to check that

$$\int_\Omega |\delta(h\eta_k)|v\,dx \le \int_\Omega \eta_k|\delta h|v\,dx + \int_\Omega h|\delta\eta_k|v\,dx$$

and that $|\delta\eta_k| \le k$ if $kd \le 1$ while $|\delta\eta_k| = 0$ if $kd > 1$. The monotone convergence theorem implies that

$$\int_\Omega \eta_k|\delta h|v\,dx \to \int_\Omega |\delta h|v\,dx,$$

and, because $\partial\Omega \in C^2$, we have

$$k\int_{\Omega_k} hv\,dx \to \int_{\partial\Omega} hv\,ds$$

as $k \to \infty$, where $\Omega_k$ is the subset of $\Omega$ on which $d \le 1/k$.

We therefore obtain

$$\left(\int_\Omega h^{n/(n-1)} v\,dx\right)^{1/\kappa} \le C(n)\left(\int_\Omega (|\delta h| + h|H|)v\,dx + \int_{\partial\Omega} hv\,ds\right). \tag{10.46}$$

The proof is completed by combining this inequality with (10.45a) and the proof of Theorem 5.8. □

Next, we show that, if $u$ satisfies the boundary condition (10.1b) and $A$ and $\psi$ satisfy appropriate structure conditions, then condition (10.44) is satisfied.

**Lemma 10.10.** *Suppose $u \in C^1(\overline{\Omega})$ and $\partial\Omega \in C^1$. Suppose also that $u$ satisfies* (10.1b). *If there are constants $\beta_1$, $\beta_2$, and $\tau_0$ with $0 \le \beta_1, \beta_2 < 1$ and $\tau_0 \ge 1$ such that*

$$\beta_1 |A| \le \nu \cdot A, \tag{10.47a}$$

$$|\psi| \le \beta_2 \nu \cdot A \tag{10.47b}$$

*on $\Sigma(\tau_0)$, then*

$$|\nu \cdot \gamma| \le \left(1 + \frac{(1-\beta_2)^2}{\beta_1^2}\right)^{-1/2} \tag{10.48}$$

*on $\Sigma(\tau_0)$.*

***Proof.*** Set

$$F_0 = (\gamma - (\gamma \cdot \nu)\nu) \cdot (A + \psi\gamma).$$

From the Schwarz inequality and condition (10.47a), we conclude that

$$|F_0| \le |\gamma - (\gamma \cdot \nu)\nu||A + \psi\gamma| \le \beta_1 (1 - (\gamma \cdot \nu)^2)^{1/2} \nu \cdot A.$$

Since $A \cdot \gamma + \psi = 0$, we conclude that

$$F_0 = -(\gamma \cdot \nu)(A \cdot \nu + \psi(\gamma \cdot \nu)),$$

so

$$|F_0| \ge |\gamma \cdot \nu|(A \cdot \nu)(1 - \beta_2)$$

by virtue of (10.47b). Combining these two inequalities and dividing by $A \cdot \nu$, we obtain

$$(1 - \beta_2)|\gamma \cdot \nu| \le \beta_1 (1 - (\gamma \cdot \nu)^2)^{1/2},$$

and simple algebra yields (10.48). □

Our final step is to derive a differentiated weak form of (10.1).

**Lemma 10.11.** *Let $u \in C^2(\overline{\Omega})$ be a weak solution of* (10.1) *with $\partial\Omega \in C^2$. Suppose that $A$ is differentiable with respect to $x$, $z$, and $p$. Suppose also that there are functions $C^i_k$ and $D^i_k$ satisfying* (10.11) *and $D^i_k$ is differentiable with respect to $x$, $z$, and $p$. If $\eta \in C^1(\overline{\Omega})$ is a vector field with $\eta \cdot \gamma = 0$ on $\partial\Omega$, then*

$$\int_\Omega [(a^{ij} D_{jk} u + C^i_k) D_i \eta^k - D_i(D^i_k)\eta^k] \, dx = \int_{\partial\Omega} g_k \eta^k \, ds, \tag{10.49}$$

*where*

$$g_k = p_k \frac{\partial \psi}{\partial z} + \frac{\partial \psi}{\partial x^k} + A^i c^{ij} D_j \gamma_k - \gamma_i D^i_k. \tag{10.50}$$

***Proof.*** We first take a $C^2(\overline{\Omega})$ vector field $\theta$ (with no restrictions on its normal component on $\partial\Omega$). With $D_k\theta^k$ as our test function, we see that

$$\int_\Omega (A^i D_{ik}\theta^k - BD_k\theta^k)\,dx = \int_{\partial\Omega} \psi D_k\theta^k\,ds,$$

and then the divergence theorem gives

$$\int_\Omega A^i D_{ik}\theta^k\,dx = -\int_\Omega D_k A^i D_i\eta^k\,dx - \int_{\partial\Omega} A^i D_i\theta^k\gamma_k\,ds.$$

We now rewrite the boundary integrands.

Since

$$\begin{aligned} A^i D_i\theta^k\gamma_k &= A^i c^{ij} D_j\theta^k\gamma_k - \psi\gamma_i D_i\theta^k\gamma_k \\ &= A^i c^{ij} D_j(\theta\cdot\gamma) - A^i c^{ij}\theta^k D_j\gamma_k - \psi\gamma_i D_i\theta^k\gamma_k \end{aligned}$$

and

$$\psi D_k\theta_k = \psi c^{ij} D_i\theta^j + \psi\gamma_i D_i\theta^k\gamma_k,$$

we have

$$\begin{aligned} \int_\Omega (D_k A^i + \delta^i_k B) D_i\theta^k\,dx &= \int_{\partial\Omega} (-\psi c^{ij} D_i\theta^j - A^i c^{ij} D_j(\theta\cdot\gamma))\,ds \\ &\quad + \int_{\partial\Omega} A^i c^{ij} D_j\gamma_k\theta_k\,ds. \end{aligned}$$

Now, we replace $\theta$ by a sequence of $C^2(\overline{\Omega})$ vectors which converge in $C^1(\overline{\Omega})$ to $\eta$ and note that $c^{ij}D_j(\eta\cdot\gamma) = 0$ on $\partial\Omega$ to infer that

$$\int_\Omega (D_k A^i + \delta^i_k B) D_i\eta^k\,dx = \int_{\partial\Omega} (-\psi c^{ij} D_i\eta^j + A^i c^{ij} D_j\gamma_k\eta^k)\,ds.$$

We then rewrite

$$D_k A^i + \delta^i_k B = a^{ij} D_{jk}u + C^i_k + D^i_k$$

and integrate by parts to obtain

$$\int_\Omega [(a^{ij} D_{jk}u + C^i_k) D_i\eta^k - D_i(D^i_k)\eta^k]\,dx = I,$$

where

$$I = \int_{\partial\Omega} (-\psi c^{ij} D_i\eta^j + A^i c^{ij} D_j\gamma_k\eta^k - \gamma_i D^i_k\eta^k)\,ds.$$

The proof is completed by observing that we can integrate by parts on $\partial\Omega$ to obtain

$$\int_{\partial\Omega} -\psi c^{ij} D_i\eta^j\,ds = \int_{\partial\Omega} c^{ij} D_i\psi\eta^j\,ds$$

and using the definition of $g_k$. □

## 10.4 Gradient bounds for the conormal problem

We are now ready to prove some gradient bounds in $C^2$ domains. In outline, our proof is a simple modification of the method from Section 10.2, but we shall present the proof in several steps so that the changes are easier to digest. Most of the changes (other than the obvious one of using $v_1$ in place of $v$) appear in the proof of the energy estimate, and we break that proof down into three steps.

The first step is to estimate the terms which come from $a^{ij}D_{jk}uD_j\eta^k$ with a suitable choice of $\eta$. (In our previous case, this estimate was simple because $D_i\nu^k = g^{ik}/v$ and $D_iv = D_{ij}u\nu^j$.) We begin by noting that, if there are positive decreasing functions $\varepsilon_x$ and $\varepsilon_z$ satisfying (10.34) and a positive constant $\beta_0$ such that

$$|A_z(x,u,p)\cdot\gamma| + |\psi_z(x,u)| \le \varepsilon_z(v)va^{ij}(x,u,p)\gamma_i\gamma_j, \tag{10.51a}$$

$$|A_x(x,u,p)\cdot\gamma| + \frac{1}{R_0}|A(x,u,p)| + |\psi_x(x,u)| \le \varepsilon_x(v)v^2a^{ij}(x,u,p)\gamma_i\gamma_j, \tag{10.51b}$$

$$\sum_i |a^{ij}(x,u,p)\gamma_j| \le \beta_0 a^{ij}(x,u,p)\gamma_i\gamma_j \tag{10.51c}$$

for all $(x,p) \in \partial\Omega\times\mathbb{R}^n$ with $A(x,u,p)\cdot\gamma + \psi(x,u) = 0$ and $v \ge \tau_0$, then $G(x,z,p) = A(x,z,p)\cdot\gamma + \psi(x,z)$ satisfies the hypotheses of Lemma 10.7, and hence there is a function $v_1$ as in Lemma 10.8; in what follows, we define $\nu_1$ and $b^{km}$ from that lemma. We also define

$$\mathcal{E}_1 = a^{ij}D_iv_1D_jv_1, \quad \mathfrak{C}_1^2 = a^{ij}D_{ik}uD_{jk}u. \tag{10.52}$$

**Lemma 10.12.** *Let $u \in C^2(\overline{\Omega})$ and let $A$ satisfy* (10.47). *Suppose also that $u$ satisfies* (10.1b). *Suppose further that there are positive decreasing functions $\varepsilon_x$ and $\varepsilon_z$ satisfying* (10.34) *and a positive constant $\beta_0$ such that* (10.51) *holds for all $(x,p) \in \partial\Omega\times\mathbb{R}^n$ with $A(x,u,p)\cdot\gamma+\psi(x,u) = 0$ and $v \ge \tau_0$. Suppose finally that there are functions $\bar\mu$ and $\Lambda_0$ along with a nonnegative constant $\beta_3$ such that*

$$va^{ij}\xi_i\bar\xi_j \le (a^{ij}\xi_i\xi_j)^{1/2}(\bar\mu|\bar\xi|^2)^{1/2}, \tag{10.53a}$$

$$va^{ij}\bar\xi_i\xi_j \le (a^{ij}\xi_i\xi_j)^{1/2}(\bar\mu|\bar\xi|^2)^{1/2}, \tag{10.53b}$$

$$\varepsilon_z(v)va^{ij}\xi_iD_ju \le \beta_3(a^{ij}\xi_i\xi_j)^{1/2}(\Lambda_0)^{1/2}, \tag{10.53c}$$

$$\varepsilon_z(v)va^{ij}D_iu\xi_j \le \beta_3(a^{ij}\xi_i\xi_j)^{1/2}(\Lambda_0)^{1/2} \tag{10.53d}$$

*and*

$$\left(\frac{1}{R_0} + \varepsilon_x(v)v\right)^2\bar\mu \le \beta_3^2\Lambda_0 \tag{10.54}$$

*for all $\xi$ and $\bar{\xi}$ in $\mathbb{R}^n$, as long as $v \geq \tau_0$ and $a^{ij}$ is evaluated at $(x, u, Du)$. Let $\chi$ be an increasing function defined on $[\tau, \infty)$ with $\tau$ sufficiently large that $v_1 \geq \tau$ implies $v \geq \tau_0$, and set*

$$\eta^k = \chi(v_1)(v_1 - \tau)_+ \zeta^2 \nu_1^k. \tag{10.55}$$

*Then $\eta \cdot \gamma = 0$ on $\partial\Omega$, and there are constants $c_2$ and $c_3$, determined only by $n$, $\beta_0$, $\beta_1$, and $\beta_2$ such that*

$$\begin{aligned} a^{ij} D_{jk} u D_i \eta^k \geq c_2 &\left( [(v_1 - \tau)\chi' + \chi]\mathcal{E}_1 + (1 - \frac{\tau}{v_1})\chi \mathcal{C}_1^2 \right) \zeta^2 \\ &- c_3 [(v_1 - \tau)\chi' + \chi]\beta_3^2 \Lambda_0 \zeta^2 - c_3 \bar{\mu} |D\zeta|^2 \chi \end{aligned} \tag{10.56}$$

*on $\Omega_\tau$, the subset of $\Omega$ on which $v_1 \geq \tau$.*

***Proof.*** First, because $N = 0$ on $\Sigma(\tau_0)$, we immediately conclude that $\eta \cdot \gamma = 0$ on $\partial\Omega$.

Next, a direct calculation gives

$$a^{ij} D_{jk} u D_i \eta^k = H_1 + H_2 + H_3,$$

where

$$\begin{aligned} H_1 &= a^{ij} D_{jk} u \nu_1^k D_i v_1 [(v_1 - \tau) + \chi]\zeta^2, \\ H_2 &= a^{ij} D_{jk} u D_i \nu_1^k (v_1 - \tau)\chi\zeta^2, \\ H_3 &= a^{ij} D_{jk} u \nu_1^k D_i \zeta 2\zeta \chi (v_1 - \tau). \end{aligned}$$

Each one of these terms needs special care in its estimation. First, we set

$$a_i = \frac{1}{v_1}(\frac{1}{2} D_i(c^{km}) p_k p_m + \varepsilon N N_i),$$

so

$$D_i v_1 = D_{ij} u \nu_1^j + a_i + \frac{\varepsilon}{v_1} N N_z D_i u.$$

It follows that

$$H_1 = [(v_1 - \tau)\chi' + \chi]\zeta^2 (\mathcal{E}_1 - a^{ij} D_i v_1 a_j - \frac{\varepsilon}{v_1} N N_z a^{ij} D_i v_1 D_j u).$$

By applying (10.53a) (10.53c), and (10.54), we then infer that

$$\begin{aligned} H_1 &\geq [(v_1 - \tau)\chi' + \chi]\zeta^2 (\mathcal{E}_1 - C\beta_3 (\mathcal{E}_1)^{1/2} \Lambda_0^{1/2}) \\ &\geq [(v_1 - \tau)\chi' + \chi]\zeta^2 (\frac{1}{2}\mathcal{E}_1 - C\beta_3^2 \Lambda_0) \end{aligned}$$

In addition, if we set

$$\tilde{\mathcal{E}} = a^{ij} D_{ik} u \nu_1^k D_{jm} u \nu_1^m,$$

then

$$H_1 = [(v_1 - \tau)\chi' + \chi]\zeta^2(\tilde{\mathcal{E}} + a^{ij}D_{ik}u\nu_1^k a_j + \frac{\varepsilon}{v_1}NN_z a^{ij}D_{ik}u\nu_1^k D_j u).$$

and hence

$$H_1 \geq [(v_1 - \tau)\chi' + \chi]\zeta^2(\tilde{\mathcal{E}} - C\beta_3(\tilde{\mathcal{E}})^{1/2}\Lambda_0^{1/2})$$

by virtue of (10.53b) (10.53d), and (10.54). Another application of Cauchy's inequality gives

$$H_1 \geq [(v_1 - \tau)\chi' + \chi]\zeta^2((1-\theta)\tilde{\mathcal{E}} - \frac{C}{\theta}\beta_3^2\Lambda_0)$$

for any $\theta \in (0,1)$ and hence (by choosing $\theta$ so that $(1-\theta)(1-c_1/4) = 1 - c_1/2$),

$$H_1 \geq [(v_1 - \tau)\chi' + \chi]\zeta^2(\frac{c_1}{8}\theta\mathcal{E}_1 + (1 - \frac{c_1}{2})\tilde{\mathcal{E}} - C\beta_3^2\Lambda_0).$$

Then we have

$$H_2 = J_1 + J_2 + J_3 + J_4,$$

where

$$J_1 = \left(1 - \frac{\tau}{v_1}\right)\chi\zeta^2 a^{ij}b^{km}D_{ik}uD_{jm}u,$$

$$J_2 = -\left(1 - \frac{\tau}{v_1}\right)\chi\zeta^2\tilde{\mathcal{E}},$$

$$J_3 = (v_1 - \tau)\chi\zeta^2 a^{ij}D_{jk}u\frac{\partial\nu_1^k}{\partial x^i},$$

$$J_4 = (v_1 - \tau)\chi\zeta^2 a^{ij}D_{jk}u\frac{\partial\nu_1^k}{\partial z}D_i u.$$

(Of course, here, we consider $\nu_1^k$ as a function of the variables $x, z, p$.) From (10.43), it follows that

$$J_1 \geq c_1\left(1 - \frac{\tau}{v_1}\right)\chi\zeta^2\mathfrak{C}_1^2,$$

and the estimate $\tilde{\mathcal{E}} \leq \mathfrak{C}_1^2$ implies that

$$J_2 \geq -\frac{1}{2}c_1\left(1 - \frac{\tau}{v_1}\right)\chi\zeta^2\mathfrak{C}_1^2 + \left(\frac{1}{2}c_1 - 1\right)\chi\tilde{\mathcal{E}}\zeta^2.$$

It's easy to check that

$$\left|\frac{\partial\nu}{\partial x}\right| \leq C\left(\frac{1}{R_0} + \varepsilon_x(v)v\right)$$

and

$$\left|\frac{\partial \nu}{\partial z}\right| \le C\varepsilon_z(v),$$

so

$$J_3 + J_4 \ge -C\beta_3 \left(1 - \frac{\tau}{v_1}\right) \mathcal{C}_1 \Lambda_0^{1/2}.$$

Finally,

$$H_3 \ge -C\left(1 - \frac{\tau}{v_1}\right) \mathcal{C}_1 \zeta(\bar{\mu}|D\zeta|^2)^{1/2},$$

and the proof is completed by combining all these inequalities with Cauchy's inequality. Note that $c_2 = c_1/8$. □

We are now ready to present our energy estimate.

**Lemma 10.13.** *Let $u \in C^2(\overline{\Omega})$ be a solution of (10.1) and suppose that $\partial\Omega \in C^2$ with conditions (10.47) and (10.51) satisfied on $\Sigma(\tau_0)$ and conditions (10.53) satisfied wherever $v \ge \tau_0$. Suppose also that*

$$C_k^i \nu_1^k \xi_i \le (a^{ij}\xi_i\xi_j)^{1/2}\beta_3\Lambda_0^{1/2} \tag{10.57a}$$

*for all $\xi \in \mathbb{R}^n$,*

$$C_k^i \eta_i^k \le (a^{ij}\eta_i^k\eta_j^k)^{1/2}\beta_3\Lambda_0^{1/2}, \tag{10.57b}$$

$$v\nu_1^k \frac{\partial D_k^i}{\partial p_j}\eta_{ij} \le (a^{ij}\eta_{ik}\eta_{jk})^{1/2}\beta_3\Lambda_0^{1/2} \tag{10.57c}$$

*for all matrices $\eta$,*

$$\nu_1^k p_i \frac{\partial D_k^i}{\partial z} + \nu_1^k \frac{\partial D_k^i}{\partial x^i} \le \beta_3^2 \Lambda_0 \tag{10.57d}$$

*wherever $v \ge \tau_0$. Suppose further that there is a function $\lambda_0$ such that*

$$\lambda_0 g^{ij}\xi_i\xi_j \le a^{ij}\xi_i\xi_j \tag{10.58}$$

*wherever $v \ge \tau_0$. Now let $\tau$ be a positive constant such that $v_1 \ge \tau$ implies $v \ge \tau_0$ and suppose that there is a $C^1([\tau,\infty))$ function $\Lambda_1$ such that*

$$g_k\nu_1^k \le \Lambda_1(v_1) \tag{10.59}$$

*on the portion of $\partial\Omega$ where $v_1 \ge \tau$ and that*

$$(v_1\Lambda_1'(v_1))^2 + \Lambda_1(v_1)^2 \le \beta_3^2\lambda_0\Lambda_0, \tag{10.60a}$$

$$\Lambda_1(v_1)v_1 \le \beta_3\Lambda_0 \tag{10.60b}$$

*wherever* $v_1 \geq \tau$. *If* $\chi$ *is a* $C^1([\tau,\infty))$ *function satisfying* (10.15) *for all* $s > \tau$, *then there is a constant* $c_4$, *determined only by* $n$, $\beta_0$, $\beta_1$, *and* $\beta_2$ *such that*

$$\int_{\Omega_\tau} [\left(1-\frac{\tau}{v_1}\right)\mathcal{C}_1^2 + \mathcal{E}_1]\chi\zeta^2\,dx \\ \leq c_4(1+c_\chi)^2 \int_{\Omega_\tau} ((\beta_3+1/R_0)^2\Lambda_0\zeta^2 + \bar{\mu}|D\zeta|^2)\chi\,dx \tag{10.61}$$

*for any* $C^1$ *function* $\zeta$ *which vanishes wherever* $d \geq R_0$.

***Proof.*** Using $\eta$ from Lemma 10.12 and arguing as in that lemma, we find that

$$\int C_k^i D_i\eta^k - D_i(D_k^i)\eta^k\,dx \leq \frac{c_2}{4}\int_{\Omega_\tau} [(v_1-\tau)\chi' + \chi]\mathcal{E}_1 + \left(1-\frac{\tau}{v_1}\right)\chi\mathcal{C}_1^2)\zeta^2\,dx \\ + C(1+c_\chi)\int_{\Omega_\tau} \beta_3^2\Lambda_0\zeta^2 + \bar{\mu}|D\zeta|^2\,dx.$$

Using this $\eta$ in (10.49) and recalling also the result of Lemma 10.12, we find that

$$\int_{\Omega_\tau} ([(v_1-\tau)\chi' + \chi]\mathcal{E}_1 + \left(1-\frac{\tau}{v_1}\right)\chi\mathcal{C}_1^2)\zeta^2\,dx \\ \leq C(1+c_\chi)\int_{\Omega_\tau} \beta_3^2\Lambda_0\zeta^2 + \bar{\mu}|D\zeta|^2\,dx + C\int_{\partial\Omega} g_k\eta^k\,ds.$$

To proceed, we estimate this boundary integral. From (10.59) and (10.39), we have

$$\int_{\partial\Omega} g_k\eta^k\,ds \leq C\int_{\partial\Omega} \Lambda_1(v_1)\left(1-\frac{\tau}{v_1}\right)_+ \chi\zeta^2 v\,ds,$$

and then (10.45a) implies that

$$\int_{\partial\Omega} g_k\eta^k\,ds \leq C\int_{\Omega_\tau} |\delta(\Lambda_1(v_1)(1-\frac{\tau}{v_1})\chi)|\zeta^2 v\,dx \\ + C\int_{\Omega_\tau} (\Lambda_1\chi\zeta|\delta\zeta| + \lambda_1(1-\frac{\tau}{v_1})\chi\zeta^2|H| + (1/R_0)\Lambda_1\chi\zeta^2)v\,dx.$$

These integrals are estimated straightforwardly. First, we differentiate to see that

$$|\delta(\Lambda_1(v_1)(1-\frac{\tau}{v_1})\chi)| = |\Lambda_1'(v_1)(1-\frac{\tau}{v_1})\chi + \Lambda_1\frac{\tau}{v_1^2}\chi + \Lambda_1(1-\frac{\tau}{v_1})\chi'||\delta v_1|$$

and

$$|\Lambda_1'(v_1)(1-\frac{\tau}{v_1})\chi + \Lambda_1\frac{\tau}{v_1^2}\chi + \Lambda_1(1-\frac{\tau}{v_1})\chi'| \le \frac{1+c_\chi}{v_1}\beta_3\lambda_0^{1/2}\Lambda_0^{1/2}$$

by (10.15) and (10.60a). It then follows from (10.39) and (10.58) that

$$\int_{\Omega_\tau} |\delta(\Lambda_1(v_1)(1-\frac{\tau}{v_1})\chi)|\zeta^2 v\,dx \le C(1+c_\chi)\int_{\Omega_\tau} \mathcal{E}_1^{1/2}\beta_3\Lambda_0^{1/2}\,dx.$$

Similarly, by invoking (10.10), (10.53a), and (10.39), we see that

$$\begin{aligned}\int_{\Omega_\tau} (\Lambda_1\chi\zeta|\delta\zeta| + \lambda_1(1-\frac{\tau}{v_1})\chi\zeta^2|H| + (1/R_0)\Lambda_1\chi\zeta^2)v\,dx \\ \le C\int_{\Omega_\tau} ((1-\frac{\tau}{v_1})\mathcal{C}_1^2)^{1/2}\beta_3\Lambda_0^{1/2}\zeta^2\,dx \\ + C\int_{\Omega_\tau} (\beta_3^2\Lambda_0\zeta^2 + \bar\mu|D\zeta|^2)\,dx.\end{aligned}$$

The proof is completed by combining these inequalities and applying Cauchy's inequality. □

We are now in a position to prove our gradient estimates. The first estimate (for the maximum of the gradient in terms of an integral) is proved by making three simple replacements in the proof of Lemma 10.4: we use $v_1$ in place of $v$, we use our energy inequality Lemma 10.13 in place of Lemma 10.3 and we use the Sobolev inequality Lemma 10.9 in place of Lemma 10.2.

**Lemma 10.14.** *Let $u \in C^2(\overline{\Omega})$ be a solution of* (10.1) *and suppose that $\partial\Omega \in C^2$ with conditions* (10.47) *and* (10.51) *satisfied on $\Sigma(\tau_0)$ and conditions* (10.53) *satisfied wherever $v \ge \tau_0$. Suppose also that conditions* (10.57) *and* (10.58) *hold wherever $v \ge \tau_0$, that condition* (10.59) *holds on the portion of $\partial\Omega$ on which $v_1 \ge \tau$, and that conditions* (10.60) *hold wherever $v_1 \ge \tau$. Suppose further that there are functions $\Lambda \in C(\overline{\Omega})$ and $w$, $\lambda$, and $\tilde\lambda$ in $C^1([\tau,\infty))$ such that*

$$\Lambda_0 \le \Lambda v, \tag{10.62a}$$

$$\lambda(v_1) \le \Lambda(x), \tag{10.62b}$$

$$\lambda(v_1)\left(1+\left(\frac{v_1\lambda'(v_1)}{\lambda}\right)^2\right)g^{ij}\xi_i\xi_j \le va^{ij}\xi_i\xi_j, \tag{10.62c}$$

$$(\Lambda(x)/\lambda(v_1))^{N/2}/\Lambda(x) \le \tilde\lambda(v_1) \tag{10.62d}$$

*wherever $v_1 \ge \tau$. Suppose finally that conditions* (10.17) *and* (10.19) *hold. Then there is a constant $c_5$, determined only by $n$, $N$, $\beta$, $\beta_1$, $\beta_2$, and $\beta_3 R$, such that*

$$\sup_{\Omega_\tau(y,R/2)}\left(1-\frac{\tau}{v_1}\right)^N w(v_1) \le c_5 R^{-n}\int_{\Omega_\tau(y,R)} w(v_1)\Lambda\tilde\lambda v\,dx \tag{10.63}$$

*for any $y \in \Omega$ and $R > 0$ such that $d \le R_0$ in $\Omega[y, R]$.*

The integral in (10.63) is estimated by essentially the same technique that was used to estimate the integral in (10.20). We assume that there are positive constants $\beta_4$ and $\beta_5$ such that

$$\Lambda(x)\tilde{\lambda}(v_1)v \le \beta_4 w^{\beta_5} Du \cdot A, \tag{10.64}$$

so Lemma 10.14 reduces the gradient estimate to one on

$$\int w^q Du \cdot A\, dx$$

for $q = 1 + \beta_5$. To estimate this integral, we set $\psi_0 = \sup |\psi(x, u)|$ and we assume that there are positive constants $\beta_6$, $\beta_7$ and $\beta_8$ such that

$$w' A \cdot \xi \le \beta_6 (a^{ij}\xi_i\xi_j)^{1/2} (Du \cdot A)^{1/2}, \tag{10.65a}$$
$$w' \psi_0 \gamma \cdot \xi \le \beta_6 (a^{ij}\xi_i\xi_j)^{1/2} (Du \cdot A)^{1/2} \tag{10.65b}$$

for all $\xi \in \mathbb{R}^n$,

$$w|A| \le \beta_6 Du \cdot A, \tag{10.65c}$$
$$\Lambda_0 \le \varepsilon(v_1) w^2 Du \cdot A, \tag{10.65d}$$
$$\bar{\mu} \le \beta_7 Du \cdot A, \tag{10.65e}$$
$$B \le \beta_8 Du \cdot A \tag{10.65f}$$

wherever $v \ge \tau_0$. (These conditions are essentially the same as (10.23).) We also assume that there is a constant $\beta_9$ such that

$$\psi_0 \le \beta_9 |A| \tag{10.66}$$

wherever $v \ge \tau_0$. Note that this inequality is almost the same as (10.47b); in practice, this inequality with $\beta_9 < 1$ is considered the basic assumption even though it is somewhat stronger than (10.47b).

**Lemma 10.15.** *Let $u \in C^2(\overline{\Omega})$ be a solution of* (10.1) *and suppose that $\partial\Omega \in C^2$ with conditions* (10.47) *and* (10.51) *satisfied on $\Sigma(\tau_0)$. Suppose also that conditions* (10.53), (10.57), (10.58), (10.65), *and* (10.66) *are satisfied wherever $v \ge \tau_0$. Let $\tau$ be a positive constant such that $v_1 \ge \tau$ implies $v \ge \tau_0$ and suppose conditions* (10.17), (10.59), *and* (10.60) *are also satisfied. Let $q \ge 2$, set $\sigma = \operatorname{osc} u$, and $E = \exp(\beta_8\sigma)$. If there is a $\tau_1 \ge \tau$ such that*

$$16c_4(1 + \beta q)^2 \left(\beta_3 + \frac{1}{R}\right)^2 \beta_6^2 \sigma^2 q^2 E^2 \varepsilon(\tau_1) \le 1, \tag{10.67}$$

*then there is a constant $C$, determined only by $\beta$, $\beta_0$, $\beta_1$, $\beta_2$, $\beta_7$, $\beta_9$, and $q$, such that*

$$\int_{\Omega_\tau(y,R/2)} w^q Du\cdot A\,dx \le CE^q\left(w(\tau_1)+\frac{\beta_6\sigma}{R}\right)^q\int_{\Omega_\tau(y,R)} Du\cdot A\,dx \tag{10.68}$$

*for any $y\in\Omega$ and $R>0$ such that $d\le R_0$ in $\Omega[y,R]$.*

***Proof.*** The proof is essentially the same as for Lemma 10.5. With $m=\inf_{\Omega(R)} u$, we define $f_1$ and $f$ by

$$f_1(s)=\exp(\beta_8(s-m)),\quad f(s)=(s-m)f_1(s),$$

and we set

$$I_1=\int_\Omega (w(v_1)^q-w(\tau)^q)_+\zeta^q f'(u)Du\cdot A\,dx,$$

$$I_2=-\int_{\Omega_\tau} qf(u)w^{q-1}\zeta^q w' Dv_1\cdot A\,dx,$$

$$I_3=-\int_\Omega qf(u)(w^q-w(\tau)^q)_+\zeta^{q-1}D\zeta\cdot A\,dx,$$

$$I_4=-\int_\Omega (w^q-w(\tau)^q))_+\zeta^q f(u)\operatorname{div} A\,dx,$$

$$I_5=\int_\Omega (w^q-w(\tau)^q))_+\zeta^q f_1(u)Du\cdot A\,dx,$$

$$I_6=\int_\Omega (w^q-w(\tau)^q))_+\zeta^q\beta_8 f(u)Du\cdot A\,dx,$$

$$I_7=\int_{\partial\Omega} (w^q-w(\tau)^q)_+\zeta^q f(u)A\cdot\gamma\,ds.$$

Integration by parts gives

$$I_1=I_2+I_3+I_4+I_7,$$

the explicit form of $f$ implies that $I_1=I_5+I_6$, and the argument in Lemma 10.5 shows that $I_4\le I_6$. It follows that

$$I_5\le I_2+I_3+I_7.$$

We now estimate $I_7$. First,

$$\begin{aligned} I_7 &\le E\psi_0\sigma\int_{\partial\Omega}(w^q-w(\tau)^q)_+\zeta^q\,ds\\ &= E\psi_0\sigma\int_{\partial\Omega}(w^q-w(\tau)^q)_+\zeta^q\gamma\cdot\gamma\,ds\\ &= E\psi_0\sigma[J_1+J_2+J_3],\end{aligned}$$

where

$$J_1 = -\int_{\Omega_\tau} q w^{q-1} w' \gamma \cdot D v_1 \zeta^q \, dx,$$

$$J_2 = -\int_{\Omega} q(w^q - w(\tau)^q)_+ \zeta^{q-1} \gamma \cdot D\zeta \, dx,$$

$$J_3 = -\int_{\Omega} (w^q - w(\tau)^q)_+ \zeta^q \operatorname{div} \gamma \, dx.$$

By invoking (10.65a), (10.65b), (10.65c), and (10.66), we infer that

$$\begin{aligned} I_5 &\le 2E\sigma q \beta_6 \int_{\Omega_\tau} w^{q-1} \zeta^q \mathcal{E}_1^{1/2} (Du \cdot A) \, dx \\ &+ (1+\beta_9)\beta_6 E\sigma \int_{\Omega_\tau} q w^{q-1} \zeta^{q-1} |D\zeta| Du \cdot A \, dx \\ &+ \beta_9 E \frac{\sigma}{R} \int_{\Omega_\tau} Du \cdot A \frac{w^q}{v} \zeta^q \, dx. \end{aligned}$$

We now choose $\zeta$ as in Lemma 10.5 so $|D\zeta| \le 2/R$. The Cauchy-Schwarz inequality along with Lemma 10.13 and conditions (10.65d) and (10.65e) imply that

$$\begin{aligned} 2E\sigma q\beta_6 &\int_{\Omega_\tau} w^{q-1} \zeta^q \mathcal{E}_1^{1/2} (Du \cdot A) \, dx \\ &\le \frac{1}{2} \int_{\Omega_\tau} w^q \zeta^q Du \cdot A \, dx + C E^2 \beta_6^2 \frac{\sigma^2}{R^2} \int_{\Omega_\tau} w^{q-2} \zeta^{q-2} Du \cdot A \, dx \end{aligned}$$

provided $\tau \ge \tau_1$. From (10.65c) and the inequality $Du \cdot A \le v|A|$, we conclude that $w \le \beta_6 v$, so

$$\begin{aligned} (1+\beta_9)\beta_6 E\sigma &\int_{\Omega_\tau} q w^{q-1} \zeta^{q-1} |D\zeta| Du \cdot A \, dx + \beta_9 E \frac{\sigma}{R} \int_{\Omega_\tau} Du \cdot A \frac{w^q}{v} \zeta^q \, dx \\ &\le C E \beta_6 \frac{\sigma}{R} \int_{\Omega_\tau} w^{q-1} \zeta^{q-1} Du \cdot A \, dx. \end{aligned}$$

Combining these inequalities as in the proof of Lemma 10.5 with Young's inequality completes the proof. □

Finally, we estimate $\int Du \cdot A \, dx$. To simplify notation, we note that, if $\Omega$ is a bounded Lipschitz domain, then there is a constant $K^*(\Omega)$ such that

$$|\partial\Omega \cap B(y, R)| \le K^* R^{n-1}$$

for any $y \in \overline{\Omega}$ and any $R > 0$, where $|\partial\Omega \cap B(y, R)|$ denotes $(n-1)$-dimensional measure.

**Lemma 10.16.** *Let $u$ be a bounded weak solution of* (10.1) *and suppose that there are a decreasing function $\varepsilon_1$ and a nonnegative constant $\beta_6$ such that*

$$|A| \le \varepsilon_1(v) Du \cdot A, \tag{10.69a}$$

$$B \le \beta_8 Du \cdot A \tag{10.69b}$$

*for $v \ge \tau_0$. Set $\sigma = \operatorname{osc}_{\Omega[y,R]} u$ and $E = \exp(\beta_6\sigma)$ and let $\tau \ge \tau_0$. If there is a constant $\tau_2 \ge \tau$ such that*

$$4\varepsilon_1(\tau_2)\frac{\sigma}{R} \le 1, \tag{10.70}$$

*then*

$$\int_{\Omega_\tau(y,R/2)} Du \cdot A\, dx \le C(n)R^n E\Delta_1 + 2K^*\psi_0 E\sigma R^{n-1}, \tag{10.71}$$

*where*

$$\Delta_1 = \sup_{S_0}\left(\sigma(B - \beta_8 Du\cdot A)_+ + (Du\cdot A)_- + \frac{\sigma|A|}{R}\right)$$

*and $S_0$ is the subset of $\Omega[y,R]$ on which $v < \tau_2$.*

***Proof.*** By following the proof of Lemma 10.6 with $\beta_8$ replacing $\beta_6$ and $\Omega[y,R]$ replacing $B(y,R)$, we find that

$$\int_{\Omega_\tau(y,R/2)} Du\cdot A\, dx \le C(n)R^n E\Delta_1 + 2\int_{\partial\Omega}\eta(x)\psi\, ds,$$

where $\eta$ is a function satisfying the conditions $0 \le \zeta \le E\sigma$ in $\Omega$ and $\zeta = 0$ outside $B(y,R)$. (The proof of Lemma 10.6 requires more of $\eta$ but these are the only properties of relevance to the estimate of the boundary integral.) It follows that

$$\int_{\partial\Omega}\eta(x)\psi\, ds \le E\sigma K^* R^{n-1},$$

which then implies (10.71). □

It is a simple matter to verify that our examples are covered by these results.

In our first example, we have $A = \nu$ and $B(x,z,p) = B^*(x,z,p) + B_0(x,z,p)$ satisfies the conditions

$$|B^*| + v|B_0| \le \theta_1, \quad v(|B^*_p| + |B^*_p\cdot p|) \le \theta_1, \quad vB^*_z \le \theta_1^2, \quad |B^*_x| \le \theta_1^2, \quad B \le \theta_2$$

for some nonnegative constants $\theta_1$ and $\theta_2$. If we assume that there are nonnegative constants $\psi_0$, $\psi_1$, and $\psi_2$ with $\psi_0 < 1$ such that

$$|\psi| \le \psi_0, \quad |\psi_x| \le \psi_1, \quad -\psi_2 \le \psi_z \le 0,$$

then the hypotheses of Lemmata 10.14, 10.15, and 10.16 are satisfied with

$$\beta_0 = \frac{1}{1-\psi_0^2}, \quad \beta_1 = \frac{1}{2}, \quad \beta_2 = \frac{1+\psi_0}{2},$$
$$\beta_3 = C(n, N, \psi_0)(\theta_1 + \psi_1 + \psi_2 + \frac{1}{R_0}),$$
$$\beta_6 = C(n, \psi_0), \quad \beta_7 = 2, \quad \beta_8 = \beta_{10} = \theta_2,$$
$$q = 1,$$
$$\varepsilon_z(v) = \frac{\psi_2}{1-\psi_0^2}, \quad \varepsilon_x(v) = \frac{\psi_1 + 1/R_0}{v(1-\psi_0^2)},$$
$$\Lambda_0 = v, \quad \lambda_0 = \frac{1}{v}, \quad \Lambda_1 = 2\psi_1 + 2\theta_1 + \frac{2n}{R_0}, \quad \bar{\mu} = v,$$
$$\varepsilon(v) = \frac{2}{(\ln v)^2}, \quad \tau_1 = C\left(\frac{\sigma}{R} + 1\right),$$
$$\lambda = \Lambda = \tilde{\lambda} = 1, \quad \varepsilon_1(v) = \frac{2}{v}',$$

$w = (\ln v)^2$ in Lemma 10.14 and $w = \ln v$ in Lemma 10.15. Hence

$$v(y) \le \exp\left(C(n, \psi_0, [\theta_1 + \theta_2 + \psi_1 + \psi_2]R)\left(1 + \frac{\sigma}{R}\right)^{3/2}\right).$$

For uniformly elliptic equations, we assume that $\Psi$ satisfies (10.30) and that $A$ and $B$ satisfy (10.31). In addition, we assume that there are nonnegative constants $\psi_0$, $\psi_1$, and $\psi_2$ such that

$$|\psi| \le \psi_0, \quad |\psi_x| \le \psi_1, \quad |\psi_z| \le \psi_2. \tag{10.72}$$

(Note that, this time, we do not assume that $\psi_z \le 0$.) If we take $\tau_0$ so large that $\tau_0\Psi(\tau_0) \ge 1 + 2\theta_1$ and $\Psi(\tau_0) \ge 2\psi_0$, then the hypotheses of Lemmata 10.14, 10.15, and 10.16 are satisfied with

$$\beta_0 = \theta_3, \quad \beta_1 = \frac{1}{2\theta_2}, \quad \beta_2 = \frac{1}{2}$$
$$\beta_3 = C(\theta_0, \theta_1, \theta_2, \theta_3)(\theta_4 + \psi_1 + \psi_2 + \frac{1}{R_0}),$$
$$\beta_6 = 4\theta_2, \quad \beta_7 = 4\theta_3, \quad \beta_8 = 2\theta_4,$$
$$q = N + 2,$$
$$\varepsilon_z(v) = 2\theta_4 + \psi_2, \quad \varepsilon_x(v) = \theta_4 + \psi_2 + \frac{1}{R_0},$$
$$\Lambda_0 = v^3\Psi(v), \quad \lambda_0 = \frac{\Psi(v)}{v}, \quad \Lambda_1 = C(\theta_0, \theta_1, \theta_2, \theta_3)(\psi_1 + \Psi_2 + \frac{1}{R_0})v_1\Psi(v_1),$$
$$\bar{\mu} = \theta_3 v\Psi(v),$$

$$\Lambda = (1+\theta_3)v^2\Psi(v), \quad \lambda = C(\theta_0,\theta_1,\theta_2,\theta_3)\Psi(v_1), \quad \tilde\lambda = C(\theta_0,\theta_1,\theta_2,\theta_3)v_1^{N-1},$$
$$\varepsilon_1(v) = \frac{2\theta_2}{v}, \quad w = v$$

provided $R$ is sufficiently small.

Finally, we examine the exponential case. We consider (10.1) when $A = \exp(\theta v^2)p$ for some positive constant $\theta$ and $B$ satisfies the condition

$$|B(x,z,p)| \le \theta_1 \exp(\theta v^2)v^2$$

for some nonnegative constant $\theta_1$, and we assume that $\psi_0$, $\psi_1$, And $\psi_2$ are nonnegative constants such that (10.72) holds. By observing that

$$a^{ij}\gamma_j = 2\theta\exp(v^2)\gamma^i - 2\theta\psi p_i$$

wherever $A\cdot\gamma + \psi = 0$, we see that the hypotheses of Lemma 10.14 are satisfied with

$$\beta_0 = 2, \quad \beta_1 = \beta_2 = \frac{1}{2},$$
$$\varepsilon_z = \frac{2\theta\psi_1}{v}, \quad \varepsilon_x = \frac{2\theta}{v}[\psi_2 + \frac{1}{R_0}],$$
$$\beta_3 = e(\psi_1+\psi_2+\theta_1+\frac{1}{R_0}), \quad \beta_8 = 2\theta_1,$$
$$\beta = 1,$$
$$\Lambda = \Lambda_0 = \exp(\theta v^2)v^3, \quad \bar\mu = \exp(\theta v^2)v^2, \quad \lambda = \exp(\theta v_1^2)v_1^{-3},$$
$$\Lambda_1 = (\psi_1+\psi_2+\frac{1}{R_0})\exp(\theta v_1^2)v_1, \quad \lambda_0 = \exp(\theta v^2), \quad \tilde\lambda = v_1^{-2},$$
$$w = v.$$

We then obtain

$$\sup_{\Omega_\tau(y,R/4)} \left(1 - \frac{\tau}{v\,1}\right)^{N+2} v \le CR^{-n} \int_{\Omega_\tau(y,R/2)} \exp(\theta v^2)v^2\,dx.$$

The hypotheses of Lemma 10.16 are satisfied with $\varepsilon_1(v) = 1/(2v)$ and $\beta_8 = 2\theta_1$, so we obtain

$$v(y) \le C\left(1+\frac{\sigma}{R}\right)^2 \exp(\theta\left(1+\frac{\sigma}{R}\right)^2).$$

## Notes

As mentioned in Chapter 9, the basic idea for deriving gradient estimates is due to Bernstein [9]. The corresponding result for uniformly elliptic equations in divergence form was developed by Ladyzhenskaya and Ural'tseva

[91] and extended to the minimal surface equation by Bombieri, De Giorgi, and Miranda [12]. Further extensions can be found in [92] and (for general prescribed mean curvature equations) in [13]. In this chapter, we follow the arguments in [170], which combines arguments from the uniformly elliptic case and the prescribed mean curvature case. In particular, the Sobolev-type inequality (Lemma 10.1), which is taken directly from [131], is a generalized version of Teorema 2 from [137] although these works use very different methods of proof.

The interior gradient estimates in Section 10.2 are taken from [170] with a few slight modifications: our structure conditions are written without direct reference to the minimum eigenvalue of $[a^{ij}]$, we incorporate the decomposition in terms of $C^i_k$ and $D^i_k$ whereas [170] assumes that one of these two matrices is zero, and our $a^{ij}$ differs from that in [170] by a factor of $1/v$. (In addition, we use slightly different structure functions; in particular [170] does not distinguish between $\Lambda_0$ and what would be $\Lambda v$ in our notation or between $\lambda$ and $v\lambda_0$.) Specifically, our Lemma 10.3 is the same as (2.11) of [170], Lemma 10.4 is the same as Lemma 1 of [170], Lemma 10.5 is the same as Lemma 2 of [170], and Lemma 10.6 is the same as Lemma 3 of [170]. As a further refinement, [12] gives the gradient estimate for the prescribed mean curvature equation in the form

$$v(y) \leq c_1 \exp(c_2 \sup(u - u(y))/d(y))$$

with $c_1$ and $c_2$ determined only by $n$ and $d(y)\theta_1$. (As mentioned in the notes to Chapter 9, such an estimate was first proved by Finn [45] but only for $n = 2$; moreover, Finn showed there that this form cannot be improved.) Theorem 2 of [170] shows that the method in this chapter can be further modified to obtain an interior estimate of this form, and it is possible (see Section 4 in [106]) to prove corresponding estimates for the conormal problems in this chapter but, due to the length of this chapter, we omit even the statements of such estimates.

Moreover, results similar to those in [170] were proved later by Marcellini [125]. The primary difference between the estimates in [170] and those in [125] is a technical one; Marcellini shows directly that weak solutions (which are not assumed to have second derivatives) have bounded gradient while Simon assumes initially more smoothness of the solutions (although a proof that weak solutions of uniformly elliptic equations have appropriate derivatives is given on pages 845 and 846 of [170]), and Simon's structure conditions are slightly more general.

A gradient estimate for oblique derivative problems was announced by Ural′tseva in [182] and proved by Ladyzhenskaya and Ural′tseva in Section

X.2 of [91] under slightly stronger assumptions than given here; it was briefly pointed out in Section V.7 of [90] that the assumptions on the first derivatives of $B$ and the second derivatives of $\psi$ can be eliminated, and it is easy to modify the proof in [91] to handle theses weaker hypotheses. The proof given there was an integral version of the method from Chapter 9: after a suitable change of variable so that $\partial\Omega$ is replaced by $B^0(R)$ and $\Omega$ is replaced by $B^+(R)$, the authors estimate $|D'u|$, and the remaining derivative is estimated as in Chapter 9 (but using integral methods).

A major advance was made by Ural′tseva in [183]: she introduced the function $v_2$, defined by

$$v_2(x) = F(x, u, Du) + \bar{\psi}(x, u)Du \cdot \bar{\gamma},$$

when $A$ behaves in appropriate sense like $\nu$ and $F$ depends only on $p$. In that work, $\psi$ was assumed to be identically zero and $\Omega$ was assumed convex. These two restrictions were removed in two subsequent works [184] and [185] with $\psi$ assumed constant but nonzero and $\Omega$ assumed convex in [184]; the general case of nonconstant $\psi$ and not necessarily convex $\Omega$ (but $\partial\Omega \in C^2$) was studied in [185] specifically for the case $A = \nu$. A further generalization, allowing $F$ to depend also on $x$ and $z$, was given in [186]. The results of [185] were also proved in [61] using very similar techniques; as Gerhardt points out in [61], the main justification of that work is that Ural′tseva's work was not readily available in the West at the time. (Although [184] was published in 1973, the English translation did not appear until 1979, and [185] was published in 1975 with the English version not appearing until 1980. Gerhardt's work appeared in 1976.) All of these results were extended to general variational conormal problems in smooth domains by the present author in [97] (using a straightforward combination of ideas in [170], [184], and [61]); we mention in particular that our (10.45a) and its proof come from Lemma 1.4 of [61] and that our (10.46) and its proof come from Lemma 1.1 of [61]. In addition, the proof of our Lemma 10.10 is based on the proof of Lemma 1 in [184]. Corresponding results in nonsmooth domains were proved in [113] by combining the ideas developed in [97] and [84], and we refer the reader to those works for more details. We also mention [83], in which the authors prove that the graph of the solution of the capillary problem is a Lipschitz surface even when the contact angle is zero; their results also give gradient bound for the capillary problem when the contact angle is bounded away from zero.

The gradient estimates in Sections 10.3 and 10.4 were first presented in [106] which was primarily concerned with parabolic conormal problems.

The main new ingredient in that work is the function $v_1$, which was suggested by [38]; this work studies oblique derivative problems for fully nonlinear parabolic equations and it shows that, under suitable hypotheses, the function

$$\bar{v}_1 = (D'u)^2 + \varepsilon(Du \cdot \gamma - g(x, u, Du))^2$$

satisfies a nice boundary value problem. However, this function is only useful under stronger hypotheses than seem appropriate. First, in order to derive the differential equation, one must assume that $g$ is twice differentiable with respect to its arguments (whereas our $v_1$ only uses first derivatives of $g$), and, second, some of these structure conditions seem to be too strong for even the simple case of linear boundary conditions. The second objection is overcome with a slightly different function:

$$\hat{v}_1 = \max\{\bar{v}_1, \varepsilon_1 |Du|^2\}$$

by Ural′tseva in [187] and [188], and we refer to those works for further discussion of the use of $\hat{v}_1$ in place of $\bar{v}_1$. The more elaborate construction of $v_1$ in Lemmata 10.7 and 10.8 eliminates the need for any structure conditions on the second derivatives of the boundary condition, even for problems in nondivergence form. Details of this statement have, so far, only been carried out for uniformly parabolic problems in Theorem 13.13 of [114].

We point out in addition, that Nezhinskaya [147, 148] has derived gradient bounds for conormal problems under a slightly different set of hypotheses than allowed here by obtaining suitable integral estimates for the quantity denoted by $w_T$ in Chapter 9; this approach is based on that in [91], and it would be interesting to know if either this approach can be generalized to include the results given here or if the approach given here can be generalized to include the results in [147, 148].

## Exercises

**10.1** Show that condition (10.23e) can be replaced by

$$\frac{\partial A^i}{\partial z} D_i u + \frac{\partial A^i}{\partial x^i} \le \beta_6 Du \cdot A$$

in Lemma 10.5 if (10.24) is suitably adjusted. Also prove the corresponding modification of Lemma 10.15.

**10.2** Show that Lemma 10.6 remains valid if $B$ is replaced by $-B$ in (10.28b).

## Chapter 11

# Higher Order Estimates and Existence of Solutions for Quasilinear Oblique Derivative Problems

### Introduction

Our next step in studying the oblique derivative problem is to obtain estimates on higher derivatives. Specifically, we prove a Hölder estimate for the first and second derivatives of $u$ assuming that we have an *a priori* bound for the maximum modulus of the gradient of $u$. We also use these estimates to prove a number of existence results. Our first existence results will apply to a number of problems with sufficiently smooth data, and we shall use these results along with suitable estimates to prove the existence of solutions to oblique derivative problems with nonsmooth data. For the nonsmooth data, we shall prove some existence results in which the solutions are known to have some global smoothness (typically $C^{1,\alpha}(\overline{\Omega})$) and some results when the solutions are not known to be so smooth.

### 11.1 The Hölder gradient estimate for conormal problems

The easiest of our estimates is the one for the Hölder norm of the gradient of the solution of a conormal problem. This estimate is proved using the observation that the tangential derivatives of the solution are themselves solutions of suitable linear conormal problems. In fact, for the interior Hölder gradient estimate, all the derivatives of the solution are solutions of suitable linear equations in divergence form.

**Lemma 11.1.** *Let $u \in C^2(B(R))$ for some ball $B(R)$ and suppose $u$ is a solution of the equation*

$$\operatorname{div} A(x,u,Du) + B(x,u,Du) = 0 \text{ in } B(R). \tag{11.1}$$

*If there are positive constants $K$, $\lambda_K$, $\Lambda_K$, and $\mu_K$ such that*

$$|Du| \le K, \tag{11.2a}$$

$$a^{ij}\xi_i\xi_j \ge \lambda_K|\xi|^2 \tag{11.2b}$$

*for all* $\xi \in \mathbb{R}^n$,

$$|a^{ij}| \le \Lambda_K, \tag{11.2c}$$

$$|A_x| + K|A_z| + |B| \le \mu_K \tag{11.2d}$$

*in* $B(R)$, *where* $a^{ij} = \partial A^i/\partial p_j$, *and* $A$, $A_p$, $A_x$, $A_z$, *and* $B$ *are evaluated at* $(x, u, Du)$, *then there are positive constants* $\alpha$ *and* $C$, *determined only by* $n$ *and* $\Lambda_K/\lambda_K$ *such that*

$$\operatorname*{osc}_{B(r)} Du \le C\left(\frac{r}{R}\right)^\alpha \left(\operatorname*{osc}_{B(R)} Du + \frac{\mu_K}{\lambda_K}R\right) \tag{11.3}$$

*for all* $r \in (0, R)$.

***Proof.*** For $k = 1, \dots, n$, set $w_k = D_k u$ and

$$f_k^i = \frac{\partial A^i}{\partial x^k} + D_k u\frac{\partial A^i}{\partial z} + \delta_k^i B.$$

The weak form of (11.1) says that

$$\int_{B(R)} [A^i D_i\zeta - B\zeta]\,dx = 0$$

for any $C^1(\overline{B(R)})$ function $\zeta$ with compact support. If we let $\varphi$ be an arbitrary $C^2(\overline{B(R)})$ function with compact support, set $\zeta = D_k\varphi$, and integrate by parts, we obtain

$$\int_{B(R)} [a^{ij}D_j w_k + f_k^i]D_i\varphi\,dx = 0.$$

The desired estimate then follows from Theorem 5.45. □

A similar argument provides a Hölder gradient estimate up to the boundary.

**Theorem 11.2.** *Let* $\partial\Omega \in C^2$ *and suppose* $u \in C^2(\overline{\Omega})$ *satisfies*

$$\operatorname{div} A(x, u, Du) + B(x, u, Du) = 0 \text{ in } \Omega \cap B(R), \tag{11.4a}$$

$$A(x, u, Du)\cdot\gamma + \psi(x, u) = 0 \text{ on } \partial\Omega \cap B(R). \tag{11.4b}$$

*Suppose also that there are positive constants* $K$, $\lambda_K$, $\Lambda_K$, *and* $\mu_K$ *such that* (11.2) *holds and that*

$$|\psi_x(x, z)| + |\psi_z(x, z)|K \le \mu_K$$

*for* $x \in \partial\Omega$ *and* $|z| \le \sup|u|$. *Then there are positive constants* $\alpha < 1$ *and* $C$, *both determined by* $\Lambda_K/\lambda_K$ *and* $n$, *such that*

$$\operatorname*{osc}_{\Omega\cap B(r)} Du \le C\left(\frac{r}{R}\right)^\alpha (\operatorname*{osc}_{\Omega\cap B(R)} Du + \frac{\mu_K}{\lambda_K}R) \tag{11.5}$$

*for all* $r \in (0,R)$.

***Proof.*** Since the hypotheses are invariant under a $C^2$ change of variables, there is no loss of generality in assuming that $\Omega \cap B(R) = B^+(R)$ and $\partial\Omega \cap B(R) = B^0(R)$. With $w_k = D_k u$ as before, we see that

$$D_i(a^{ij}D_j w_k + f^i_k) = 0$$

in $B^+(R)$. Moreover, for $k < n$, we also have

$$a^{nj}D_j w_k + f^n_k + \psi_k = 0$$

on $B^0(R)$ with

$$\psi_k = \frac{\partial\psi(x,u)}{\partial x^k} + \frac{\partial\psi(x,u)}{\partial z}D_k u.$$

It follows from Theorem 5.45 that

$$\operatorname*{osc}_{B^+(r)} D_k u \le C\left(\frac{r}{R}\right)^\alpha (\operatorname*{osc}_{B^+(R)} D_k u + \frac{\mu_K}{\lambda_K}R).$$

In addition, for any $\rho \in (0,R)$, we have

$$\begin{aligned}\operatorname*{osc}_{B^0(\rho)} D_n u &\le C\left(\sum_{k<n}\operatorname*{osc}_{B^0(\rho)} D_k u + \frac{\mu_K}{\lambda_K}\rho\right)\\ &\le C\left(\frac{\rho}{R}\right)^\alpha\left(\sum_{k<n}\operatorname*{osc}_{B^+(R)} D_k u + \frac{\mu_K}{\lambda_K}R\right),\end{aligned}$$

and Theorem 5.46 implies that

$$\operatorname*{osc}_{B^+(r)} D_n u \le C\left(\frac{r}{R}\right)^\alpha (\operatorname*{osc}_{B^+(R)} D_n u + \frac{\mu_K}{\lambda_K}R).$$

The proof is completed by combining this estimate with the corresponding one for $D_k u$, $k < n$. □

## 11.2 A solvability theorem

In this section, we prove a solvability theorem for operators on subsets of Banach spaces which will be used repeatedly in our examination of existence of solutions for nonlinear boundary value problems. To simplify writing of the statement of our theorem we make the following definition. Let $\mathbb{B}$ be a Banach space with norm $\|\cdot\|$. Then for any $\eta \in (0,1)$ and $v_0 \in \mathbb{B}$, we write $S(v_0,\eta)$ for the set of all $v \in \mathbb{B}$ such that

$$\|v-(1-\delta)v_0\| \le \eta\delta\|v_0\|$$

for some $\delta \in [0,1]$.

**Theorem 11.3.** *Let $\mathbb{B}$ and $\mathbb{B}_1$ be Banach spaces, let $U$ be a subset of $\mathbb{B}_1$, let $J\colon U \to \mathbb{B}$, and let $\eta \in (0,1)$. Suppose that for every $u \in U$, there are $u_1 \in U$ and $\varepsilon \in (0,1)$ such that*

$$\|Ju_1-(1-\varepsilon)Ju\| \le \eta\varepsilon\|Ju\|. \tag{11.6}$$

*Suppose further that there is a topology $\mathcal{T}$ on $\mathbb{B}$ such that if $(v_n)$ is a sequence in $\mathbb{B}$ with a uniform bound $M$ on $\|v_n\|$ which converges in $\mathcal{T}$ to some $v \in \mathbb{B}$, then $\|v\| \le M$. Suppose finally that there is $u_0 \in U$ such that $JU \cap S(Ju_0,\eta)$ is sequentially compact in $\mathcal{T}$. Then there is $u \in U$ such that $Ju = 0$.*

***Proof.*** If $Ju_0 = 0$, there is nothing to prove, so we suppose that $Ju_0 \ne 0$. To simplify writing, we abbreviate $S = S(Ju_0,\eta)$, and we observe that $JU \cap S$ is non-empty because it contains $u_0$.

Next, let $v \in JU$ and choose any $u \in U$ such that $Ju = v$. By hypothesis, there are $u_1 \in U$ and $\varepsilon \in (0,1)$ (which will depend on $u$) such that (11.6) is satisfied. It follows from the definition of $S$ that $v_1 = Ju_1$ satisfies

$$\|v_1-(1-\varepsilon)v\| \le \eta\varepsilon\|v\|,$$

so

$$\|v_1\| \le (1-\varepsilon+\eta\varepsilon)\|v\|,$$

which means that $\|v_1\| < \|v\|$. The same argument shows that $\|v\| \le \|v_0\|$ for any $v \in S$.

Moreover,

$$\begin{aligned}\|v_1-(1-\delta)(1-\varepsilon)Ju_0\| &\le \|v_1-(1-\varepsilon)Ju\| + (1-\varepsilon)\|Ju-(1-\delta)Ju_0\| \\ &\le \eta\varepsilon\|Ju\| + (1-\varepsilon)\eta\delta\|Ju_0\|.\end{aligned}$$

Since $\|Ju\| \le \|Ju_0\|$ and

$$(1-\delta)(1-\varepsilon) = 1 - [\varepsilon + (1-\varepsilon)\delta] \in (0,1),$$

it follows that $v_1 \in JU \cap S$. In brief, given any $v \in JU \cap S$, there is $v_1 \in JU \cap S$ such that $\|v_1\| < \|v\|$.

Now, we set

$$I = \inf\{\|v\| : v \in JU \cap S\},$$

and we let $(v_n)$ be a sequence in $JU \cap S$ such that $\|v_n\| \to I$. By hypothesis, there is a subsequence $(v_{n_k})$ which converges in $\mathcal{T}$ to some $v \in JU \cap S$. Moreover, given any $\varepsilon > 0$, there is a $K$ such that $\|v_{n_k}\| < I + \varepsilon$ if $k > K$, and hence $\|v\| \le I + \varepsilon$. Since this inequality is true for any $\varepsilon$, it follows that $\|v\| \le I$, and the definition of $I$ implies that $\|v\| \ge I$. Therefore $\|v\| = I$. If $I > 0$, then we could find a $v_1 \in JU \cap S$ with $\|v_1\| < \|v\|$, so $I = 0$. There is a $v \in JU \cap S$ with $\|v\| = 0$, and hence $0 \in JU \cap S$, as desired. □

Condition (11.6) is easily connected to the concept of Gâteaux variation, which we now define. Let $\mathbb{B}_1$ and $\mathbb{B}$ be Banach spaces, and let $V$ be a subset of $\mathbb{B}_1$. For a function $J\colon V \to \mathbb{B}$, we define the *Gâteaux variation of $J$ at $u \in V$ in the direction $\psi \in \mathbb{B}_1$* to be

$$J_u(\psi) = \lim_{\varepsilon \to 0} \frac{J(u+\varepsilon\psi) - Ju}{\varepsilon}.$$

Note that this limit is undefined unless $u + \varepsilon\psi \in V$ for all sufficiently small $\varepsilon$, but it does not require that $u$ be an interior point of $V$. Since the choice of $V$ depends on the specific problem that we study, we shall only give a particularly simple example of the Gâteaux variation here. Let $\Omega$ be a bounded domain in $\mathbb{R}^n$ with $\partial\Omega \in H_{1+\alpha}$ for some $\alpha \in (0,1)$, and define $Q$ and $N$ by

$$Qu = a^{ij}(x,u,Du)D_{ij}u + a(x,u,Du) \text{ in } \Omega, \tag{11.7a}$$

$$Nu = b(x,u,Du) \text{ on } \partial\Omega. \tag{11.7b}$$

We suppose that the functions $a^{ij}$, $a$, and $b$ are differentiable with respect to $z$ and $p$ and that these functions, along with their derivatives with respect to $z$ and $p$, are in $H_\alpha(S)$ for any compact subset of $\overline{\Omega} \times \mathbb{R} \times \mathbb{R}^n$ (or $\partial\Omega \times \mathbb{R} \times \mathbb{R}^n$ for $b$ and its derivatives). If we take $\mathbb{B}_1 = H_{2+\delta}^{(-1-\delta)}$ for any $\delta \in (0,\alpha)$ and define $J$ by $Ju = (Qu, Nu)$ (with $\mathbb{B} = H_\delta^{(1-\delta)} \times H_\alpha(\partial\Omega)$), then

$$J_u\psi = (a^{ij}D_{ij}\psi + (a_p^{ij}D_{ij}u + a_p)\cdot D\psi + (a_z^{ij}D_{ij}u + a_z)\psi,\ b_p \cdot D\psi + b_z\psi),$$

where $a^{ij}$, $a$, $b$ and their derivatives are all evaluated at $(x,u,Du)$.

The topology $\mathcal{T}$ will always be chosen in the same way, which we now describe. We start by taking $\Omega$ to be a bounded subset of $\mathbb{R}^n$ with $\partial\Omega \in H_{k+\alpha}$ for some positive integer $k$ and some $\alpha \in (0,1)$, and we take $\mathbb{B}$ to be one of the following spaces

(1) $\mathbb{B} = H_a(\Omega)$ for some non-integer $a \in (0, k+\alpha)$,
(2) $\mathbb{B} = H_b(\partial\Omega)$ for some non-integer $b \in (0, k+\alpha)$,
(3) $\mathbb{B} = H_a(\Omega) \times H_b(\partial\Omega)$ for some non-integers $a$ and $b$ in $(0, k+\alpha)$.

We then take the topology $\mathcal{T}$ to be the one induced by the $L^\infty(S)$ norm, with $S = \Omega$ in the first and third cases and $S = \partial\Omega$ in the second case. Now suppose that $(v_m)$ is a sequence in $\mathbb{B}$ with a uniform bound $M$ on $\|v_m\|$ which converges in $\mathcal{T}$ to some $v \in \mathbb{B}$. If $\mathbb{B} = H_a(\Omega)$, then the Arzela-Ascoli theorem implies that $D^j v_m$ converges uniformly to $D^j v$ for any integer $j < a$, so $\|v\| \le M$. Similar arguments show that $\|v\| \le M$ in the other two cases.

## 11.3 Existence results and estimates for linear equations and nonlinear boundary conditions in spherical caps

In our study of the oblique derivative problem for quasilinear elliptic equations, we shall use some results on the existence and regularity of solutions to related, simpler problems. It will be convenient to study problems in a special geometry first, namely, the spherical cap from Section 2.6. With $\mu_0$ a fixed positive constant and $\kappa$ a fixed constant in $(\mu_0/(1+\mu_0^2)^{1/2}, 1)$, we define, for any $r > 0$,

$$\Gamma(r) = \{x \in \mathbb{R}^n : |x'|^2 + |x^n + \kappa r|^2 < r^2, x^n > 0\},$$

and we write $\Gamma^0(r)$ and $\Gamma^*(r)$ for the subsets of $\partial\Gamma(r)$ on which $|x'|^2 + |x^n + \kappa r|^2 < r^2$ and $|x'|^2 + |x^n + \kappa r|^2 = r^2$, respectively. We also take $[a^{ij}]$ to be a continuous positive-definite matrix-valued functions with eigenvalues in the interval $[\lambda, \mu\lambda]$ for some positive numbers $\lambda$ and $\mu$, and we take $g$ to be a continuous function defined on $\Gamma^0(r) \times \mathbb{R}^{n-1}$ such that

$$|g(x,p') - g(x,q')| \le \mu_0 |p' - q'|$$

for all $x \in \Gamma^0(r)$ and all $p'$ and $q'$ in $\mathbb{R}^{n-1}$.

We now estimate the Hölder norm of the gradient for solutions of various simpler problems. To write our estimates more simply, we abbreviate

$$[\cdot]_{\alpha,\rho} = [\cdot]_{\alpha,\Gamma(\rho)}$$

for any $\rho \in (0, r)$ and $\alpha \in (0, 1)$.

**Lemma 11.4.** *Suppose that $a^{ij}$ and $g$ are independent of $x$. If $v$ satisfies*

$$a^{ij}D_{ij}v = 0 \text{ in } \Gamma(r), \quad D_n v = g(D'v) \text{ on } \Gamma^0(r), \tag{11.8}$$

*then there are positive constants $C$ and $\sigma_0$, both determined only by $\mu_0$, $n$, and $\mu$, such that*

$$[Dv]_{\alpha,r/2} \le Cr^{-1-\alpha}|v|_0 \tag{11.9}$$

*for any $\alpha \in (0, \sigma_0]$.*

***Proof.*** First, for $k \in \{1, \dots, n-1\}$ and $h \in (0, r/8)$, we set

$$v_{k,h}(x) = \frac{1}{h}[v(x + e_k) - v(x)].$$

Then there is a vector field $\beta_{k,h}$ with $\beta^n_{k,h} \equiv 1$ and $|\beta'_{k,h}| \le \mu_0$ such that

$$a^{ij}D_{ij}v_{k,h} = 0 \text{ in } \Gamma(7r/8), \quad \beta_{k,h} \cdot Dv_{k,h} = 0 \text{ on } \Gamma^0(7r/8).$$

The local Hölder estimate Theorem 1.26 then gives a positive constant $\sigma_1$ determined only by $\mu_0$, $n$, $\Lambda/\lambda$ such that

$$[v_{k,h}]_{\alpha,3r/4} \le Cr^{-\alpha}|v_{k,h}|_{0,7r/8}$$

if $\alpha \in (0, \sigma_1)$. Sending $h \to 0$, we obtain

$$[D_k v]_{\alpha,3r/4} \le Cr^{-\alpha}|D_k v|_{0,7r/8}.$$

To continue, we observe that

$$a^{ij}D_{ij}D_n v = 0 \text{ in } \Gamma(3r/4)$$

and

$$[D_n v]_{\alpha,\Gamma^0(3r/4)} \le C\sum_{k=1}^{n-1}[D_k v]_{\alpha,3r/4}.$$

It follows from Corollary 1.29 that there is a constant $\sigma_2$, determined only by $\Lambda/\lambda$ and $n$, such that

$$[D_n v]_{\alpha,r/2} \le C[D_n v]_{\alpha,\Gamma^0(3r/4)}$$

if $\alpha \in (0, \sigma_2)$. With this further restriction on $\alpha$, we have

$$[Dv]_{\alpha,r/2} \le Cr^{-\alpha}|Dv|_0.$$

To continue, we use weighted norms with respect to $d^*$, the distance to $\Gamma^*(r)$. It follows that

$$[v]^{(0)}_{1+\alpha} \le C[v]^{(0)}_1,$$

and the interpolation inequality Lemma 2.3 implies that

$$[v]^{(0)}_1 \le C([v]^{(0)}_{1+\alpha} + |v|_0)^{1/(1+\alpha)}(|v|_0)^{\alpha/(1+\alpha)}.$$

From these two inequalities, we conclude that

$$[v]^{(0)}_{1+\alpha} \le C|v|_0,$$

and (11.9) follows easily from this estimate with $\sigma_0 = \min\{\sigma_1, \sigma_2\}$. □

To proceed, we now present a key extension lemma.

**Lemma 11.5.** *Let $\alpha \in (0,1)$, and suppose there is an nonnegative increasing function $\mu_1$ such that*

$$|g(x,p') - g(y,p')| \le \mu_1(M)|x-y|^\alpha$$

*for all $x$ and $y$ in $\Gamma^0(r)$ and all $p' \in \mathbb{R}^{n-1}$ with $|p'| \le M$. Then, for all $\varphi \in H_{1+\alpha}(\Gamma^*(r))$, there is a function $\varphi_0 \in H_2^{(-1-\alpha)}(\Gamma(r)) \cap C^2(\Gamma(r))$ such that*

$$\varphi_0 = \varphi, \tag{11.10a}$$

$$D\varphi_0 \cdot \tau = D\varphi \cdot \tau \tag{11.10b}$$

*for any vector field $\tau$ tangential to $\Gamma^*(r)$,*

$$D_n\varphi_0 = g(D'\varphi_0) \tag{11.10c}$$

*on $\Gamma^{**}(r)$. Moreover, there is a constant $C$, determined only by $\mu_0$, $\kappa$, $\alpha$, $n$, $\mu_1(|\varphi|_1)$, and $r$ such that*

$$|\varphi_0|_2^{(-1-\alpha)} \le C|\varphi|_{1+\alpha}.$$

***Proof.*** We begin by defining $K$ to be the set of all points $x$ with $|x'| = \sqrt{1-\kappa^2}r$. Then, for any function $f^* \in H_\alpha(\Gamma^{**}(r))$, there is a function $\varphi^* \in H_{1+\alpha}(K)$ such that $\varphi^* = \varphi$ and $D_n\varphi^* = f^*$ on $\Gamma^{**}(r)$. Furthermore, we may take $\varphi^*$ so that $|\varphi^*|_1 \le |\varphi|_1$ and $|\varphi^*|_{1+\alpha} \le C|\varphi|_{1+\alpha}$.

To define a suitable function $f^*$, we fix a point $x_0 \in \Gamma^{**}(r)$ and rotate axes so that, in the new coordinates, $x_0 = (0,\dots,0,\sqrt{1-\kappa^2}r,0)$. Then the vector $(0,\dots,0,-\kappa,\sqrt{1-\kappa^2})$ is tangential to $\Gamma^*(r)$ at $x_0$, and there is a Lipschitz function $G$ with Lipschitz constant no greater than $\mu_0$ such that the equation $v_n = g(x_0,v')$ in the old coordinates is equivalent to $v_n = G(v')$ in the new coordinates. (For simplicity, we suppress the dependence of $G$ on $x_0$ here.) Write $h$ for the directional derivative of $\varphi$ in the direction of $\tau$ at $x_0$, and, for $k = 1,\dots,n-2$, write $v_k$ for the directional derivative of $\varphi$ in the direction of the positive $x_k$ axis. We then define the function $H$ by

$$H(s) = s - G\left(v'', h + \frac{s\sqrt{1-\kappa^2}}{\kappa}\right).$$

A simple calculation shows that $H'$ has a positive lower bound and hence there is a unique solution $t$ of the equation $H(t) = 0$; we define $f^*(x_0)$ to be this number $t$. It is straightforward to verify that $|f^*|_\alpha \le C|\varphi|_{1+\alpha}$.

We now observe that, at $x_0$, the vector $(0,\dots,-1,0)$ is normal to $K$, and the vector $v = (v'', h + t\sqrt{1-\kappa^2}/\kappa, t)$ (where $t$ is as above) satisfies the conditions $v \cdot \tau = D\varphi \cdot \tau$ at $x_0$ for any $\tau$ tangential to $\Gamma^*(r)$ at $x_0$ and $v_n = G(v')$. In other words, $v$ is the appropriate gradient of our extension function $\varphi_0$. We now define $f^{**}$ on $K$ by

$$f^{**}(x', x^n) = D_\tau\varphi(x', 0) + \frac{f^*(x', 0)\sqrt{1-\kappa^2}}{\kappa},$$

where $\tau$ (which depends on $x'$) is the vector tangent to $\Gamma^*(r)$ at $(x', 0)$ which lies in the plane containing the normal to $K$ at $(x', 0)$ and the point $(0, -\kappa r)$. It follows that $f^{**} \in H_\alpha(K)$ with $|f^{**}|_\alpha \le C|\varphi|_{1+\alpha}$.

Finally, we write $J$ for the set of all points $x$ with $|x'| < \sqrt{1-\kappa^2}r$. The proof is completed by taking $\varphi_0$ to be an $H_2^{(-1-\alpha)}(J) \cap C^2(J)$ extension of $\varphi^*$ with normal derivative equal to $f^{**}$. □

Our next step is a conditional Hölder gradient estimate, that is, we prove such an estimate under the additional assumption that a certain class of boundary value problems has a solution. This conditional estimate will be used later to show that the associated problems really do have a solution.

**Lemma 11.6.** *Let $r > 0$ be given, and suppose that there are positive constants $\mu_1$ and $\mu_2$ such that*

$$|g(x, p') - g(y, p')| \le [\mu_1 + \mu_2|p'|]|x - y|^\alpha \tag{11.11}$$

*for all $x$ and $y$ in $\Gamma^0(r)$ and all $p' \in \mathbb{R}^{n-1}$. Suppose also that $g(x, 0) = 0$ for all $x \in \Gamma^0(r)$. Then there is a positive constant $\sigma_1$, determined only by $\mu_0$, $\tau$, $n$, and $\Lambda/\lambda$ such that if the problem*

$$a^{ij}(x_0)D_{ij}\tilde{w} = 0 \text{ in } \Gamma(r), \tag{11.12a}$$

$$\tilde{w} = \tilde{\varphi} \text{ on } \Gamma^*(r), \tag{11.12b}$$

$$D_n\tilde{w} = g(x_0, D'\tilde{w}) \text{ on } \Gamma^0(r) \tag{11.12c}$$

*is solvable for all $x_0 \in \Gamma^0(r)$, all $\tilde{\varphi} \in H_{1+\alpha}$ and some $\alpha \in (0, \sigma_1)$, then there is a constant $C$, determined only by $\mu$, $\mu_0$, $\kappa$, $\alpha$, $n$, $r$, and the modulus of continuity of $a^{ij}$ such that any solution $v$ of the problem*

$$a^{ij}D_{ij}v = f \text{ in } \Gamma(r), \tag{11.13a}$$

$$v = \varphi \text{ on } \Gamma^*(r), \tag{11.13b}$$

$$D_nv = g(D'v) + \psi \text{ on } \Gamma^0(r) \tag{11.13c}$$

*satisfies the estimate*

$$|v|_{1+\alpha} \le C(|f/\lambda|_0^{(1-\alpha)} + |\varphi|_{1+\alpha} + \mu_1 + \mu_2^{(1+\alpha)/\alpha} + |\psi|_\alpha). \tag{11.14}$$

*Moreover, for any $\theta \in (0, 1)$, there is a constant $C$, determined also by $\theta$ such that*

$$|v|_{1+\alpha,\theta r} \le C(|f/\lambda|_0^{1-\alpha} + |v|_0 + \mu_1 + \mu_2^{(1+\alpha)/\alpha} + |\psi|_\alpha). \tag{11.15}$$

***Proof.*** We first suppose that $\mu_2 = 0$ and we define $\hat{d}$ by

$$\hat{d}(x) = r - \frac{|x'|^2}{r} - \frac{(x^n + \kappa r)^2}{r}. \tag{11.16}$$

There are several main estimates to be proved. The first is an estimate of the maximum of $v$, and we introduce the function

$$w_0 = (2 + 2A_0)r^{1+\alpha} - \hat{d}^{1+\alpha} - A_0(x^n)^{1+\alpha} - A_0 r^\alpha x^n$$

with $A_0$ a constant at our disposal to effect this estimate. Elementary calculations show that, by choosing $A_0$ sufficiently large (depending only on $\mu$, $\mu_0$, and $\alpha$), we have

$$a^{ij}D_{ij}w_0 \le -\lambda\alpha(1+\alpha)d^{\alpha-1} \text{ in } \Gamma(r),$$
$$w_0 \ge 0 \text{ on } \Gamma^*(r),$$
$$D_n w_0 \le -\mu_0(1+\alpha)r^\alpha, \quad |D'w_0| \le (1+\alpha)r^\alpha \text{ on } \Gamma^0(r)$$

If we set $F = |f/\lambda|_0^{(1-\alpha)}$, $\Phi_0 = |\varphi|_0$, and $\Psi_0 = |\psi|_0$, and define

$$w = Fw_0 + \Psi_0[r - x^n] + \Phi_0,$$

it follows that

$$a^{ij}D_{ij}(w \pm v) \le -\lambda\alpha(1+\alpha)d^{\alpha-1} \text{ in } \Gamma(r),$$
$$w \pm v \ge 0 \text{ on } \Gamma^*(r),$$
$$\beta \cdot D(w \pm v) \le 0 \text{ on } \Gamma^0(r)$$

for some vector $\beta$ satisfying $\beta^n = 1$ and $|\beta'| \le \mu_0$, so the maximum principle implies that

$$|v| \le w,$$

and the definition of $w$ gives us

$$|v|_0 \le C[F + \Psi_0 + \Phi_0]. \tag{11.17}$$

Our next estimate is on the rate at which $v$ grows near $\Gamma^{**}$, the set of all $x \in \mathbb{R}^n$ with $x^n = 0$ and $|x'|^2 = (1-\kappa^2)r^2$. As in Section 3.6, we write $d^{**}$ for the distance to $\Gamma^{**}$. For convenience, we abbreviate $|\varphi|_{1+\alpha}$ to $\Phi$ and $|\psi|_\alpha$ to $\Psi$. Then, we use Lemma 11.5 to infer that there is a function $\varphi_0 \in C^2(\Gamma(r)) \cap C^1(\overline{\Gamma(r)})$ such that

$$\varphi_0 = \varphi \text{ on } \Gamma^{**},$$
$$D\varphi_0 = Dv \text{ on } \Gamma^{**},$$
$$|D^2\varphi_0| \le C_0\Phi(d^{**})^{\alpha-1}.$$

for some constant $C_0$. Further, we note that the constant $\alpha_1 = \alpha_1(\arctan\kappa, n, \mu_0, \mu)$ from Proposition 3.17 is greater than 1. For $\alpha \in (0, \alpha_1 - 1)$ (to be further specified), we use $\tilde{w}_{1+\alpha}$, the function from Proposition 3.17, and $\varepsilon$, the constant from that proposition. With $A_1$, $A_2$, and $A_3$ positive constants at our disposal, we set

$$w_1 = A_1\tilde{w}_{1+\alpha} - A_2(\hat{d}\,)^{1+\alpha} - A_3(x^n)^{1+\alpha}.$$

Since $a^{ij}D_{ij}\tilde{w}_{1+\alpha} \le 0$, we can choose $A_2$ and $A_3$ so that

$$a^{ij}D_{ij}w_1 \le -C_0 d^{\alpha-1}$$

in $\Gamma(r)$. On $\Gamma^*(r)$, we have $d^* = 0$ and $0 \le x^n \le Cd^{**}$, so we have

$$w_1 \ge C_0(d^{**})^{1+\alpha}$$

on $\Gamma^*(r)$ for all sufficiently large $A_1$. In addition, where $d^{**} = \varepsilon r$, we have $d^* \le Cd^{**}$ and $x^n \le Cd^{**}$, so we also have

$$w_1 \ge C_0(d^{**})^{1+\alpha}$$

where $d^{**} < \varepsilon r$ for $A_1$ sufficiently large. Finally, we have $|D[(d^*)^{1+\alpha}]| \le C(d^{**})^{\alpha}$ on $\Gamma^0(r)$, so we also have

$$\beta \cdot Dw_1 \le -C_0(d^{**})^{\alpha}$$

on the subset of $\Gamma^0(r)$ where $d^{**} \le \varepsilon r$. It follows from the maximum principle that

$$|v - \varphi_0| \le C(F + \Phi + \Psi)w_1 \le C(F + \Phi + \Psi)(d^{**})^{1+\alpha}$$

as long as $d^{**} \le \varepsilon r$, and hence, for any $x_0 \in \Gamma^{**}(r)$, we have

$$|v(x) - v(x_0) - Dv(x_0) \cdot (x - x_0)| \le C(F + \Phi + \Psi)|x - x_0|^{1+\alpha}.$$

Now we fix $x_0 \in \Gamma^0(r)$ and set $R_0 = d^{**}(x_0)$; we also set

$$\Gamma(\rho, x_0) = \{x \in \mathbb{R}^n : x - x_0 \in \Gamma(\rho)\}$$

for $\rho > 0$, with corresponding definitions for $\Gamma^*(\rho, x_0)$ and $\Gamma^0(\rho, x_0)$. Further, we note that there is a positive constant $\varepsilon_1(\kappa)$ such that $\Gamma(\varepsilon_1 R_0, x_0) \subset \Gamma(r)$. Choosing $x_1 \in \Gamma^*(r)$ so that $|x_1 - x_0| = R_0$, we infer that

$$|v(x) - v(x_0) - Dv(x_1) \cdot (x - x_0)| \le C\Phi|x - x_0|^{1+\alpha}$$

if $|x - x_0| \ge \varepsilon_1 R_0$.

If $|x - x_0| < \varepsilon_1 R_0$, we define $v_0$ by $v_0(x) = v(x) - Dv(x_1) \cdot (x - x_0)$, so $v_0$ satisfies the conditions

$$a^{ij}D_{ij}v_0 = f \text{ in } \Gamma(r), \quad D_n v_0 = g_0(x_0, D'v) + \psi_0 \text{ on } \Gamma^0(r),$$

with $g_0$ defined by

$$g_0(p') = g(x, p' + D'v(x_1)) - g(x, D'v(x_1)),$$

and $\psi_0$ defined by

$$\psi_0(x) = \psi(x) - \psi(x_1).$$

With $\delta_1 \in (0,1)$ and $\tau \in (0,1)$ to be chosen, we set $\rho_k = \tau^k \delta_1 R_0$ for any nonnegative integer $k$, and we define three sequences of functions $(v_k)$, $(g_k)$, and $(\psi_k)$, and a sequence of linear polynomials $(L_k)$ inductively as follows. Given $v_k$ and $g_k$, we write $w_k$ to be the solution of

$$\begin{aligned} a^{ij}(x_0)D_{ij}w_k &= 0 \text{ in } \Gamma(\rho_k, x_0), \\ D_n w_k &= g_k(x_0, D'w_k) \text{ on } \Gamma^0(\rho_k, x_0), \\ w_k &= v_k \text{ on } \Gamma^*(\rho_k, x_0). \end{aligned}$$

Such a function exists by virtue of our assumptions because we can take

$$w_k = \tilde{w} - g_{k-1}(x_0, D'L_{k-1})x^n + D'L_{k-1} \cdot x,$$

where $\tilde{w}$ is the solution of

$$\begin{aligned} a^{ij}(x_0)D_{ij}\tilde{w} &= 0 \text{ in } \Gamma(\rho_k, x_0), \\ D_n\tilde{w} &= g_{k-1}(x_0, D'\tilde{w}) \text{ on } \Gamma^0(\rho_k, x_0), \\ \tilde{w} &= v_k + g(x_0, D'L_{k-1})x^n - D'L_{k-1} \cdot x \text{ on } \Gamma^*(\rho_k, x_0) \end{aligned}$$

given by our assumption. We then take $L_k$ to be the first degree Taylor polynomial for $w_k$ centered at $x_0$ and we set $\psi_k(x) = \psi_{k-1}(x) - \psi_{k-1}(x_0)$, noting that $\psi_k(x_0) = 0$ if $k \geq 1$. Using Lemma 11.4 to estimate $w_k$, we can imitate the proof of Theorem 4.5 to find constants $\tau \in (0,1)$ and $\sigma \in (0,1)$ such that

$$\sup_{\Gamma(\tau\rho_k, x_0)} |v_k - L_k| \leq \tau^{1+\sigma} \sup_{\Gamma(\rho_k, x_0)} |v_k| + C\rho_0^{1+\alpha}(F + \Phi + \Psi + \mu_1)$$

for $L_k = Dw_k(x_0)$ if $\delta_1$ is sufficiently small. Proceeding inductively and arguing as in Theorem 4.5, we obtain that, for every $\rho \in (0,r)$, there is a linear polynomial $P_1(\rho, x_0)$ such that

$$\sup_{\Gamma(r)\cap B(x_0,\rho)} |v - P_1(\rho, x_0)| \leq C(|v_0| + F)\frac{\rho^{1+\alpha}}{r^{1+\alpha}}. \tag{11.18}$$

Similar arguments show that this estimate is valid for any $x_0 \in \Gamma^*(r)$ and then for any $x_0 \in \Gamma(r)$. The estimate (11.14) then follows from Lemma 2.9 in this case. By noting that (11.11) implies that

$$|g(x,p') - g(y,p')| \leq (\mu_1 + \mu_2 M)|x-y|^\alpha$$

for all $x$ and $y$ in $\Gamma^0(r)$ and all $p'$ with $|p'| \le M$, we infer from the case $\mu_2 = 0$ that

$$|v|_{1+\alpha} \le C(|v|_0 + F + \Psi + \Phi + \mu_1 + \mu_2|v|_1).$$

The interpolation inequality Lemma 2.4 then implies (11.14).

To complete the proof, we observe that the process used to prove (11.18) implies that, if there is a constant $\mu_3$ such that

$$|g(x,p') - g(y,p')| \le \mu_3 \frac{|x-y|^\alpha}{d^*(x)}$$

for all $x$ and $y$ in $\Gamma^0(r)$ with $|x-y| \le \frac{1}{2}d^*(x)$ and all $p' \in \mathbb{R}^{n-1}$ in place of (11.11), then

$$|v|^{(0)}_{1+\alpha} \le C(|v|_0 + \mu_3 + F + |\psi|^{(1)}_{1+\alpha}).$$

It follows that

$$|v|^{(0)}_{1+\alpha} \le C(|v|_0 + \mu_1 + \mu_2|v|^{(0)}_1 + F + |\psi|^{(1)}_{1+\alpha}),$$

and the proof is completed by using Lemma 2.5. □

We are now ready to prove an existence result for linear equations with nonlinear boundary conditions, based on a variation of the method of continuity. Although its primary application is to show that (11.12) is solvable, we state a more general result because that result is critical to the proof of the solvability of (11.12).

**Lemma 11.7.** *Let $\sigma_1$ be the constant from Lemma 11.6, let $\alpha \in (0,\sigma_1)$, and suppose that there are constants $\mu_1$ and $\mu_2$ such that (11.11) holds. If $a^{ij} \in H^{(0)}_\delta$ for some $\delta \in (0,1)$, then, for any $f \in H^{(1-\alpha)}_\delta(\Gamma(r))$, $h \in H_\alpha(\Gamma^0(r))$ and continuous $\varphi$, there is a solution of*

$$a^{ij}D_{ij}w = f \text{ in } \Gamma(r), \tag{11.19a}$$

$$D_n w = g(x, D'w) + h \text{ on } \Gamma^0(r), \tag{11.19b}$$

$$w = \varphi \text{ on } \Gamma^*(r). \tag{11.19c}$$

***Proof.*** Suppose first that $g \in C^2(\overline{\Gamma^0(r)} \times \mathbb{R}^{n-1})$ and that $g$ is independent of $x$. We write $T$ for the set of all $t \in [0,1]$ such that the problem

$$a^{ij}D_{ij}w = f \text{ in } \Gamma(r),$$

$$D_n w = tg(D'w) + h \text{ on } \Gamma^0(r),$$

$$w = \varphi \text{ on } \Gamma^*(r).$$

has a solution for every $f$, $h$, and $\varphi$ as in the hypotheses of this lemma. From the linear theory, we know that $0 \in T$, and that $w$ is in $H_{2+\delta}^{(-1-\alpha)}$ if $\varphi$ is in $H_{1+\alpha}$.

With $\varepsilon$ a positive constant to be further determined, let $t \in T$, let $s \in [0,1]$, and suppose that $|s-t| \le \varepsilon$. For fixed $f \in H_{\delta}^{(1-\alpha)}$ and $\varphi \in H_{1+\alpha}$, we set $\mathbb{B}_1 = H_{2+\delta}^{(-1-\alpha)}$ and $\mathbb{B} = H_\alpha(\Gamma^0(r))$, and we write $U$ for the subset of all $u \in \mathbb{B}_1$ such that

$$a^{ij}D_{ij}u = f \text{ in } \Gamma(r), \quad u = \varphi \text{ on } \Gamma^*(r).$$

We then define the operator $J\colon U \to \mathbb{B}$ by setting

$$Ju = D_n u - sg(D'u) - h.$$

Our goal is to show that $Ju = 0$ for some $u \in U$, and we shall do so by appropriate application of Theorem 11.3. As a first step, we note that, if $\psi \in H_{2+\delta}^{(-1-\alpha)}$ with

$$a^{ij}D_{ij}\psi = 0 \text{ in } \Gamma(r), \quad u = 0 \text{ on } \Gamma^0(r),$$

then $u + \varepsilon\psi \in U$ for any $\varepsilon$ and that, in this case, we have

$$J_u\psi = D_n\psi - s\frac{\partial g}{\partial p'}(D'u)\cdot D'\psi.$$

From the linear theory, it follows that there is a function $\psi \in \mathbb{B}_1$ such that

$$\begin{aligned} a^{ij}D_{ij}\psi &= 0 \text{ in } \Gamma(r), \\ \psi &= 0 \text{ on } \Gamma^*(r), \\ D_n\psi - s\frac{\partial g}{\partial p'}(D'u)\cdot D'\psi &= -Ju \text{ on } \Gamma^0(r), \end{aligned}$$

and hence, for every $u \in U$, there are $u_1 \in U$ and $\varepsilon \in (0,1)$ such that (11.6) is satisfied with $\eta = 1/2$.

Our next step is to observe that the definition of $J$ implies that

$$D_n u - tg(D'u) = (s-t)g(D'u) + Ju,$$

so Lemma 11.6 implies that

$$\|u\|_{\mathbb{B}_1} \le C[|f|_{\delta}^{(1-\alpha)} + |\varphi|_{1+\alpha} + |(s-t)g(D'u) + Ju|_\alpha).$$

It follows that

$$\|u\|_{\mathbb{B}_1} \le C_1\varepsilon\|u\|_{\mathbb{B}_1} + C_2$$

with $C_1$ determined only by $\mu$, $n$, $r$, $\alpha$, $|a^{ij}|_{\delta}^{(0)}$, and the modulus of continuity of $a^{ij}$, and $C_2$, determined also by $f$, $h$, and $\varphi$. If $C_1\varepsilon \le 1/2$, we conclude that

$$\|u\|_{\mathbb{B}_1} \le 2C_2. \tag{11.20}$$

As mentioned in the previous section, we now take $\mathcal{T}$ to be the topology induced by the $L^\infty(\Gamma^0(r))$ norm, and we take $u_0$ to be the solution of

$$a^{ij}D_{ij}u_0 = f \text{ in } \Gamma(r), u_0 = \varphi \text{ on } \Gamma^*(r), D_n u_0 = 0 \text{ on } \Gamma^0(r).$$

If $(v_m)$ is a sequence in $JU \cap S(Ju_0, 1/2)$, then it's bounded in $\mathbb{B}$, and hence it contains a subsequence which converges in $\mathcal{T}$ to some function $v \in \mathbb{B}$. For each $m$, there is a $u_m \in U$ such that $Ju_m = v_m$ and (11.20) implies that $(u_m)$ is bounded in $\mathbb{B}_1$. Hence, by the Arzela-Ascoli theorem, this sequence contains a subsequence which converges in $H_{1+\alpha/2}(\Gamma(r))$ to some $u \in \mathbb{B}_1$, and hence $Ju_m \to Ju$ uniformly on $\Gamma^0(r)$. It's easy to check that $u \in U$ and that $Ju \in S(u_0, 1/2)$, so $JU \cap S(u_0, 1/2)$ is sequentially compact in $\mathcal{T}$. Theorem 11.3 then implies that there is $u \in U$ such that $Ju = 0$, which means that $u$ solves

$$a^{ij}D_{ij}u = f \text{ in } \Gamma(r), \tag{11.21a}$$

$$D_n u = sg(D'u) + h \text{ on } \Gamma^0(r), \tag{11.21b}$$

$$u = \varphi \text{ on } \Gamma^*(r). \tag{11.21c}$$

We now observe that, if $\varphi$ is only assumed continuous, then there is a sequence of functions $(\varphi_m)$ in $H_{1+\alpha}$ converging uniformly to $\varphi$. If we write $u_m$ for the solution of (11.21) with $\varphi_m$ in place of $\varphi$, then the maximum principle shows that $(u_m)$ is uniformly Cauchy, so it has a limit $v$. From (11.15) and the interior Schauder estimate, we infer that $Dv_m \to Dv$ uniformly on compact subsets of $\Gamma^0(r)$ and $D^2v_m \to D^2v$ uniformly on compact subsets of $\Gamma(r)$, so $s \in T$ if $|s-t| \le \varepsilon$. Since $\varepsilon$ is a constant independent of $t$, it follows that $T = [0,1]$ and therefore (11.12) has a solution.

When $g$ depends on $x$, we repeat the above argument, but we have to use the case of $g$ independent of $x$ to infer (11.15).

To remove the assumption that $g \in C^2$, we observe that there is a sequence of $C^2$ functions $(g_m)$ which converges uniformly on compact subsets of $\Gamma^0(r) \times \mathbb{R}^{n-1}$ to $g$ such that

$$|g_m(x,p') - g_m(x,q')| \le \mu_0|p'-q'|,$$

$$|g_m(x,p') - g_m(y,p')| \le 2[\mu_1 + \mu_2|p'|]|x-y|^\alpha$$

for all $x$ and $y$ in $\Gamma^0(r)$ and all $p'$ and $q'$ in $\mathbb{R}^{n-1}$. If $\varphi \in H_{1+\alpha}$, then the solutions of

$$a^{ij}D_{ij}w_m = f \text{ in } \Gamma(r),$$

$$D_n w_m = g_m(x, D'w_m) + h \text{ on } \Gamma^0(r),$$

$$w_m = \varphi \text{ on } \Gamma^*(r)$$

are uniformly bounded in $H_{2+\delta}^{(-1-\alpha)}$, so there is a subsequence that converges to the unique solution of (11.12). If $\varphi$ is only continuous, then we take a sequence of $H_{1+\alpha}$ functions $(\varphi_m)$ converging uniformly to $\varphi$ and argue as before to conclude that (11.12) has a unique solution, which is the limit of the solutions to

$$\begin{aligned} a^{ij}D_{ij}v_m &= f \text{ in } \Gamma(r), \\ D_n v_m &= g(x, D'v_m) + h \text{ in } \Gamma^0(r), \\ v_m &= \varphi_m \text{ on } \Gamma^*(r). \end{aligned}$$

□

In fact, the same result is true if the equation has lower order terms. To prove this more general result, we use the following version of the Leray-Schauder fixed point, which reduces the existence question to a verification of an *a priori* estimate; we refer to Theorem 11.6 of [64] for a proof.

**Theorem 11.8.** *Let $\mathbb{B}$ be a Banach space, and let $T\colon \mathbb{B} \times [0,1] \to \mathbb{B}$ be a compact mapping with $T(x,0) = 0$ for all $x \in \mathbb{B}$. If there is a constant $M$ such that*

$$\|x\|_{\mathbb{B}} < M$$

*for any $x \in \mathbb{B}$ satisfying $x = T(x,t)$ for some $s \in [0,1]$, then there is a fixed point of $T(\cdot,1)$.*

The following generalization of Lemma 11.7 to linear equations with lower order terms will be used several times in the rest of this book.

**Proposition 11.9.** *Let $\sigma_1$ be the constant from Lemma 11.6, let $\alpha \in (0,\sigma_1)$, and suppose that there are constants $\mu_1$ and $\mu_2$ such that (11.11) holds. Suppose also that there is a nonnegative constant $B_0$ such that $|b| \le \lambda B_0 (x^n)^{\alpha-1}$. Let $a^{ij} \in H_\delta^{(0)}$, $b^i \in H_\delta^{(1-\alpha)}$, and $c \in H_\delta^{(1-\alpha)}$ for some $\delta \in (0,1)$. If $c \le 0$ in $\Gamma(r)$, then, for any $f \in H_\delta^{(1-\alpha)}(\Gamma(r))$, $h \in H_\alpha(\Gamma^0(r))$ and continuous $\varphi$, there is a solution of*

$$a^{ij}D_{ij}w + b^i D_i w + cw = f \text{ in } \Gamma(r), \tag{11.22a}$$

$$D_n w = g(x, D'w) \text{ on } \Gamma^0(r), \tag{11.22b}$$

$$w = \varphi \text{ on } \Gamma^*(r), \tag{11.22c}$$

*and this solution satisfies the estimate*

$$|w|_{1+\alpha} \le C(|f/\lambda|_0^{(1-\alpha)} + |\varphi|_{1+\alpha} + |g(\cdot,0)|_0 + \mu_1 + \mu_2^{(1+\alpha)/\alpha}). \tag{11.23}$$

*Moreover (even without the assumption $c \le 0$), there is a constant $C$, determined only by $n$, $r$, $\alpha$, $\mu$, $\mu_0$, $\kappa$, $B_0$, $|c|_0^{(1-\alpha)}/\lambda$, and the modulus of continuity of $a^{ij}$ such that*

$$|w|_{1+\alpha,r/2} \le C(|w|_0 + |f/\lambda|_0^{(1-\alpha)} + \mu_1 + \mu_2^{(1+\alpha)/\alpha}). \tag{11.24}$$

***Proof.*** Suppose first that $\varphi \in H_{1+\alpha}$. We introduce the Banach space $\mathbb{B} = H_1(\Omega)$, and we define the function $T\colon \mathbb{B} \times [0,1] \to \mathbb{B}$ by saying $u = T(v,t)$ if $u$ solves the boundary value problem

$$\begin{aligned} a^{ij}D_{ij}u + tb^iD_iv + tcv &= tf \text{ in } \Gamma(r), \\ D_nu &= tg(x, D'u) \text{ on } \Gamma^0(r), \\ u &= t\varphi \text{ on } \Gamma^*(r). \end{aligned}$$

It follows from Lemma 11.7 that this problem has a solution in $H_{1+\alpha}$ for any $(v,s) \in \mathbb{B} \times [0,1]$ and hence $T$ is a compact mapping. The linear theory implies that $T(v,0) = 0$ for all $v \in \mathbb{B}$, so we just have to provide a uniform bound on $\|u\|_1$ for all solutions of

$$\begin{aligned} a^{ij}D_{ij}u + tb^iD_iu + tcu &= tf \text{ in } \Gamma(r), \\ D_nu &= tg(x, D'u) \text{ on } \Gamma^0(r), \\ u &= t\varphi \text{ on } \Gamma^*(r). \end{aligned}$$

Our first step is an estimate on the maximum of $u$. To obtain this estimate, we define the function $h$ by

$$h(s) = \frac{1}{B_0}s - \frac{1+B_0}{B_0}\int_0^s \exp(B_0\sigma^\alpha/\alpha)\,d\sigma.$$

Since $h''(s) - B_0s^{\alpha-1}h'(s) = -s^{\alpha-1}$ for all $s > 0$ and $h'(0) = -1$, it follows that $w_0 = h(x^n)$ satisfies

$$\begin{aligned} a^{ij}D_{ij}w_0 + tb^iD_iw_0 &\le -(x^n)^{\alpha-1} \text{ in } \Gamma(r), \\ D_nw_0 \le -1,\ D'w_0 &= 0 \text{ on } \Gamma^0(r), \\ w_0 &\ge -\frac{1+B_0}{B_0}r\exp(B_0r^\alpha/\alpha) \text{ on } \Gamma^*(r). \end{aligned}$$

We now set $F = |f/\lambda|_0^{(1-\alpha)} + |g(\cdot,0)|$ and conclude that there are constants $A_1$ and $A_2$ such that $w_1 = F[A_1 + A_2w_0 - \hat{d}^{1+\alpha}]$ satisfies

$$\begin{aligned} a^{ij}D_{ij}w_1 + tb^iD_iw_1 + tcw_1 &\le f \text{ in } \Gamma(r), \\ D_nw_1 &< g(x, D'w) \text{ on } \Gamma^0(r), \\ w_1 &\ge t\varphi \text{ on } \Gamma^*(r). \end{aligned}$$

It follows that $u \le w_1$, and a similar argument gives a lower bound for $u$, so

$$|u|_0 \le CF. \tag{11.25}$$

Then Lemmata 11.6 and 11.7 imply that

$$|u|_{1+\alpha} \le C(|f/\lambda|_0 + \mu_1 + \mu_2^{(1+\alpha)/\alpha} + |b|_0^{(1-\alpha)}|u|_1),$$

and interpolation, in this case Lemma 2.4, along with (11.25), gives (11.23). It then follows from Lemma 2.14 that

$$|u|^{(-1-\alpha)}_{2+\delta} \le C(|f/\lambda|^{(1-\alpha)}_{\delta} + \mu_1 + \mu_2^{(1+\alpha)/\alpha}).$$

In particular, this implies an upper bound on $|u|_1$, so Theorem 11.8 implies that (11.22) has a solution.

When $\varphi$ is only assumed continuous, we argue as in Lemma 11.7 to infer (11.24) and solvability of (11.22). □

## 11.4 Estimates and existence results for linear equations and nonlinear boundary conditions in general domains

From these preliminaries, we can prove a local Hölder gradient estimate for solutions of such problems in arbitrary $H_{1+\alpha}$ domains.

**Theorem 11.10.** *Let $\omega \in H_{1+\alpha}(B_{n-1}(0,R))$ for some $\alpha \in (0,1)$ and some $R > 0$, and suppose $\omega(0) = 0$. Write $\Omega[R]$ and $\Sigma[R]$ for the subsets of $\mathbb{R}^n$ with $|x| < R$ such that $x^n > \omega(x')$ or $x^n = \omega(x')$, respectively, and suppose that $g$ is a continuous function defined on $\Sigma[R] \times \mathbb{R}^{n-1}$. Let $a^{ij} \in H^{(0)}_{\delta}$, $b^i \in H^{(1-\alpha)}_{\delta}$, $c \in H^{(1-\alpha)}_{\delta}$ with $a^{ij}$ continuous on $\overline{\Omega}$ with eigenvalues in the interval $[\lambda, \mu\lambda]$. (Here the weighted norms are respect to distance from $\Sigma[R]$.) Suppose finally that there are nonnegative constants $\mu_0$, $\mu_1$, and $\mu_2$ such that*

$$|g(x,p') - g(x,q')| \le \mu_0|p' - q'|,$$
$$g(x,p') - g(y,p')| \le [\mu_1 + \mu_2|p'|]|x-y|^{\alpha}$$

*for all $x$ and $y$ in $\partial\Omega$ and all $p$ and $q$ in $\mathbb{R}^{n-1}$. Then there are constants $\alpha_1$, determined only by $\mu$, $\mu_0$, $n$, and $\max|D\omega|$, and $C$, determined only by $|\omega|_{1+\alpha}$, $R$, $|b|^{(1-\alpha)}_0$, and $|c|^{(1-\alpha)}_0$ such that any solution of*

$$a^{ij}D_{ij}v + b^iD_iv + cv = f \text{ in } \Omega[R],$$
$$D_nv = g(x, D'v) \text{ on } \Sigma[R]$$

*satisfies the estimate*

$$|v|_{1+\alpha,\Omega[R/2]} \le C(|v|_0 + |f/\lambda|^{(1-\alpha)}_0 + \mu_1 + \mu_2^{(1+\alpha)/\alpha})$$

*provided $\alpha \le \alpha_1$.*

***Proof.*** We note that, after a suitable $H^{(-1-\alpha)}_{2+\delta}$ change of variable, we may assume that $\omega \equiv 0$ (with appropriate change in the constants) and then this estimate follows from (11.23). □

We also obtain an existence result in $H_{1+\alpha}$ domains.

**Theorem 11.11.** *Let $\partial\Omega \in H_{1+\alpha}$ for some $\alpha \in (0,1)$, and suppose that $g$ is a continuous function defined on $\partial\Omega \times \mathbb{R}^n$ such that $g(x,p) = g(x,q)$ whenever $p - q$ is parallel to $\gamma(x)$. Let $a^{ij} \in H_\delta^{(0)}$, $b^i \in H_\delta^{(1-\alpha)}$, $c \in H_\delta^{(1-\alpha)}$ with $a^{ij}$ continuous on $\overline{\Omega}$ with eigenvalues in the interval $[\lambda, \mu\lambda]$. Suppose finally that there are nonnegative constants $\mu_0$, $\mu_1$, and $\mu_2$ such that*

$$|g(x,p) - g(x,q)| \le \mu_0|p' - q'|,$$
$$|g(x,p) - g(y,p)| \le [\mu_1 + \mu_2|p'|]|x - y|^\alpha$$

*for all $x$ and $y$ in $\partial\Omega$ and all $p$ and $q$ in $\mathbb{R}^n$. (Here $p' = p - \gamma(x)p \cdot \gamma(x)$.) Suppose finally that $c \le -\lambda$. Then, for any $f \in H_\delta^{(1-\alpha)}$, there is a solution of*

$$a^{ij}D_{ij}v + b^iD_iv + cv = f \text{ in } \Omega, \tag{11.26a}$$
$$\gamma \cdot Dv = g(x, D'v) \text{ on } \partial\Omega. \tag{11.26b}$$

*Then there are constants $\alpha_1$, determined only by $\mu$, $\mu_0$, $n$, and $\Omega$, and $C$, determined also by $\Omega$, $|b|_0^{(1-\alpha)}$, and $|c|_0^{(1-\alpha)}$, such that any solution of (11.26) satisfies the estimate*

$$|v|_{1+\alpha} \le C(|f/\lambda|_0^{(1-\alpha)} + \mu_1 + \mu_2^{(1+\alpha)/\alpha}) \tag{11.27}$$

*provided $\alpha \le \alpha_1$.*

***Proof.*** For ease of notation, we write $L$ for the operator defined by $Lw = a^{ij}D_{ij}w + b^iD_iw + cw$.

As in Lemma 11.7, we first assume that $g \in C^2(\partial\Omega \times \mathbb{R})$. Instead of the method of continuity argument used in that lemma, we directly apply Theorem 11.3, so we set $\mathbb{B}_1 = H_{2+\delta}^{(-1-\alpha)}(\Omega)$ and $\mathbb{B} = H_\alpha(\partial\Omega)$, and, for a fixed $f \in H_\delta^{(1-\alpha)}$, we write $U$ for the set of all $u \in \mathbb{B}_1$ such that $Lu = f$ in $\Omega$. Also, we define the operator $J\colon U \to \mathbb{B}$ by

$$Ju = \gamma \cdot Du - g(x, D'u).$$

Now, for any $\psi \in H_{2+\delta}^{(-1-\alpha)}$ with $L\psi = 0$ in $\Omega$, we find that

$$J_u\psi = \gamma \cdot D\psi - \frac{\partial g}{\partial p'} \cdot D'\psi,$$

so, just as in Lemma 11.7, for any $u \in U$, there is $u_1 \in U$ such that (11.6) holds with $\eta = 1/2$. Therefore, Theorem 11.3 reduces the proof of the existence of solutions to the establishment of (11.27).

To verify this estimate, we fix $v \in U$, and let $\rho$ be a regularized distance for $\Omega$ such that $|D\rho| \geq 1/\sqrt{2}$ in some neighborhood of $\partial\Omega$, say the set on which $\rho < \delta_0$. With

$$B_0 = \sup_{\{\rho<\delta_0\}} (a^{ij}D_{ij}\rho + b^i D_i\rho)\rho^{1-\alpha}/\lambda,$$

$h$ as in the proof of Proposition 11.9, and

$$F = |f/\lambda|_0^{(1-\alpha)} + \sup|g(\cdot, 0)| + \sup|Jv|,$$

we find that there is a positive constant $A_0$ such that $w_0 = F[A_0 + h(\rho)]$ satisfies

$$Lw_0 \leq f \text{ in } \Omega,$$
$$\gamma \cdot Dv < \tau g(x, D'v) + Jv \text{ on } \partial\Omega.$$

Hence, we obtain $|v|_0 \leq CF$, and then (11.27) (for sufficiently small $\alpha$) follows from Theorem 11.10 and the interior estimates for the linear equation $Lv = f$. From this estimate, the existence of solutions for (11.26) with $g$ not necessarily $C^2$ follows by a simple approximation argument. □

## 11.5 Mixed boundary value problems for simple quasilinear differential equations and nonlinear boundary conditions in spherical caps

From our study of linear differential equations with nonlinear boundary conditions, we now obtain corresponding results when the equations are quasilinear. In this section, we look only at a special class of quasilinear equations. The choice of this class is made to facilitate our analysis, but it can be used in conjunction with our gradient estimates to obtain results for a much wider class of equations; we provide this application in a later section.

We now take $\mu_0$, $\kappa$, $r$, $\Gamma(r)$, $\Gamma^0(r)$, $\Gamma^*(r)$, and $g$ as in Section 11.3, and we write $[a^{ij}]$ for a matrix-valued function defined on $\Gamma(r) \times \mathbb{R}^n$ which is differentiable with respect to the second variable and which satisfies the hypotheses of Section 11.3, that is, $a^{ij}$ is continuous with respect to both variables and its eigenvalues lie in the interval $[\lambda, \mu\lambda]$ for positive constants $\lambda$ and $\mu$.

Our first step is a local Hölder gradient estimate, which we prove in a slightly different geometric setting.

**Lemma 11.12.** *Suppose that $a^{ij}$ and $g$ are independent of $x$, and suppose that there are constants $K$ and $A_0$ such that*

$$\left|\frac{\partial a^{ij}(p)}{\partial p}\right| \le A_0\lambda$$

*for all $p \in \mathbb{R}^n$ with $|p| \le K$. If $v \in C^3(B^+(R)) \cap C^2(B^+(R) \cup B^0(R))$ satisfies*

$$a^{ij}(Dv)D_{ij}v = 0 \text{ in } B^+(R), \quad D_n v = g(D'v) \text{ on } B^0(R) \tag{11.28}$$

*with $|Dv| \le K$ for some $R > 0$, there are positive constants $C$ and $\sigma$, determined only by $\mu_0$, $n$, $\mu$, $K$, and $A_0$, such that*

$$\operatorname*{osc}_{B^+(\rho)} Dv \le C\left(\frac{\rho}{R}\right)^\sigma \operatorname*{osc}_{B^+(R)} Dv \tag{11.29}$$

*for all $\rho \in (0, R)$.*

***Proof.*** We use the function $N$ from Lemma 10.8, noting that $N$ is a function of $p$ only, and fixing a positive constant $\varepsilon$ which is less than the number $\varepsilon_0$ from that lemma. Moreover, it follows from the proof of that lemma that $N(p) = 0$ whenever $p_n = g(p')$. We also define $V = |p'|^2 + \varepsilon N^2$. (In the notation of Lemma 10.8, we have $V = v_1^2 - 1$.)

Now, with $\theta$ a positive constant at our disposal, we define (for $k = 1, \dots, n$)

$$w_k^\pm = \pm D_k v + \theta V,$$

and we define

$$b = \frac{\partial a^{ij}(Dv)}{\partial p} D_{ij}v.$$

A simple calculation shows that

$$a^{ij}D_{ij}w_k^\pm + b^i D_i w_k^\pm = 2\theta a^{ij} b^{mr} D_{im}v D_{jr}v,$$

where $[b^{km}]$ is the matrix from Lemma 10.8, and hence there is a positive constant $c_1$ such that

$$a^{ij}D_{ij}w_k^\pm + b^i D_i w_k^\pm \ge c_1\theta\lambda|D^2v|^2.$$

Since $|b| \le A_0\lambda|D^2v|$, we conclude that

$$a^{ij}D_{ij}w_k^\pm \ge -\frac{c_2}{\theta}\lambda|Dw_k^\pm|^2$$

for $c_2 = A_0^2/c_1$. The quadratic gradient term is then eliminated by a standard exponential substitution. First, we define

$$W_k^\pm = \sup_{B^+(R)} w_k^\pm.$$

Then we define

$$h_k^\pm = \frac{\theta}{c_2}\left(\exp\left(\frac{c_2}{\theta}(W_k^\pm - w_k^\pm)\right) - 1\right).$$

It follows that

$$a^{ij}D_{ij}h_k^\pm \le 0, \quad h_k^\pm \ge 0 \text{ in } B^+(R). \tag{11.30}$$

Our next step is to prove a suitable weak Harnack inequality for $h_k^\pm$. The case $k < n$ is quite straightforward, so we investigate this one first. To this end, we set

$$\beta = \left(-\frac{\partial g}{\partial p}(D'v), 1\right)$$

and then we have $\beta \cdot DD_k v = 0$. Since $V = |D'v|^2$ and $D_n V = D_n(|D'v|^2)$ on $B^0(R)$, it follows that $\beta \cdot Dw_k^\pm = 0$ on $B^0(R)$, and therefore

$$\beta \cdot Dh_k^\pm = 0 \text{ on } B^0(R).$$

We then infer from Theorem 1.20 that there are positive constants $P$ and $C$ such that

$$\left(R^{-n}\int_{B^+(R/2)} (h_k^\pm)^P\,dx\right)^{1/P} \le C \inf_{B^+(R/2)} h_k^\pm. \tag{11.31}$$

For $k = n$, we define $H^\pm = \inf_{B^0(R)} h_n^\pm$ and

$$\mathcal{H}^\pm(x) = \begin{cases} \min\{h_n^\pm(x), H^\pm\} & \text{if } x \in B^+, \\ H^\pm & \text{if } x \notin B^+. \end{cases}$$

We now apply Theorem 1.28 with $\Omega = \mathbb{R}^n$, $\omega(x') = |x'|$, $\Sigma[R] = \emptyset$, and $G = 0$ to infer that

$$\left(R^{-n}\int_{B^*(0,R/2)} (\mathcal{H}^\pm)^P\right)^{1/P} \le C \inf_{B(0,R/2)} \mathcal{H}^\pm,$$

where $B^*(R/2)$ is the set of all $x \in \mathbb{R}^n$ with $|x| < R$ and $x^n > -|x'|$. Since $|B^*(R/2) \setminus B^+(R/2)| = C(n)R^n$ and $\mathcal{H}^\pm = H^\pm$ on $B^*(R/2) \setminus B^+(R/2)$, it follows that

$$H^\pm \le C \inf_{B(0,R/2)} \mathcal{H}^\pm.$$

Since $\mathcal{H}^\pm \le h_n^\pm$, it follows that

$$\inf_{B^0(R)} h_n^\pm \le C \inf_{B(0.R/2)} h_n^\pm. \tag{11.32}$$

We next observe that

$$W_k^\pm - w_k^\pm \le h_k^\pm \le \exp\left(\frac{c_2}{\theta}(W_k^\pm - w_k^\pm)\right)(W_k^\pm - w_k^\pm).$$

The assumed gradient bound implies that there is a constant $C_\theta$ (determined by the known quantities and $\theta$) such that

$$\exp\left(\frac{c_2}{\theta}(W_k^\pm - w_k^\pm)\right) \le C_\theta.$$

Hence

$$W_k^\pm - w_k^\pm \le h_k^\pm \le C_\theta(W_k^\pm - w_k^\pm).$$

This inequality allows us to rewrite our estimates for $h_k^\pm$ as estimates for $w_k^\pm$. Specifically, we infer that

$$\left(R^{-n}\int_{B_+(R/2)}(W_k^\pm - w_k^\pm)^P\right)^{1/P} \le C_\theta \inf_{B^+(R/2)}(W_k^\pm - w_k^\pm)$$

if $k < n$ and

$$\inf_{B^0(R/2)}(W_n^\pm - w_n^\pm) \le C_\theta \inf_{B^+(R/2)}(W_n^\pm - w_n^\pm).$$

We now introduce some notation to connect these inequalities to oscillations. Specifically, we define

$$\omega_k(\rho) = \operatorname*{osc}_{B^+(\rho)} w_k^+ + \operatorname*{osc}_{B^+(\rho)} w_k^-$$

(for $k = 1, \ldots, n$) and

$$\omega_0(\rho) = \operatorname*{osc}_{B^+(\rho)} V.$$

We now note that $\inf_{B^+(R/2)} w \ge \inf_{B^+(R)} w$ for any function $w$, and, hence, if we set $W = \sup_{B^+(R)} w$, then

$$\begin{aligned}\inf_{B^+(R/2)}(W - w) &= \sup_{B^+(R)} w - \sup_{B^+(R/2)} w \\ &\le \sup_{B^+(R)} w - \inf_{B^+(R)} w + \inf_{B^+(R/2)} w - \sup_{B^+(R/2)} w \\ &= \operatorname*{osc}_{B^+(R)} w - \operatorname*{osc}_{B^+(R/2)} w.\end{aligned}$$

It follows that

$$\inf_{B^+(R/2)}(W_k^+ - w_k^+) + \inf_{B^+(R/2)}(W_k^- - w_k^-) \le \omega_k(R) - \omega_k(R/2).$$

In addition, we have $a^P+b^P \ge C(P)(a+b)^P$ for any nonnegative $a$ and $b$, so

$$\left(\int_S (W_k^+ - w_k^+ + W_k^- - w_k^-)^P\,dx\right)^{1/P} \le \left(\int_S (W_k^+ - w_k^+)^P\,dx\right)^{1/P} + \left(\int_S (W_k^- - w_k^-)^P\,dx\right)^{1/P}.$$

Moreover,

$$\begin{aligned} W_k^+ - w_k^+ + W_k^- - w_k^- &= \sup_{B^+(R)} w_k + \sup_{B^+(R)} w_k^- - 2\theta V \\ &\ge \operatorname*{osc}_{B^+(R)} D_k v - 2\theta\omega_0(R), \end{aligned}$$

so our previous estimates (in particular, (11.31)) give

$$\operatorname*{osc}_{B^+(R)} D_k v \le C_\theta[\omega_k(R) - \omega_k(R/2)] + 2\theta\omega_0(R). \tag{11.33}$$

Next, we observe that

$$\begin{aligned} \inf_{B^0(R/2)} (W_n^+ - w_n^+) &+ \inf_{B^0(R/2)} (W_n^- - w_n^-) \\ &\ge \operatorname*{osc}_{B^+(R)} D_n v - \operatorname*{osc}_{B^0(R/2)} D_n v - 2\theta\omega_0(R). \end{aligned}$$

From the boundary condition, we have that

$$\operatorname*{osc}_{B^0(R/2)} D_n v \le \mu_1 n \sum_{k=1}^{n-1} \operatorname*{osc}_{B^0(R/2)} D_k v,$$

and we know that $\operatorname{osc}_{B^0(R/2)} D_k v \le \operatorname{osc}_{B^+(R)} D_k v$. Hence

$$\begin{aligned} \inf_{B^0(R/2)} (W_n^+ - w_n^+) &+ \inf_{B^0(R/2)} (W_n^- - w_n^-) \\ &\ge \operatorname*{osc}_{B^+(R)} D_n v - n\mu_0 \sum_{k=1}^{n-1} \operatorname*{osc}_{B^+(R/2)} D_k v - 2\theta\omega_0(R). \end{aligned}$$

Using this inequality along with (11.32), we conclude that

$$\operatorname*{osc}_{B^+(R)} D_n v \le \mu_0 n \sum_{k=1}^{n-1} \operatorname*{osc}_{B^+(R)} D_k v + 2\theta\omega_0(R) + C_\theta(\omega_n(R) - \omega_n(R/2)). \tag{11.34}$$

We now set $\bar\omega = \sum_{j=1}^n \omega_j$ and we define

$$\omega(\rho) = \sum_{j=1}^n \operatorname*{osc}_{B^+(\rho)} D_j v.$$

Summing our inequalities (11.33) on $k$ and adding the resultant inequality to (11.34), we find that

$$\omega(R) - 2n(1+\mu_0)\theta\bar{\omega}_0(R) \le C_\theta[\bar{\omega}(R) - \bar{\omega}(R/2)].$$

Now there is a positive constant $K_1$ such that $\omega_0(R) \le K_1\bar{\omega}(R)$ and hence, if we take $\theta = 1/(2n(1+\mu_0)K_1)$, we have

$$\omega(R) \le C[\bar{\omega}(R) - \bar{\omega}(R/2)]$$

and also

$$\frac{1}{2}\bar{\omega}(\rho) \le \omega(\rho) \le \frac{3}{2}\bar{\omega}(\rho)$$

for any $\rho \in (0, R)$. Hence,

$$\bar{\omega}(R) \le C[\bar{\omega}(R) - \bar{\omega}(R/2)],$$

which implies that

$$\bar{\omega}(R/2) \le \delta\bar{\omega}(R)$$

with $\delta = (C-1)/C \in (0,1)$.

We can repeat this argument with $R$ replaced by any $\rho \in (0, R)$ and hence we obtain

$$\bar{\omega}(\tau\rho) \le \tau^\alpha\bar{\omega}(\rho)$$

with $\tau = 1/2$ and $\alpha = \log_{1/2}\delta > 0$. Lemma 1.25 then implies that

$$\bar{\omega}(\rho) \le C\left(\frac{\rho}{R}\right)^\alpha \bar{\omega}(R),$$

and therefore

$$\omega(\rho) \le C\left(\frac{\rho}{R}\right)^\alpha \omega(R).$$

Simple algebra allows us to infer (11.29). □

Note that the constant $C$ in (11.29) depends exponentially on the gradient bound for $v$, and hence gradient bounds cannot be obtained from this one via interpolation. For this reason, we need to modify our approach to various estimates somewhat compared to the methods in the previous sections. In addition, we need to use the following result of Krylov, related to the Hölder gradient estimate for solutions of Dirichlet problems for linear equations up to the boundary. To state this lemma, we fix a point $x_0 \in \Gamma^*(r)$ and we define (for $s > 0$ and $\hat{d}$ defined by (11.16))

$$G(s) = \{x \in \mathbb{R}^n : |x' - x_0'| < s, 0 < \hat{d}(x) < \frac{s}{32n\mu}\},$$

$$G'(s) = \{x \in \mathbb{R}^n : |x' - x_0'| < 2s, \frac{s}{32n\mu} < \hat{d}(x) < \frac{s}{16n\mu}\}.$$

**Lemma 11.13.** *Let $u$ be a nonnegative $C^2(G(2s))$ function for some $s > 0$ so small that $G(2s) \subset \Gamma(r)$, and suppose that there are positive constants $\alpha < 1$ and $F$ such that*

$$a^{ij}D_{ij}u \le \lambda F\hat{d}^{\alpha-1} \text{ in } G(2s). \tag{11.35}$$

*Then*

$$\inf_{G'(s)} \frac{u}{\hat{d}} \le 2 \inf_{G(s)} \frac{u}{\hat{d}} + \frac{16}{\alpha} F s^\alpha. \tag{11.36}$$

***Proof.*** We set $A = \inf_{G'(s)} u/\hat{d}$ and introduce functions

$$w_1 = \left(-3 - 8n\mu\frac{\hat{d}}{s} + \frac{|x' - x_0'|^2}{s^2}\right)\hat{d},$$

$$w_2 = \frac{1}{\alpha}[\left(\frac{16s}{n\mu}\right)^\alpha - \hat{d}^\alpha]\hat{d},$$

$$w = u + \frac{A}{4}w_1 + Fw_2.$$

Direct calculation shows that

$$\begin{aligned} a^{ij}D_{ij}w_1 = &\left(1 - \frac{16n\mu\hat{d}}{s} + \frac{|x' - x_0'|^2}{s^2}\right)a^{ij}D_{ij}\hat{d} - 4a^{ij}D_{ij}\hat{d} \\ &+ \frac{2}{s^2}\mathcal{T}'\hat{d} + \frac{4}{s^2}\sum_{i,j=1}^{n-1} a^{ij}(x^i - x_0^i)D_j\hat{d} + \frac{2}{s^2}\sum_{i=1}^{n-1} a^{in}(x^i - x_0^i)D_n\hat{d} \\ &- \frac{16}{s}\mu a^{ij}D_i\hat{d}D_j\hat{d}, \end{aligned}$$

where $\mathcal{T}' = \sum_{i=1}^{n-1} a^{ii}$. Since $a^{ij}D_{ij}\hat{d} = -2\mathcal{T}/r$ and

$$1 - \frac{16n\mu\hat{d}}{s} + \frac{|x' - x_0'|^2}{s^2} \ge 0$$

in $G(2s)$, we conclude that

$$a^{ij}D_{ij}w_1 \le \frac{\lambda}{s}(\frac{1}{8} + 6\mu|D\hat{d}| - 16\mu|D\hat{d}|^2) + 8n\mu\lambda/r$$

in $G(2s)$. In addition, $|D\hat{d}|^2 = 4 - 4\hat{d}/r$, so $|D\hat{d}| \ge 1$ in $G(2s)$ and hence $a^{ij}D_{ij}w_1 \le 0$ in $G(2s)$.

Next,

$$\begin{aligned} a^{ij}D_{ij}w_2 = &\frac{1}{\alpha}\left(\left(\frac{s}{16n\mu}\right)^\alpha - \hat{d}^\alpha\right)a^{ij}D_{ij}\hat{d} \\ &- \hat{d}^{\alpha-1}(\hat{d}a^{ij}D_{ij}\hat{d} + (1+\alpha)a^{ij}D_i\hat{d}D_j\hat{d}). \end{aligned}$$

The explicit expressions for $a^{ij}D_{ij}\hat{d}$ and $|D\hat{d}|^2$ allow us to infer that

$$a^{ij}D_{ij}w_2 \le \lambda\hat{d}^{\alpha-1}[-4-4\alpha+\frac{\hat{d}}{r}(2n\mu+4+4\alpha)] \le -\lambda\hat{d}^{\alpha-1}.$$

Therefore $a^{ij}D_{ij}w \le 0$ in $G(2s)$ and $w \ge 0$ on $\partial G(2s)$, so the maximum principle implies that $w \ge 0$ in $G(2s)$. In particular $w \ge 0$ in $G(s)$. Since $w_1 \le -2\hat{d}$ and $w_2 \le 8s^\alpha\hat{d}/\alpha$ in $G(s)$, we have $u \ge ((A/2)-(8F/\alpha))\hat{d}$ there. Simple rearrangement gives (11.36). □

This lemma leads in a standard way to an oscillation estimate for $u/\hat{d}$.

**Lemma 11.14.** *With $x_0$ and $\alpha$ as in Lemma 11.13, let $u \in C^2(G(2R))$ for some $R > 0$ such that $G(2R) \subset \Gamma(r)$, and suppose that $|u/\hat{d}|$ is bounded in $G(2R)$. Then there is a constant $\theta_0 \in (0,1]$ determined by $n$ and $\mu$, and, for any $\theta$ and $\alpha$ satisfying $0 < \alpha < \theta < \theta_0$, there is a positive constant $C$, determined also by $\alpha$ and $\theta$, such that, if there is a nonnegative constant $F$ with*

$$|a^{ij}D_{ij}u| \le \lambda F\hat{d}^{\alpha-1} \text{ in } G(2R), \tag{11.37}$$

*then*

$$\operatorname*{osc}_{G(s)}\frac{u}{\hat{d}} \le C[\Big(\frac{s}{R}\Big)^\theta \sup_{G(2R)}\left|\frac{u}{\hat{d}}\right| + Fs^\alpha] \tag{11.38}$$

*for any $s \in (0,2R)$.*

***Proof.*** Note from Lemma 1.25 that it suffices to show that there is a constant $\delta(n,\mu) \in (0,1)$ such that

$$\operatorname*{osc}_{G(s)}\frac{u}{\hat{d}} \le \delta \operatorname*{osc}_{G(2s)}\frac{u}{\hat{d}} + Fs^\alpha + \frac{s}{R}\sup_{G(2R)}\left|\frac{u}{\hat{d}}\right| \tag{11.39}$$

for all $s \in (0,R/2)$.

To prove (11.39), we set

$$m_i = \inf_{G(is)}\frac{u}{\hat{d}}, \quad M_i = \sup_{G(is)}\frac{u}{\hat{d}}$$

for $i = 1,2$, and

$$M = \sup_{G(2R)}\left|\frac{u}{\hat{d}}\right|.$$

We then apply the weak Harnack inequality (Theorem 1.16) to $u - m_2\hat{d}$ in

$$\Sigma = \{x \in \mathbb{R}^n : |x'-x_0'| < \frac{3}{2}s, \ \frac{s}{64n\mu} < \hat{d} < \frac{s}{8n\mu}\},$$

obtaining

$$\left(\frac{1}{|G'(s)|}\int_{G'(s)}(u-m_2\hat{d})^\kappa\,dx\right)^{1/\kappa} \le C[\inf_{G'(s)}(w-m_2)+s^\alpha F+\frac{s^2}{r}|m_2|].$$

It then follows from Lemma 11.13 that

$$\left(\frac{1}{|G'(s)|}\int_{G'(s)}(u-m_2\hat{d})^\kappa\,dx\right)^{1/\kappa} \le Cs[m_1-m_2+s^\alpha F+\frac{s}{R}M],$$

and similar reasoning with $M_2\hat{d}-u$ in place of $u-m_2\hat{d}$ yields

$$\left(\frac{1}{|G'(s)|}\int_{G'(s)}(M_2\hat{d}-u)^\kappa\,dx\right)^{1/\kappa} \le Cs[M_2-M_1+s^\alpha F+\frac{s}{R}M].$$

Adding these inequalities and using the inequality $a^\kappa+b^\kappa \le C(\kappa)(a+b)^\kappa$ for all nonnegative $a$ and $b$ then implies that

$$(M_2-m_2)s \le Cs[m_1-m_2+M_2-M_1+s^\alpha F+\frac{s}{R}M],$$

and simple rearrangement yields (11.39) with $\delta=(C-1)/C$. □

From this lemma, we easily infer a Taylor expansion result for $v$.

**Proposition 11.15.** *There is a constant $\theta_0$ determined only by $n$ and $\mu$, and, for any $\alpha$ and $\theta$ satisfying $0<\alpha<\theta<\theta_0$, there is a constant $C$, determined also by $\alpha$ and $\theta$, such that, if $v \in C^2(\Gamma(r)\cap B(x_0,R))$ for some $x_0 \in \Gamma^*(r)$ and $R \in (0,x_0^n)$ satisfies the conditions*

$$a^{ij}D_{ij}v = 0 \text{ in } \Gamma(r)\cap B(x_0,R), \tag{11.40a}$$

$$v=\varphi \text{ on } \Gamma^*(r)\cap B(x_0,R) \tag{11.40b}$$

*for some $\varphi \in H_{1+\alpha}$, then there is a linear polynomial $L$ (determined also by $v$) such that*

$$\sup_{\Gamma(r)\cap B(x_0,\tau R)}|v-L| \le C[\tau^{1+\theta}\sup_{\Gamma(r)\cap B(x_0,R)}|v|+[\varphi]_{1+\alpha}R^{1+\alpha}) \tag{11.41}$$

*for all $\tau \in (0,1/2)$.*

***Proof.*** For simplicity of notation, we set $\Phi=[\varphi]_{1+\alpha}$. Since $\Gamma(r)\cap B(x_0,s) \subset G(16n\mu s)$ and $G(s) \subset \Gamma(r)\cap B(x_0,2s)$, it suffices to find a linear polynomial $L$ such that

$$\sup_{G(\tau R)}|v-L| \le C[\tau^{1+\theta}\sup_{G(R/2)}|v|+\Phi R^{1+\alpha}] \tag{11.42}$$

for all $\tau \in (0,1/2)$.

Without loss of generality, we may assume that $\varphi$ has been extended to a $C^2(\Gamma(r) \cap B(x_0, R)) \cap C^{1,\alpha}(\overline{\Gamma(r) \cap B(x_0, R)})$, also denoted by $\varphi$ with $|D^2\varphi| \le C\Phi\hat{d}^{\alpha-1}$, and we set $\hat{v} = v - \varphi$.

From Lemma 11.14, we infer that

$$k_0 = \lim_{h\to 0^+} \frac{\hat{v}(x_0 + hD\hat{d}(x_0))}{h}$$

exists and that

$$\sup_{G(\tau R)} \left| \frac{\hat{v}}{\hat{d}} - k_0 \right| \le C[\tau^\theta \sup_{G(R/4)} \left| \frac{\hat{v}}{\hat{d}} \right| + \Phi R^\alpha].$$

In addition, the proof of Lemma 1.31 (with $[R^\alpha \hat{d} - \hat{d}^{1+\alpha}]/\alpha$ in place of $Ry^n - (y^n)^2$) implies that

$$\sup_{G(R/4)} \left| \frac{\hat{v}}{\hat{d}} \right| \le C \left( \frac{1}{R} \sup_{G(R/2)} |\hat{v}| + \Phi R^\alpha \right).$$

Finally, the inequality $0 < \hat{d} < \tau R$ in $G(\tau R)$ implies that

$$\sup_{G(\tau R)} |\hat{v} - k_0\hat{d}| \le \tau R \sup_{G(\tau R)} \left| \frac{\hat{v}}{\hat{d}} - k_0 \right|.$$

By combining these three estimates, we conclude that

$$\sup_{G(\tau R)} |\hat{v} - k_0\hat{d}| \le C[\tau^{1+\theta} \sup_{G(R/2)} |\hat{v}| + \Phi R^{1+\alpha}).$$

We now set $L(x) = \varphi(x_0) + [D\varphi(x_0) + k_0 D\hat{d}(x_0)] \cdot (x - x_0)$. The proof of (11.42) is completed by observing that

$$|v - L| \le |\hat{v} - k_0\hat{d}| + C\Phi R^{1+\alpha} + Ck_0 \frac{\tau^2 R^2}{r}$$

in $G(\tau R)$, that $\sup |h| \le \sup |v|$, and that $|k_0| \le C \sup |v|/R$. □

Our next step is to prove an existence theorem for the problem

$$a^{ij}(Dv)D_{ij}v = 0 \text{ in } \Gamma(r), \tag{11.43a}$$

$$v = \varphi \text{ on } \Gamma^*(r), \tag{11.43b}$$

$$D_n v = g(D'v) \text{ on } \Gamma^0(r). \tag{11.43c}$$

(So here, we explicitly indicated that $a^{ij}$ and $g$ are independent of $x$.)

Our existence result is now very straightforward.

**Lemma 11.16.** *Suppose there is a constant $A_1$ such that*

$$(1 + |p|) \left| \frac{\partial a^{ij}}{\partial p}(x, p) \right| \le A_1$$

*for all $(x, p) \in \Gamma(r) \times \mathbb{R}^n$. Then, for any $\varphi \in C(\Gamma^*(r))$, there is a unique solution of* (11.43).

***Proof.*** We assume initially that $g$ and $a^{ij}$ are $C^2$ functions and that $\varphi \in H_{1+\alpha}$.

In this case, we take $\mathbb{B}$ in Theorem 11.8 to be $H_{1+\alpha}(\Gamma(r))$, and we define the mapping $T$ by $v = T(w,s)$ means that $v$ is the solution of

$$a^{ij}(Dw)D_{ij}v = 0 \text{ in } \Gamma(r),$$
$$v = s\varphi \text{ on } \Gamma^*(r),$$
$$D_n v = sg(D'v) \text{ on } \Gamma^0(r)$$

given by Lemma 11.7. It's immediate from the maximum principle that $T(w,0) = 0$ for any $w \in \mathbb{B}$.

The proof of the uniform bound is a combination of previous estimates; however, due to the regularity in the hypotheses of these estimates, we must first show that any solution of $v = T(v,s)$ is sufficiently smooth in $\Gamma(r) \cap \Gamma^0(r)$. To this end, let $R \in (0,r)$ and set $r' = (R+r)/2$. Then, for all sufficiently small $\eta$ and $k = 1,\dots,n-1$, the functions $v_{k,\eta}$, defined by

$$v_{k,\eta}(x) = \frac{v(x+\eta e_k) - v(x)}{\eta},$$

are solutions of linear problems

$$a^{ij}_{k,\eta}(x)D_{ij}v_{k,\eta} = a_{k,\eta}(x) \text{ in } \Gamma(r'),$$
$$\beta_{k,\eta}(x)\cdot Dv_{k,\eta} = \psi_{k,\eta}(x) \text{ on } \Gamma(r')$$

with the functions $a^{ij}_{k,\eta}$, $\beta_{k,\eta}$, and $\psi_{k,\eta}$ uniformly bounded in $H_\alpha$ and $|a_{k,\eta}|(x^n)^{1-\alpha}$ also uniformly bounded in $L^\infty$. Moreover, the eigenvalues of $[a^{ij}_{k,j}]$ are in $[\lambda,\mu\lambda]$ and we have $\beta^n_{k,\eta} = 1$ and $|\beta'_{k,\eta}| \le \mu_0$. We then obtain a uniform bound on $|v_{k,\eta}|_{1+\alpha;R}$ which implies that $D_k v \in H_{1+\alpha}(\Gamma(R))$. By also using the differential equation to estimate $D_{nn}v$, we see that $v \in H_{2+\alpha}(\Gamma(R))$ and hence $v \in C^2(\Gamma(r)\cap\Gamma^0(r))$. A similar difference quotient argument shows that $v \in C^3(\Gamma(r))$.

We are now ready to estimate solutions of the equation $v = T(v,s)$, which means that $v$ is a solution of

$$a^{ij}(Dv)D_{ij}v = 0 \text{ in } \Gamma(r),$$
$$v = s\varphi \text{ on } \Gamma^*(r),$$
$$D_n v = sg(D'v) \text{ on } \Gamma^0(r).$$

The maximum principle immediately implies that $\sup|v| \le s\sup|\varphi| \le \sup|\varphi|$. The proof of the growth of $v$ near $\Gamma^{**}(r)$ from Lemma 11.6 then shows there is a constant $C$ such that

$$|v(x) - v(x_0)| \le C|x - x_0|$$

for all $x \in \Gamma(r)$ and all $x_0 \in \Gamma^{**}(r)$, and the modification of Lemma 1.31 from Proposition 11.15 provides another constant $C$ such that this estimate holds for all $x \in \Gamma(r)$ and all $x_0 \in \Gamma^*(r)$ as well. It then follows from Theorem 9.9 that $|Dv|$ is uniformly bounded (independent of $s$). Finally, we obtain a global Hölder gradient estimate by following the argument in Lemma 11.6 but with Lemma 11.12 in place of Lemma 11.4 and Proposition 11.15 in place of Lemma 1.35. It then follows that (11.43) has a solution under the stronger hypotheses that $a^{ij}$ and $g$ are $C^2$.

To remove these additional assumptions, we first note that the $H_{1+\alpha}$ estimates depend on $a^{ij}$ and $g$ only through the constants $A_1$, $\lambda$, $\mu$, and $\mu_0$. It follows that (11.43) has a solution under the original hypotheses on $a^{ij}$ and $g$, and the extension to merely continuous $\varphi$ is done as before. □

## 11.6 Hölder gradient estimates for quasilinear equations

We are now ready to prove a Hölder gradient estimate for general quasilinear equations and nonlinear boundary conditions in a spherical cap. Since we will assume that we have a gradient bound for the solution in question, there is no loss of generality in assuming that $a^{ij}$ and $g$ are independent of $z$, and that $a^{ij}$ is independent of $p$ if $|p|$ is larger than some fixed number $K$, which is no smaller than the assumed bound for the gradient of the solution. We prove this fact as part of our Hölder gradient estimate.

We also recall the definition of a generalized half-ball from Section 4.4: An open subset $G$ of $\mathbb{R}^n$ is a *generalized half-ball* if there is an open subset $G'$ of $\mathbb{R}^{n-1}$ such that every point $x = (x', x^n)$ in $G$ satisfies the conditions $x' \in G'$ and $x^n > 0$. We also write $G^0 = G' \times \{0\}$.

**Theorem 11.17.** *Let $G$ be a generalized half-ball, let $[a^{ij}]$ be a matrix-valued function defined on $G \times \mathbb{R}^n$, and let $g$ be a real-valued function defined on $G^0 \times \mathbb{R}^{n-1}$. Suppose that there are positive constants $\alpha < 1$, $\mu$, $\lambda$, $\mu_0$, $\mu_1$, $A_1$, $K$, such that*

$$\lambda|\xi|^2 \le a^{ij}(x,p)\xi_i\xi_j \le \mu\lambda|\xi|^2, \tag{11.44a}$$

$$\left|\frac{\partial a^{ij}(x,p)}{\partial p}\right| \le A_1\lambda \tag{11.44b}$$

*for all $(x,p) \in G \times \mathbb{R}^n$ with $|p| \le K$ and all $\xi \in \mathbb{R}^b$, and*

$$|g(x,p') - g(y,q')| \le \mu_0|p' - q'|, \tag{11.45a}$$

$$|g(x,p') - g(y,p')| \le \mu_1|x - y|^\alpha \tag{11.45b}$$

*for all $x$ and $y$ in $G^0$ and all $p'$ and $q'$ in $\mathbb{R}^{n-1}$ with $|p'|, |q'| \le K$. Suppose also that there is a nonnegative, increasing function $\theta$, defined on $[0, 1/2]$ with $\theta(0) = 0$, such that*

$$|a^{ij}(x,p) - a^{ij}(y,p)| \le \theta\left(\frac{|x-y|}{d^*(x)}\right) \tag{11.46}$$

*for all $x$ and $y$ in $G$ with $|x - y| \le \frac{1}{2}d^*(x)$ and all $p \in \mathbb{R}^n$ with $|p| \le K$. Then is a constant $\alpha' \in (0, 1)$, determined only by $K$, $\alpha$, $A_1$, $n$, $\mu$, and $\mu_0$, and, for any $A_2 \ge 0$, there is a constant $C$, determined also by $A_2$, $\mu_1$, and $G_0 = \sup_{x\in G} |g(x, 0)|$ such that any function $u \in C^2(G) \cap C^1(\overline{G})$ satisfying the conditions $|Du| \le K$ in $G$ and*

$$|a^{ij}(x, Du)D_{ij}u| \le A_2 d^*(x)^{-\alpha-1}(x^n)^{\alpha-1} \text{ in } G, \tag{11.47a}$$

$$D_n u = g(x, D'u) \text{ on } G^0 \tag{11.47b}$$

*obeys the estimate*

$$|Du|_{\alpha'}^{(0)} \le C.$$

***Proof.*** As a first step, we create extensions of $a^{ij}$ and $g$ which satisfy the appropriate conditions for all $p$, $p'$, and $q'$. The argument in Lemma 1.3 gives us a matrix-valued function $[\tilde{a}^{ij}]$ defined on $G \times \mathbb{R}^n$ such that $\tilde{a}^{ij}(x,p) = a^{ij}(x,p)$ for all $(x,p) \in G \times \mathbb{R}^n$ with $|p| \le K$,

$$|\tilde{a}^{ij}(x,p) - \tilde{a}^{ij}(x,q)| \le A_1|p-q|,$$

$$|\tilde{a}^{ij}(x,p) - \tilde{a}^{ij}(y,p)| \le \theta\left(\frac{|x-y|}{d^*(x)}\right)$$

for all $x \in G$, all $p$ and $q$ in $\mathbb{R}^n$ and all $y \in G$ with $|x - y| \le \frac{1}{2}d^*(x)$. It follows that the eigenvalues of $[\tilde{a}^{ij}]$ are in the interval $[\lambda/2, 2\mu\lambda]$ if $|p| \le K + (2A_1)^{-1}$. We then let $\eta$: $[0, \infty] \to [0, 1]$ be a fixed Lipschitz function which is 1 on $[0, K]$ and 0 on $[K + (2A_1)^{-1}, \infty)$ and set

$$A^{ij} = \eta\tilde{a}^{ij} + (1-\eta)\lambda\delta^{ij}.$$

It follows that $A^{ij}$ has eigenvalues in the range $[\lambda/2, 2\mu\lambda]$ and satisfies the conditions

$$\left|\frac{\partial A^{ij}(x,p)}{\partial p}\right| \le C(A_1, K)\lambda,$$

for all $(x,p) \in G \times \mathbb{R}^n$ and

$$|A^{ij}(x,p) - A^{ij}(y,p)| \le \theta\left(\frac{|x-y|}{d^*(x)}\right)$$

for all $x$ and $y$ in $G$ with $|x-y| \le \frac{1}{2}d^*(x)$ and all $p \in \mathbb{R}^n$. Moreover, we have that $A^{ij}$ is independent of $p$ if $|p| \ge K + (2A_1)^{-1}$. For ease of notation, we denote this extension also by $a^{ij}$. Similarly, if we define $g$ for $|p'| \ge K$ by

$$g(x,p') = \eta(|p'|)g(x, Kp'/|p'|),$$

we find that $g$ satisfies (11.45) for all $p'$ and $q'$.

We need to obtain the usual iterative inequality involving the oscillation of the difference between $u$ and suitable linear functions. Specifically, we fix $x_0 \in G^0$ and with $\delta_1 \in (0,1/2)$ to be chosen, we set $R = \delta_1 d^*(x_0)$. With $\tau \in (0,1/2)$ also to be chosen, we define $R_k = \tau^k R$ and we wish to create a sequence of functions $(u_k)$ and a sequence of linear polynomials $(L_k)$ inductively. As a preliminary step, we set $L_{-1}(x) = g(x_0,0)(x^n - x_0^n)$, $u_0 = u - L_{-1}$, $a_0^{ij}(x,p) = a^{ij}(x, p + DL_{-1})$, and $g_0(x,p') = g(x,p') - g(x_0,0)$. Then $u_0$ satisfies the conditions

$$|a_0^{ij}(x, Du_0)D_{ij}u_0| \le A_2(x^n)^{\alpha-1} \text{ in } \Gamma(x_0,R),$$
$$D_n u_0 = g_0(x, D'u_0) \text{ on } \Gamma^0(x_0,R).$$

Note that conditions (11.44), (11.45), and (11.46) hold if we we replace $a^{ij}$ by $a_0^{ij}$ and $g$ by $g_0$. In addition, $a_0^{ij}$ is independent of $p$ if $|p| \ge K + G_0$ and $g_0(x_0,0) = 0$.

We then define $\tilde{v}_k$ to be the solution of

$$a_0^{ij}(x_0, D\tilde{v}_k)D_{ij}\tilde{v}_k = 0 \text{ in } \Gamma(x_0,R_k),$$
$$\tilde{v}_k = u_k \text{ on } \Gamma^*(x_0,R_k),$$
$$D_n\tilde{v}_k = g_0(x_0, D'\tilde{v}_k) \text{ on } \Gamma^0(x_0,R_k)$$

given by Lemma 11.16, and we write $\tilde{L}_k$ for the first order Taylor polynomial for $\tilde{v}_k$ centered at $x_0$. We also define $v_k = \tilde{v}_k - \tilde{L}_k$ and we write $L_k$ for the first order Taylor polynomial for $v_k$ centered at $x_0$. Finally, we define $u_k$ inductively by $u_{k+1} = u_k - L_k$.

To proceed, we note from Theorem 9.9 (and the assumed bound $|Du_0| \le K + G_0$) that, for any $\delta \in (0,1)$, there is a constant $C(\delta)$ such that $|D\tilde{v}_k| \le C(\delta)$ in $\Gamma(x_0,(1-\delta)R_k)$. From Lemma 11.12 applied to $\tilde{v}_k$ in $\Gamma(x_0,R_k/2)$, we infer that there is a constant $\sigma$ such that

$$\sup_{\Gamma(x_0,R_k)} |v_k - L_k| \le C\tau^{1+\sigma} \sup_{\Gamma(x_0,R_k)} |v_k|. \tag{11.48}$$

Then, Corollary 3.13 gives a constant $\zeta$ such that

$$\sup_{\Gamma(x_0,R_k)\setminus\Gamma(x_0,(1-\delta)R_k)} |u_k - v_k| \le C\delta^\zeta[\sup_{\Gamma(x_0,R_k)} |u_k| + (A_2+\mu_1)R_k^{1+\alpha}]$$

for any $\delta \in (0,1)$. (Note that the gradient estimate for $u$ is not used here.)

To estimate the supremum of $|u_k - v_k|$ over $\Gamma(x_0,(1-\delta)R_k)$, we need to estimate the second derivatives of $v_k$, and this estimate is more delicate than before. From the gradient bound for $\tilde{v}_k$ and Lemma 11.12, we conclude that

$$[a_0^{ij}(x_0, D\tilde{v}_k)]^{(0)}_{\sigma,S} \le C(\delta)$$

for any open subset $S$ of $\Gamma(x_0,(1-\delta/2)R_k)$.

By invoking the interior Schauder estimate, Proposition 2.13, we conclude that

$$|D^2\tilde{v}_k(x)| \le C(\delta)R_k^{-\sigma}(x^n)^{\sigma-1}$$

and

$$|D^2\tilde{v}_k| \le C(\delta)R_k^{-1-\sigma}\sup|v_k|(x^n)^{\sigma-1}$$

for all $x \in \Gamma(x_0,(1-\delta)R_k)$. We now write

$$a_0^{ij}(x, Du_0)D_{ij}\tilde{v}_k = b_k^i D_i w_k + f_0,$$

with

$$w_k = \tilde{v}_k - u_0,$$

$$b_k(x) = \begin{cases} \frac{a_0^{ij}(x,Du_0(x)) - a_0^{ij}(x,D\tilde{v}_k(x))}{|Dw_k(x)|} D_{ij}\tilde{v}_k(x)\frac{Dw_k(x)}{|Dw_k(x)|} & \text{if } Dw_k(x) \ne 0, \\ 0 & \text{if } Dw_k(x) = 0, \end{cases}$$

$$f_0 = [a_0^{ij}(x, D\tilde{v}_k) - a_0^{ij}(x_0, D\tilde{v}_k)]D_{ij}\tilde{v}_k.$$

Then

$$|b| \le C(\delta)R_k^{-\sigma}(x^n)^{\sigma-1}$$

and

$$|f_0| \le C(\delta)\theta(\delta_1)R_k^{-1-\sigma}\sup|v_k|(x^n)^{\sigma-1}.$$

We now set

$$M_k = C\delta^{\zeta}[\sup_{\Gamma(x_0,R_k)}|u_k| + (A_2+\mu_1)R_k^{1+\alpha}]$$

and

$$B_k = C(\delta)\theta(\delta_1)R_k^{-1-\sigma}\sup|v_k| + A_2.$$

It follows that there is a vector $\beta_k$ such satisfying $\beta_k^n = 1$ and $|\beta'| \le \mu_0$ such that

$$|a^{ij}D_{ij}w_k + b_k^i D_i w_k| \le \lambda B_k(x^n)^{\sigma-1} \text{ in } \Gamma(x_0,(1-\delta)R_k),$$
$$|w_k| \le M_k \text{ on } \Gamma^*(x_0,(1-\delta)R_k),$$
$$|\beta\cdot Dw_k| \le \mu_1 R_k^{\sigma} \text{ on } \Gamma^0(x_0,(1-\delta)R_k).$$

A simple calculation shows that $h_k$, defined by

$$h_k(x) = [\mu_1 + B_k]R_k^\sigma \int_{x^n}^{R_k} \exp\left(\frac{[C(\delta)+1]s^\sigma}{\sigma R_k^\sigma}\right) ds + M_k,$$

satisfies

$$\begin{gathered} a^{ij}D_{ij}h_k + b_k^i D_i h_k \le -\lambda B_k (x^n)^{\sigma-1} \text{ in } \Gamma(x_0,(1-\delta)R_k), \\ h_k \ge M_k \text{ on } \Gamma^*(x_0,(1-\delta)R_k), \\ \beta \cdot Dh_k \le -\mu_1 R_k^\sigma \text{ on } \Gamma^0(x_0,(1-\delta)R_k), \end{gathered}$$

and then the maximum principle implies that $\pm w_k \le h_k$ in $\Gamma(x_0,(1-\delta)R_k)$. Since

$$\begin{aligned} \int_{x^n}^{R_k} \exp\left(\frac{[C(\delta)+1]s^\sigma}{\sigma R_k^\sigma}\right) ds &\le \int_0^{R_k} \exp\left(\frac{[C(\delta)+1]s^\sigma}{\sigma R_k^\sigma}\right) ds \\ &\le R_k \exp\left(\frac{C(\delta)+1}{\sigma}\right) \end{aligned}$$

and $u_k - v_k = w_k$, it follows that

$$|u_k - v_k| \le C(\delta)R_k^{1+\alpha} + (C(\delta)\theta(\delta_1) + C\delta^\varsigma) \sup_{\Gamma(x_0,R_k)} |u_k|$$

in $\Gamma(x_0,R_k)$. It follows that

$$\sup_{\Gamma(x_0,R_{k+1})} |u_k - L_k| \le C(\delta)R_k^{1+\alpha} + (C(\delta)\theta(\delta_1) + C\delta^\varsigma + C\tau^{1+\alpha}) \sup_{\Gamma(x_0,R_k)} |u_k|.$$

Just as in the proof of Theorem 4.5, we can choose first $\tau$, then $\delta$, and finally $\delta_1$ in a suitable way and then conclude that, for each $s \in (0, \frac{1}{2}d^*(x_0))$, there is a linear polynomial $P(x_0,s)$ such that

$$\sup_{\Gamma(x_0,s)} |u - P(x_0,s)| \le C\frac{s^{1+\sigma}}{r^{1+\sigma}} \sup|u| + Cs^{1+\alpha}.$$

A similar estimate is true if $x_0 \in G$, so the desired estimate follows by virtue of Lemma 2.9. □

We then infer a Hölder gradient estimate near a $C^{1,\alpha}$ portion of $\partial\Omega$.

**Proposition 11.18.** *Let $\omega \in H_{1+\alpha}(B_{n-1}(x_0,R))$ for some $\alpha \in (0,1)$, $x_0 \in \mathbb{R}^n$ with $x_0^n = \omega(x_0')$ and some $R > 0$, let $[a^{ij}]$ be a matrix-valued function defined on $\Omega[R] \times \mathbb{R}^n$, and let $b$ be a real-valued function defined on $\Sigma[R] \times \mathbb{R}^n$. Suppose that there are positive constants $\mu$, $\lambda$, $\mu_0$, $\mu_1$, $A_1$,*

*$K$, and $\chi$ such that (11.44) holds for all $(x,p) \in \Omega[R] \times \mathbb{R}^n$ with $|p| \le K$ and all $\xi \in \mathbb{R}^n$, and*

$$\left|\frac{\partial b(x,p)}{\partial p}\right| \le \mu_0 \frac{\partial b(x,p)}{\partial p} \cdot \gamma, \tag{11.49a}$$

$$\frac{\partial b(x,p)}{\partial p} \cdot \gamma \ge \chi, \tag{11.49b}$$

$$|b(x,p) - b(y,p)| \le \mu_1 \chi |x-y|^\alpha \tag{11.49c}$$

*for all $(x,p) \in \Omega[R] \times \mathbb{R}^n$ with $|p| \le K$ and all $y \in \Omega[R]$. Suppose also that there is a nonnegative, increasing function $\theta$, defined on $[0,1/2]$ with $\theta(0) = 0$ such that (11.46) for all $x$ and $y$ in $\Omega[R]$ with $|x-y| \le \frac{1}{2}d^*(x)$ and all $p \in \mathbb{R}^n$ with $|p| \le K$. Then is a constant $\alpha' \in (0,1)$, determined only by $K$, $\alpha$, $A_1$, $n$, $\mu$, and $\mu_0$, and, for any $A_2 \ge 0$, there is a constant $C$, determined also by $A_2$, $\mu_1$, and $G_0 = \sup_{x\in\Sigma[R]} |b(x,0)|$ such that any function $u \in C^2(\Omega[R]) \cap C^1(\overline{\Omega[R]})$ satisfying the conditions $|Du| \le K$ in $\Omega[R]$ and*

$$|a^{ij}(x,Du)D_{ij}u| \le A_2 d^*(x)^{-\alpha-1} d(x)^{\alpha-1} \text{ in } \Omega[R], \tag{11.50a}$$

$$b(x,Du) = 0 \text{ on } \Sigma[R] \tag{11.50b}$$

*obeys the estimate*

$$|Du|^{(0)}_{\alpha'} \le C.$$

***Proof.*** Note that the boundary condition can be rewritten as

$$Du \cdot \gamma = g(x, D'u)$$

with $D'u = Du - (Du \cdot \gamma)\gamma$ and the hypotheses (11.49) imply that

$$|g(x,p') - g(x,q')| \le C(\mu_0, n)|p' - q'|, \tag{11.51a}$$

$$|g(x,p') - g(y,p')| \le C(\mu_0, \mu_1, n, \alpha)|x-y|^\alpha \tag{11.51b}$$

for all $x$ and $y$ in $\Sigma[R]$ and all $p'$ and $q'$ with $|p'|, |q'| \le K$.

Since all of the hypotheses of this proposition (with (11.51) in place of (11.49)) are invariant under an $H_2^{(-1-\alpha)}$ change of variables, the conclusion follows from Theorem 11.17. □

This local estimate leads to a global Hölder gradient estimate, which we state in a form for applications to existence results.

**Theorem 11.19.** *Let $\partial\Omega \in H_{1+\alpha}$ for some $\alpha \in (0,1)$, let $[a^{ij}]$ be a matrix-valued function defined on $\Omega \times \mathbb{R}^n$, and let $b$ be a real-valued function defined on $\partial\Omega \times \mathbb{R}^n$. Suppose that there are positive constants $\mu$, $\lambda$, $\mu_0$, $\mu_1$, $A_1$,*

*$K$, and $\chi$ such that (11.44) holds for all $(x,p) \in \Omega \times \mathbb{R}^n$ with $|p| \le K$ and all $\xi \in \mathbb{R}^n$, and (11.49) holds for all $(x,p) \in \Omega \times \mathbb{R}^n$ with $|p| \le K$ and all $y \in \Omega$. Suppose also that there is a nonnegative, increasing function $\theta$, defined on $[0,1/2]$ with $\theta(0) = 0$ such that*

$$|a^{ij}(x,p) - a^{ij}(y,p)| \le \theta(|x-y|)$$

*for all $x$ and $y$ in $\Omega$ with $|x-y| \le \frac{1}{2}$ and all $p \in \mathbb{R}^n$ with $|p| \le K$. Then is a constant $\alpha' \in (0,1)$, determined only by $K$, $\alpha$, $A_1$, $n$, $\mu$, $\mu_0$, and $\Omega$, and, for any $A_2 \ge 0$, there is a constant $C$, determined also by $A_2$, $\mu_1$, and $G_0 = \sup_{x\in\partial\Omega} |b(x,0)|$ such that any function $u \in C^2(\Omega) \cap C^1(\overline{\Omega})$ satisfying the conditions $|Du| \le K$ in $\Omega$ and*

$$|a^{ij}(x,Du)D_{ij}u| \le A_2 d(x)^{\alpha-1} \text{ in } \Omega, \tag{11.52a}$$

$$b(x,Du) = 0 \text{ on } \partial\Omega \tag{11.52b}$$

*obeys the estimate*

$$|Du|_{\alpha'} \le C.$$

## 11.7 A basic existence theorem for quasilinear elliptic equations with nonlinear boundary conditions

Our global Hölder gradient estimate leads to a relatively simple but useful existence theorem.

**Theorem 11.20.** *Let $\partial\Omega \in H_{1+\alpha}$ for some $\alpha \in (0,1)$, and let $a^{ij}$ and $a$ be Hölder continuous functions defined on $\overline{\Omega} \times \mathbb{R} \times \mathbb{R}^n \times [0,1]$ such that $a^{ij}$ is Lipschitz with respect to $p$. Let $b$ be a Hölder continuous function defined on $\partial\Omega \times \mathbb{R} \times \mathbb{R}^n \times [0,1]$ such that $b$ is Lipschitz with respect to $p$. Suppose $[a^{ij}]$ is positive definite and that $b$ is oblique. Suppose finally that $a(x,z,p;0) = 0$ for all $(x,z,p) \in \Omega \times \mathbb{R} \times \mathbb{R}^n$ and $b(x,z,p;0) = b(x,w,p;0)$ for all $(x,p) \in \partial\Omega \times \mathbb{R}^n$ and all $z$ and $w$ in $\mathbb{R}$ with $b(x,0,0;0) = 0$ for all $x \in \partial\Omega$. If there is a constant $M$ such that any function $u \in C^2(\Omega) \cap C^1(\overline{\Omega})$ satisfying*

$$a^{ij}(x,u,Du;s)D_{ij}u + (s-1)u + a(x,u,Du;s) = 0 \text{ in } \Omega, \tag{11.53a}$$

$$b(x,u,Du;s) = 0 \text{ on } \partial\Omega \tag{11.53b}$$

*for some $s \in [0,1]$ obeys the estimate*

$$|u|_1 \le M, \tag{11.54}$$

*then there is a solution* $u \in C^2(\Omega) \cap C^1(\overline{\Omega})$ *of*

$$a^{ij}(x,u,Du;1)D_{ij}u + a(x,u,Du;1) = 0 \text{ in } \Omega, \tag{11.55a}$$

$$b(x,u,Du;1) = 0 \text{ on } \partial\Omega. \tag{11.55b}$$

*Moreover this solution is in* $H_{2+\delta}^{(-1-\delta)}$ *for some* $\delta \in (0,1)$.

***Proof.*** With $\delta \in (0,\alpha)$ to be chosen, we take $\mathbb{B} = H_{1+\delta}$, and we define the operator $T\colon \mathbb{B} \times [0,1] \to \mathbb{B}$ by writing $u = T(v,s)$ if $u$ is the solution, given by Theorem 11.11, of the problem

$$a^{ij}(x,v,Dv;s)D_{ij}u + sv - u + a(x,v,Dv;s) = 0 \text{ in } \Omega,$$

$$b(x,v,Du;s) = 0 \text{ on } \partial\Omega.$$

This solution is in $H_{1+\alpha'}$ for some $\alpha' \in (0,1)$, so $T$ will be compact if $\delta < \alpha'$. It is also easy to see that $T(v,0) = 0$ for all $v \in \mathbb{B}$ and that $u = T(u,s)$ for some $s$ if and only if $u$ satisfies (11.53).

Finally, Theorem 11.19 gives a constant $\alpha''$ such that

$$|u|_{1+\alpha''} \le C.$$

If we choose $\delta$ so that $\delta \le \alpha''$ and $\delta \in (0,\alpha')$, it follows from Theorem 11.8 that the operator $T(\cdot,1)$ has a fixed point, which will be a solution of (11.55).

The Hölder gradient estimate just quoted shows that $u \in H_{1+\delta}$, and then the functions $\tilde{a}^{ij}$ and $\tilde{a}$, defined by

$$\tilde{a}^{ij}(x) = a^{ij}(x,u,Du;1), \quad \tilde{a}(x) = a(x,u,Du;1), \tag{11.56}$$

are in $H_\delta$ and hence the interior Schauder estimate implies that $u \in H_{2+\delta}^{(-1-\delta)}$. □

As stated, our existence theorem has an important limitation which prevents us from applying it directly to the examples from the previous chapters. The gradient estimate from Chapters 9 and 10 always assume the solution of our boundary value problem to be in $C^2(\overline{\Omega})$ but the existence theorem refers to solutions that are only assumed to be $H_{1+\delta}$ for some $\delta < 1$. The purpose of the next section is to prove that, under simple assumptions on the data, any $H_{1+\delta}$ solution is actually in $C^2$.

## 11.8 Second derivative Hölder estimates

We now show that solutions of the oblique derivative problem have globally Hölder continuous second derivatives if the data are smooth enough. By adapting the proof of the first part of the proof of Lemma 11.6, we can show that $u \in H_{2+\alpha}$ if $\tilde{a}^{ij}$ and $\tilde{a}$ (defined by (11.56)) are Hölder continuous, $b$ has Hölder continuous first derivatives and $\partial\Omega \in H_{2+\alpha}$, but we wish to obtain this regularity even when $\partial\Omega \in H_{1+\alpha}$. Our proof is a variant of that for Lemma 4.39, and, in this way, we can make the quasilinear regularity theory closer to that of the linear theory.

As before, we need only consider solutions of linear differential equations with nonlinear boundary conditions.

**Proposition 11.21.** *Let $\omega \in H_{1+\alpha}(B_{n-1}(x_0', R))$ for some $\alpha \in (0,1)$, some $x_0 \in \mathbb{R}^n$ with $x_0^n = \omega(x_0')$ and some $R \in (0,1)$. Let $a^{ij} \in H_\alpha^{(0)}(\Omega[R])$, let $a \in H_\alpha^{(2)}(\Omega[R])$, and $b \in H_{1+\alpha}^{(0)}(\Sigma[R] \times B(0,K))$ for some positive constant $K$. Suppose also that $a^{ij}$ has its eigenvalues in the interval $[\lambda, \mu\lambda]$ for some positive constants $\lambda$ and $\mu$, and $b$ is uniformly oblique on $\partial\Omega$. Specifically, suppose that there are nonnegative constants $\chi > 0$, $A_1$, and $B_1$ such that*

$$|a^{ij}|_\alpha^{(0)} + |a|_\alpha^{(2)} \le A_1\lambda, \tag{11.57a}$$

$$\frac{\partial b}{\partial p}(x,p)\cdot\gamma \ge \chi, \tag{11.57b}$$

$$|Db(x,p) - Db(y,q)| \le B_1\chi[|p-q|^\alpha + d^*(x)^{-1-\alpha}|x-y|^\alpha] \tag{11.57c}$$

*for all $(x,p)$ and $(y,q)$ in $\Sigma[R] \times \mathbb{R}$ with $|x-y| \le \frac{1}{2}d^*(x)$ and $|p|, |q| \le K$. Here $Db$ denotes the vector*

$$\left(\frac{\partial b}{\partial x}, \frac{\partial b}{\partial p}\right).$$

*Then, there is a constant $C$, determined only by $\alpha$, $A_1$, $B_1$, $\mu$, $n$, and $|\omega|_{1+\alpha}$ such that*

$$|u|_{2+\alpha}^{(0)} \le C \tag{11.58}$$

*for any solution $u \in H_1(\Omega[R]) \cap C^2(\Omega[R])$ of*

$$a^{ij}D_{ij}u + a = 0 \text{ in } \Omega[R], \tag{11.59a}$$

$$b(x, Du) = 0 \text{ on } \Sigma[R] \tag{11.59b}$$

*with $|Du| \le K$ in $\Omega[R]$.*

***Proof.*** From Proposition 11.18, there is a constant $\alpha'$ such that

$$|u|_{1+\alpha'}^{(0)} \le C.$$

We then take $\delta = \min\{\alpha, \alpha'\}$, and suppose that $u \in H_{2+\delta}^{(0)}$.

We then fix $y_0 \in \Sigma[R]$ and set $R_0 = \theta d^*(y_0)$ with $\theta \in (0,1)$ to be further specified. We now set $E = \Omega[y_0, R_0]$, $E^0 = \Sigma[y_0, R_0]$ and $E^* = \sigma[y_0, R_0]$, and we note that

$$|a^{ij}|^{(0)}_{\delta;E} + |a|^{(2)}_{\delta;E} \le CA_1\lambda.$$

Moreover, we have $\beta \cdot Du = g$ on $E^0$ with $\beta = b_p(y_0, Du(y_0))$ and

$$g(x) = \beta \cdot Du - b(x, Du).$$

A quick calculation shows that

$$\begin{aligned} D_i g(x) - D_i g(y) = {} & [\beta^k - b^k(x, Du(x))][D_{ik}u(x) - D_{ik}u(y)] \\ & + [b^k(y, Du(y)) - b^k(x, Du(x))]D_{ik}u \\ & + [b_i(x, Du(x)) - b_i(y, Du(y))], \end{aligned}$$

and hence

$$|g|^{(1/2)}_{1+\delta;E^0} \le C + C\theta^\delta |u|^{(-1/2)}_{2+\delta;E}.$$

It follows from Lemma 4.39 that

$$|u|^{(-1/2)}_{2+\delta:E} \le C + C\theta^\delta |u|^{(-1/2)}_{2+\delta;E},$$

and hence, if $C\theta^\delta \le 1$, we infer that

$$|u|^{(-1/2)}_{2+\delta;E} \le C.$$

Since this estimate is valid for any $y_0 \in \Sigma[R]$ and a corresponding inequality holds for any ball in $\Omega[R]$, we have

$$|u|^{(0)}_{2+\delta,\Omega[R]} \le C. \tag{11.60}$$

To see that, in fact, $u \in H_{2+\delta}$, we first fix $y \in \Sigma[R]$ and imitate the proof of Lemma 11.7 to conclude that there are a point $y^* \in \mathbb{R}^n$ and a positive number $r$ such that $|y - y^*| < r$, $d^*(y) < R$, and the problem

$$\begin{aligned} a^{ij}D_{ij}v &= a \text{ in } \Omega \cap B(y^*, r), \\ b(x, Dv) &= 0 \text{ on } \partial\Omega \cap B(y^*, r), \\ v &= u \text{ on } \Omega \cap \partial B(y^*, r) \end{aligned}$$

has a solution in $H^{(-1-\alpha')}_{2+\delta}(\Omega \cap B(y^*, r))$. Here the weighted norms are with respect to distance to $\partial B(y^*, r)$, and we have used the preceding argument to estimate $|v|^{(-1-\alpha')}_{2+\delta}$. By the maximum principle $v = u$ and hence $u \in H^{(0)}_{2+\delta}(\Omega[R'])$ for any $R' \in (0, R)$. Since the constant $C$ in (11.60) with $R'$ replaced by $R$ is independent of $R'$, it follows that (11.60) holds. The proof is completing by noting that, by interpolation, we may assume that $\alpha' = 1$.

□

## 11.9 Existence theorems for our examples

In this section, we collect existence theorems for our examples from previous chapters.

For our conormal problems, we recall the abbreviation $v = \sqrt{1+|p|^2}$ Our first example is the capillary problem:

$$\operatorname{div}\left(\frac{Du}{\sqrt{1+|Du|^2}}\right) + B(x,u,Du) = 0 \text{ in } \Omega, \tag{11.61a}$$

$$\frac{Du\cdot\gamma}{\sqrt{1+|Du|^2}} + \psi(x,u) = 0 \text{ on } \partial\Omega. \tag{11.61b}$$

We prove the existence of solutions for this problem under suitable, standard hypotheses on $B$, $\psi$, and $\Omega$.

**Theorem 11.22.** *Suppose that*

$$\lim_{|z|\to\infty} \sup_{(x,p)\in(\Omega\times\mathbb{R}^n)} (\operatorname{sgn} z)B(x,z,p) = -\infty \tag{11.62a}$$

*and that there are functions $B^*$ and $B_0$, along with an increasing nonnegative function $\theta_1$ such that $B = B^* + B_0$ and*

$$|B^*(x,z,p)| + v|B_0(x,z,p)| \le \theta_1(|z|), \tag{11.62b}$$

$$v(|B^*_p(x,z,p)| + |p\cdot B^*_p(x,z,p)|) \le \theta_1(|z|), \tag{11.62c}$$

$$vB^*_z(x,z,p) \le \theta_1^2(|z|), \quad |B^*_x(x,z,p)| \le \theta_1^2(|z|). \tag{11.62d}$$

*Suppose also that there are a constant $c_0 \in (0,1)$ and an increasing, nonnegative function $\psi_0$ with $\psi_0(t) < 1$ for all $t > 0$, such that*

$$\operatorname{sgn} z\psi(x,z) \le c_0, \tag{11.63a}$$

$$|\psi(x,z)| \le \psi_0(|z|), \tag{11.63b}$$

$$\psi_z(x,z) \le 0 \tag{11.63c}$$

*for all $(x,z) \in \partial\Omega\times\mathbb{R}$. In addition, suppose that there is a constant $\alpha \in (0,1)$ such that $B \in H_\alpha(K)$ for any bounded subset $K$ of $\Omega\times\mathbb{R}\times\mathbb{R}^n$, $\psi \in H_{1+\alpha}(K')$ for any bounded subset $K'$ of $\partial\Omega\times\mathbb{R}$, and $\partial\Omega \in H_{2+\alpha}$. Then there is a solution of (11.61) in $C^0(\overline{\Omega})\cap C^2(\Omega)$. Moreover $u \in H_{2+\alpha'}(\Omega)$ for any $\alpha' \in (0,\alpha)$.*

***Proof.*** In this case, we define the coefficients in (11.53) by setting

$$a^{ij}(x,z,p;s) = \frac{\delta^{ij} - \nu^i\nu^j}{v},$$

$$a(x,z,p;s) = sB(x,z,p),$$

$$b(x,z,p;s) = \frac{p\cdot\gamma}{v} + s\psi(x,z),$$

where $\nu = p/v$, so conditions (11.62) and (11.63) are satisfied with $B^*(x,z,p;s) = (s-1)z + sB^*(x,z,p)$, $B_0(x,z,p;s) = sB_0(x,z,p)$. In particular, the limit behavior in (11.62a) is independent of $s$. Hence we may assume for simplicity that $s = 1$ in obtaining our estimates.

We first observe that conditions (8.21) are satisfied with $a_1 = 1$, $b_0 = 0$, $R = 1$, $g \equiv 1$, $G_1(t) = t$, and any $M \geq 1$, so

$$\sup |u| \leq C[\int_\Omega |u|\,dx + M].$$

We now choose $q \geq 1$ so that $c_0 \leq q/(q+1)$ and, with $K_0$ from Lemma 5.17, we set

$$b_2 = 4(q + 1 + K_0).$$

Then there is a positive constant $M$ (determined by the limit behavior in (11.62a)) such that conditions (8.25) are satisfied with $a_1 = 1$, $b_0 = 0$, and $\varepsilon = 1/2$. In this way we obtain from Lemma 8.6 that

$$\int_\Omega |u|\,dx \leq 2^{1/q} qM|\Omega|.$$

In conjunction with our previous estimate, this inequality implies a uniform estimate on $\sup |u|$, and the combination of Lemmata 10.4, 10.5, and 10.6 provides a uniform estimate on $\sup |Du|$. Existence of a solution of (11.61) then follows from Theorem 11.20 and Proposition 11.21. An additional application of Proposition 11.21 gives the desired regularity.

□

Our next example is a uniformly elliptic conormal problem, that is, a problem of the form (10.1):

$$\operatorname{div} A(x,u,Du) + B(x,u,Du) = 0 \text{ in } \Omega, \tag{11.64a}$$

$$A(x,u,Du) \cdot \gamma + \psi(x,u) = 0 \text{ on } \partial\Omega, \tag{11.64b}$$

with $A$, $B$ and $\psi$ satisfying suitable conditions, which we define in term of a positive $C^1(0,\infty)$ function $\Psi$, satisfying the inequality

$$\frac{\Psi(s)}{\theta_0} \leq s\Psi'(s) \leq \theta_0 \Psi(s) \tag{11.65}$$

for some positive constant $\theta_0$ and all $s > 0$, and an increasing function $g$, satisfying

$$g(t) \leq \Psi(t), \quad G(t) \geq \delta t g(t)$$

for all sufficiently large $t$, where $G$ is defined by

$$G(s) = \int_0^s \Psi(t)\,dt.$$

(Note that an easy example of $g$ is $g = \Psi$, but $g = \Psi^{\mu}$ for $\mu \in (\delta, 1)$ is also acceptable.)

**Theorem 11.23.** *Suppose that there are nonnegative constants $\hat{a}_1$, $\hat{c}_0$, and $M_0$ such that*

$$p \cdot A(x, z, p) \geq v\Psi(v) - \hat{a}_1 G(|z|) \tag{11.66}$$

*for all $(x, z, p) \in \Omega \times \mathbb{R} \times \mathbb{R}^n$ with $|z| \geq M_0$ and*

$$z\psi(x, z) \leq \hat{c}_0 G(|z|) \tag{11.67}$$

*for all $(x, z) \in \Omega \times \mathbb{R}$ with $|z| \geq M_0$. Suppose also that there is a nonnegative constant $b_0$ such that, for any positive constant $\hat{b}_2$, there is a constant $M_1$ with*

$$zB(x, z, p) \leq b_0 p \cdot A(x, z, p) - \hat{b}_2 G(|z|) \tag{11.68}$$

*for all $(x, z, p) \in \Omega \times \mathbb{R} \times \mathbb{R}^n$ with $|z| \geq M_1$. Suppose further that there are increasing functions $\theta_2$, $\theta_3$, $\theta_4$, $\psi_0$, and $\psi_1$ such that*

$$|A(x, z, p)| \leq \theta_2(|z|)\Psi(v), \tag{11.69a}$$

$$\frac{\Psi(v)}{v}|\xi|^2 \leq \frac{\partial A^i}{\partial p_j}(x, z, p)\xi_i\xi_j \leq \theta_3(|z|)\frac{\Psi(v)}{v}|\xi|^2, \tag{11.69b}$$

$$v|A_z(x, z, p)| + |A_x(x, z, p)| + |B(x, z, p)| \leq \theta_4(|z|)v\Psi(v) \tag{11.69c}$$

*for all $(x, z, p) \in \Omega \times \mathbb{R} \times \mathbb{R}^n$. Suppose finally that there is a constant $\alpha \in (0, 1)$ such that $B \in H_\alpha(K)$ for any bounded subset $K$ of $\Omega \times \mathbb{R} \times \mathbb{R}^n$, $\partial\Omega \in H_{2+\alpha}$, and $\psi \in H_{1+\alpha}(K')$ for any bounded subset $K'$ of $\partial\Omega \times \mathbb{R}$. Then there is a solution $u \in C^0(\overline{\Omega}) \cap C^2(\Omega)$ of (11.64). In fact , $u \in H_{2+\alpha'}(\Omega)$ for any $\alpha' \in (0, \alpha)$.*

***Proof.*** This time, we define the coefficients in (11.53) by

$$\begin{aligned}
a^{ij}(x, z, p; s) &= s\frac{\partial A^i}{\partial p_j}(x, z, p) + (1 - s)\left(\Psi(v)\delta^{ij} + \frac{\Psi'(v)}{v}p_i p_j\right), \\
a(x, z, p; s) &= s[A^i_z(x, z, p)p_i + A^i_i(x, z, p) + B(x, z, p)], \\
b(x, z, p; s) &= s[A(x, z, p) \cdot \gamma + \psi(x, z)] + (1 - s)\Psi(v)p \cdot \gamma.
\end{aligned}$$

For our maximum estimate on $u$, we invoke Lemmata 8.5 and 8.6 with $g$ replaced by the function $g_s$, defined by

$$g_s(t) = sg(t) + (1 - s),$$

$\theta = 0$, and the following prescription for $a_1$, $c_0$, $b_1$, $b_2$, $R$, and $M$. First, we choose $R \geq 1$ so that

$$\hat{c}_0 \leq \frac{1}{2}R^{-1/\delta}.$$

Since

$$\frac{G(t)}{t^{1/\delta}} \le \frac{G(\tau)}{\tau^{1/\delta}}$$

for any $\tau \le t$, we have (8.21) satisfied with $a_1 \ge \hat{a}_1 R^{1/\delta}/\delta$, $b_1 = 0$, and $c_0 = 1/2$ provided $M \ge M_0$. Finally, set $q = \max\{1, b_0 - 1\}$ and choose $b_2$ so that

$$a_1 \max\{q, q+1-b_0\} + \frac{1}{2}(q+1+K_0)R \le \frac{1}{2}b_2.$$

Then there is a constant $\hat{b}_2$ such that

$$\hat{b}_2 G(|z|) \ge b_2 G(|z|/R);$$

with $M_2$ the constant corresponding to this $\hat{b}_2$, we set $M = \max\{M_1, M_2, R\}$, and then infer a maximum estimate for $|u|$. (Note that, although the function $g_s$ depends on $s$, the constants in the estimates are independent of $s$.)

Then Lemma 10.14, Theorem 8.8, Lemma 10.15, Lemma 10.16 provide the gradient bound, and the proof is completed as before. □

Our third and final conormal problem is for the false mean curvature equation:

$$\operatorname{div}(\exp(\theta v^2)Du) + B(x,u,Du) = 0 \text{ in } \Omega, \tag{11.70a}$$

$$\exp(\theta v^2)Du \cdot \gamma + \psi(x,u) = 0 \text{ on } \partial\Omega \tag{11.70b}$$

for some $\theta > 0$ with $B = O(|p|^2 \exp(\theta |p|^2))$. We also require suitable conditions on $B$ and $\psi$ to derive a bound on $u$.

**Theorem 11.24.** *Let $\theta > 0$ and suppose there are nonnegative constants $b_0$, $c_0 < 1$, and $M_0$, along with an increasing function $b_2$ such that*

$$zB(x,z,p) \le b_0 \exp(\theta v^2)|p|^2 - b_2(|z|)\exp(\theta z^2) \tag{11.71a}$$

*for all $(x,z,p) \in \Omega \times \mathbb{R} \times \mathbb{R}^n$ with $|z| \ge M_0$,*

$$z\psi(x,z) \le \frac{c_0}{4\theta^2 z^2}\exp(\theta z^2) \tag{11.71b}$$

*for all $(x,z) \in \partial\Omega \times \mathbb{R}$ with $|z| \ge M_0$. Suppose also that*

$$\lim_{s\to\infty} b_2(s) = \infty. \tag{11.72}$$

*Suppose further that there is an increasing function $\theta_1$ such that*

$$|B(x,z,p)| \le \theta_1(|z|)\exp(\theta v^2)v^2 \tag{11.73}$$

*for all $(x,z,p) \in \Omega \times \mathbb{R} \times \mathbb{R}^n$. If there is a constant $\alpha \in (0,1)$ such that $B \in H_\alpha(K)$ for any bounded subset $K$ of $\Omega \times \mathbb{R} \times \mathbb{R}^n$, $\partial\Omega \in H_{2+\alpha}$, and $\psi \in H_{1+\alpha}(K')$ for any bounded subset $K'$ of $\partial\Omega \times \mathbb{R}$. Then there is a solution $u \in C^0(\overline{\Omega}) \cap C^2(\Omega)$ of (11.70). In fact, $u \in H_{2+\alpha'}(\Omega)$ for any $\alpha' \in (0,\alpha)$.*

***Proof.*** Now Lemmata 8.5 with

$$a_1 = 0, \quad b_1 = 0, \quad R = 1, M = M_0,$$
$$g(s) = s\exp(\theta s^2), \quad G_1(s) = \frac{1}{4\theta^2 s^2}\exp(\theta s^2),$$

provides a bound on $u$ in terms of $\int G(|u|)\,dx$, and 8.6 (with the same $a_1$, $g$, and $G_1$) bounds this integral if we take $q$ so large that $c_0 \le q/\max\{q, q+1-b_0\}$ and $M$ so large that

$$c_0(q+1+K_0) \le \frac{1}{2}\hat{b}_2(M).$$

(Of course, we also take $b_2 = \hat{b}_2(M)$.) The proof is completed as before, taking into account the gradient bound estimate at the end of Section 10.4. □

We also observe that, by taking $g$ to be a power function (or, more generally, a function satisfying $G(s) \ge \delta s g(s)$ for some $\delta > 0$), we have a similar result. For the reader's convenience, we state the theorem here.

**Theorem 11.25.** *Let $\theta > 0$ and suppose there are nonnegative constants $b_0$, $c_0$, $m$, and $M_0$, along with an increasing function $b_2$ such that*

$$zB(x,z,p) \le b_0\exp(\theta v^2)|p|^2 - b_2(|z|)|z|^m \tag{11.74a}$$

*for all $(x,z,p) \in \Omega \times \mathbb{R} \times \mathbb{R}^n$ with $|z| \ge M_0$,*

$$z\psi(x,z) \le c_0|z|^m \tag{11.74b}$$

*for all $(x,z) \in \partial\Omega \times \mathbb{R}$ with $|z| \ge M_0$. Suppose also that* (11.72) *holds and that there is an increasing function $\theta_1$ such that* (11.73) *holds. If there is a constant $\alpha \in (0,1)$ such that $B \in H_\alpha(K)$ for any bounded subset $K$ of $\Omega \times \mathbb{R} \times \mathbb{R}^n$, $\partial\Omega \in H_{2+\alpha}$, and $\psi \in H_{1+\alpha}(K')$ for any bounded subset $K'$ of $\partial\Omega \times \mathbb{R}$. Then there is a solution $u \in C^0(\overline{\Omega}) \cap C^2(\Omega)$ of* (11.70). *In fact, $u \in H_{2+\alpha'}(\Omega)$ for any $\alpha' \in (0,\alpha)$.*

For problems which are not of conormal form, our existence theorems are still straightforward.

For capillary-type problems, we recall that $g^{ij} = \delta^{ij} - \nu^i\nu^j$, where $\nu = Du/\sqrt{1+|Du|^2}$, and our boundary value problem can be written as

$$g^{ij}D_{ij}u + a(x,u,Du) = 0 \text{ in } \Omega, \tag{11.75a}$$
$$b(x,u,Du) = 0 \text{ on } \partial\omega. \tag{11.75b}$$

Our existence result then takes the following form.

**Theorem 11.26.** *Suppose that $\partial\Omega \in C^3$ and that there are positive constants $\mu_1$ and $M_1$ such that*

$$\mu_1\rho_2 + (\operatorname{sgn} z)a(x,z,p) < 0 \tag{11.76a}$$

*for all $(x,z,p) \in \Omega \times \mathbb{R} \times \mathbb{R}^n$ with $|z| \geq M_1$ and $|p| \leq \mu_1$,*

$$b(x,z,-(\operatorname{sgn} z)\mu_1\gamma) < 0 \tag{11.76b}$$

*for all $(x,z,p) \in \partial\Omega \times \mathbb{R} \times \mathbb{R}^n$ with $|z| \geq M_1$ and $|p| \leq \mu_1$. Suppose also that*

$$\delta a,\ \delta_T a,\ (\bar\delta - 1)a \leq O(1), \tag{11.77a}$$
$$|a_p| = O(1), \tag{11.77b}$$
$$\delta_T b \leq O(b_p \cdot \gamma), \tag{11.77c}$$
$$|b_p| = O(b_p \cdot \gamma), \tag{11.77d}$$
$$\bar\delta b \leq o(b_p \cdot \gamma), \tag{11.77e}$$

*where the limit behavior is uniform on bounded subsets of $\Omega \times \mathbb{R}$. If there is a constant $\alpha \in (0,1)$ such that $b \in H_{1+\alpha}(K')$ for any bounded subset $K'$ of $\partial\Omega \times \mathbb{R} \times \mathbb{R}^n$, then there is a solution $u \in C^0(\overline{\Omega}) \cap C^2(\Omega)$ of (11.75). In fact, $u \in H_{2+\alpha'}$ for any $\alpha' \in (0,\alpha)$.*

***Proof.*** We first note that conditions (11.76a) are just the hypotheses of Lemma 8.3, so we have a bound on $u$.

A quick calculation shows that (9.57) is satisfied with $\Lambda \equiv 1$ and $\mu_1 = 1$. In addition, $\mathcal{E} = |p|^2/v^2$, so

$$\delta_T \mathcal{E} \leq O(v^{-3}),$$
$$|\bar\delta g^{ij}|,\ \bar\delta\mathcal{E} = O(v^{-2}),$$
$$|\delta_T g^{ij}|,\ |\delta_N g^{ij}| = O(v^{-1}).$$

Hence the hypotheses of Theorem 9.12 and Lemma 9.13 are satisfied. The existence and regularity results then follow from Theorem 11.20 and Proposition 11.21. □

For uniformly elliptic problems, we have the most general class of possible boundary conditions.

**Theorem 11.27.** *Suppose that $\partial\Omega \in C^3$ and that there are positive constants $\mu_1$ and $M_1$ such that*

$$\mu_1\rho_2\Lambda(x,z,p) + (\operatorname{sgn} z)a(x,z,p) < 0 \tag{11.78}$$

*for all* $(x, z, p) \in \Omega \times \mathbb{R} \times \mathbb{R}^n$ *with* $|z| \geq M_1$ *and* $|p| \leq \mu_1$ *while* (11.76b) *holds for all* $(x, z, p) \in \partial\Omega \times \mathbb{R} \times \mathbb{R}^n$ *with* $|z| \geq M_1$ *and* $|p| \leq \mu_1$. *Suppose also that*

$$\Lambda = O(\lambda), \quad |a| = O(\lambda|p|^2), \tag{11.79a}$$

$$|p||a_p^{ij}| + |a_z^{ij}| + |a_x^{ij}| = O(\lambda), \tag{11.79b}$$

$$|p||a_p| + |a_z| + |a_x| = O(\lambda|p|^2), \tag{11.79c}$$

*with the limit behavior uniform on bounded subsets of* $\Omega \times \mathbb{R}$ *as* $|p| \to \infty$, *and*

$$|p||b_p| + |b_z| + |b_x| = O(|p|b_p \cdot \gamma) \tag{11.80}$$

*with the limit behavior uniform on bounded subsets of* $\partial\Omega \times \mathbb{R}$ *as* $|p| \to \infty$.

*If there is a constant* $\alpha \in (0, 1)$ *such that* $b \in H_{1+\alpha}(K')$ *for any bounded subset* $K'$ *of* $\partial\Omega \times \mathbb{R} \times \mathbb{R}^n$, *then there is a solution* $u \in C^0(\overline{\Omega}) \cap C^2(\Omega)$ *of* (11.55). *In fact,* $u \in H_{2+\alpha'}$ *for any* $\alpha' \in (0, \alpha)$.

***Proof.*** Now our bounds follow from Lemmata 8.3, 8.4 and Theorems 9.1 and 9.10. □

Finally, we write the false mean curvature equation in the form

$$\Delta u + \theta D_i u D_j u D_{ij} u + a(x, u, Du) = 0$$

for some positive constant $\theta$.

**Theorem 11.28.** *Suppose that* $\partial\Omega \in C^3$ *and that there are positive constants* $\mu_1$ *and* $M_1$ *such that* (11.78) *for all* $(x, z, p) \in \Omega \times \mathbb{R} \times \mathbb{R}^n$ *with* $|z| \geq M_1$ *and* $|p| \leq \mu_1$ *while* (11.76b) *holds for all* $(x, z, p) \in \partial\Omega \times \mathbb{R} \times \mathbb{R}^n$ *with* $|z| \geq M_1$ *and* $|p| \leq \mu_1$. *Suppose also that*

$$|a| = o(|p|^3), \tag{11.81a}$$

$$|p||a_p| + |\delta_T a| = O(|p|^4), \tag{11.81b}$$

$$\delta a, \ (\bar{\delta} - 1)a \leq O(|p|^4) \tag{11.81c}$$

*and*

$$\delta_T b, \ \bar{\delta} b \leq o(|p|b_p \cdot \gamma). \tag{11.82}$$

*If there is a constant* $\alpha \in (0, 1)$ *such that* $b \in H_{1+\alpha}(K')$ *for any bounded subset* $K'$ *of* $\partial\Omega \times \mathbb{R} \times \mathbb{R}^n$, *then there is a solution* $u \in C^0(\overline{\Omega}) \cap C^2(\Omega)$ *of*

$$\Delta u + \theta D_i u D_j u D_{ij} u + a(x, u, Du) = 0 \text{ in } \Omega, \tag{11.83a}$$

$$b(x, u, Du) = 0 \text{ on } \partial\Omega. \tag{11.83b}$$

*In fact,* $u \in H_{2+\alpha'}$ *for any* $\alpha' \in (0, \alpha)$.

***Proof.*** As before, conditions (11.78) and (11.76b) provide a bound on $u$.

Then conditions (9.46) (9.47), and (9.48) are satisfied with $a_*^{ij} = \delta^{ij}$ and $\tau \equiv \theta$, so the gradient bound follows from Theorem 9.8. □

## Notes

The results in Section 11.1 are due to Ladyzhenskaya and Ural′tseva (see Chapter X of [91]).

Theorem 11.3 and the discussion in Section 11.2 come from [119]. The form of Theorem 11.3 given here is a slight restatement of the version of Theorem 1 in [119], which is based on [96]. In turn, [96] is based on ideas of Kirk and Caristi (see [28, 80]) and a result from Miranda [136]. We refer to the notes for Chapter XII of [114] for a more detailed discussion of this situation, but we point out here that the application of the Gâteaux variation is taken directly from [80]. We also point out that our self-contained proof is considerably shorter (and more elementary) than that of its earlier form, Lemma 8.5 of [114]. Furthermore, the nonlinear method of continuity from Section 17.2 of [64] can be used in place of Theorem 11.3 everywhere in this book via appropriate modification of the various proofs.

The approach to the Hölder gradient estimate given here is a more careful working of the method described in Sections XIII.5 and XIII.6 of [114]. As mentioned in the notes to Chapter XIII of [114], a simpler form of the Hölder gradient estimate for solutions of equations in divergence form, but with general oblique boundary conditions, was first given as Theorem 5.2 of [54]. This estimate was extended to nonlinear oblique derivative problems with equations in general form (but with more regularity assumed on the coefficients $a^{ij}$, $a$, and $b$) in Lemma 2.4 of [98]. A variant of the method used in these papers was presented in Theorem 4.1 of [120], but the current method is based on ideas in [111]. Alternative approaches (which also influenced [111]) are given in Theorem 4.5 of [38] and Theorem 2 of [187].

The existence results in Section 11.9 come from several sources. Existence results for the capillary problem have been proved many times with various methods under varying hypotheses; we mention Chapter 7 of [46] as a more complete bibliographic reference for this topic, but list some of the more relevant theorems here. In [183], Ural′tseva proved the existence of a solution for the problem

$$\operatorname{div} A(Du) = \psi(x,u) \text{ in } \Omega, \quad A(Du)\cdot\gamma = 0 \text{ on } \partial\Omega$$

if $A$ behaves (in a suitable way) like $\nu$, if $\psi_z$ has a positive lower bound, and if $\Omega$ is convex. She allowed the boundary condition to have the form $A(Du)\cdot\gamma = \cos\theta$ (for some constant $\theta \in (0,\pi)$) in [184], and, in [185], she considered nonconstant $\theta$ and she removed the convexity hypothesis on $\partial\Omega$.

As previously mentioned, Gerhardt [61] reproduced (via slightly different methods) the results in [185] although he only considered explicitly the case that $A = \nu$; in addition, he studied the obstacle problem. Theorem 11.22 first appears (although in slightly different form) as a brief remark at the end of [97]; see also Theorem 7.6 in [106]. Existence of a solution for the capillary problem under slightly different hypotheses was proved by Spruck [174] in two dimensions and by Simon and Spruck [171] in higher dimensions; these papers allow the contact angle $\theta$ to be zero or $\pi$ on some portions of the boundary, in which case the solution will have an unbounded gradient.

Existence of a solution for uniformly elliptic conormal problems is discussed briefly in Section 10.2 of [91] under more restrictive hypotheses than those of Theorem 11.23 although the method of proof there can be adapted to prove our theorem. A more complete version of Theorem 11.23 is given in [97] (for variational problems) and in [106] (for non-variational problems) although the exact statement here is new; it requires the Hölder estimate Theorem 8.8, which wasn't available at the time of [97, 106].

An existence theorem for the conormal problem with exponential growth first appears in [106] (see p. 62 of that work).

Problems for nondivergence problems (or, as a special case, problems with the equation in divergence form but with boundary conditions that are not conormal) were first discussed by Fiorenza ([53–55]) but he did not provide the estimates necessary to prove an existence theorem. Theorem 11.26 seems not to have been stated before, but it is an easy corollary of the gradient estimate in [107]. A restricted version of Theorem 11.27 first appears in Section 5 of [98], and the exact form given here was first proved as Theorem 7.6 in [120]. Theorem 11.28 is new, but it is an easy corollary of the gradient estimate in [101].

We also mention that a number of existence and regularity results are available under hypotheses not covered by the methods in this chapter. In particular, the method used for proving Hölder gradient estimates for discontinuous $\beta$ in Sections 4.5 and 4.6 has a nonlinear variant, described in [108] for equations in divergence form (focused on the capillary problem). Estimates for nonlinear problems in corners in other types of function spaces have also been extensively studied. We mention here [94] which provides a complete description of the regularity of solutions of the capillary problem near a corner, depending on the data of the problem near that corner.

In addition, by using the refined gradient bounds from [113], it was shown (see page 62 of [113]) that Lipschitz solutions exist for conormal problems in bounded Lipschitz domains satisfying a uniform exterior sphere condition. A variant of this result was proved using different techniques in [30] for convex domains and uniformly elliptic conormal problems; the most important improvement here is that some of the assumptions on the vector field $A$ are relaxed. In addition, [30] proves some gradient bounds for a subclass of $C^1$ domains.

When the boundary condition is linear, then a Hölder gradient estimate was proved, even for fully nonlinear differential equations, in [117]. The proof is essentially that given for linear boundary conditions with linear differential equations in Section 4.1; it would of great interest to prove a corresponding result for nonlinear boundary conditions, but I know of no such results other than Theorem 1.1 from [108] (and its precursor Theorem 1 from [105]), which is only valid when $n = 2$.

## Chapter 12

# Oblique Derivative Problems for Fully Nonlinear Elliptic Equations

## Introduction

We now examine the oblique derivative problem for elliptic equations which are fully nonlinear, that is, equations of the form

$$F(x, u, Du, D^2u) = 0 \tag{12.1}$$

for some function $F$ defined on $\Gamma = \Omega \times \mathbb{R} \times \mathbb{R}^n \times \mathbb{S}^n$, where $\mathbb{S}^n$ denotes the set of all real-valued symmetric matrices. We write $(x, z, p, r)$ for a typical point in $\Gamma$ and we define the operator $F$ by

$$F[u](x) = F(x, u, Du, D^2u). \tag{12.2}$$

We say that the function $F$ is elliptic at a point $(x, z, p, r) \in \Gamma$ if

$$F(x, z, p, r + \eta) > F(x, z, p, r)$$

for any positive definite matrix $\eta$, and we say that the operator $F$ is elliptic at a function $u$ if the function $F$ is elliptic at $(x, u(x), Du(x), D^2u(x))$ for every $x \in \Omega$. If, in addition there are positive constants $\Lambda$ and $\lambda$ such that

$$\lambda \operatorname{tr} \eta \le F(x, z, p, r + \eta) - F(x, z, p, r) \le \Lambda \operatorname{tr} \eta \tag{12.3}$$

for all $(x, z, p, r)$ in some subset $\Sigma$ of $\Gamma$ and all positive definite matrices $\eta$, we say that $F$ is *uniformly elliptic* on $\Sigma$. In this chapter, we shall be concerned primarily with uniformly elliptic equations. A brief discussion of various nonuniformly elliptic equations will be given in the Notes to this chapter.

Note that, if $F$ is continuously differentiable with respect to $r$, then $F$ is elliptic if and only if $F_r$ is positive definite and $F$ is uniformly elliptic if and only if the eigenvalues of $F_r$ are between $\lambda$ and $\Lambda$. For notational convenience, we write

$$F^{ij} = \frac{\partial F}{\partial r_{ij}}.$$

An important subclass of uniformly elliptic equations is given via the set of Bellman operators; we define a Bellman operator from a indexed family of linear operators $(L_\nu)_{\nu\in\mathcal{V}}$ and an indexed family of functions $(f_\nu)_{\nu\in\mathcal{V}}$ for some index set $V$. In this case, we define

$$F[u](x) = \inf_{\nu\in\mathcal{V}}(L_\nu u(x) - f_\nu(x)).$$

Such an operator will be uniformly elliptic if we can write each operator in the form

$$L_\nu w = a_\nu^{ij} D_{ij} w + a_\nu^i D_i w + a_\nu w$$

and if there are positive constants $\lambda$ and $\Lambda$ such that

$$\lambda|\xi|^2 \le a_\nu^{ij}\xi_i\xi_j \le \Lambda|\xi|^2$$

for all $\nu \in \mathcal{V}$ and all $\xi \in \mathbb{R}^n$. This class of operators includes as a special case the Pucci minimal operator $m_\mu$ for $\mu \ge 1$, in which the indexed family of linear operators is the set of all linear operators of the form $Lw = a^{ij}D_{ij}u$ with $[a^{ij}]$ a constant matrix with eigenvalues in the interval $[1,\mu]$. This operator is related to the operator $\mathcal{M}_\mu$ defined in Chapter 3 via the equation $\mathcal{M}_\mu(-u) = -m_\mu(u)$.

We also write our boundary condition in the form

$$b(x,u,Du) = 0, \tag{12.4}$$

and we define the operator $N$ by

$$N[u] = b(x,u,Du). \tag{12.5}$$

As before, the most general definition of obliqueness is that the function $N^*$, defined by

$$N^*(\tau) = b(x,z,p+\tau\beta)$$

is increasing for some oblique vector field $\beta$. For the most part, we shall assume here that $N$ is $C^1$ with respect to $p$ and that $\partial\Omega \in C^1$, so that obliqueness is equivalent to the condition

$$\frac{\partial N}{\partial p}(x,z,p)\cdot\gamma > 0. \tag{12.6}$$

## 12.1 Maximum estimates, comparison principles, and a uniqueness theorem

It is easy to extend the maximum estimates from Chapter 8 so that they apply to fully nonlinear problems, assuming that $F$ is differentiable with respect to $p$. First, we write

$$F[u] = a^{ij}(x, u, Du)D_{ij}u + a(x, u, Du),$$

with $a^{ij}$ and $a$ defined by

$$a^{ij}(x, z, p) = \int_0^1 F^{ij}(x, z, p, tD^2u(x))\,dt,$$
$$a(x, z, p) = F(x, z, p, 0).$$

We also define $\mathcal{E}$ by

$$\mathcal{E}(x, z, p, r) = F^{ij}(x, z, p, r)p_ip_j.$$

From Lemma 8.1, we have the following result.

**Lemma 12.1.** *Let $u \in C^1(\overline{\Omega}) \cap C^2(\Omega)$ be a solution of*

$$F[u] = 0 \text{ in } \Omega, \quad N[u] = 0 \text{ on } \partial\Omega, \tag{12.7}$$

*with $F$ and $N$ given by (12.2) and (12.5). Suppose there are nonnegative constants $\mu_1$, $\mu_2$, $M_1$, and $R$ such that* $\operatorname{diam}\Omega \le R$ *and*

$$\frac{(\operatorname{sgn} z)F(x, z, p, 0)}{\mathcal{E}(x, z, p, r)} \le \frac{\mu_1|p| + \mu_2}{|p|^2} \tag{12.8a}$$

*for all $(x, z, p, r) \in \Omega \times \mathbb{R} \times \mathbb{R}^n \times \mathbb{S}^n$ with $|z| \ge M_1$ and $0 < |p| \le \mu_2 \exp((1+\mu_1)R)$,*

$$(\operatorname{sgn} z)b(x, z, p) < 0 \tag{12.8b}$$

*for all $(x, z, p) \in \partial\Omega \times \mathbb{R} \times \mathbb{R}^n$ with $|z| \ge M_1$ and $|p| \le \mu_2 \exp((1 + \mu_1)R)$. Then*

$$\sup_\Omega |u| \le M_1 + \frac{\mu_2}{1 + \mu_1} \exp((1 + \mu_2)R).$$

In a similar vein, Lemma 8.3 gives the following result if we write $\Lambda(x, z, p, r)$ for the maximum eigenvalue of the matrix $F_r(x, z, p, r)$.

**Lemma 12.2.** *Suppose $\partial\Omega \in C^2$. Suppose also that there are positive constant $\mu_1$ and $M_1$ such that*

$$\mu_1\rho_2\Lambda(x, z, p, r) + (\operatorname{sgn} z)a(x, z, p) < 0 \tag{12.9a}$$

*for all* $(x,z,p) \in \Omega \times \mathbb{R} \times \mathbb{R}^n$ *with* $|z| \geq M_1$ *and* $|p| \leq \mu_1$,

$$b(x,z,-(\operatorname{sgn} z)\mu_1\gamma) < 0 \tag{12.9b}$$

*for all* $(x,z,p) \in \partial\Omega \times \mathbb{R} \times \mathbb{R}^n$ *with* $|z| \geq M_1$ *and* $|p| \leq \mu_1$. *Then*

$$\sup_{\Omega} |u| \leq M_1 + R_0\mu_1.$$

In a slightly different vein, we infer from Theorem 1.9 the following estimate for equations of Monge-Ampère type, which is essentially the same as Lemma 8.2.

**Lemma 12.3.** *Let* $u \in C^2(\Omega) \cap C^1(\overline{\Omega})$ *be a solution of* (12.7) *with* $F$ *given by* (12.64). *Suppose that there are positive, measurable functions* $g \in L^n_{loc}(\mathbb{R}^n)$ *and* $h \in L^n(\Omega)$ *such that*

$$\int_{\Omega} h(x)^n\,dx < \int_{\mathbb{R}^n} g(p)^n\,dx \tag{12.10}$$

*Suppose also that*

$$-(\operatorname{sgn} z)\psi(x,z,p) \leq \frac{h(x)}{g(p)} \tag{12.11a}$$

*for all* $(x,z,p) \in \Omega \times \mathbb{R} \times \mathbb{R}^n$ *and that there is a continuous, increasing function* $M_1$ *such that*

$$(\operatorname{sgn} z)b(x,z,p) < 0 \tag{12.11b}$$

*for all* $(x,z,p) \in \partial\Omega \times \mathbb{R} \times \mathbb{R}^n$ *with* $|z| \geq M_1(|p|)$. *Then there is a number* $t > 0$, *determined only by* $h$ *and* $g$ *such that*

$$\int_{\Omega} h(x)^n\,dx = \int_{\{|p|\leq t\}} g(p)^n\,dx. \tag{12.12}$$

*Moreover,*

$$\sup |u| \leq M_1(t) + t\operatorname{diam}\Omega. \tag{12.13}$$

It is also easy to prove a fully nonlinear comparison theorem, which we shall use to derive a uniqueness result. In fact, we present two slightly different comparison theorems, analogous to the two maximum principles, Lemmata 1.2 and 1.6.

**Theorem 12.4.** *Let* $\partial\Omega \in C^1$, *and let* $u$ *and* $v$ *be in* $C^2(\Omega) \cap C^1(\overline{\Omega})$ *with* $F[u] \geq F[v]$ *in* $\Omega$ *and* $N[u] \geq N[v]$ *on* $\partial\Omega$. *Suppose that* $F$ *is elliptic at* $tu + (1-t)v$ *for all* $t \in [0,1]$ *and decreasing in* $z$ *for each* $(x,p,r) \in \Omega \times \mathbb{R}^n \times \mathbb{S}^n$. *Suppose also that* $F$ *is locally uniformly Lipschitz with respect to* $z$, $p$, *and* $r$ *with* $|F_p|/\lambda$ *locally bounded in* $\Omega \times \mathbb{R} \times \mathbb{R}^n \times \mathbb{S}^n$. *Suppose further that* $N$ *is oblique at* $u$ *and strictly decreasing in* $z$ *for each* $(x,p) \in \partial\Omega \times \mathbb{R}^n$. *Then* $u \leq v$ *in* $\overline{\Omega}$.

***Proof.*** We argue as in Lemma 1.2. Let $w = u - v$, fix $\varepsilon > 0$ and $x_0 \in \Omega$, and set $M_0 = \sup_{\partial\Omega} w$. Then there is a compact set $K \subset \Omega$ such that $x_0$ is an interior point of $K$ and $w \le M_0 + \varepsilon$ on $\partial K$. Since $|F_p|/\lambda$ is locally uniformly bounded, there is a constant $K_1$ such that $|F_p| \le K_1\lambda$ in $K$ and, with $\gamma = 1 + K_1$, we define the function $\varphi$ by

$$\varphi(x) = \exp(-\gamma x^1).$$

We now write $\Omega^+$ for the subset of $K$ on which $w > 0$, and suppose that $w + \varepsilon\varphi$ has a positive maximum at some $x_1 \in \overline{\Omega^+}$. If $x_1$ is an interior point of $\Omega^+$, then we have $Du + \varepsilon D\varphi = Dv$ and $D^2u + \varepsilon D^2\varphi \ge D^2v$ at $x_1$, so

$$F(x_1, u(x_1), Du(x_1) + \varepsilon D\varphi(x_1), D^2u(x_1) + \varepsilon\varphi(x_1)) \le F[v](x_1).$$

Using the conditions on $F$, we find that

$$\begin{aligned} F(x_1, u(x_1), Du(x_1) &+ \varepsilon D\varphi(x_1), D^2u(x_1) + \varepsilon\varphi(x_1)) \\ &\ge F[u](x_1) + \varepsilon\lambda[\Delta\varphi(x_1) - |D\varphi(x_1)], \end{aligned}$$

and direct computation gives

$$\Delta\varphi(x_1) - |D\varphi(x_1)| > 0,$$

so $F[u](x_1) < F[v](x_1)$. Hence $w$ must attain its maximum over $\overline{\Omega^+}$ at a boundary point. Now the proof of Lemma 1.2 implies that, if $w$ has a positive maximum, then it must occur at some $x_2 \in \partial\Omega$. But then, $Du(x_2) = Dv(x_2) + \tau\gamma(x_2)$ for some nonpositive constant $\tau$, so

$$N[u](x_2) \le b(x_2, u(x_2), Dv(x_2))$$

because $N$ is oblique and

$$b(x_2, u(x_2), Dv(x_2)) < N[v](x_2)$$

because $N$ is strictly decreasing. These inequalities contradict the assumption that $N[u] \ge N[v]$, and hence $w$ can't have a positive maximum so $u \le v$ in $\Omega$. □

**Theorem 12.5.** *Let $\Omega$ be a bounded Lipschitz domain, and let $u$ and $v$ be in $C^2(\Omega) \cap C^1(\overline{\Omega})$ with $F[u] \ge F[v]$ in $\Omega$ and $N[u] \ge N[v]$ on $\partial\Omega$. Suppose that $F$ is elliptic at $tu + (1-t)v$ for all $t \in [0,1]$ and decreasing in $z$ for each $(x, p, r) \in \Omega \times \mathbb{R}^n \times \mathbb{S}^n$. Suppose also that $F$ is locally uniformly Lipschitz with respect to $z$, $p$, and $r$ with $|F_p|/|F_z|$ and $|F_r|/|F_z|$ bounded in $\Omega \times \mathbb{R} \times \mathbb{R}^n \times \mathbb{S}^n$ and $F_z < 0$. Suppose further that $N$ is oblique at $u$ and decreasing in $z$ for each $(x, p) \in \partial\Omega \times \mathbb{R}^n$ and that there is a constant $\varepsilon \in (0, 1)$ such that the vector field $\partial b(x, z, p)/\partial p$ has modulus of obliqueness at least $\varepsilon$ for any $(z, p) \in \mathbb{R} \times \mathbb{R}^n$. Then $u \le v$ in $\overline{\Omega}$.*

***Proof.*** From Lemma 1.5, there is a function $\rho$ such that

$$\frac{\partial N}{\partial p}(x,z,p)\cdot D\rho(x) > 0$$

for all $x \in \partial\Omega$. With $\delta > 0$ at our disposal, we now set

$$w = u - v - \delta\exp(-\delta\rho),$$

and we suppose that $w$ has a positive maximum at $x_0 \in \overline{\Omega}$.

If $x_0 \in \Omega$, then, at $x_0$, we have

$$F[u](x_0) \le F(x_0, v+\delta\psi, Dv - \delta^2\psi D\rho, D^2v - \delta^2\psi D^2\rho + \delta^3\psi D\rho\otimes D\rho),$$

where we have written $\psi$ for $\exp(-\delta\rho)$ and $D\rho\otimes D\rho$ is the matrix $[D_i\rho D_j\rho]$. The hypotheses on $F$ imply that

$$\begin{aligned} F(x_0, v+\delta\psi, Dv - \delta^2\psi D\rho, D^2v - \delta^2\psi D^2\rho + \delta^3\psi D\rho\otimes D\rho) \\ \le F[v](x_0) + \delta\psi F_z[x_0,\Xi](1 - C_1\delta) \end{aligned}$$

for $\Xi$ some point on the line segment in $\mathbb{R}\times\mathbb{R}^n\times\mathbb{S}^n$ joining $(v, Dv, D^2v)$ and $(v+\delta\psi, Dv - \delta^2\psi D\rho, D^2v - \delta^2\psi D^2\rho + \delta^3\psi D\rho\otimes D\rho)$ and $C_1$ some positive constant. It follows that $x_0$ can't be in $\Omega$ if $C_1\delta < 1$.

Similarly, if $x_0 \in \partial\Omega$, then

$$N[u](x_0) \le b(x_0, v(x_0), Du) \le N[v](x_0) + \beta\cdot D(u-v)$$

for some vector field $\beta$ with modulus of obliqueness at least $\varepsilon$, and hence

$$N[u](x_0) \le N[v](x_0) - \delta\psi\beta\cdot D\rho < N[v](x_0).$$

(Here, we only use the inequality $\delta > 0$.) Therefore $w$ can't have a positive maximum if $C_1\delta < 1$. Sending $\delta \to 0$ completes the proof. □

From the comparison principle, we infer an immediate uniqueness statement.

**Corollary 12.6.** *If $F$, $N$ and $\Omega$ satisfy the hypotheses of Theorem* 12.4 *or Theorem* 12.5, *then the problem* (12.7) *has at most one solution in $C^2(\Omega)\cap C^1(\overline{\Omega})$.*

## 12.2 Second derivative Hölder estimates

Our proof of second derivative Hölder estimates is a combination of standard ideas which we identify more precisely in the Notes. The key to these estimates is an algebraic lemma due to Motzkin and Wasow [141]; we refer the reader to Lemma 17.13 of [64] (or Lemma 14.5 of [114]) for a proof.

**Lemma 12.7.** *Let $\lambda \le \Lambda$ be positive constants and let $n$ be a positive integer. Then there is a positive constant $\lambda^*$, determined only by $n$, $\lambda$, and $\Lambda$, and a finite set $\{\xi^1, \dots, \xi^K\}$ of unit vectors such that any matrix $[A^{ij}]$ satisfying*

$$\lambda|\xi|^2 \le A^{ij}\xi_i\xi_j \le \Lambda|\xi|^2 \tag{12.14}$$

*for all $\xi \in \mathbb{R}^n$ can be written as*

$$A^{ij} = \sum_{k=1}^{K} \beta_k \xi_i^k \xi_j^k \tag{12.15}$$

*for constants $\beta_1, \dots, \beta_k$ in $[\lambda^*, \Lambda]$. Moreover, the set $\{\xi^1, \dots, \xi^K\}$ can be taken to include the vectors $e_1, \dots, e_n$ and $(e_i \pm e_j)/\sqrt{2}$ for all $i \ne j$.*

From this lemma, we first obtain an estimate in a very special circumstance but under very weak assumptions about the regularity of the functions involved.

**Lemma 12.8.** *Let $F\colon \mathbb{S}^n \to \mathbb{R}$ satisfy*

$$\lambda \operatorname{tr} \eta \le F(r + \eta) - F(r) \le \Lambda \operatorname{tr} \eta \tag{12.16}$$

*for some positive constants $\lambda \le \Lambda$ and all positive definite matrices $\eta$. Suppose also that $F$ is concave. Then there are constants $C$ and $\alpha$ determined only by $n$ and $\Lambda/\lambda$ such that any solution $u \in C^2(B(R))$ of $F(D^2u) = 0$ in $B(R)$ satisfies the estimate*

$$\operatorname*{osc}_{B(\rho)} D^2u \le C\left(\frac{\rho}{R}\right)^{\alpha} \operatorname*{osc}_{B(R)} D^2u \tag{12.17}$$

*for any $\rho \in (0, R)$.*

***Proof.*** For $h \in (0, R/4)$ and $\xi \in \mathbb{R}^n$ a unit vector, we define $u_h$ by

$$u_h(x, \xi) = \frac{1}{2}[u(x + h\xi) + u(x - h\xi)] - u(x).$$

Since $F$ is concave, we have

$$F\left(\frac{1}{2}[u(x + h\xi) + u(x - h\xi)]\right) \ge \frac{1}{2}[F(u(x + h\xi)) + F(x - h\xi)] = Fu.$$

It then follows from (12.16) that there is a matrix-valued function $[a^{ij}]$ (determined by $F$, $h$, and $\xi$) with eigenvalues in the interval $[\lambda, \Lambda]$ such that

$$a^{ij}D_{ij}u_h \ge 0 \text{ in } B(3R/4).$$

Setting $H_{s,h} = \sup_{B(sR/4)} u_h$ for $s = 1, 2$, we infer from the interior weak Harnack inequality Theorem 1.16 that there are positive constants $C$ and $\kappa$ such that

$$\left(R^{-n}\int_{B(R/4)} (H_{2,h} - u_h)^\kappa \, dx\right)^{1/\kappa} \le C(H_{2,h} - H_{1,h}).$$

We now divide this inequality by $h^2/2$, send $h \to 0$, and define

$$w(x,\xi) = D_{ij}u(x)\xi_i\xi_j, \quad M_s = \sup_{B(sR/4)} w$$

to infer that

$$\left(R^{-n}\int_{B(R/4)} (M_2 - w)^\kappa \, dx\right)^{1/\kappa} \le C(M_2 - M_1). \tag{12.18}$$

Now, for every $x$ and $y$ in $B(R/2)$, there is a matrix $[A^{ij}]$ satisfying (12.14) such that

$$A^{ij}[D_{ij}u(x) - D_{ij}u(y)] = F[u](x) - F[u](y) = 0,$$

and hence Lemma 12.7 gives unit vectors $\xi^1, \dots, \xi^K$, a constant $\lambda^*$ and functions $\beta_1, \dots, \beta_K$ with values in $[\lambda^*, \Lambda]$ such that

$$\sum_{k=1}^K \beta_k(x,y)[w(x,\xi^k) - w(y,\xi^k)] = 0. \tag{12.19}$$

We further abbreviate

$$w^{(k)} = w(\cdot,\xi^k), \quad M_s^{(k)} = \sup_{B(sR/4)} w^{(k)}, \quad m_s^{(k)} = \inf_{B(sR/4)} w^{(k)}$$

for $s = 1, 2$ and set

$$\omega(sR/4) = \sum_{k=1}^K M_s^{(k)} - m_s^{(k)}.$$

With $j$ fixed, we now sum (12.18), with $\xi$ replaced by $\xi_k$, for $k \ne j$. Since $m_2^{(k)} \le m_1^{(k)}$, we conclude that

$$\left(R^{-n}\int_{B(R/4)} \sum_{k\ne j} (M_2^{(k)} - w^{(k)})^\kappa \, dx\right)^{1/\kappa} \le C(\omega(R/2) - \omega(R/4)).$$

From (12.19), we conclude that

$$\beta_j(x,y)[w^{(j)}(y) - w^{(j)}(x)] = \sum_{k\ne j} \beta_k[w^{(k)}(x) - w^{(k)}(y)],$$

and hence (by taking the infimum over all $x \in B(R/2)$)

$$\lambda^*(w^{(j)}(y) - m_2^{(j)}) \le \Lambda \sum_{k \ne j} [M_2^{(k)} - w^{(k)}(y)].$$

We now raise this inequality to the power $\kappa$ and sum on $j$; similarly, we sum (12.18) with $\xi$ replaced by $\xi^{(k)}$ on $k$. Adding the resulting inequalities yields

$$\omega(R/2) \le C[\omega(R/2) - \omega(R/4)],$$

so

$$\omega(R/4) \le \delta\omega(R/2)$$

for some constant $\delta \in (0,1)$. Iteration yields

$$\omega(\rho) \le C\left(\frac{\rho}{R}\right)^\alpha \omega(R)$$

for constants $C$ and $\alpha$ as in the statement of this lemma, and (12.17) follows from this estimate because

$$\omega(\rho) \ge \operatorname*{osc}_{B(\rho)} D^2u$$

and

$$\omega(R) \le K \operatorname*{osc}_{B(R)} D^2u.$$

□

For our further analysis of problems with fully nonlinear equations, we now study the interior Hölder estimate for second derivatives of solutions of (12.1) when $F$ has a particular form. A key step is the following weighted estimate in a ball.

**Lemma 12.9.** *Assume that $F$ is as in Lemma 12.8 and that $F \in C^2$. Let $\eta$ be a $C^2$ function defined on $\mathbb{R}^n$ with support in $B = B(0, R_0)$ for some $R_0 > 0$ with $0 \le \eta \le 1$ in $\mathbb{R}^n$, let $g \in C^2(B)$, and define $\tilde{F}$ by*

$$\tilde{F}(x,r) = \eta F(r) + (1-\eta)\lambda \operatorname{tr} r - \lambda g(x).$$

*If $v \in C^4(B) \cap L^\infty(B)$ satisfies $\tilde{F}(x, D^2v) = 0$ in $B$, then there are positive constants $\theta < 1$ and $C$, determined only by $\Lambda/\lambda$, $n$, and $\eta$, such that*

$$|u|^{(0)}_{2+\theta} \le C(|u|_0 + |g|^{(2)}_2). \tag{12.20}$$

***Proof.*** To simplify notation, we assume without loss of generality that $\lambda = 1$. If we differentiate the equation $\tilde{F}(x, D^2v) = 0$ twice in the direction of any unit vector $\xi$ and use the concavity of $F$, we conclude that $h(x,\xi)$, defined by $h(x,\xi) = D_{ij}u(x)\xi^i\xi^j$, satisfies the inequality

$$\tilde{F}^{ij} D_{ij} h(x,\xi) + 2\xi \cdot D\eta A^{ij} D_{ijk} v\xi^k + A \geq 0,$$

where

$$A^{ij} = 2\xi \cdot D\eta[F^{ij}(D^2v) - \delta^{ij}]$$

$$A = (F(D^2v) - \lambda\Delta v)\frac{\partial^2\eta}{\partial x^i \partial x^j}\xi^i\xi^j - \lambda D_{ij} g\xi^i\xi^j.$$

Hence, if we fix a point $y \in B$ and set $M_2 = \sup_{B(y,d(y)/2)} |D^2v|$ and $G_2 = \sup_{B(y,d(y)/2)} |D^2g|$, it follows that there are constants $A_0$ and $B_0$, determined only by $\Lambda/\lambda$, $n$, and $\eta$ such that

$$\tilde{F}^{ij} D_{ij} h(x,\xi) \geq -\lambda[A_0|D^3v| + B_0M_2 + G_2]$$

in $B(y, d(y)/2)$. With $\xi_1, \dots, \xi_K$ the vectors from Lemma 12.7, $h^{(k)} = h(\cdot, \xi_k)$, and

$$\tilde{v} = \frac{1}{M_2}\sum_{k=1}^{K}(h^{(k)})^2,$$

it's easy to check that

$$4\sum_{k=1}^{K} |Dh^{(k)}|^2 \geq |D^3u|^2,$$

and hence

$$\tilde{F}^{ij} D_{ij}\tilde{v} \geq -C(B_0 + A_0^2)\lambda M_2 - CG_2.$$

For $k = 1, \dots, K$ and $\varepsilon \in (0,1)$ a constant to be chosen, we now set $w^{(k)} = h^{(k)} + \varepsilon\tilde{v}$, and conclude that

$$\tilde{F}^{ij} D_{ij} w^{(k)} \geq -C\bar{\mu}M_2 - CG_2,$$

with

$$\bar{\mu} = B_0 + \frac{A_0^2}{\varepsilon}.$$

We now fix $R \in (0, d(y)/4)$ and we further define

$$W_i^{(k)} = \sup_{B(y,iR/4)} w^{(k)},$$

$$M_i^{(k)} = \sup_{B(y,iR/4)} h^{(k)}, \quad m_i^{(k)} = \inf_{B(y,iR/4)} h^{(k)},$$

$$\omega(s) = \sum_{k=1}^{K} \operatorname*{osc}_{B(y,s)} h^{(k)}$$

for $i = 1, 2$, and $s \in (0, R)$. Then the weak Harnack inequality implies that

$$\left(R^{-n}\int_{B(y,R/4)}(W_2^{(k)} - w^{(k)})^\kappa\,dx\right)^{1/\kappa} \le C[W_2^{(k)} - W_1^{(k)} + \bar\mu M_2R^2 + G_2R^2).$$

Since

$$W_2^{(k)} - w^{(k)} \ge M_2^{(k)} - h^{(k)} - 2\varepsilon\omega(R/2)$$

and

$$W_2^{(k)} - W_1^{(k)} \le M_2^{(k)} - M_1^{(k)} + 2\varepsilon\omega(R/2),$$

we conclude that, for any $j \in \{1, \dots, K\}$,

$$\left(R^{-n}\int_{B(y,R/4)}\sum_{k\ne j}(M_2^{(k)} - h^{(k)})^\kappa\,dx\right)^{1/\kappa}$$
$$\le C[(1+\varepsilon)\omega(R/2) - \omega(R/4) + \bar\mu M_2R^2 + G_2R^2).$$

To obtain a corresponding inequality for $h^{(k)} - m_2^{(k)}$, we note that, for any $x_1$ and $x_2$ in $B(y, R)$, there are a matrix $[a^{ij}]$ and a constant $D_0$ such that

$$\lambda|\xi|^2 \le a^{ij}\xi_i\xi_j \le \Lambda|\xi|^2$$

for any $\xi \in \mathbb{R}^n$ and

$$a^{ij}[D_{ij}u(x_1) - D_{ij}u(x_2)] \le M_2\lambda D_0|x_1 - x_2| + G_1|x_1 - x_2|,$$

where $G_1 = \sup_{B(y,d(y)/2)}|Dg|$. From Lemma 12.7 and the argument in the proof of Lemma 12.8, we infer that

$$h^{(j)}(x) - m_2^{(j)} \le C[D_0M_2R + G_1R + \sum_{k\ne j}(M_2^{(k)} - h^{(k)}(x))],$$

and hence

$$\left(R^{-n}\int_{B(y,R/4)}(h^{(k)} - m_2^{(k)})^\kappa\,dx\right)^{1/\kappa}$$
$$\le C[(1+\varepsilon)\omega(R/2) - \omega(R/4) + M_2(\bar\mu R^2 + D_0R) + G_1R + G_2R^2].$$

Adding these inequalities then yields

$$\omega(R/2) \le C[(1+\varepsilon)\omega(R/2) - \omega(R/4) + \bar\mu M_2(R^2 + D_0R) + G_1R + G_2R^2].$$

By rearranging this inequality and setting $\delta = 1 - 1/C$, we see that

$$\omega(R/4) \le (\delta + \varepsilon)\omega(R/2) + M_2(D_0 R + \bar{\mu} R^2) + G_1 R + G_2 R^2.$$

We now fix $\varepsilon$ so that $\delta + \varepsilon < 1$ to conclude that there are positive constants $\theta$, determined only by $n$ and $\Lambda/\lambda$, and $K$, determined also by $\eta$ such that

$$\omega(R) \le \left(\frac{R}{d(y)}\right)^\theta C(M_2 + G_1 R + G_2 R^2)$$

for any $R \in (0, d(y))$. It follows that

$$\operatorname*{osc}_{B(y,R)} D^2 v \le C(M_2 + G_1 R + G_2 R^2) \left(\frac{R}{d(y)}\right)^\theta,$$

so

$$|u|^{(0)}_{2+\theta} \le C(|u|^{(0)}_2 + |g|^{(2)}_2).$$

The desired estimate (12.20) now follows from this one via Lemma 2.3. □

In fact, this second derivative Hölder estimate can be extended to study solutions of (12.1) assuming suitable conditions on the second derivatives of $F$ with respect to all arguments, but we shall only use the simpler version described here. Our next step is to prove a corresponding existence result.

**Lemma 12.10.** *Suppose $F$ satisfies the hypotheses of Lemma 12.9, and fix $R_0 > 0$. Then, for any $\varphi \in C(\overline{B(0, R_0)})$, there is a unique solution of*

$$F(D^2 v) = 0 \text{ in } B(0, R_0), \quad v = \varphi \text{ on } \partial B(0, R_0). \tag{12.21}$$

***Proof.*** Suppose first that $\varphi \in H_\delta$ for some $\delta \in (0, 1)$, fix a function $\eta$ as in Lemma 12.9, and define $\tilde{F}$ by

$$\tilde{F}(x, r) = \eta F(r) + (1 - \eta)\lambda \operatorname{tr} r.$$

With $\theta \in (0, 1)$ arbitrary, we set $\mathbb{B}_1 = H^{(-\delta)}_{4+\theta}$ and $\mathbb{B} = H^{(2-\delta)}_{2+\theta}$, and we define the map $J\colon \mathbb{B}_1 \to \mathbb{B}$ by $J(u)(x) = \tilde{F}(x, D^2 u(x))$. Recalling the definition of the Gâteaux variation $J_u(\psi)$, from Section 11.2, it is easy to check that

$$J_u(\psi) = \tilde{F}^{ij}(x, D^2 u) D_{ij}\psi,$$

and hence, if $U$ denotes the subset of all $u \in \mathbb{B}_1$ such that $u = \varphi$ on $\partial B$, it follows that, for any $u \in U$, there are $u_1 \in U$ and $\varepsilon \in (0, 1)$ such that

$$\|Ju_1 - (1 - \varepsilon)Ju\| \le \frac{1}{2}\varepsilon\|Ju\|,$$

which is just (11.6) with $\eta = 1/2$. By combining the linear theory with Lemma 12.9, we infer that there is a constant $C_0$ (depending on $R$, $\theta$, $\delta$, and $\eta$) such that

$$\|u\|_{4+\theta}^{(-\delta)} \le C_0(\|\tilde{F}(x, D^2u)\|_{2+\alpha}^{(2-\delta)} + \|\varphi\|_\delta)$$

for all $u \in U$. It then follows from Theorem 11.3 that there is a solution $u \in H_{4+\theta}^{(-\delta)}$ of

$$\tilde{F}(x, Du) = 0 \text{ in } B, \quad u = \varphi \text{ on } \partial B,$$

and the proof of Corollary 12.6 is easily modified to verify that this solution is unique.

Now let $(\eta_k)$ be an increasing sequence of functions as above such that

$$\lim_{k\to\infty} \eta_k(x) = 1$$

for all $x \in B$, let $(\varphi_k)$ be a sequence of $H_\delta$ functions converging uniformly to $\varphi$, write $\tilde{F}_k$ for the function (denoted above by $\tilde{F}$) with $\eta_k$ in place of $\eta$, and write $u_k$ for the solution of

$$\tilde{F}_k(x, Du_k) = 0 \text{ in } B, \quad u_k = \varphi_k \text{ on } \partial B$$

that we have just proved to exist. Lemma 12.9 gives a locally uniformly second derivative Hölder estimate for the sequence $(u_k)$ while Theorem 1.26 and Corollary 1.29 give a uniform global modulus of continuity estimate. In addition, the maximum principle shows that the sequence $(u_k)$ is uniformly bounded, so we can extract a uniformly convergent subsequence which, by virtue of the second derivative Hölder estimate, converges to a solution of the differential equation $F(D^2u) = 0$ in $B$. □

We can also obtain Hölder estimates for solutions of (12.1) under fairly weak conditions on the function $F$. As usual, we prove these estimates via our existence result.

**Theorem 12.11.** *Let $B = B(x_0, R)$ for some $x_0 \in \mathbb{R}^n$ and some $R > 0$, let $F\colon B \times \mathbb{S}^n \to \mathbb{R}$ and suppose that there are positive constants $\lambda \le \Lambda$, $\sigma$, $A_1$, and $A_2$ such that*

$$\lambda \operatorname{tr} \eta \le F(x, r+\eta) - F(x, r) \le \Lambda \operatorname{tr} \eta \tag{12.22a}$$

*for all $x \in B$ and all positive definite matrices $\eta$, and*

$$|F(x, r) - F(y, r)| \le (A_1 d(x)^{-2} + A_2|r|)\lambda|x-y|^\sigma d(x)^{-\sigma} \tag{12.22b}$$

*for all $x$ and $y$ in $B$ with $|x-y| \le \frac{1}{2}d(x)$. Suppose also that $F$ is concave with respect to $r$. If $\sigma < \alpha$, the constant from Lemma 12.8, then there is a constant $C$, determined by $n$, $\Lambda/\lambda$, $\sigma$, and $A_2$ such that any solution $u \in C^2(B) \cap L^\infty(B)$ of $F(x, D^2u) = 0$ in $B$ satisfies the estimate*

$$[u]_{2+\sigma}^{(0)} \le C([u]_2^{(0)} + A_1).$$

***Proof.*** Without loss of generality, we may assume that $F$ is $C^2$ with respect to $r$. We then follow a familiar path. We fix a point $y \in B$, we set $\rho_0 = \frac{1}{2}d(y)$, and we take $\tau \in (0,1)$ to be further determined. We then define $\rho_k$, for any positive integer $k$, to be $\tau^{-k}\rho_0$, and we abbreviate $B(y,\rho_k)$ to $B(k)$. We take $\tilde{v}_k$ to be the solution of

$$F(y, D^2\tilde{v}_k) = 0 \text{ in } B(k), \quad \tilde{v}_k = u \text{ on } \partial B(k)$$

given by Lemma 12.10, and we write $\tilde{Q}_k$ to be the second order Taylor polynomial of $\tilde{v}_k$ centered at $y$. Next, we define $u_k$ and $Q_k$ inductively by setting $u_0 = u$, $Q_0 = 0$,

$$Q_k = \tilde{Q}_k + \sum_{j=0}^{k-1} Q_j,$$

$$u_{k+1} = u_k - Q_k.$$

For notational convenience, we also set

$$v_k = \tilde{v}_k + \sum_{j=0}^{k-1} Q_j.$$

We infer from Lemma 12.8 that

$$\sup_{B(k+1)} |v_k - Q_k| \le \operatorname*{osc}_{B(k+1)} \tilde{v}_k \le C\tau^{2+\alpha} \sup_{B(k)} \tilde{v}_k \le C\tau^{2+\alpha} \sup_{B(k)} |v_k|.$$

In addition, there is a matrix $[a^{ij}]$ with eigenvalues in the interval $[\lambda, \Lambda]$ such that

$$a^{ij}D_{ij}(u - \tilde{v}_k) = F(y, D^2u) - F(y, D^2\tilde{v}_k)$$

(which will depend on $k$ but we suppress this dependence from the notation). Using the differential equations for $u$ and $\tilde{v}_k$ along with (12.22b), we infer that

$$|F(y, D^2u) - F(y, D^2\tilde{v}_k)| \le \rho_k^\sigma A_3$$

in $B(k)$, where

$$A_3 = \rho_0^{-2}(A_1 + A_2[u]_2^{(0)}).$$

In addition, since $u_k - v_k = u - \tilde{v}_k$, the maximum principle implies that

$$|u_k - v_k| \le CA_3\rho_k^{2+\sigma}.$$

Combining all these estimates gives

$$\sup_{B(k+1)} |u_{k+1}| \le C\tau^{2+\alpha} \sup_{B(k)} |u_k| + C\rho_k^{2+\alpha}A_3.$$

If we now set

$$A_4 = \rho_0^{-2}(A_1 + (A_2 + 1)[u]_2^{(0)}),$$

then our iteration scheme implies that, for every $y \in B$ and every $\rho \in (0, d(y))$, there is a second degree polynomial $P_2(y,\rho)$ such that

$$\sup_{B(y,\rho)} |u - P_2(y,\rho)| \le C\rho^{2+\sigma}A_4,$$

and the proof is completed by invoking Lemma 2.8. □

## 12.3 Second derivative Hölder estimates for solutions of oblique derivative problems

We now show how to prove estimates for solutions of oblique derivative problems. Our first step is an estimate near the boundary with a simple differential equation and a constant coefficient linear boundary condition. Although this estimate is not, by itself, a Hölder estimate for the second derivatives of the solutions, it contains the important analytical aspect of the proof of our desired estimate.

**Lemma 12.12.** *Let $\omega \in H_{1+\alpha}(B_{n-1}(0,R_0))$ for some $\alpha \in (0,1)$ and $R_0 \in (0,1]$ with $\omega(0) = 0$ and $D\omega(0) = 0$, and, for $s \in (0, R_0]$, set*

$$\Omega[s] = \{x \in \mathbb{R}^n : |x| < s,\ x^n > \omega(x')\}, \tag{12.23a}$$

$$\Sigma[s] = \{x \in \mathbb{R}^n : |x| < s,\ x^n = \omega(x')\}. \tag{12.23b}$$

*Suppose $F\colon \Omega[R_0]\times\mathbb{S}^n \to \mathbb{R}$ is differentiable with respect to $x$ and $r$, concave with respect to $r$ and that there are positive constants $\lambda \le \Lambda$ and $A_0$ such that*

$$\lambda \operatorname{tr}\eta \le F(x, r+\eta) - F(x,r) \le \Lambda \operatorname{tr}\eta, \tag{12.24a}$$

*for all positive definite matrices $\eta$ and all $(x,r) \in \Omega[R_0]\times\mathbb{S}^n$,*

$$|F(x,r) - F(y,r)| \le A_0 d(x)^{\alpha-1}|x-y| \tag{12.24b}$$

*for all $x$ and $y$ in $\Omega[R_0]$ with $|x-y| \le \frac{1}{2}d(x)$ and all $r \in \mathbb{S}^n$. Let $\beta$ be a unit vector such that*

$$|\beta'| \le \frac{1-\varepsilon}{\omega_0}\beta^n \tag{12.25}$$

*for some positive constants $\varepsilon < 1$ and $\omega_0 \ge \sup|D\omega|$. Then there are constants $\tau_0 \in (0,1)$, determined only by $\varepsilon$ and $\omega_0$, $\theta \in (0,1)$, determined only by $n$, $\varepsilon$, $\Lambda/\lambda$, and $\omega_0$ and $C$, determined also by $[D\omega]_\alpha$ such that, if $u \in H_1(\Omega[R_0]) \cap C^2(\Omega[R_0])$ satisfies*

$$F(x, D^2u) = 0 \text{ in } \Omega[R_0], \quad \beta\cdot Du = 0 \text{ on } \Sigma[R_0] \tag{12.26}$$

*and if $\alpha < \theta$, then, for any $R \in (0, R_0/2)$, there is a quadratic polynomial $Q_R$ such that*

$$\operatorname*{osc}_{\Omega[\tau R]} |u - Q_R| \le C(\tau^{2+\theta} \operatorname*{osc}_{\Omega[R]} |u-Q| + A_0R^{2+\alpha} + \frac{R^{2+\alpha}}{R_0^{1+\alpha}} \sup_{\Omega[R]} |\beta\cdot Du|) \tag{12.27}$$

*for any quadratic polynomial $Q$ and any $\tau \in (0,\tau_0)$.*

***Proof.*** Suppose first that $\beta = e_n$, and set $H = D_{nn}u(0)$ and $M = \sup_{\Omega[R_0]} |D_n u|$. Let $\rho$ be a regularized distance from Lemma 4.9, and note from (4.24) with $\varepsilon = 1/2$, that we may assume that $|D\rho| \geq 1/2$ in $\Omega[R_0]$. Moreover, the construction shows that $D\rho(0) = e_n$, and there is a constant $A_1$, determined only by $n$, $[D\omega]_\alpha$, and $\alpha$ such that $|D^2\rho| \leq A_1 \rho^{\alpha-1}$. We now observe that

$$\hat{\rho} = \int_0^\rho \exp\left(-\frac{4A_1\Lambda}{\alpha\lambda} t^\alpha\right) dt$$

satisfies $a^{ij} D_{ij}\hat{\rho} \leq 0$ and $C\rho \leq \hat{\rho} \leq \rho$ in $\Omega[R_0]$ for any matrix $[a^{ij}]$ with eigenvalues in the interval $[\lambda, \Lambda]$.

It then follows from the proof of Proposition 11.15, with $\hat{\rho}$ replacing $\hat{d}$, that there is a constant $\theta > 0$ such that

$$\left|\frac{D_n u(x)}{\hat{\rho}(x)} - H\right| \leq C[A_0|x|^\alpha + \frac{|x|^\theta}{R_0^\theta} \sup_{\Omega[3R_0/4]} \left|\frac{D_n u}{\hat{\rho}}\right|]$$

for any $x \in \Omega[R_0/2]$ and that

$$\sup_{\Omega[3R_0/4]} \left|\frac{D_n u}{\hat{\rho}}\right| \leq C(A_0 R_0^\alpha + \frac{M}{R_0}).$$

From this inequality, we immediately conclude that

$$H \leq C(A_0 R_0^\alpha + \frac{M}{R_0}). \tag{12.28}$$

In addition, $D\hat{\rho}(0) = e_n$, and hence

$$|D_n u(x) - H[x^n - \omega(x')]| \leq C(A_0|x|^{1+\alpha} + M\frac{|x|^{1+\theta}}{R_0^{1+\theta}}) \tag{12.29}$$

for all $x \in \Omega[R_0/2]$.

We next observe that there is a constant $\tau_1 \in (0, 1/4)$ such that $B((R/2)e_n, \tau_1 R) \subset \Omega[R]$ for any $R \in (0, R_0/2)$. In what follows, we always assume that $\tau \in (0, \tau_1/2)$, or equivalently, $\tau_0 \leq \tau_1/2$. We then define $\bar{x}_1 = (R/2)e_n$ and we define the polynomial $P$ by

$$P(x) = Du(\bar{x}_1) \cdot (x - \bar{x}_1) + \frac{1}{2} D_{ij}u(\bar{x}_1)(x - \bar{x}_1)^i (x - \bar{x}_1)^j$$

and infer from Theorem 12.11 that

$$\operatorname*{osc}_{B(\bar{x}_1, \tau R)} |u - P| \leq C[\tau^{2+\theta} \operatorname*{osc}_{\Omega[R]} |u - Q| + A_0(\tau R)^{2+\alpha}] \tag{12.30}$$

for any quadratic polynomial $Q$. We now define

$$b_{ij} = \begin{cases} D_{ij}u(\bar{x}_1) & \text{if } i, j < n, \\ D_{ij}u(0) & \text{if } i = n \text{ or } j = n \end{cases}$$

and we set

$$Q_R(x) = Du(\bar{x}_1)\cdot(x-\bar{x}_1) + \frac{1}{2}b_{ij}(x-\bar{x}_1)^i(x-\bar{x}_1)^j.$$

Since $b_{in} = b_{ni} = 0$ if $i < n$, we have

$$\begin{aligned} Q_R(x) - P(x) &= \frac{1}{2}\sum_{i=1}^{n-1}[D_{in}u(\bar{x}_1) - b_{in}](x-\bar{x}_1)^i(x-\bar{x}_1)^n \\ &\quad + \frac{1}{2}\sum_{j=1}^{n}[D_{nj}u(\bar{x}_1) - b_{nj}](x-\bar{x}_1)^j(x-\bar{x}_1)^n. \end{aligned}$$

We now define, for $i = 1, \ldots, n$,

$$f_i = \frac{D_n u(\bar{x}_1 + \tau R e_i) - D_n u(\bar{x}_1)}{\tau R}.$$

When $i < n$, we have $b_{in} = 0$, and hence

$$D_{in}u(\bar{x}_1) - b_{in} = (D_{in}u(\bar{x}_1) - f_i) + f_i.$$

The mean value theorem then gives a point $y_i$ on the line segment joining $\bar{x}_1$ and $\bar{x}_1 + \tau R e_i$ (and hence in $B(\bar{x}_1, \tau R)$) such that $f_i = D_{in}u(y_i)$. Theorem 12.4 then implies that

$$|D_{in}u(\bar{x}_1) - f_i| \le \operatorname*{osc}_{B(\bar{x}_1,\tau R)} D_{in}u \le C\left(\tau^\sigma \frac{\operatorname{osc}_{\Omega[R]}(u-Q)}{R^2} + A_0\tau R^\alpha\right).$$

Furthermore, (12.29) implies that

$$|f_i| \le \frac{C}{\tau R}[(\tau R)^{1+\alpha}H + A_0(\tau R)^{1+\alpha} + M\left(\frac{R}{R_0}\right)^{1+\theta}]$$

because $(\bar{x}_1 + \tau R e_i)^n = \bar{x}_1^n$, $\omega(\bar{x}_1') = \omega(0) = 0$, and $|\omega(\bar{x}_1 + \tau R e_i)| \le C(\tau R)^{1+\alpha}$. Using the estimate (12.28) and the inequalities $0 \le R_0 \le 1$, we find that

$$|f_i| \le C[A_0(\tau R)^{1+\alpha} + M\left(\frac{R}{R_0}\right)^{1+\theta}].$$

To estimate the final term in the expansion of $Q_R - P$, we observe that

$$D_{nn}u(\bar{x}_1) - D_{nn}u(0) = D_{nn}u(\bar{x}_1) - f_n + f_n - H.$$

It follows as before that

$$|D_{nn}u(\bar{x}_1) - f_n| \le C\left(\tau^\sigma \frac{\operatorname{osc}_{\Omega[R]}(u-Q)}{R^2} + A_0\tau R^\alpha\right),$$

and simple algebra yields

$$f_n - H = \frac{D_n(\bar{x}_1 + \tau R e_n) - (1+\tau R)H}{\tau R} + \frac{H - D_n u(\bar{x}_1)}{\tau R}.$$

Since $\bar{x}_1' = (\bar{x}_1 + \tau R e_n)' = 0$ and $\omega(0) = 0$, we conclude from (12.29) that

$$|D_{nn}u(\bar{x}_1) - D_{nn}u(0)| \le \frac{C}{\tau R}[A_0 \tau R^{1+\alpha} + M\frac{R^{1+\theta}}{R_0^{1+\theta}}].$$

Combining these inequalities, we find that

$$|Q_R(x) - P(x)| \le C[\tau^{2+\theta} \operatorname*{osc}_{\Omega[R]}(u - Q) + A_0 R^{2+\alpha} + M\frac{R^{2+\alpha}}{R_0^{1+\alpha}}]. \qquad (12.31)$$

In combination with (12.30), this inequality implies that

$$\operatorname*{osc}_{B(\bar{x}_1,\tau R)}(u - Q_R) \le C\left[\tau^{2+\theta} \operatorname*{osc}_{\Omega[R]}(u - Q) + A_0 R^{2+\alpha} + M\frac{R^{2+\alpha}}{R_0^{1+\alpha}}\right]. \qquad (12.32)$$

Now let $x \in \Omega[\tau R]$ and set $\bar{x} = x + \bar{x}_1$, $w = u - Q_R$. It follows that

$$w(x) - w(0) = w(\bar{x}) - w(\bar{x}_1) + w(x) - w(\bar{x}) + w(\bar{x}_1) - w(0),$$

and

$$|w(\bar{x}) - w(\bar{x}_1)| \le \operatorname*{osc}_{B(\bar{x}_1,\tau R)}(u - Q_2).$$

In addition,

$$w(x) - w(\bar{x}) + w(\bar{x}_1) - w(0) = \frac{R}{2}\int_0^1 \left[D_n w\left(\frac{sR}{2}e_n\right) - D_n w\left(x + \frac{sR}{2}e_n\right)\right] ds$$

and

$$D_n w\left(\frac{sR}{2}e_n\right) - D_n w\left(x + \frac{sR}{2}e_n\right)$$
$$= D_n u\left(\frac{sR}{2}e_n\right) - H\frac{sR}{2} - D_n u\left(x + \frac{sR}{2}e_n\right) + H\left[x^n + \frac{sR}{2}\right].$$

Using (12.29) again, we find that

$$\left|D_n w\left(\frac{sR}{2}e_n\right) - D_n w\left(x + \frac{sR}{2}e_n\right)\right| \le H|\omega(x')| + C(A_0 R^{1+\alpha} + \frac{MR^{1+\theta}}{R_0^{1+\theta}}).$$

Since

$$H|\omega(x')| \le C[A_0 R_0^\alpha R^{1+\alpha} + \frac{MR^{1+\alpha}}{R_0}],$$

it follows (after recalling that $R_0 \le 1$ and $\alpha \le \theta$) that

$$|w(x) - w(\bar{x}) + w(\bar{x}_1) - w(0)| \le C[A_0 R^{2+\alpha} + \frac{MR^{2+\alpha}}{R_0^{1+\alpha}}],$$

and hence

$$\operatorname*{osc}_{\Omega[\tau R]}(u - Q_2) \le C\left[\operatorname*{osc}_{B(\bar{x}_1,\tau R)}(u - Q_2) + A_0 R^{2+\alpha} + \frac{MR^{2+\alpha}}{R_0^{1+\alpha}}\right].$$

Combining this inequality with (12.32) and (12.31) yields (12.27) in the special case that $\beta = e_n$.

In the case of general $\beta$, we begin by defining the linear change of variables $T$ by saying that $y = Tx$ if

$$y^i = \begin{cases} x^i - (\beta^i/\beta^n)x^n & \text{if } i < n, \\ x^n & \text{if } i = n. \end{cases}$$

We also define $\tilde{u} = u \circ T^{-1}$ and we define the function $\tilde{\omega}$ implicitly in a neighborhood of 0 to satisfy the equation

$$\tilde{\omega}(y') = \omega\left(y' + \frac{\beta'}{\beta^n}\tilde{\omega}(y')\right);$$

hence $x^n = \omega(x')$ if and only if $x^n = \tilde{\omega}((Tx)')$. We then write $\tilde{\Omega}[s]$ (for suitable $s > 0$) for the set of all $y \in \mathbb{R}^n$ with $|y| < s$ and $y^n > \tilde{\omega}(y')$, and we write $\tilde{\Sigma}[s]$ for the set of all $y \in \mathbb{R}^n$ with $|y| < s$ and $y^n = \tilde{\omega}(y')$. It follows that

$$\frac{\partial \tilde{u}}{\partial y^n} = 0$$

on $T(\Sigma[R_0])$ and that there are positive constants $c_1$ and $c_2$, determined only by $\omega_0$ and $\varepsilon$, such that $|D\tilde{\omega}| \le c_1$, $[D\tilde{\omega}]_\alpha \le c_1[D\omega]_\alpha$,

$$T(\Omega[s]) \subset \tilde{\Omega}[c_1 s]$$

for all $s \in (0, R_1)$ (where $R_1$ is chosen so that $\tilde{\Omega}[c_1 R_0] \subset T(\Omega[R_0])$; we can take $R_1 = c_2 R_0$),

$$\tilde{\Omega}[c_2 s] \subset T(\Omega[s])$$

for all $s \in (0, R_0)$, and $D\tilde{\omega}(0) = 0$. Further, there is a function $\tilde{F}\colon T(\Omega[R_0]) \times \mathbb{S}^n \to \mathbb{R}$ such that

$$c_2\lambda \operatorname{tr}\eta \le \tilde{F}(y, r + \eta) - \tilde{F}(y, r) \le c_1\Lambda \operatorname{tr}\eta$$

for any positive definite matrix $\eta$ and all $(y, r) \in T(\Omega[R_0]) \times \mathbb{S}^n$,

$$|\tilde{F}(x, r) - \tilde{F}(y, r)| \le c_1 A_0 \tilde{d}(x)^{\alpha-1}|x - y|$$

for all $x$ and $y$ in $T(\Omega[R_0])$ and all $r \in \mathbb{S}^n$, where $\tilde{d}$ denotes distance to $\tilde{\Sigma}[R_0]$, and

$$\tilde{F}(y, D^2\tilde{u}) = 0$$

in $T(\Omega[R_0])$. Our previous step shows that $\tilde{u}$ satisfies an estimate like (12.27) but with $\tilde{\Omega}[s]$ in place of $\Omega[s]$ for $s = \tau R, R$, and the proof is completed by pointing out that the class of quadratic polynomials is unchanged under the mapping $T$. □

At this point, we jump to the relevant existence result with an attendant estimate.

**Lemma 12.13.** *Let $\omega \in H_{1+\alpha}(B_{n-1}(0,R))$ for some $\alpha \in (0,1)$ and $R \in (0,1]$ with $\omega(0) = 0$, for $s \in (0,R]$, define $\Omega[s]$ and $\Sigma[s]$ by* (12.23), *and define*

$$\sigma[s] = \{x \in \mathbb{R}^n : |x| = s, x^n \geq \omega(x')\}. \tag{12.33}$$

*Suppose $F$ is as in Lemma* 12.8 *and let $\beta$ be a unit vector satisfying* (12.25) *for some positive constants $\varepsilon < 1$ and $\omega_0 \geq \sup|D\omega|$. Then, for any function $\varphi \in C(\sigma[R])$, there is a unique solution $v \in C^2(\Sigma[R]) \cap C(\overline{\Sigma[R]})$ of*

$$F(D^2 v) = 0 \text{ in } \Omega[R], \quad \beta \cdot Dv = 0 \text{ on } \Sigma[R], \quad v = \varphi \text{ on } \sigma[R]. \tag{12.34}$$

*Moreover, there are positive constants $\theta$, determined only by $\Lambda/\lambda$, $n$, $\varepsilon$, and $\omega_0$, and $C$, determined also by $[D\omega]_\alpha$, such that if $\alpha < \theta$, then*

$$\sup_{\Omega[\tau R]} |v - \bar{Q}| \leq C\tau^{2+\alpha} \sup_{\Omega[R]} |v| \tag{12.35}$$

*for any $\tau \in (0, 1/2)$, where $\bar{Q}$ is the second degree Taylor polynomial for $v$ centered at* 0.

***Proof.*** First, we note that Lemma 12.12 remains valid even without the assumption that $D\omega(0) = 0$ via a simple rotation of axes. Then, as in Section 12.2, we fix a nonnegative $C^\infty$ function $\eta\colon \mathbb{R}^n \to [0,1]$ with compact support in $B(0,R)$, define $\tilde{F}$ by

$$\tilde{F}(x,r) = \eta F(r) + (1-\eta)\operatorname{tr} r,$$

and we assume that $\varphi \in H_\delta$ for some $\delta \in (0,1)$. The combination of that lemma with the argument of Theorem 12.11 shows that any solution of the problem

$$\tilde{F}(x, D^2 u) = g(x) \text{ in } \Omega[R], \tag{12.36a}$$

$$\beta \cdot Du = 0 \text{ on } \Sigma[R], \tag{12.36b}$$

$$u = \varphi \text{ on } \sigma[R] \tag{12.36c}$$

with $g \in H^{(2-\delta)}_{2+\alpha}$ satisfies the estimate

$$|u|^{(-\delta)}_{4+\alpha} \leq C(|u|^{(-\delta)}_2 + |g|^{(2-\delta)}_{2+\alpha} + |\varphi|^{(-\delta)}_\delta).$$

Lemma 2.3 shows that

$$|u|^{(-\delta)}_{4+\alpha} \leq C(|u|^{(-\delta)}_0 + |g|^{(2-\delta)}_{2+\alpha} + |\varphi|^{(-\delta)}_\delta),$$

and a simple application of the maximum principle yields

$$|u|^{(-\delta)}_{4+\alpha} \le C(|g|^{(2-\delta)}_{2+\alpha} + |\varphi|^{(-\delta)}_{\delta} + F(0)).$$

It then follows from Theorem 11.3 that (12.36) is solvable in $H^{(-\delta)}_{4+\alpha}$ for any $g$, and the approximation arguments used in the proof of Theorem 12.11 yield our existence result.

From Lemmata 12.12, 2.8, and 2.3, we have

$$|v|^{(0)}_{2+\alpha} \le C|v|_0$$

and then integration yields (12.35). □

A straightforward modification of our prior perturbation arguments leads to a Hölder estimate for second derivatives of solutions of fully nonlinear oblique derivative problems.

**Theorem 12.14.** *Let $\alpha$ and $\alpha'$ be positive numbers with $\alpha < \alpha' \le 1$ and set $\delta = \alpha/(2\alpha' - \alpha)$. Let $\Omega$ be a domain in $\mathbb{R}^n$ with a $C^{1,\alpha}$ boundary portion $T$, and write $d^*$ for the distance to $\partial\Omega\setminus T$. Suppose $F$ is a real-valued function defined on $\Omega \times \mathbb{S}^n$ which is concave with respect to $r$ and which satisfies* (12.24a) *for all positive definite matrices $\eta$ and all $(x,r) \in \Omega \times \mathbb{S}^n$, and*

$$|F(x,r) - F(y,r)| \le [A_1 d^*(x)^{-2} + A_2|r|]\lambda \frac{|x-y|^\alpha}{(d^*(x))^\alpha}$$

*for all $x$ and $y$ in $\Omega$ with $|x-y| \le d^*(x)/2$ and all $r \in \mathbb{S}^n$. Let $b$ be a real-valued function defined on $T \times \mathbb{R}^n$ and set $\chi = \partial b/\partial p$. Suppose that there are positive constants $b_0$ and $G_1$ such that*

$$|b_p(x,p)| \le b_0\chi(x,p), \tag{12.37a}$$

$$|b_x(x,p)| \le G_1\chi(x,p)/d^*(x) \tag{12.37b}$$

*for all $(x,p) \in \Sigma[R^*] \times \mathbb{R}^n$. Suppose further that there is a nonnegative constant $G_2$ such that*

$$|b_x(x,p) - b_x(y,q)| \le \left[G_1 \frac{|x-y|^\alpha}{(d^*(x))^{2+\alpha}} + G_2 G_1^{1/(1+\alpha)} \frac{|p-q|^\alpha}{d^*(x)^{2-\alpha}}\right] \chi(x,p), \tag{12.38a}$$

$$|b_p(x,p) - b_p(y,q)| \le \left[G_1^{\alpha/(1+\alpha)} \frac{|x-y|^\alpha}{d^*(x)^\alpha} + G_2|p-q|^{\alpha'}\right] \chi(x,p) \tag{12.38b}$$

*for all $(x,p)$ and $(y,q)$ in $T \times \mathbb{R}^n$ with $|x-y| \le d^*(x)/2$. If $u \in H^{(0)}_2(\Omega) \cap C^2(\Omega) \cap H_{1+\delta}(\Omega)$ (with the weight in the $H^{(0)}_2$ norm being with respect to $d^*$) satisfies*

$$F(x, D^2u) = 0 \text{ in } \Omega, \quad b(x, Du) = 0 \text{ on } T, \tag{12.39}$$

*then there are positive constants $\theta$, determined only by $n$, $\Lambda/\lambda$, $b_0$, and $\Omega$, and $C$, determined also by $A_2$ and $G_2$, such that*

$$|u|^{(0)}_{2+\alpha} \le C(|u|_0 + A_1 + G_1 + [Du]_\delta^{2(1+\alpha)(2\alpha'-\alpha)/\alpha} + ([u]_1^{(0)})^{2+\alpha}). \quad (12.40)$$

***Proof.*** By arguing as in Theorem 12.11, we may assume that $A_2$ is zero. We begin by fixing a point $x_0 \in T$. We then rotate and translate so that $\gamma(x_0) = e_n$ and $x_0 = 0$. Now we choose $R_0 \in (0,1]$ so that $d^*(x_0) \ge R_0$. By further decreasing $R_0$, we may assume that there is a function $\omega \in H_{1+\alpha}(B_{n-1}(0,R_0))$ such that $\omega(0) = 0$ and

$$\Omega \cap B(x_0, R_0) = \Omega[R_0], \quad T \cap B(x_0, R_0) = \Sigma[R_0],$$

using the definitions (12.23). Moreover, there are constants $\omega_0 > 0$ and $\varepsilon \in (0,1)$ such that $|D\omega| \le \omega_0$ and $\beta = b_p(0, Du(0))$ satisfies

$$|\beta'| \le \frac{1-\varepsilon}{\omega_0}\beta^n.$$

Using (12.37) and (12.38), we infer that there is a constant $C$ such that

$$|\beta \cdot (Du - Du(0)) + b_x(0, Du(0)) \cdot x| \le C(I_1 + I_2 + I_3 + I_4),$$

where

$$\begin{aligned}
I_1 &= G_1 \frac{|x|^{1+\alpha}}{R_0^{2+\alpha}}, \\
I_2 &= G_1^{1/(1+\alpha)} \frac{|Du - Du(0)|^\alpha |x|}{R_0^{2-\alpha}}, \\
I_3 &= G_1^{\alpha/(1+\alpha)} \frac{|x|^\alpha |Du - Du(0)|}{R_0^\alpha}, \\
I_4 &= |Du - Du(0)|^{1+\alpha'}.
\end{aligned}$$

(Strictly speaking, $b_x(0, Du(0))$ denotes a tangential derivative along $\partial\Omega$ but we can take it to be a vector with $n$th component equal to zero.) Now we set $U_\delta = [Du]_\delta$, $U_2 = [u]_2^{(0)}$, and $\zeta = (2+\alpha)/(2+2\alpha')$, and we note that

$$|Du - Du(0)| \le R_0^{-2} U_2 |x|, \quad (12.41a)$$

$$|Du - Du(0)| \le U_\delta^{1-\zeta} U_2^\zeta R_0^{-2\zeta} |x - y|^{\zeta + (1-\zeta)\delta}. \quad (12.41b)$$

Using (12.41a), we conclude that

$$I_2 \le G_1^{1/(1+\alpha)} U_2^\alpha \frac{|x|^{1+\alpha}}{R_0^{2+\alpha}}$$

and

$$I_3 \le G_1^{\alpha/(1+\alpha)} U_2 \frac{|x|^{1+\alpha}}{R_0^{2+\alpha}}.$$

To estimate $I_4$, we use (12.41b) and the easily checked identities

$$\zeta(1+\alpha') = 1 + \frac{\alpha}{2}, \quad (\zeta + (1-\zeta)\delta)(1+\alpha') = 1+\alpha$$

to infer that

$$I_4 \le U_\delta^{(1-\zeta)(1-\alpha')} U_2^{1+\alpha/2} \frac{|x|^{1+\alpha}}{R_0^{2+\alpha}}.$$

Combining these inequalities with Young's inequality and setting

$$U = G_1 + U_\delta^{2(1+\alpha)(2\alpha'-\alpha)/\alpha} + U_2^{1+\alpha/2},$$

we conclude that

$$I_1 + I_2 + I_3 + I_4 \le CU \frac{|x|^{1+\alpha}}{R_0^{2+\alpha}}.$$

We now write $\hat{\beta}$ for the vector perpendicular to $\beta$ with $\hat{\beta}' = b_x(0, Du(0))$ and define $u_0$ by

$$u_0(x) = u(x) - Du(0) \cdot x + \frac{1}{|\beta|^2}(\beta \cdot x)(\hat{\beta} \cdot x).$$

Since

$$\beta \cdot D((\beta \cdot x)(\hat{\beta} \cdot x)) = |\beta|^2 \hat{\beta} \cdot x,$$
$$|\hat{\beta} \cdot x - b_x(0, Du(0)) \cdot x| = |\hat{\beta}^n||x^n|,$$

and $|x^n| = \omega(x') \le C|x'|^{1+\alpha}$ on $\Sigma[R_0]$, it follows that

$$|\beta \cdot Du_0| \le C\beta^n \left( \frac{|x|^{1+\alpha}}{R_0^{2+\alpha}} + \frac{|Du - Du(0)|^{1+\alpha}}{R_0^\alpha} \right).$$

We therefore infer that

$$|\beta \cdot Du_0| \le \beta^n C_4 U \frac{|x|^{1+\alpha}}{R_0^{2+\alpha}}. \tag{12.42}$$

Next, we write $r^0$ for the matrix with entries $r^0_{ij} = \beta^i \hat{\beta}^j$, and we define $F_0$ by $F_0(x, r) = F(x, r - r^0)$.

With $\tau \in (0,1)$ a constant at our disposal, we define $R_k = \tau^k$ for any nonnegative integer $k$. We also define $F_k$, $v_k$, $r^k$, and $Q_k$ inductively as follows. First, $v_k$ is the solution of

$$F_k(0, D^2 v_k) = 0 \text{ in } \Omega[R_k],$$
$$\beta \cdot Dv_k = 0 \text{ on } \Sigma[R_k],$$
$$v_k = u_k \text{ on } \sigma[R_k].$$

We then write $Q_k$ for the second order Taylor polynomial of $v_k$ centered at 0, and $r^k = D^2v_k(0)$. Finally, we set $u_{k+1} = u_k - Q_k$ and we define $F_{k+1}$ by $F_{k+1}(x, r) = F_k(x, r - r^k)$. It follows from Lemma 12.13 that there are constants $\theta$ (as in the statement of this theorem) and $C_1$ such that

$$\sup_{\Omega[R_{k+1}]} |v_k - Q_k| \leq C_1\tau^{2+\alpha} \sup_{\Omega[R_k]} |v_k| \tag{12.43}$$

for $\alpha \leq \theta$. In addition, the maximum principle implies that

$$|v_k - u_k| \leq C_2 \left( A_1 \frac{R_k^{2+\alpha}}{R_0^{2+\alpha}} + \sup_{\Sigma[R_k]} \frac{|\beta \cdot Du_k|}{\beta^n} R_k \right). \tag{12.44}$$

On $\Sigma[R_k]$, we have

$$\beta \cdot Du_k = \beta \cdot Du_{k-1} + \beta \cdot DQ_{k-1}$$

and it's easy to check that

$$\beta \cdot DQ_k = \beta^i D_{in}v_k(0)x^n.$$

By arguing as in Proposition 11.15, we find that

$$|\beta^i D_{in}v_k(0)| \leq C\beta^n \sup |v_k|R_k^{-2},$$

and we have $|x^n| \leq C|x|^{1+\alpha}$ on $\Sigma[R_k]$, so

$$|\beta \cdot DQ_k| \leq C_3\beta_n R_k^{-2}|x|^{1+\alpha} \sup_{\Omega[R_k]} |v_k|.$$

We now set

$$B_k(x) = \frac{|\beta \cdot Du_k(x)|}{\beta^n}, \quad V_k = \sup_{\Omega[R_k]} |v_k|,$$

and infer that

$$B_k(x) \leq B_{k-1}(x) + C_3V_{k-1}R_{k-1}^{-2}|x|^{1+\alpha}.$$

A straightforward induction yields

$$B_k(x) \leq B_0(x) + C_3|x|^{1+\alpha} \sum_{j=0}^{k-1} R_j^{-2}V_j$$

and hence, by virtue of (12.42), we have

$$R_kB_k(x) \leq C_4U\tau^{(2+\alpha)k} + C_3R_0^\alpha\tau^{(2+\alpha)k} \sum_{j=0}^{k-1} \frac{V_j}{\tau^{2j}} \tag{12.45}$$

for $x \in \Sigma[R_k]$. Inserting this inequality into (12.44) gives

$$|v_k - u_k| \leq (C_2A_1 + C_2C_4U)\tau^{(2+\alpha)k} + C_3R_0^\alpha\tau^{(2+\alpha)k} \sum_{j=0}^{k-1} \frac{V_j}{\tau^{2j}},$$

so (12.43) implies that

$$V_{k+1} \le (C_1 + C_2C_3R_0^\alpha)\tau^{2+\alpha}V_k + H\tau^{(2+\alpha)k} + 2C_2C_3R_0^\alpha\tau^{(2+\alpha)k}\sum_{j=0}^{k-1}\frac{V_j}{\tau^{2j}}.$$

with $H = 2(C_2A_1 + C_2C_4U)$. We now choose $\theta \in (0,\alpha)$ and set $\sigma = (\theta+\alpha)/2$. In addition, we take $R_0$ so small that $C_2C_3R_0^\alpha \le 1$ and choose $\tau$ so small that $(C_1+1)\tau^{2+\alpha} \le \tau^{2+\sigma}$. Then our induction inequality becomes

$$V_{k+1} \le \tau^{2+\sigma}V_k + H\tau^{(2+\theta)k} + 2C_2C_3R_0^\alpha\tau^{(2+\theta)k}\sum_{j=0}^{k-1}\frac{V_j}{\tau^{2j}}.$$

For suitable choices of $\tau$ and $R_0$, it follows from the argument deriving (4.19) from (4.18) that

$$V_k \le \tau^{(2+\theta)k}V_0 + 2H\tau^{(2+\alpha)(k-1)}. \tag{12.46}$$

We now use this inequality, along with the summability of the infinite series

$$\sum_{j=0}^{\infty}\tau^{\theta j}$$

and (12.45) to infer that

$$R_kB_k(x) \le C[U + V_0 + H]\tau^{(2+\theta)k}$$

on $\Sigma[R_k]$. It then follows from (12.44) and (12.46) that

$$\sup_{\Omega[R_k} |u_k| \le [U + U_0 + H]\tau^{(2+\theta)k}$$

with

$$U_0 = \sup_{\Omega[R_0]} |u_0|.$$

Hence, for any $\rho \in (0,R)$, there is a second order polynomial $P(\rho)$ such that

$$\sup_{\Omega[\rho]}|u - P(\rho)| \le C(U + U_0 + H)\rho^{2+\theta}.$$

Combining this inequality with the corresponding interior estimates, we find that

$$[u]_{2+\alpha}^{(0)} \le C(|u|_0 + U + U_0 + H),$$

which implies that

$$[u]_{2+\alpha} \le C(|u|_0 + A_1 + G_1 + [Du]_\delta^{2(1+\alpha)(2\alpha'-\alpha)/\alpha}) + C([u]_2^{(0)})^{1+\alpha/2}.$$

The estimate (12.40) follows from this inequality via Lemma 2.3. □

## 12.4 Uniformly elliptic fully nonlinear problems

Our next step is to derive suitable *a priori* estimates for solutions of uniformly elliptic problems.

The Hölder estimate for the solution is particularly simple.

**Theorem 12.15.** *Let $\Omega$ be a bounded $C^1$ domain in $\mathbb{R}^n$ and let $u \in C^2(\Omega) \cap C(\overline{\Omega})$ be a solution of (12.7). Suppose that there are positive constants $\lambda \leq \Lambda$ and an increasing function $A_0$ such that*

$$\lambda \operatorname{tr} \eta \leq F(x, z, p, r + \eta) - F(x, z, p, r) \leq \Lambda \operatorname{tr} \eta \tag{12.47a}$$

*for all $(x, z, p, r) \in \Omega \times \mathbb{R} \times \mathbb{R}^n \times \mathbb{S}^n$ and all positive definite matrices $\eta$,*

$$|F(x, z, p, 0)| \leq A_0(|z|)\lambda(1 + |p|^2) \tag{12.47b}$$

*for all $(x, z, p) \in \Omega \times \mathbb{R} \times \mathbb{R}^n$. Suppose also that*

$$|b(x, z, p')| \leq A_0(|z|)\chi(1 + |p'|) \tag{12.48}$$

*for all $(x, z, p) \in \partial\Omega \times \mathbb{R} \times \mathbb{R}^n$ with $b(x, z, p) = 0$, where $p' = p - (p \cdot \gamma)p$. Then there are constants $\alpha \in (0, 1)$ and $C > 0$, determined only by $n$, $\Lambda/\lambda$, $\sup|u|$ and $A_0(\sup|u|)$ such that*

$$|u|_\alpha \leq C. \tag{12.49}$$

***Proof.*** With $a^{ij}$ and $a$ as in the remarks before Lemma 12.1, we see that $u$ satisfies the differential equation $a^{ij} D_{ij}u + a = 0$ in $\Omega$ with $a^{ij}$ having eigenvalues in the range $[\lambda, \Lambda]$ and $|a| \leq A_0(\sup|u|)\lambda(1 + |p|^2)$. Similarly, $u$ satisfies the boundary condition $\beta \cdot Du + g = 0$ for some vector field $\beta$ having a modulus of obliqueness determined by $A_0(\sup|u|)$ and $\Omega$ and $g/|\beta|$ is bounded by a constant determined by the same quantities. The Hölder estimate then follows from Lemma 8.4. □

If we impose suitable assumptions on the derivatives of $F$ and $b$ with respect to $(x, z, p, r)$, then we can imitate the arguments in Chapter 9 to obtain a bound on the maximum of the gradient. As in that chapter, an important part in the derivation of this bound is played by a suitable interior gradient bound, which we now present. It will be convenient to introduce some additional notation here.

First, we set $\mathcal{E}_2 = F^{ij} r_{ik} r_{jk}$ and we use the operators $\delta$ and $\delta$ from Chapter 9. In other words,

$$\delta g = g_z + |p|^{-2} p \cdot g_p, \quad \bar{\delta} g = p \cdot g_p.$$

**Lemma 12.16.** *Let $u \in C^3(\Omega)$ be a solution of $F[u] = 0$ in $\Omega$, and suppose that there are nonnegative constants $\mu$, $\mu_1$, and $M$ such that*

$$\Lambda|p|^2 \leq \mu\mathcal{E}, \quad |p||F_p|,\ \delta F \leq \frac{\mu_1}{\varepsilon} + \frac{\varepsilon}{|p|^2}\mathcal{E}_2 \tag{12.50}$$

*for all $|p| \geq M$ and $\varepsilon \in (0,1)$. Then there are constants $C$ and $\theta$, determined only by $n$, $\mu$, and $\mu_1$, such that, for any $y \in \Omega$ and $R \in (0, d(y))$ such that*

$$\operatorname*{osc}_{B(y,R)} u \leq \theta,$$

*we have*

$$|Du(y)| \leq C\frac{\operatorname{osc}_{B(y,R)} u}{R} + M. \tag{12.51}$$

***Proof.*** We set $\alpha = \operatorname{osc}_{B(y,R)} u$ and assume first that $\alpha \leq 1/(2\mu_1)$. We then define $\eta$ by $\eta(x) = (1 - |x-y|^2)/R^2$, and $h$ by

$$h(z) = \exp((z - \sup_{B(y,R)} u)/\alpha).$$

With $\beta \in (0, 1/2)$ to be further determined, we set

$$M_1 = \sup_{B(y,R)} \eta^2|Du|^2, \quad w = \eta^2|Du|^2 + \beta M_1 h(u).$$

If we apply the operator $D_k u D_k$ to the equation $F[u] = 0$, then a straightforward calculation yields

$$F^{ij}D_{ij}w + B^iD_iw = |Du|^2H_0 + \frac{\beta}{\alpha}M_1hJ_0 + 2\eta^2\mathcal{E}_2,$$

where

$$\begin{aligned}
B^i &= F^i - \frac{4}{\eta}F^{ij}D_j\eta,\\
H_0 &= 2\eta F^{ij}D_{ij}\eta - 6F^{ij}D_i\eta D_j\eta - 2\eta F^iD_i\eta - \eta^2\delta F,\\
J_0 &= F^iD_iu - \frac{4}{\eta}F^{ij}D_i\eta D_ju + F^{ij}D_{ij}u + \frac{1}{\alpha}\mathcal{E}.
\end{aligned}$$

Since $|D\eta| \leq 2/R$ and $F^{ij}D_{ij}\eta \geq -4n\Lambda R^{-2}$, we infer from (12.50) with $\varepsilon = 2\alpha\mu_1$ that

$$F^{ij}D_{ij}w + B^iD_iw \geq H_1\mathcal{E} + J_1\mathcal{E}_2,$$

where

$$\begin{aligned}
H_1 = &- 36\mu R^{-2} - 2\eta|Du|/(\alpha R) - \eta^2|Du|^2/(2\alpha) - \beta M_1h/(2\alpha^2)\\
&- 8\beta M_1h\mu/(\alpha\eta|Du|R) - \beta^2M_1^2h/(2\alpha^2\eta^2|Du|^2) + \frac{\beta}{\alpha^2}M_1h
\end{aligned}$$

and

$$J_1 = -8\mu_1\alpha\eta/(R|Du|) - 2\mu_1\alpha\eta^2 - 2\mu_1 M_1 h\beta/|Du|^2 + 3\eta/2.$$

Now let $x_0$ be a point at which $w$ attains its maximum. Since $\beta \in (0, 1/2]$ and $h(u) \in (0,1)$, it follows that

$$M_1 \le w(x_0) \le \eta^2(x_0)|Du(x_0)|^2 + \frac{M_1}{2},$$

and hence $\eta^2(x_0)|Du(x_0)|^2 \ge M_1/2$. It follows that, at $x_0$, we have

$$F^{ij}D_{ij}w \ge (H_2 + H_3 h\beta/\alpha^2)\mathcal{E} + J_2\eta^2\mathcal{E}_2$$

with

$$\begin{aligned} H_2 &= -(4n+32)\mu/R^2 - 2M_1^{1/2}/(\alpha R) - M_1/\alpha, \\ H_3 &= -8\mu\alpha M_1^{1/2}/R - \beta n\mu M_1 + M_1/2, \\ J_2 &= -8\mu_1\alpha/(RM_1^{1/2}) - 2\alpha\mu_1 - 4\beta\mu_1 + 3/2. \end{aligned}$$

We now take

$$\beta = \min\{1/2, 1/(8\mu_1), 1/(8n\mu)\}$$

and note that $J_2 > 0$ if

$$(16\mu_1\alpha/R)^2 < M_1, \quad \alpha \le 1/(4\mu_1).$$

Moreover, if we have

$$((16(\mu_+\mu_1)\alpha/R)^2 < M_1,$$

it follows that

$$F^{ij}D_{ij}w \ge \mathcal{E}[\beta M_1/(24\alpha)^2 - (4n+32)\mu/R^2 - 2M_1^{1/2}/(\alpha R) - M_1/\alpha]$$

at $x_0$ because $h(u) \ge e^{-1} \ge 1/3$. It follows that there is a constant $C_n$ determined only by $n$ such that $F^{ij}D_{ij}w(x_0) > 0$ if

$$\alpha \le \frac{C_n}{\mu + \mu_1}, \tag{12.52}$$

$$M_1 > \left(\frac{(\mu+\mu_1)\alpha}{C_n R}\right)^2. \tag{12.53}$$

Since we must have $F^{ij}D_{ij}w(x_0) \le 0$, it follows that

$$M_1 \le \max\left\{M^2, \left(\frac{(\mu+\mu_1)\alpha}{C_n R}\right)^2\right\}$$

if $\alpha \le \theta$ with $\theta = C_n/(\mu + \mu_1)$, and this inequality implies (12.51). □

Note that (12.50) follows from the assumptions (12.47a) and

$$(1+|p|)|F_p|,\ |F_z|, |F_x|/(1+|p|) \le \lambda A_1(|z|)(1+|p|^2+|r|) \qquad (12.54)$$

for some increasing function $A_1$.

Our next step is to prove a gradient bound near a flat boundary.

**Lemma 12.17.** *Let $R \in (0,1)$, suppose $F \in C^1(B^+(0,R)\times\mathbb{R}\times\mathbb{R}^n\times\mathbb{S}^n)$, and suppose that $b \in C^1(B^0(0,R)\times\mathbb{R}\times\mathbb{R}^n)$ with $b^n > 0$. Suppose also that there are positive constants $\Lambda$, $\lambda$, and an increasing function $A_1$ such that conditions* (12.47a) *and* (12.54) *are satisfied for all $(x,z,p,r) \in B^+(0,R)\times\mathbb{R}\times\mathbb{R}^n\times\mathbb{S}^n$. Suppose also that there is an increasing function $B_1$ such that*

$$(1+|p|^2)|b_p| + (1+|p|)|b_z| + |b_x| \le B_1(|z|)(1+|p|^2)b^n \qquad (12.55)$$

*for all $(x,z,p) \in B^0(0,R)\times\mathbb{R}\times\mathbb{R}^n$ with $b(x,z,p)=0$ If $u \in C^3(B^+(0,R))\cap C^2(\overline{B^+(0,R)})$ satisfies*

$$F[u] = 0 \text{ in } B^+(0,R), \quad N[u] = 0 \text{ on } B^0(0,R), \qquad (12.56)$$

*then there are positive constants $\theta$ and $C$, determined only by $\Lambda/\lambda$, $A_1(\sup|u|)$, $B_1(\sup|u|)$, and $n$ such that, if*

$$\operatorname*{osc}_{B^+(0,R)} u \le \theta,$$

*then*

$$\sup_{B^+(0,R/4)} |Du| \le C\left(\frac{\operatorname{osc}_{B^+(0,R)} u}{R} + 1\right). \qquad (12.57)$$

***Proof.*** Set $\alpha = \operatorname{osc}_{B^+(0,R)} u$, $\rho = \min\{R/4, \alpha\}$, and write $S$ for the set of all $x \in B^+(0,R/2)$ with $x^n < \rho$. Similarly to the proof of Lemma 12.16, we define functions $\eta$ and $h$ by

$$\eta(x) = 1 - \frac{4|x|^2}{R^2}, \quad h(z) = \exp\left(\frac{z - \sup_{B^+(0,R)} u}{\alpha}\right).$$

Our next step is to note that a simple modification of the proof of Lemma 10.10 gives a function $N \in C^2(B^+(0,R)\times\mathbb{R}\times\mathbb{R}^n)$ and a constant $C$, determined only by $B_1(\sup|u|)$ and $n$ such that

$$\begin{gathered}
\frac{1}{2} \le N^n \le \frac{3}{2}, \quad |N_p| \le 4B_1,\\
|p||N_z| + |N_x| \le C|p|^2,\\
|p|^2|NN_{pp}| + |p||NN_{pz}| + |NN_{px}| \le C|p|^2,\\
|p|^2|NN_{zz}| + |p||NN_{zx}| + |NN_{xx}| \le C|p|^2,
\end{gathered}$$

for $|p| \geq C$ and $b(x, u, Du) = 0$ on $B^0(0, R)$ wherever $|Du| \geq C$. (Note that the expressions $NN_{pp}$, etc., are not typographical errors.) By taking $\varepsilon$ sufficiently small and setting

$$v_1 = (1 + |p'|^2 + \varepsilon N^2)^{1/2}$$

(where $p' = (p_1, \dots, p_{n-1}, 0)$), we have that $v\sqrt{\varepsilon}/2 \leq v_1 \leq 2v$, and $\nu_1 = \partial v_1/\partial p$ satisfies

$$|\nu_1| \leq 2, \quad \nu_1 \cdot Du \geq v_1/8$$

for $v_1 \geq 3$. Finally, there is a constant $C_1$ such that

$$\left(\nu_1 \frac{\partial^2 v_1}{\partial p_i \partial p_j} + \nu_1^i \nu_1^j\right) \xi_i \xi_j \geq C_1 |\xi|^2$$

for all $x \in \mathbb{R}^n$.

We now set $M_1 = \sup_{B^+(0,R/2)} \eta^2 v_1^2$ and, with $\alpha_1$ and $\alpha_2$ positive constants to be further specified, we set

$$w = \eta^2 v_1^2 + \alpha_1 M_1 h(u) + \alpha_2 M_1 v x^n,$$

and we write $x_0$ for a point in $\overline{S}$ at which $w$ attains its maximum. By using Lemma 12.16, we find that $vx^n \leq C_2 \alpha$ in $B^+(0, R/2)$ with $C_2$ determined only by $n$, $\Lambda/\lambda$ and $A_1(\sup|u|)$ if $\alpha$ is sufficiently small. Hence, if $\alpha_1 \leq 1/4$ and

$$C_2 \alpha \alpha_2 \leq 1/4, \tag{12.58}$$

we conclude that $\eta^2 v_1^2 \geq M_1/2$ and $\eta^2 v^2 \geq M_1/8$ at $x_0$.

If $x_0 \in B^+(0, R/2)$ and $|Du(x_0)| \geq C$, then we have

$$\begin{aligned} b^i D_i w &= 2v_1^2 \eta b^i D_i \eta - 2\eta^2 (b_z |D'u|^2 + b_x \cdot D'u) + \frac{\alpha_1}{\alpha} M_1 h Du \cdot b_p + \alpha_2 M_1 v b^n \\ &\geq b^n M_1 v \left(\alpha_2 - \frac{2B_0}{\eta v R} - \frac{B_0 \alpha_1}{\alpha} - 4B_1\right). \end{aligned}$$

If $M_1 \geq (8\alpha/(\alpha_1 R))^2$ and

$$\alpha_2 = 4B_1 + \frac{3B_0 \alpha_1}{\alpha},$$

then we conclude that $b^i D_i w(x_0) > 0$ which contradicts the assumption that $w$ has a maximum at $x_0$. We also observe that our choice of $\alpha_2$ implies that (12.58) follows from the inequality $C_1 M(4B_1 \alpha + 3B_0 \alpha_1) \leq 1/4$, and this inequality holds if $\alpha$ and $\alpha_1$ are sufficiently small.

If $x_0 \in S$ and $|Du(x_0)| \geq C$, then we define the operator $L$ by

$$Lw = F^{ij} D_{ij} w + (F^i - \frac{4}{\eta} F^{ij} D_j \eta) D_i w.$$

It follows that there are positive constants $C_2$ and $C_3$ such that

$$Lw(x_0) \geq M_1\alpha^{-2}h(u)\mathcal{E}(\alpha_1 - C_2\alpha_1^2 - C_3\alpha^2)$$

so we can't have $x_0 \in S$ if $|Du(x_0)| \geq C$ provided

$$\alpha_1 \leq \frac{1}{2C_2}, \quad \alpha \leq \frac{1}{(2C_3)^{1/2}}.$$

It follows that the maximum of $w$ occurs either where $|Du| \leq C$ or where $x^n \geq \rho$. Since Lemma 12.16 gives a bound on $Du(x_0)$ if $x^n \geq \rho$, we obtain (12.57) by straightforward algebra. □

By combining this estimate with Theorem 12.15, Lemma 12.16, and a $C^3$ change of variables, we infer the following global gradient bound.

**Theorem 12.18.** *Let $\Omega$ be a bounded domain in $\mathbb{R}^n$ with $\partial\Omega \in C^3$. Suppose conditions* (12.47), (12.48), (12.54) *are satisfied. Suppose also that there is an increasing function $B_1$ such that*

$$(1+|p|^2)|b_p| + (1+|p|)|b_z| + |b_x| \leq B_1(|z|)(1+|p|^2)b_p \cdot \gamma \tag{12.59}$$

*for all $(x,z,p) \in \partial\Omega \times \mathbb{R} \times \mathbb{R}^n$ with $b(x,z,p) = 0$. Then, for any solution $u \in C^3(\Omega) \cap C^2(\overline{\Omega})$ of* (12.7), *there is a constant $C$, determined only by $\Lambda/\lambda$, $A_0(\sup|u|)$, $A_1(\sup|u|)$, $B_0(\sup|u|)$, $B_1(\sup|u|)$, and $\Omega$ such that*

$$\sup_\Omega |Du| \leq C. \tag{12.60}$$

A Hölder gradient estimate then follows by straightforward modification of the proof of Lemma 11.12.

**Theorem 12.19.** *Let $\partial\Omega \in C^3$ and suppose that there are positive constants $\lambda$, $\Lambda$ such that* (12.47) *is satisfied. Suppose also that there is an increasing function $\mu_1$ such that*

$$|F_p| + |F_z| + |F_x| \leq \mu_1(|z| + |p|)(1+|r|) \tag{12.61}$$

*for all $(x,z,p,r) \in \Omega \times \mathbb{R} \times \mathbb{R}^n \times \mathbb{S}^n$ and*

$$|b_p| + |b_z| + |b_x| \leq \mu_1(|z| + |p|)b_p \cdot \gamma \tag{12.62}$$

*for all $(x,z,p) \in \Omega \times \mathbb{R} \times \mathbb{R}^n$ with $b(x,z,p) = 0$. If $u \in C^3(\Omega) \cap C^2(\overline{\Omega})$ solves* (12.7), *then there is are positive constants $C$ and $\delta$, determined only by $n$, $\Omega$, $\Lambda/\lambda$, and $\mu_1(|u|_1)$ such that*

$$[Du]_\delta \leq C. \tag{12.63}$$

***Proof.*** We only prove a local Hölder gradient estimate near a point in $\partial\Omega$ since the corresponding interior estimate is proved similarly. In addition, since the hypotheses are invariant under a $C^3$ change of independent variable, we only may assume that $u$ satisfies (12.56). With $N$ as in the proof of Lemma 12.17 and $\varepsilon > 0$ sufficiently small, we set $V = |D'u|^2 + \varepsilon N^2$. Then with $\theta$ a positive constant to be further determined, we define, for $k = 1, \ldots, n$,

$$w_k^{\pm} = \pm D_k u + \theta V.$$

Using the given structure conditions, we find that there are constants $c_2$ and $c_3$ (determined only by $n$, $\Lambda/\lambda$, and $\mu_1(|u|_1)$ such that

$$F^{ij} D_{ij} w_k^{\pm} \geq -\frac{c_2}{\theta}\lambda |Dw_k^{\pm}|^2 - c_3\lambda.$$

The proof of the Hölder gradient estimate at 0 is now almost identical to that of Lemma 11.12; the only change is that the functions $h_k^{\pm}$ in that proof satisfy inhomogeneous differential equations. □

We note that second derivative Hölder estimates for $u$ follow immediately from Theorem 12.14, so we obtain the following existence result very easily.

**Theorem 12.20.** *Let* $\partial\Omega \in C^3$, *let* $F \in C^1(\Omega \times \mathbb{R} \times \mathbb{R}^n \times \mathbb{S}^n)$, *and let* $b \in C^{1,\alpha}(\partial\Omega \times \mathbb{R} \times \mathbb{R}^n)$ *for some* $\alpha \in (0,1)$. *Suppose $F$ and $b$ satisfy conditions* (12.9), (12.47), (12.48), (12.54), *and* (12.59). *Suppose also that* $F_z(x,z,p,r) < 0$ *for all* $(x,z,p,r) \in \Omega \times \mathbb{R} \times \mathbb{R}^n \times \mathbb{S}^n$ *and* $b_z(x,z,p) \leq 0$ *for all* $(x,z,p) \in \partial\Omega \times \mathbb{R} \times \mathbb{R}^n$. *Then there is a unique solution of* (12.7).

## Notes

The basic material in Sections 12.1 and 12.2 comes from Chapter 17 of [64], and we refer the reader there for detailed references. We only mention that Lemma 12.7 is the lemma on page 255 of [141], and that Lemma 12.8 was first proved simultaneously (using essentially the method described here) by Evans [42] and Krylov [86].

The first study of oblique derivative problems for fully nonlinear elliptic equations was [121]; the authors looked at equations of Bellman type and linear boundary conditions. Shortly thereafter, Trudinger and I, in [120] extended the results of [121] to general uniformly elliptic equations and nonlinear boundary conditions. Unfortunately, our second derivative

Hölder estimates required an estimate on the Hölder norm of the derivatives of $F$ with respect to the variables $x, z, p, r$, although (as pointed out in the Supplementary remarks to [120]) it was shown in [179] that the second derivative Hölder estimate can be proved without this condition. Further variations on the second derivative Hölder estimate are given in [164] and [117]. The proof in [164] is stated for nonlinear equations with linear boundary conditions, but it can be modified to include nonlinear boundary conditions like the ones studied here. In broad outline, the goal in [164] is to reduce the estimate to one with the boundary condition $D_n u = 0$ on $\partial\Omega$ and then to show that the estimate in this case follows from an estimate for the same boundary condition but with $\partial\Omega$ being locally the hyperplane $x^n = 0$. Our proof is based on the one in [117], but we have included more details. We have followed [120] here in most other respects except that we use the Hölder gradient estimate method described in Chapter 11, and we have used the function $N$ (from Lemma 10.8) to deduce the gradient bound. We refer the reader to [120] for further examination of uniformly elliptic oblique derivative problems; in particular, we mention Corollary 7.10 in [120] which removes the monotonicity assumptions (with respect to $z$) on $F$ and $b$ in our Theorem 12.20.

For nonuniformly elliptic problems, the study of oblique derivative problems is much more complicated and not yet in a final form. An important class of nonuniformly elliptic equations is based on Monge-Ampère operators, which have the form

$$F[u] = \det D^2 u - \psi(x, u, Du). \tag{12.64}$$

Since $F^{ij}(x, z, p, r) = r^{ij}$, where $r^{ij}$ is the cofactor of $r_{ij}$, it follows that these operators are elliptic only when $r$ is positive definite, which means that solutions must be strictly convex. In this case, we also know that $\det D^2 u > 0$, so we must restrict attention to positive functions $\psi$. This class of operators is important because it is elliptic only for some functions, and the theory presented so far has been concerned with operators that are elliptic for all functions. The Neumann problem for Monge-Ampère equations, that is, the problem

$$\det D^2 u = \psi(x, u, Du) \text{ in } \Omega, \quad \gamma \cdot Du = \varphi(x, u) \text{ on } \partial\Omega,$$

was studied in detail by Lions, Trudinger, and Urbas in [122]. They showed proved the existence of a solution to this problem under fairly simple hypothesis. They assumed that $\psi \in H_2(K)$ for any bounded subset $K$ of $\Omega \times \mathbb{R} \times \mathbb{R}^n$, $\varphi \in H_3(K')$ for any compact subset $K'$ of $\partial\Omega \times \mathbb{R}$, and that

$\partial\Omega \in H_4$. For structure conditions, they assumed $\varphi_z$ to be bounded from below by a positive constant, and that there are functions $g \in L^1(\Omega)$ and $h \in L^1_{\mathrm{loc}}(\mathbb{R}^n)$, and a constant $N$, such that

$$\psi(x, N, p) \leq \frac{g(x)}{h(p)}$$

for all $(x, p) \in \Omega \times \mathbb{R}^n$ with

$$\int_\Omega g(x)\, dx < \int_{\mathbb{R}^n} h(p)\, dp.$$

They also proved solvability of some related boundary value problems, in which the differential equation is slightly modified. The estimates on $u$ and $Du$ in [122] are also valid, with the proofs given there, for a more general class of boundary conditions, but the second derivative bounds rely heavily on the exact form of the boundary condition.

The significance of the normal derivative condition for the Monge-Ampère equation was examined in great detail in [192] and [95]. Although the two papers use slightly different methods and obtain slightly different results, the most important point we wish to make here is that both papers prove second derivative estimates and, hence, existence theorems when $b(x, z, p)$ is sufficiently close to $\gamma \cdot p - \varphi(x, z)$; specifically they assume that $|b_p - \gamma|$ is sufficiently small in an appropriate norm. Moreover, when $|b_p - \gamma|$ is large, then the estimates fail.

The extension of these results to more general differential equations is still not well studied. Urbas [189, 190] studied (12.7) with more general differential equations when $b$ is close to $\gamma \cdot p - \varphi$ but only when $n = 2$.

An alternative boundary condition, first studied by Pogorelov (see pages 41–44 of [154]), is called the second boundary value problem. In the present situation, this problem is to solve the differential equation subject to the constraint $Du(\Omega) = \Omega^*$ for a given domain $\Omega^*$. For the Monge-Ampère equation, the convexity of the solution implies that this constraint is equivalent to the pointwise condition $h(Du) = 0$ on $\partial\Omega$, where $h$ is a defining function for $\Omega^*$. In other words $h \in C^1(\mathbb{R}^n)$, $\Omega$ is the set on which $h$ is positive, and $|Dh| + |h|$ never vanishes in $\mathbb{R}^n$. When $\Omega$ and $\Omega^*$ are convex, Urbas [191] showed that such a function $h$ must be oblique on $\partial\Omega$. Hence the second boundary value problem is an oblique boundary value problem in this case, and the estimates in [122] were adapted in [191] to prove second derivative bounds for solutions of the second boundary value problem for Monge-Ampère equations. In addition, the existence of solutions was proved there. Extensions of this result to a slightly larger class of differential equations was given in [193]. We mention also [181] for a discussion of the problem of optimal transportation in this context and [197] for an extension of the results in [181] to some more general differential equations. Both papers examine the second boundary problem $Du(\Omega) = \Omega^*$ for different differential equations using the idea of conversion to an oblique boundary value problem.

# Bibliography


[1] Agmon, S., Douglis, A. and Nirenberg, L. (1959). Estimates near the boundary for solutions of elliptic partial differential equations satisfying general boundary conditions. I, *Comm. Pure Appl. Math.* **12**, pp. 623–727.

[2] Aleksandroff, A. D. (1963). Uniqueness conditions and estimates for the solution of the Dirichlet problem, *Vestnik Leningrad Univ.* **18**, 3, pp. 5–29, [Russian]. English transl. in Amer. Math. Soc. Transl. (2) **68** (1968), 89–119.

[3] Alexandroff, A. D. (1939). Almost everywhere existence of the second differential of a convex function and some properties of convex surfaces connected with it, *Leningrad State Univ. Annals [Uchenye Zapiski] Math. Ser.* **6**, pp. 3–35, [Russian].

[4] Athanassenas, M. and Finn, R. (2006). Compressible fluids in a capillary tube, *Pacific J. Math.* **224**, pp. 201–229.

[5] Bakel′man, I. Y. (1961). Theory of quasilinear elliptic equations, *Sibirsk. Mat. J̌.* **2**, pp. 179–186, [Russian].

[6] Barles, G. (1993). Fully nonlinear Neumann type boundary conditions for second-order elliptic and parabolic equations, *J. Differential Equations* **106**, pp. 90–106.

[7] Barles, G. (1999). Nonlinear Neumann boundary conditions for quasilinear degenerate elliptic equations and applications, *J. Differential Equations* **154**, pp. 191–224.

[8] Barles, G. and Da Lio, F. (2006). Local $C^{0,\alpha}$ estimates for viscosity solutions of Neumann-type boundary problems, *J. Differential Equations* **225**, pp. 202–241.

[9] Bernstein, S. (1910). Sur la gènèralisation du problème de Dirichlet. ii, *Math. Ann.* **69**, pp. 82–136.

[10] Bers, L. and Nirenberg, L. (1955). On linear and non-linear elliptic boundary value problems in the plane, in *Convegno Internazionale sulle Equazioni Lineari alle Derivate Parziali, Trieste, 1954* (Edizioni Cremonese, Roma), pp. 141–167.

[11] Bhattacharya, T. and Leonetti, F. (1991). A new Poincaré inequality and its application to the regularity of minimizers of integral functionals with

nonstandard growth, *Nonlinear Anal.* **17**, pp. 833–839.

[12] Bombieri, E., De Giorgi, E. and Miranda, M. (1969). Una maggiorazione a priori relativa alle ipersuperfici minimali non parametriche, *Arch. Rational Mech. Anal.* **32**, pp. 255–267.

[13] Bombieri, E. and Giusti, E. (1973). Local estimates for the gradient of non-parametric surfaces of prescribed mean curvature, *Comm. Pure Appl. Math.* **26**, pp. 381–394.

[14] Brandt, A. (1969a). Interior estimates for second-order elliptic differential (or finite-difference) equations via the maximum principle, *Israel J. Math.* **7**, pp. 95–121.

[15] Brandt, A. (1969b). Interior Schauder estimates for parabolic differential-(or difference-) equations via the maximum principle, *Israel J. Math.* **7**, pp. 254–262.

[16] Browder, F. E. (1959). Estimates and existence theorems for elliptic boundary value problems, *Proc. Nat. Acad. Sci. U.S.A.* **45**, pp. 365–372.

[17] Busemann, H. and Feller, W. (1936). Krümmungseigenschaften Konvexer Fläschen, *Acta Math.* **66**, pp. 1–47, [German].

[18] Byun, S. and Wang, L. (2004). Elliptic equations with BMO coefficients in Reifenberg domains, *Comm. Pure Appl. Math.* **57**, pp. 1283–1310.

[19] Byun, S. and Wang, L. (2005). The conormal derivative problem for elliptic equations with BMO coefficients on Reifenberg flat domains, *Proc. London Math. Soc.* **90**, pp. 245–272.

[20] Caccioppoli, R. (1934). Sulle equazioni ellittiche a derivate parziali con $n$ variabili indipendenti. *Atti Accad. Naz. Lincei, Rend., VI. Ser.* **19**, pp. 83–89.

[21] Caffarelli, L. A. (1989). Interior a priori estimates for solutions of fully nonlinear equations, *Ann. of Math. (2)* **130**, 1, pp. 189–213.

[22] Calderón, A. P. (1985). Boundary value problems for the Laplace equation in Lipschitzian domains, in *Recent Progress in Fourier Analysis* (Elsevier), pp. 33–48.

[23] Caldeŕon, A. P. and Zygmund, A. (1952). On the existence of certain singular integrals, *Acta Math.* **88**, pp. 85–139.

[24] Calkin, J. W. (1940). Functions of several variables and absolute continuity. I, *Duke Math. J.* **6**, pp. 170–186.

[25] Campanato, S. (1964). Proprietà di una famiglia di spazi funzionali, *Ann. Scuola Norm. Sup. Pisa (3)* **18**, pp. 137–160.

[26] Campanato, S. (1965). Equazioni elllitiche del $II^o$ ordine e spazi $\mathfrak{L}^{(2,\lambda)}$, *Ann. Mat. Pura Appl.* **(4) 69**, pp. 321–381.

[27] Canic, S., Keyftiz, B. L. and Lieberman, G. M. (2000). A proof of existence of perturbed steady transonic shocks via a free boundary problem, *Comm. Pure Appl. Math.* **53**, pp. 484–511.

[28] Caristi, J. (1976). Fixed point theorems for mappings satisfying inwardness conditions, *Trans. Amer. Math. Soc.* **215**, pp. 241–251.

[29] Chiarenza, F., Frasca, M. and Longo, P. (1993). $W^{2,p}$-solvability of the Dirichlet problem for nondivergence elliptic equations with VMO coefficients, *Trans. Amer. Math. Soc.* **336**, pp. 841–853.

[30] Cianchi, A. and Maz'ya, V. G. (2011). Global Lipschitz regularity for a class of quasilinear elliptic equations, *Comm. Partial Differential Equations* **36**, pp. 100–133.
[31] Concus, P. and Finn, R. (1975a). On capillary free surfaces in a gravitational field, *Acta Math.* **132**, pp. 207–223.
[32] Concus, P. and Finn, R. (1975b). On capillary free surfaces in the absence of gravity, *Acta Math.* **132**, pp. 177–198.
[33] Crandall, M. and Ishii, H. (1990). The maximum principle for semicontinuous functions, *Differential Integral Equations* **3**, pp. 1001–1014.
[34] Crandall, M. G., Ishii, H. and Lions, P. L. (1992). User's guide to viscosity solutions of second order partial differential equations, *Bull. Amer. Math. Soc.* **27**, pp. 1–67.
[35] De Giorgi, E. (1957). Sulla differenziabilità e l'analiticità delle estremali degli integrali multipli regolari, *Mem. Accad. Sci. Torino. Cl. Sci. Fis. Mat. Nat. (3)* **3**, pp. 25–43.
[36] DiBenedetto, E. (2002). *Real analysis*, Birkhäuser Advanced Texts: Basler Lehrbücher (Birkhäuser Boston Inc., Boston, MA).
[37] DiBenedetto, E. and Trudinger, N. S. (1984). Harnack inequalities for quasi-minima of variational integrals, *Ann. Inst. H. Poincaré Anal. Non Linéaire* **1**, 4, pp. 295–308.
[38] Dong, G. C. (1988). Initial and boundary value problems for fully nonlinear parabolic equations, *J. Partial Differential Equations Ser. A* **1**, pp. 12–42.
[39] Douglis, A. and Nirenberg, L. (1955). Interior estimates for elliptic systems of partial differential equations, *Comm. Pure Appl. Math.* **8**, pp. 503–538.
[40] Dupuis, P. and Ishii, H. (1990). On oblique derivative problems for fully nonlinear second-order partial differential equations on nonsmooth domains, *Nonlinear Anal.* **15**, pp. 1123–1138.
[41] Dupuis, P. and Ishii, H. (1991). On oblique derivative problems for fully nonlinear second-order elliptic PDEs on domains with corners, *Hokkaido Math. J.* **20**, pp. 135–164.
[42] Evans, L. C. (1982). Classical solutions of fully nonlinear, convex, second order elliptic equations, *Comm. Pure Appl. Math.* **35**, pp. 333–363.
[43] Evans, L. C. and Gariepy, R. F. (1992). *Measure theory and fine properties of functions*, Studies in Advanced Mathematics (CRC Press, Boca Raton, FL).
[44] Fichera, G. (1960). On a unified theory of boundary value problems for elliptic-parabolic equations of second order, in *Boundary problems in differential equations* (Univ. of Wisconsin Press, Madison), pp. 97–120.
[45] Finn, R. (1963). New estimates for equations of minimal surface type, *Arch. Rational Mech. Anal.* **14**, pp. 337–375.
[46] Finn, R. (1986). *Equilibrium Capillary Surfaces, Grundlehren der mathematischen Wissenschaften*, Vol. 284 (Springer-Verlag, New York-Berlin-Heidelberg-Tokyo).
[47] Finn, R. (1999). Capillary surface interfaces, *Notices Amer. Math. Soc.* **46**, pp. 770–781.
[48] Finn, R. (2002a). Eight remarkable properties of capillary surfaces, *Math.*

*Intelligencer* **24**, 3, pp. 21–33.

[49] Finn, R. (2002b). Some properties of capillary surfaces, *Milan J. Math.* **70**, pp. 1–23.

[50] Finn, R. (2004). Capillarity problems for compressible fluids, *Mem. Differential Equations Math. Phys.* **33**, pp. 47–55.

[51] Finn, R. and Luli, G. K. (2007). On the capillary problem for compressible fluids, *J. Math. Fluid Mech.* **9**, pp. 87–103.

[52] Fiorenza, R. (1959a). Sui problemi di derivata obliqua per la equazioni ellittiche non lineari in due variabili, *Ricerche Mat.* **8**, pp. 222–239.

[53] Fiorenza, R. (1959b). Sui problemi di derivata obliqua per le equazioni ellittiche, *Ricerche Mat.* **8**, pp. 83–110.

[54] Fiorenza, R. (1965). Sulla hölderianitá delle soluzioni dei problemi di derivata obliqua regolare del secondo ordine, *Ricerche Mat.* **14**, pp. 102–123.

[55] Fiorenza, R. (1966). Sui problemi di derivata obliqua per le equazioni ellittiche quasi lineari, *Ricerche Mat.* **15**, pp. 74–108.

[56] Fraenkel, L. E. (1982). On the embedding of $C^1(\bar{\Omega})$ in $C^{0,\alpha}(\bar{\Omega})$, *J. London Math. Soc. (2)* **26**, pp. 290–298.

[57] Friedman, A. (1982). *Foundations of Modern Analysis* (Dover Publications Inc., New York), reprint of the 1970 original.

[58] Friedrichs, K. O. (1944). The identity of weak and strong extensions of differential operators, *Trans. Amer. Math. Soc.* **55**, pp. 132–151.

[59] Friedrichs, K. O. (1986). *Selecta. Vol. 1, 2*, Contemporary Mathematicians (Birkhäuser Boston Inc., Boston, MA), edited and with a foreword by Cathleen S. Morawetz, With a biography by Constance Reid, With commentaries by Peter D. Lax, Tosio Kato, Fritz John, Wolfgang Wasow, Harold Weitzner, Louis Nirenberg and David Isaacson.

[60] Gagliardo, E. (1957). Caratterizzazioni delle tracce sulla frontiera relative ad alcune classi di funzioni in $n$ variabili, *Rend. Sem. Mat. Univ. Padova* **27**, pp. 284–305.

[61] Gerhardt, C. (1976). Global regularity of the solutions to the capillarity problem, *Ann. Scuola Norm. Sup. Pisa Cl. Sci. (4)* **3**, pp. 157–175.

[62] Gidas, B., Ni, W.-M. and Nirenberg, L. (1979). Symmetry and related properties via the maximum principle, *Comm. Math. Phys.* **68**, pp. 209–243.

[63] Gilbarg, D. and Hörmander, L. (1980). Intermediate Schauder estimates, *Arch. Rational Mech. Anal.* **74**, pp. 297–318.

[64] Gilbarg, D. and Trudinger, N. S. (2001). *Elliptic Partial Differential Equations of Second Order*, Classics in Mathematics (Springer-Verlag, Berlin), reprint of the 1998 edition.

[65] Giraud, G. (1933). Problèmes de valeurs à la frontière relatifs à certaines données discontinues, *Bull. Soc. Math. France* **61**, pp. 1–54.

[66] Granucci, T. (2006). Remarks on Hölder continuity for a minimum of a convex functional with nonstandard growth, *Rend. Istit. Mat. Univ. Trieste* **38**, pp. 21–43.

[67] Greco, D. (1956). Nuove formole integrali di maggiorazione per le soluzioni

di un'equazione lineare di tipo ellittico ed applicazioni alla teoria del potenziale, *Ricerche Mat.* **5**, pp. 126–149.

[68] Grisvard, P. (1985). *Elliptic problems in nonsmooth domains*, *Monographs and Studies in Mathematics*, Vol. 24 (Pitman (Advanced Publishing Program), Boston, MA).

[69] Hardy, G. H. (1925). Notes on some points in the integral calculus. LX. An inequality between integrals, *Messenger of Math.* **54**, pp. 150–156.

[70] Harnack, C.-G. A. (1887). *Die Grundlagen der Theorie des logarithmischen Potentiales und der eindeutigen Potentialfunktion in der Ebene* (Teubner, Leipzig, Gernmany).

[71] Hölder, O. L. (1882). *Beiträge zur Potentialtheorie*, Ph.D. thesis, Tübingen.

[72] Hopf, E. (1927). Elementare Bemerkungen über die Lösungen partieller Differentialgleichungen zweiter Ordnung vom elliptischen Typus. *Sitzungsberichte Akad. Berlin* **1927**, pp. 147–152, [German].

[73] Hopf, E. (1929). Zum analytischen Charakter der Lösungen regulärer zweidimensionaler Variationsprobleme, *Math. Z.* **30**, pp. 404–413.

[74] Hopf, E. (1952). A remark on linear elliptic differential equations of second order, *Proc. Amer. Math. Soc.* **3**, pp. 791–793.

[75] Hurri-Syrjänen, R. (1996). A generalization of an inequality of Bhattacharya and Leonetti, *Canad. Math. Bull.* **39**, pp. 438–447.

[76] Il′in, V. P. (1945). An immersion theorem with limiting exponent, *Dokl. Akad. Nauk SSSR* **96**, pp. 905–908.

[77] Ishii, H. (1992). Fully nonlinear oblique derivative problems for nonlinear second-order elliptic PDE's, *Duke Math. J.* **62**, pp. 633–661.

[78] Kenig, C. and Pipher, J. (1988). The oblique derivative problem on Lipschitz domains with $L^p$ data, *Amer. J. Math.* **110**, pp. 715–737.

[79] Kim, D. and Krylov, N. V. (2007). Elliptic differential equations with coefficients measurable with respect to one variable and VMO with respect to the others, *SIAM J. Math. Anal.* **39**, pp. 489–506 (electronic).

[80] Kirk, W. A. and Caristi, J. (1975). Mapping theorems in metric and Banach spaces, *Bull. Acad. Pol. Sci.* **23**, pp. 891–894.

[81] Klimov, V. S. (2000). Embedding theorems and the continuity of generalized solutions of quasilinear elliptic equations, *Differ. Uravn.* **36**, 6, pp. 784–791, 862, [Russian]. English transl. in Differential Equations **36** (2000), 870–877.

[82] Korevaar, N. (1986). An easy proof of the interior gradient bound for solutions to the prescribed mean curvature equation, *Proc. Sympos. Pure Math.* **45 Part 2**, pp. 81–89.

[83] Korevaar, N. and Simon, L. (1996). Equations of mean curvature type with contact angle boundary conditions, in J. Jost (ed.), *Geometric Analysis and the Calculus of Variations* (International Press, Cambridge, MA), pp. 175–201.

[84] Korevaar, N. J. (1988). Maximum principle gradient estimates for the capillary problem, *Comm. Partial Differential Equations* **13**, pp. 1–31.

[85] Kozlov, V. A., Maz'ya, V. G. and Rossmann, J. (1997). *Elliptic Boundary Value Problems in Domains with Point Singularities*, *Mathematical Surveys*

*and Monographs*, Vol. 52 (American Mathematical Society, Providence, RI).

[86] Krylov, N. V. (1982). Boundedly nonhomogeneous elliptic and parabolic equations, *Izv. Akad. Nauk SSSR Ser. Mat.* **46**, pp. 487–523, [Russian]. English translation in Math. USSR-Izv. **20** (1983), 459–493.

[87] Krylov, N. V. (1996). *Lectures on elliptic and parabolic equations in Hölder spaces*, *Graduate Studies in Mathematics*, Vol. 12 (American Mathematical Society, Providence, RI).

[88] Krylov, N. V. and Safonov, M. V. (1980). Certain properties of solutions of parabolic equations with measurable coefficients, *Izv. Akad. Nauk. SSSR* **40**, pp. 161–175, [Russian]. English transl. in Math. USSR Izv. **16** (1981), 151–164.

[89] Kufner, A., Maligranda, L. and Persson, L.-E. (2006). The prehistory of the Hardy inequality, *Amer. Math. Monthly* **113**, pp. 715–732.

[90] Ladyženskaja, O. A., Solonnikov, V. A. and Ural′ceva, N. N. (1968). *Linear and quasilinear equations of parabolic type* (Izdat. "Nauka", Moscow), [Russian]. English transl. published by American Mathematical Society, 1968.

[91] Ladyzhenskaya, O. A. and Ural′tseva, N. N. (1968). *Linear and Quasilinear Elliptic Equations*, Translated from the Russian by Scripta Technica, Inc. Translation editor: Leon Ehrenpreis (Academic Press, New York).

[92] Ladyzhenskaya, O. A. and Ural′tseva, N. N. (1970). Local estimates for gradients of solutions of non-uniformly elliptic and parabolic equations, *Comm. Pure Appl. Math.* **23**, pp. 677–703.

[93] Ladyzhenskaya, O. A. and Ural′tseva, N. N. (1980). Estimate of the Hölder norm of the solutions of second-order quasilinear elliptic equations of the general form, *Zap. Nauchn. Sem. Leningrad. Otdel. Mat. Inst. Steklov. (LOMI)* **96**, pp. 161–168, [Russian]. English transl. in J. Soviet Math. **21** (1983), 762–768.

[94] Lancaster, K. E. (2010). A proof of the Concus-Finn conjecture, *Pac. J. Math.* **247**, pp. 75–108.

[95] Li, S.-Y. (1999). On the oblique boundary value problems for Monge-Ampère equations, *Pacific J. Math.* **190**, pp. 155–172.

[96] Lieberman, G. M. (1982). Solvability of quasilinear elliptic equations with nonlinear boundary conditions, *Trans. Amer. Math. Soc.* **273**, pp. 753–765.

[97] Lieberman, G. M. (1983). The conormal derivative problem for elliptic equations of variational type, *J. Differential Equations* **49**, pp. 218–257.

[98] Lieberman, G. M. (1984). The nonlinear oblique derivative problem for quasilinear elliptic equations, *Nonlinear Anal.* **8**, pp. 49–65.

[99] Lieberman, G. M. (1985). Regularized distance and its applications, *Pac. J. Math.* **117**, pp. 329–352.

[100] Lieberman, G. M. (1986). Mixed boundary value problems for elliptic and parabolic differential equations of second order, *J. Math. Anal. Appl.* **113**, pp. 422–440.

[101] Lieberman, G. M. (1987a). Gradient bounds for solutions of nonuniformly elliptic oblique derivative problems, *Nonlinear Anal.* **11**, pp. 49–61.

[102] Lieberman, G. M. (1987b). Hölder continuity of the gradient of solutions

of uniformly parabolic equations with conormal boundary conditions, *Ann. Mat. Pura Appl.* **148**, pp. 77–99, 397–398.

[103] Lieberman, G. M. (1987c). Local estimates for subsolutions and supersolutions of oblique derivative problems for general second order elliptic equations, *Trans. Amer. Math. Soc.* **304**, pp. 343–353.

[104] Lieberman, G. M. (1987d). Oblique derivative problems in Lipschitz domains, *Boll. Un. Mat. Ital.* **(7)1-B**, pp. 1185–1210.

[105] Lieberman, G. M. (1987e). Two-dimensional nonlinear boundary value problems for elliptic equations, *Trans. Amer. Math. Soc.* **300**, pp. 287–295.

[106] Lieberman, G. M. (1988a). The conormal derivative problem for nonuniformly parabolic equations, *Indiana Univ. Math. J.* **37**, pp. 23–72.

[107] Lieberman, G. M. (1988b). Gradient estimates for capillary-type problems via the maximum principle, *Comm. Partial Differential Equations* **13**, pp. 33–59.

[108] Lieberman, G. M. (1988c). Hölder continuity of the gradient at a corner for the capillary problem and related results, *Pacific J. Math.* **133**, pp. 115–135.

[109] Lieberman, G. M. (1988d). Oblique derivative problems in Lipschitz domains II. Discontinuous boundary data, *J. Reine Angew. Math.* **389**, pp. 1–21.

[110] Lieberman, G. M. (1989). Optimal Hölder regularity for mixed boundary value problems , *J. Math. Appl. Anal.* **143**, pp. 572–586.

[111] Lieberman, G. M. (1990). On the Hölder gradient estimate for solutions of nonlinear elliptic and parabolic oblique boundary value problems, *Comm. Partial Differential Equations* **15**, pp. 515–523.

[112] Lieberman, G. M. (1991). The natural generalization of the natural conditions of Ladyzhenskaya and Ural′tseva, *Comm. Partial Differential Equations* **16**, pp. 311–361.

[113] Lieberman, G. M. (1992). The conormal derivative problem for equations of variational type in nonsmooth domains, *Trans. Amer. Math. Soc.* **330**, pp. 41–67.

[114] Lieberman, G. M. (1996). *Second Order Parabolic Differential Equations* (World Scientific Publishing Co. Inc., River Edge, NJ).

[115] Lieberman, G. M. (2000). The maximum principle for equations with composite coefficients, *Electron. J. Differential Equations*, pp. No. 38, 17 pp. (electronic).

[116] Lieberman, G. M. (2001). Pointwise estimates for oblique derivative problems in nonsmooth domains, *J. Differential Equations* **173**, pp. 178–211.

[117] Lieberman, G. M. (2002). Higher regularity for nonlinear oblique derivative problems in Lipschitz domains, *Ann. Scuola Norm. Sup. Pisa* **1**, pp. 111–151.

[118] Lieberman, G. M. (2008). Elliptic equations with strongly singular lower order terms, *Indiana Univ. Math. J.* **57**, pp. 2097–2135.

[119] Lieberman, G. M. (2012). A new, simple existence theorem for fully nonlinear elliptic equations, *Comm. Appl. Nonlinear Anal.* **19**, 3, pp. 1–13.

[120] Lieberman, G. M. and Trudinger, N. S. (1986). Nonlinear oblique boundary value problems for nonlinear elliptic equations, *Trans. Amer. Math. Soc.* **295**, pp. 509–546.

[121] Lions, P. L. and Trudinger, N. S. (1986). Linear oblique derivative problems for the uniformly elliptic Bellman operator, *Math. Z.* **191**, pp. 1–15.

[122] Lions, P.-L., Trudinger, N. S. and Urbas, J. I. E. (1986). The Neumann problem for equations of Monge-Ampère type, *Comm. Pure Appl. Math.* **39**, pp. 539–563.

[123] Lipschitz, R. (1868–1869). Disamina della possibilità d'integrare completamente un dato sistema di equazioni differenziali ordinarie, *Ann. Mat. Pura Appl.* **(2) 2**, pp. 288–302.

[124] Lützen, J. (1982). *The Prehistory of the Theory of Distributions, Studies in the History of Mathematics and Physical Sciences*, Vol. 7 (Springer-Verlag, New York).

[125] Marcellini, P. (1993). Regularity for elliptic equations with general growth conditions, *J. Differential Equations* **105**, pp. 296–333.

[126] Maz′ja, V. G. and Plamenevskiĭ, B. A. (1978). Schauder estimates for the solutions of elliptic boundary value problems in domains with edges on the boundary, in *Partial differential equations (Russian), Trudy Sem. S. L. Soboleva, No. 2*, Vol. 1978 (Akad. Nauk SSSR Sibirsk. Otdel. Inst. Mat., Novosibirsk), pp. 69–102, 143, [Russian]. English transl. in Amer Math. Soc. Transl. **123** (1984), 141–169.

[127] Maz'ya, V. (2011). *Sobolev spaces with applications to elliptic partial differential equations, Grundlehren der Mathematischen Wissenschaften*, Vol. 342, augmented edn. (Springer, Heidelberg).

[128] McShane, E. J. (1934). Extension of range of functions, *Bull. Amer. Math. Soc.* **40**, pp. 837–842.

[129] Michael, J. H. (1977). A general theory for linear elliptic partial differential equations, *J. Differential Equations* **23**, pp. 1–29.

[130] Michael, J. H. (1981). Barriers for uniformly elliptic equations and the exterior cone condition, *J. Math. Anal. Appl.* **79**, pp. 203–217.

[131] Michael, J. H. and Simon, L. M. (1973). Sobolev and mean-value inequalities on generalized submanifolds of $\mathbb{R}^n$, *Comm. Pure Appl. Math.* **26**, pp. 361–379.

[132] Milakis, E. and Silvestre, L. E. (2006). Regularity for fully nonlinear elliptic equations with Neumann boundary data, *Comm. Partial Differential Equations* **31**, pp. 1227–1252.

[133] Miller, K. (1967a). Barriers on cones for uniformly elliptic operators, *Ann. Mat. Pura Appl.* **76**, pp. 93–105.

[134] Miller, K. (1967b). Existence theory for certain ordinary differential equations with a monotone singularity, *Proc. Amer. Math. Soc.* **18**, pp. 1006–1008.

[135] Miller, K. (1971). Extremal barriers on cones with Phragmèn-Lindelöf theorems and other applications, *Ann. Mat. Pura Appl.* **90**, pp. 297–329.

[136] Miranda, C. (1970). *Partial Differential Equations of Elliptic Type*, Ergebnisse der Mathematik und ihrer Grenzgebiete, Band 2 (Springer-Verlag,

New York).

[137] Miranda, M. (1967). Diseguaglianze di Sobolev sulle ipersuperfici minimali, *Rend. Sem. Mat. Univ. Padova* **38**, pp. 69–79.

[138] Morrey, C. B., Jr. (1933). A Class of Representations of Manifolds. Part I, *Amer. J. Math.* **55**, pp. 683–707.

[139] Morrey, C. B., Jr. (1940). Functions of several variables and absolute continuity, ii, *Duke Math. J.* **6**, pp. 187–215.

[140] Moser, J. (1960). A new proof of deGiorgi's theorem concerning the regularity problem for elliptic differential equations, *Comm. Pure Appl. Math.* **13**, pp. 457–468.

[141] Motzkin, T. S. and Wasow, W. (1953). On the approximation of linear elliptic differential equations by difference equations with positive coefficients, *J. Math. Physics* **31**, pp. 253–259.

[142] Nadirashvili, N. S. (1981). Lemma on the interior derivative and uniqueness of the solution of the second boundary value problem for second-order elliptic equations, *Dokl. Akad. Nauk SSSR* **261**, pp. 804–808, [Russian]. English transl. in Soviet Math. Dokl. **24** (1981), 598–601.

[143] Nadirashvili, N. S. (1983). On the question of the uniqueness of the solution of the second boundary value problem for second-order elliptic equations, *Mat. Sb. (N.S.)* **122(164)**, pp. 341–359, [Russian]. English transl. in Math. USSR-Sb. **50** (1985), 325–341.

[144] Nadirashvili, N. S. (1985). On a problem with an oblique derivative, *Mat. Sb. (N.S.)* **127(169)**, 3, pp. 398–416, [Russian]. English transl. in Math. USSR-Sb. **55** (1986), 397–414.

[145] Nadirashvili, N. S. (1988). Some estimates in a problem with oblique derivative, *Izv. Akad. Nauk SSSR* **52**, pp. 1082–1090, [Russian]. English transl. in Math. USSR-Izv. **33** (1981), 403–411.

[146] Nash, J. (1958). Continuity of solutions of parabolic and elliptic equations, *Amer. J. Math.* **80**, pp. 931–954.

[147] Nezhinskaya, I. V. (2004). On solvability of the boundary-value problem for $(p, q)$-nonlinear elliptic and parabolic equations, *Prob. Mat. Anal.* **29**, pp. 55–69, [Russian]. English transl. in J. Math. Sci. **124** (2004), 5001–5017.

[148] Nezhinskaya, I. V. (2006). Estimate for the gradient of the solution to the Neumann problem for $(p, q)$-nonlinear equations, *Prob. Mat. Anal.* **31**, pp. 47–57, [Russian]. English transl. in J. Math. Sci. **132** (2006), 428–440.

[149] Oddson, J. K. (1968). Some solutions of elliptic extremal equations in the plane, *Matematiche (Catania)* **23**, pp. 273–289.

[150] Oleĭnik, O. A. (1952). On properties of solutions of certain boundary problems for equations of elliptic type, *Mat. Sbornik N.S.* **30(72)**, pp. 695–702.

[151] Paneah, B. P. (2000). *The oblique derivative problem*, *Mathematical Topics*, Vol. 17 (Wiley-VCH Verlag Berlin GmbH, Berlin), the Poincaré-problem.

[152] Perron, O. (1923). Eine neue Behandlung der ersten Randwertaufgabe für $\Delta u = 0$, *Math. Z.* **18**, 1, pp. 42–54.

[153] Pipher, J. (1987). Oblique derivative problems for the Laplacian in Lipschitz domains, *Rev. Mat. Iberoamericana* **3**, pp. 455–472.

[154] Pogorelov, A. V. (1964). *Monge-Ampère equations of elliptic type*, Trans-

lated from the first Russian edition by Leo F. Boron with the assistance of Albert L. Rabenstein and Richard C. Bollinger (P. Noordhoff Ltd., Groningen).

[155] Popivanov, P. R. and Palagachev, D. K. (1997). *The degenerate oblique derivative problem for elliptic and parabolic equations*, *Mathematical Research*, Vol. 93 (Akademie Verlag, Berlin).

[156] Protter, M. H. and Weinberger, H. F. (1984). *Maximum Principles in Differential Equations* (Springer-Verlag, New York), corrected reprint of the 1967 original.

[157] Pucci, C. (1966). Limitazioni per soluzioni di equazioni ellittiche, *Ann. Mat. Pura Appl. (4)* **74**, pp. 15–30.

[158] Pucci, P. and Serrin, J. (2007). *The Maximum Principle*, Progress in Nonlinear Differential Equations and their Applications, 73 (Birkhäuser Verlag, Basel).

[159] Reiman, M. I. and Williams, R. J. (1988). A boundary property of semimartingales reflecting Brownian motions, *Probab. Theory Related Fields* **77**, pp. 87–97.

[160] Rešetnjak, J. G. (1968). Generalized derivatives and differentiability almost everywhere, *Mat. Sb. (N.S.)* **75(117)**, pp. 323–334, [Russian]. English transl. in Math. USSR-Sb. **4** (1968), 293–302.

[161] Riesz, F. (1916). Über lineare Funktionalgleichungen, *Acta Math.* **41**, pp. 71–98, [German].

[162] Safonov, M. V. (1980). Harnack's inequality for elliptic equations and the Hölder property of their solutions, *Zap. Nauchn. Sem. Leningrad. Otdel. Mat. Inst. Steklov. (LOMI)* **96**, pp. 272–287, [Russian]. English transl. in J. Soviet Math. **21** (1983), 851–863.

[163] Safonov, M. V. (1988). Classical solution of second-order nonlinear elliptic equations, *Izv. Akad. Nauk SSSR Ser. Mat.* **52**, pp. 1272–1287, 1328, [Russian]. English transl. in Math. USSR Izv. **33** (1988), 597–612.

[164] Safonov, M. V. (1995). On the oblique derivative problem for second order elliptic equations, *Comm. Partial Differential Equations* **20**, pp. 1349–1367.

[165] Sauvigny, F. (2006). *Partial differential equations. 1*, Universitext (Springer-Verlag, Berlin), foundations and integral representations, With consideration of lectures by E. Heinz, Translated and expanded from the 2004 German original.

[166] Schauder, J. (1930). Über lineare, vollstetige Funktionalraumen, *Studia Math.* **2**, pp. 83–96.

[167] Schauder, J. (1934a). Numerische Abschätzungen in elliptischen linearen Differentialgleichungen, *Studia Math.* **5**, pp. 34–42.

[168] Schauder, J. (1934b). Über lineare elliptische Differentialgleichungen zweiter Ordnung, *Math. Z.* **38**, 1, pp. 257–282.

[169] Serrin, J. (1971). Gradient estimates for solutions of nonlinear elliptic and parabolic equations, in *Contributions to nonlinear functional analysis (Proc. Sympos., Math. Res. Center, Univ. Wisconsin, Madison, Wis., 1971)* (Academic Press, New York), pp. 565–601.

[170] Simon, L. (1976). Interior gradient bounds for non-uniformly elliptic equa-

tions, *Indiana Univ. Math. J.* **25**, pp. 821–855.

[171] Simon, L. and Spruck, J. (1976). Existence and regularity of a capillary surface with prescribed contact angle, *Arch. Rational Mech. Anal.* **61**, pp. 19–34.

[172] Sobolev, S. L. (1938). On a theorem of functional analysis, *Math. Sb. (N. S.)* **4 (46)**, pp. 471–497, [Russian]; English transl. in Amer. Math. Soc. Transl. (2) **34** (1963), 39–68.

[173] Sobolev, S. L. (1991). *Some applications of functional analysis in mathematical physics*, *Translations of Mathematical Monographs*, Vol. 90 (American Mathematical Society, Providence, RI), translated from the third Russian edition by Harold H. McFaden, With comments by V. P. Palamodov.

[174] Spruck, J. (1975). On the existence of a capillary surface with prescribed contact angle, *Comm. Pure Appl. Math.* **28**, pp. 189–200.

[175] Stein, E. M. (1970). *Singular Integrals and Differentiability Properties of Functions* (Princeton University Press, Princeton, NJ).

[176] Taylor, M. E. (2006). *Measure Theory and Integrals* (American Mathematical Society, Providence, R.I.).

[177] Trefethen, L. and Williams, R. J. (1986). Conformal mapping solution of Laplace's equation on a polygon with oblique derivative boundary conditions, *J. Comput. Appl. Math.* **14**, pp. 227–249.

[178] Trudinger, N. S. (1980). Local estimates for subsolutions and supersolutions of general second order elliptic quasilinear equations, *Inv. Math.* **61**, pp. 67–79.

[179] Trudinger, N. S. (1984). Boundary value problems for fully nonlinear elliptic equations, *Proc. Centre Math. Anal. Austral. Nat. Univ.* **8**, pp. 65–83.

[180] Trudinger, N. S. (1985). Lectures on Nonlinear Second Order Elliptic Equations, Lecture notes, Nankai Institute of Mathematics, Tianjin, China.

[181] Trudinger, N. S. and Wang, X.-J. (2009). On the second boundary value problem for Monge-Ampère type equations and optimal transportation, *Ann. Sc. Norm. Super. Pisa Cl. Sci. (5)* **8**, pp. 143–174.

[182] Ural′ceva, N. N. (1962). Boundary-value problems for quasi-linear elliptic equations and systems with principal part of divergence type, *Dokl. Akad. Nauk SSSR* **147**, pp. 313–316, [Russian]. English transl. in Soviet Math. Dokl. **3** (1962), 1615–1618.

[183] Ural′ceva, N. N. (1971). Nonlinear boundary value problems for equations of minimal surface type, *Trudy Mat. Inst. Steklov.* **116**, pp. 217–226, 237, [Russian]. English transl. in Proc. Math. Inst. Steklov **116** (1973), 227–237.

[184] Ural′ceva, N. N. (1973). The solvability of the capillarity problem, *Vestnik Leningrad. Univ.* **1973**, 19 Mat. Meh. Astronom. Vyp. 4, pp. 54–64, 152, [Russian]. English transl. in Vestnik Lenigrad Univ. Math. **6** (1979), 363–375.

[185] Ural′ceva, N. N. (1975). The solvability of the capillarity problem. II, *Vestnik Leningrad. Univ.* **1975**, 1, Mat. Meh. Astronom. vyp. 1, pp. 143–149, 191, [Russian]. English transl. in Vestnik Lenigrad Univ. Math. **8** (1980), 151–158.

[186] Ural′tseva, N. N. (1982). Estimates for the maxima of the moduli of the gra-

dients for solutions of capillarity problems, *Zap. Nauchn. Sem. Leningrad. Otdel. Mat. Inst. Steklov. (LOMI)* **115**, pp. 274–284, [Russian]. English transl. in J. Soviet Math. **28** (1985), 806–813.
[187] Ural′tseva, N. N. (1991). A nonlinear problem with an oblique derivative for parabolic equations, *Zap. Nauchn. Leningrad. Otdel. Mat. Inst. Steklov. (LOMI)* **188**, pp. 143–158, [Russian]. English transl. in J. Math. Sci. **70** (1994), 1817–1827.
[188] Ural′tseva, N. N. (1992). Gradient estimates for solutions of nonlinear parabolic oblique derivative problem, in *Geometry and nonlinear partial differential equations (Fayetteville, AR, 1990), Contemp. Math.*, Vol. 127 (Amer. Math. Soc., Providence, RI), pp. 119–130.
[189] Urbas, J. (1995). Nonlinear oblique boundary problems for Hessian equations in two dimensions, *Ann. Inst. H. Poincaré: Anal. Non Linéaire* **12**, pp. 507–575.
[190] Urbas, J. (1996). Nonlinear oblique derivative problems for two dimensional curvature equations, *Adv. Differential Equations* **1**, pp. 301–336.
[191] Urbas, J. (1997). On the second boundary value problem for equations of Monge-Ampère type, *J. Reine Angew. Math.* **487**, pp. 115–124.
[192] Urbas, J. (1998). Oblique boundary value problems for equations of Monge-Ampère type, *Calc. Var. Partial Differential Equations* **7**, pp. 19–39.
[193] Urbas, J. (2001). The second boundary value problem for a class of Hessian equations, *Comm. Partial Differential Equations* **26**, pp. 859–882.
[194] Verchota, G. C. (2007). Counterexamples and uniqueness for $L^p(\partial\Omega)$ oblique derivative problems, *J. Funct. Anal.* **245**, pp. 413–437.
[195] Volkov, E. A. (1965a). Differential properties of solutions of boundary value problems for the Laplace equation on polygons, *Trudy Mat. Inst. Steklov* **77**, pp. 113–142, [Russian]. English transl. in Proc. Steklov Math. Inst. **77** (1968), 127–159.
[196] Volkov, E. A. (1965b). On differential properties of solutions of boundary value problems for the Laplace and Poisson equations on a rectangle, *Trudy Mat. Inst. Steklov* **77**, pp. 89–112, [Russian]. English transl. in Proc. Steklov Math. Inst. **77** (1968), 101–126.
[197] von Nessi, G. T. (2010). On the second boundary value problem for a class of modified-hessian equations, *Comm. Partial Differential Equations* **35**, pp. 745–785.
[198] Wang, L. (1992). A maximum principle for elliptic and parabolic equations with oblique derivative boundary problems, *J. Partial Differential Equations* **5**, pp. 23–27.
[199] Wang, L. (2003). A geometric approach to the Calderón-Zygmund estimates, *Acta Math. Sin. (Eng. Ser.)* **19**, pp. 381–396.
[200] Weierstrass, K. (1885). Über die analytische Darstellbarkeit sogennater willkürzer Funktionen einer reellen Veränderlichen, *Berliner Berichte* **1885**, 2, pp. 633–639, 789–805, reprinted in Mathematische Werke, Band 3, Abhandlungen III, pp. 1-37.
[201] Whitney, H. (1934). Analytic extensions of differentiable functions defined on closed sets, *Trans. Amer. Math. Soc.* **36**, pp. 63–89.

[202] Wigley, N. M. (1970). Mixed boundary value problems in plane domains with corners, *Math. Z.* **115**, pp. 33–52.

[203] Wigley, N. M. (1988). Schauder estimates in domains with corners, *Arch. Rational Mech. Anal.* **104**, pp. 271–276.

[204] Williams, R. J. (1985). Recurrence classification and invariant measure for reflected Brownian motion in a wedge, *Ann. Probab.* **13**, pp. 758–778.

[205] Yanushauskas, A. I. (1989). *The oblique derivative problem of potential theory*, Contemporary Soviet Mathematics (Consultants Bureau, New York), translated from the Russian by Norman Stein.

[206] Ziemer, W. P. (1989). *Weakly Differentiable Functions, Graduate Texts in Mathematics*, Vol. 120 (Springer-Verlag, New York).

# Index